Lecture Notes in Computer Science   16417

Founding Editors

Gerhard Goos
Juris Hartmanis

Editorial Board Members

Elisa Bertino, USA
Wen Gao, China

Bernhard Steffen, Germany
Moti Yung, USA

Advanced Research in Computing and Software Science

Subline of Lecture Notes in Computer Science

Subline Series Editors

Giorgio Ausiello, *University of Rome 'La Sapienza', Italy*
Vladimiro Sassone, *University of Southampton, UK*

Subline Advisory Board

Susanne Albers, *TU Munich, Germany*
Benjamin C. Pierce, *University of Pennsylvania, USA*
Bernhard Steffen, *University of Dortmund, Germany*
Deng Xiaotie, *Peking University, Beijing, China*
Jeannette M. Wing, *Microsoft Research, Redmond, WA, USA*

More information about this series at https://link.springer.com/bookseries/558

Yu-Fang Chen · Thomas Jensen · Ondřej Lengál
Editors

Verification, Model Checking, and Abstract Interpretation

27th International Conference, VMCAI 2026
Rennes, France, January 12–13, 2026
Proceedings

 Springer

Editors
Yu-Fang Chen
Academia Sinica
Taipei, Taiwan

Thomas Jensen
Inria
Rennes, France

Ondřej Lengál
Brno University of Technology
Brno, Czech Republic

ISSN 0302-9743   ISSN 1611-3349 (electronic)
Lecture Notes in Computer Science
ISBN 978-3-032-15699-0   ISBN 978-3-032-15700-3 (eBook)
https://doi.org/10.1007/978-3-032-15700-3

This Springer imprint is published by the registered company Springer Nature Switzerland AG
The registered company address is: Gewerbestrasse 11, 6330 Cham, Switzerland

If disposing of this product, please recycle the paper.

Preface

The 27th International Conference on Verification, Model Checking, and Abstract Interpretation (VMCAI 2026) was held in Rennes, France, from January 12–13, 2026. VMCAI serves as a premier forum for research in the broad areas of formal verification, model checking, and abstract interpretation, fostering interaction, cross-fertilization, and advancement of hybrid methods that combine these and related approaches.

VMCAI 2026 received 54 submissions, of which 18 papers were accepted (acceptance rate 33%), comprising 15 regular papers, 2 tool papers, and 1 case-study paper. Each submission was reviewed by at least three members of the Program Committee (PC). In some cases, PC members sought input from external experts. The review period was followed by an active discussion phase during which PC members and chairs deliberated on each paper. We are pleased to report that VMCAI 2026 maintained the high standard of excellence established in past editions. Each accepted paper either had a championing PC member or received overall positive scores from the reviewers.

The conference program featured three invited speakers:

- David Pichardie, Meta, France
- Mihaela Sighireanu, ENS Paris-Saclay, France
- Anthony Widjaja Lin, University of Kaiserslautern-Landau and Max-Planck Institute for Software Systems, Germany

We extend our sincere gratitude to the technical program committee and the external reviewers for their diligent work in evaluating the submissions and helping to assemble a strong technical program. The Program Committee, consisting of 51 members, represented leading researchers in core VMCAI areas including abstract interpretation, programming languages, hardware and software model checking, cyber-physical systems, formal synthesis, formal methods in AI, and concurrency. Following tradition, VMCAI adopted a single-blind review process. Each paper received reviews from at least three PC members over approximately 28 days, followed by a two-week online discussion period. Where appropriate, a PC member summarized the discussion as part of their review. We acknowledge EasyChair for providing a platform for submission and review management.

VMCAI 2025 also featured an artifact evaluation process, chaired by

- Marek Chalupa, ISTA Austria and
- Yong Li, Chinese Academy of Sciences, China

An Artifact Evaluation Committee (AEC) was formed to assess submitted artifacts. Authors were encouraged to submit artifacts alongside their papers. The artifact evaluation process consisted of two stages: an initial *smoke test* to resolve minor issues, followed by a detailed evaluation based on the criteria of availability, functionality, and reusability by three AEC members.

A total of 22 artifacts were submitted. Ten artifacts met all of the three criteria and were awarded the Available+Reproducible badge. The AEC completed its evaluations promptly, allowing their conclusions to be passed on to the main PC discussions and contribute to the final decision for accepting or rejecting a paper. We thank the AEC for their valuable service to the community.

We are also grateful to the VMCAI Steering Committee–Viktor Kunčak, Andreas Podelski, Sriram Sankaranarayanan, and Lenore Zuck–for their guidance. Special thanks go to POPL 2026 General Chair Sandrine Blazy and ACM SIGPLAN Conference Manager Neringa Young for their crucial support with the local organization. Finally, we express our appreciation to Labex CominLabs and Springer for generous financial sponsorship.

November 2025

Marek Chalupa
Yu-Fang Chen
Thomas Jensen
Ondřej Lengál
Yong Li

Organization

Program Committee Chairs

Yu-Fang Chen	Academia Sinica, Taiwan
Thomas Jensen	Inria, France
Ondřej Lengál	Brno University of Technology, Czech Republic

Artifact Evaluation Committee Chairs

Marek Chalupa	Institute of Science and Technology Austria, Austria
Yong Li	Institute of Software, Chinese Academy of Sciences, China

Steering Committee

Viktor Kunčak	École Polytechnique Fédérale de Lausanne, Switzerland
Andreas Podelski	University of Freiburg, Germany
Sriram Sankaranarayanan	University of Colorado, Boulder, USA
Lenore Zuck	University of Illinois Chicago, USA

Program Committee

Alexander Bakst	Certora, USA
Sebastien Bardin	CEA-List, France
Borzoo Bonakdarpour	Michigan State University, USA
Filip Cano	Institute of Science and Technology Austria, Austria
Milan Češka	Brno University of Technology, Czech Republic
Liqian Chen	National University of Defense Technology, China
Chih-Hong Cheng	Chalmers University of Technology, Sweden
Coen De Roover	Vrije Universiteit Brussel, Belgium

Gidon Ernst	Ludwig Maximilian University of Munich, Germany
Javier Esparza	Technical University of Munich, Germany
Grigory Fedyukovich	Florida State University, USA
Vijay Ganesh	Georgia Tech, USA
Roberto Giacobazzi	University of Arizona, USA
Simon Guilloud	Swiss Institute of Technology Lausanne, Switzerland
Ashutosh Gupta	Tata Institute of Fundamental Research, India
Lukáš Holík	Brno University of Technology, Czech Republic
Chih-Duo Hong	National Chengchi University, Taiwan
Sebastian Junges	Radboud University, Netherlands
Jan Kofroň	Charles University, Czech Republic
Alfons Laarman	Leiden University, Netherlands
Nian-Ze Lee	National Taiwan University, Taiwan
Matthieu Lemerre	CEA-List, France
Jérôme Leroux	CNRS, France
Jyun-Ao Lin	National Taipei University of Technology, Taiwan
Isabella Mastroeni	University of Verona, Italy
Nicolas Mazzocchi	Slovak University of Technology in Bratislava, Slovak Republic
Antoine Miné	LIP6, Sorbonne Université, France
Alexandre Moine	Inria, France
Benoît Montagu	Inria, France
Jorge A. Navas	Certora, USA
Luca. Negrini	Ca' Foscari University of Venice, Italy
Jakob Piribauer	TU Dresden, Germany
Adrien Pommellet	LRDE, EPITA, France
Francesco Ranzato	University of Padua, Italy
Christoph Scholl	University of Freiburg, Germany
Stephen F. Siegel	University of Delaware, USA
Mihaela Sighireanu	Université Paris-Saclay, CNRS, ENS Paris-Saclay, France
Julien Signoles	CEA-List, France
B. Srivathsan	Chennai Mathematical Institute, India
Marielle Stoelinga	University of Twente, Netherlands
Jan Strejček	Masaryk University, Czech Republic
Kohei Suenaga	Kyoto University, Japan
Tom van Dijk	University of Twente, Netherlands
Masaki Waga	Kyoto University, Japan
Kazuki Watanabe	National Institute of Informatics, Japan

Philipp Wendler	Ludwig Maximilian University of Munich, Germany
Thomas Wies	New York University, USA
Enea Zaffanella	University of Parma, Italy
Li Zhou	Max Planck Institute for Security and Privacy, Germany

Artifact Evaluation Committee

Luca Di Stefano	TU Wien, Austria
Greta Dolcetti	Ca' Foscari University of Venice, Italy
Sankalp Gambhir	École Polytechnique Fédérale de Lausanne, Switzerland
Arka Ghosh	University of Warsaw, Poland
Michal Hečko	Brno University of Technology, Czech Republic
Qifan Huang	Institute of Software, Chinese Academy of Sciences, China
David Hudák	Brno University of Technology, Czech Republic
Ayrat Khalimov	TU Clausthal, Germany
Klaus Kraßnitzer	Institute of Science and Technology Austria, Austria
Jérôme Leroux	CNRS, France
Dorian Lesbre	CEA-List, France
Zengyu Liu	National University of Defense Technology, China
Jingyi Mei	Leiden University, Netherlands
Vincent Mihalkovič	Masaryk University, Czech Republic
Sayan Mukherjee	University of Rennes, Inria, CNRS, IRISA, France
Ana Oliveira da Costa	TU Wien, Austria
Roberto Pettinau	Carl von Ossietzky University of Oldenburg, Germany
Kittiphon Phalakarn	National Institute of Informatics, Japan
Francesco Pontiggia	TU Wien, Austria
Frédéric Recoules	CEA-List, France
Diletta Rigo	University of Padua, Italy
Paul Robert	CEA-List, France
N. Ege Saraç	Institute of Science and Technology Austria, Austria
Sota Sato	Kyoto University, Japan
Gaëtan Staquet	Centrale Nantes, France
Bram Vandenbogaerde	Vrije Universiteit Brussel, Belgium

Changshun Wu	Université Grenoble Alpes, France
Hanrui Zhao	East China Normal University, China

Additional Reviewers

Konstantin Britikov	Milad Rabizadeh
David Chocholatý	Arshia Rafieioskouei
Matt Deloronzo	Franz-Xaver Reichl
Diletta Rigo	Diletta Rigo
Vincent Fischer	Daniel Riley
Vojtěch Havlena	Kenneth Rogale
Salil Kamath	Adam Rogalewicz
Valentin Krasotin	Jakob Schulz
Dorian Lesbre	Ting-Hsuan Tien
Li-Zhe Liu	Yu-Hsuan Wu
Zengyu Liu	Yi-De Wu
John Lu	Lucas Zavalia
Jingyi Mei	Ruihan Zhao
Aime Ntagengerwa	Ghiles Ziat
Milad Rabizadeh	Robin Ziemek

Invited Talks

Understanding Transformers Through the Lens of Logic and Automata

Anthony Widjaja Lin ⓘ

University of Kaiserslautern-Landau and Max Planck Institute for Software Systems

Transformers are a revolutionary neural networks architecture, which has been the backbone of our modern Large Language Models (LLMs). Despite the success of transformers in practice, we often do not know why they work (or occasionally also, why they do not work). Recent years have witnessed rapid progress in understanding transformers through the lens of logic and automata (in the community called FLaNN = Formal Languages and Neural Networks). In particular, the toolbox of logic and automata (i.e. connections to linear temporal logic) has helped us understand why PARITY (and in general "state-tracking") is difficult for transformers. I will recount some of the fundamental results in the field and open problems at the intersection of logic, automata, verification and transformers.

Current State of the Industrial-Strength INFER Static Analysis Platform

David Pichardie

Meta, France

Abstract. Static analysis plays a crucial role in improving software quality by automatically identifying potential bugs before deployment. We report on the current architecture of INFER, an industrial-strength static analysis tool developed for large-scale, production codebases. INFER employs a modular, compositional approach, leveraging separation logic and abstract interpretation to efficiently and accurately detect various classes of bugs, including null pointer dereferences, resource leaks, and memory safety issues. We detail several features that explain the tool's success and longevity.

1 Introduction

Software defects are detrimental to the user experience. Although dynamic testing is a predominant technique, its effectiveness is limited by the test cases that are executed. Static analysis offers a complementary approach, rigorously analysing the source code without execution to identify bugs that testing might miss. However, traditional static analysis tools often face two significant challenges in industrial settings: low precision (high false positive rates) and poor scalability (long analysis times for large codebases). In the long term, these tools also present engineering challenges in terms of maintainability and the ability to adapt quickly to new languages.

INFER [1–3], initially developed at Facebook (now Meta), is designed to overcome these challenges. It supports various languages, including C, C++, Objective-C, Java, Kotlin, Erlang, PHP/Hack and Python. INFER's design prioritizes speed and scalability through a compositional analysis methodology, making it suitable for continuous integration environments where fast feedback is expected.

2 INFER's Architecture and Main Features

For over ten years, INFER has proven useful for thousands of daily users and codebases containing tens of millions of lines of code. Among the many features that may explain this success, four are highlighted in this paper.

2.1 One IR to Rule Them All

A first technical challenge in supporting multiple source languages and static analyses is creating a unified analysis framework where several languages and client analyses can share the same components. INFER addresses this by translating each supported language into a single, language-agnostic Intermediate Representation (IR). This IR is a low-level, register-transfer-language-style representation designed to capture the essential semantics of memory operations and control flow, abstracting away syntax unique to the original language.

Most of our supported languages are not directly translated into INFER IR but first go through transformation toolchains that accompany these languages (Clang frontend for C/C++/Objective-C, Java bytecode for Java/Kotlin, HHVM bytecode for PHP/Hack, Python bytecode...). This is a compromise that has always been difficult to make in static analysis. Target languages like bytecode representation often have simpler semantics than their source representation, but this can sometimes make the analyzer's reasoning more difficult. In the case of INFER, this strategy has been successful, probably also thanks to the quality of the analyses implemented.

2.2 Scalable Interprocedural Analysis: Compositionality Saves You

INFER achieves its scalable interprocedural analysis through a compositional methodology that relies on building and reusing function summaries. This approach is distinct from whole-program analysis, which can be prohibitively slow for massive codebases.

For every function, INFER computes a summary. In the context of Separation Logic (SL), this summary is typically a pre/post-condition pair that describes the function's effect on the memory state (the heap) and the resources it manages. The analysis of a function is local—it only depends on the function's body and the pre-computed summaries of the functions it calls. When a function calls another, INFER uses the callee's summary to update the caller's memory state instead of re-analyzing the callee's body. Summaries need to be built in the right order. INFER follows a bottom-up approach across the call graph. Functions with no dependencies (or only dependencies on functions whose summaries are already available) are analyzed first.

To ensure rapid analysis suitable for continuous integration, INFER makes a pragmatic compromise regarding the full interprocedural analysis. INFER does not always consider a full, deep unfolding of the program's call graph. This decision is a necessary trade-off to maintain speed. However, this engineering compromise ensures the tool remains practical and fast enough to find a high volume of real-world, high-impact bugs.

2.3 Start from Anywhere

Unlike tests or whole-program analyses, INFER does not require executable entry points to start its interprocedural exploration. When integrated into a Continuous Integration (CI) pipeline, INFER's analysis typically runs against only the code that has been modified in the considered commit, along with the functions that transitively depend on this modified code (its *forward dependencies*). This drastically reduces the scope of the analysis, maintaining the necessary speed for fast developer feedback.

The bottom-up summary computation order described in the previous section is adapted to support this efficiency goal. Instead of strictly analyzing every function from the bottom of the call graph up, the analysis effectively becomes on-demand. INFER only computes summaries for functions that are determined to be **reachable** from the set of modified files. If a summary for a dependency is already available and valid, it is reused; otherwise, it is computed only when needed to analyze the changed code path. This selective, dependency-based re-analysis is essential to INFER's successful scaling to massive, continuously changing codebases.

2.4 Signal Quality Matters

Another important challenge for any industrial-strength static analysis tool stems from the theoretical limitation that finding all bugs in a program is undecidable (akin to the Halting Problem). This means every analyzer must make trade-offs, leading to two fundamental risks:

The analyzer signals a bug that is not a true defect. This typically occurs when the analysis's over-approximation of the program state is too coarse. High FP rates lead to developer fatigue and waste time triaging non-issues, significantly eroding trust in the tool.

The analyzer fails to signal a true bug. This can happen if the analysis under-approximates the program behavior, perhaps by cutting exploration short (e.g., due to time constraints) or not handling specific language features or control flow complexities perfectly.

In the industrial context, INFER's design makes a pragmatic choice: it always favors avoiding False Positives over risking False Negatives. While code quality is paramount, programmer velocity is also an important business metric. A tool that frequently flags non-bugs will be quickly dismissed and disabled by developers. Therefore, INFER is engineered to have a very low FP rate, ensuring that the warnings it does produce are highly likely to represent actionable, real-world defects. This engineering compromise is necessary for the tool's successful, long-term adoption in fast-paced development environments.

This design choice was co-developed at the same time as the fundamental theory of Incorrectness Logic [4–6]. This allows the tool to maintain a strong semantic focus while remaining pragmatic and tailored to its users.

3 Conclusion and Future Work

INFER represents a successful realization of applying advanced program analysis techniques to the challenging domain of industrial-scale software development. By prioritizing scalability, accuracy and velocity INFER provides a singular and complementary support. Its success demonstrates that sophisticated and semantically grounded static analysis can be a practical and impactful component of modern software quality assurance.

Ongoing work on INFER focuses on expanding its language support by making it compatible with the LLVM platform. We are also interested in new verification scenarios, specifically focused on validating programmes produced by generative code AI.

References

1. Calcagno, C., et al.: Moving fast with software verification. In: Havelund, K., Holzmann, G., Joshi, R. (eds.) NFM 2015. LNCS, vol. 9058, pp. 3–11. Springer, Cham (2015). https://doi.org/10.1007/978-3-319-17524-9_1
2. Distefano, D., Fähndrich, M., Logozzo, F., O'Hearn, P.W.: Scaling static analyses at Facebook. Commun. ACM **62**(8), 62–70 (2019)
3. Blackshear, S., Gorogiannis, N., O'Hearn, P.W., Sergey, I.: RacerD: compositional static race detection. Proc. ACM Program. Lang. **2**(OOPSLA), 144:1–144:28 (2018)
4. O'Hearn, P.W.: Incorrectness logic. Proc. ACM Program. Lang. **4**(POPL), 10:1–10:32 (2020)
5. Raad, A., Berdine, J., Dang, H.H., Dreyer, D., O'Hearn, P., Villard, J.: Local reasoning about the presence of bugs: incorrectness separation logic. In: Lahiri, S., Wang, C. (eds.) CAV 2020. LNCS, vol. 12225, pp. 225–252. Springer, Cham (2020). https://doi.org/10.1007/978-3-030-53291-8_14
6. Le, Q.L., Raad, A., Villard, J., Berdine, J., Dreyer, D., O'Hearn, P.W.: Finding real bugs in big programs with incorrectness logic. Proc. of ACM Program. Lang. **6**(OOPSLA1), 1–27 (2022)

TypedC: Spatial Memory Safety for Low-Level Programs by Abstract Interpretation

Mihaela Sighireanu ⓘD

Univ. Paris-Saclay, ENS Paris-Saclay, CNRS, LMF, Gif-sur-Yvette, France

Most of the software infrastructures used nowadays rely on code written in low-level programming languages like C. Such languages include idioms which are close to the underlying architectures (e.g., address values, direct stack and heap manipulation) but also provide some abstraction of various architecture-dependent features (e.g., address translation, caching). The C language is notoriously considered "memory-unsafe", meaning that it does not provide strong guarantees to prevent memory-related software bugs. Despite that, system software is often written in C because the language has a performant development infrastructure (compilers, linkers) and a well-established community of developers. Ensuring memory safety has been a leading goal of the software verification community in recent decades. To that end, several approaches have been proposed including formally and accurately capturing the memory model of C, finding vulnerabilities before a program's execution or during its execution, formally proving memory safety, or adding hardware features to enable fine-grained memory protection. Despite the significant progress in tools for memory safety verification, handling the low-level idioms (e.g., pointer arithmetic and manipulation of pointer representation, function pointers) and improving verification usability by conventional developers are still challenging for these tools. Because most developers in C are already familiar with type systems, a research trend employs various type systems (e.g., dependent types, physical types, liquid types, resource types) to introduce formal specifications of memory properties in developer's workflow and to output the detected vulnerabilities as type-checking errors. This approach is combined with memory-aware program logics (e.g., separation logic) to obtain precise specifications and error diagnosis.

In this talk, we will present a static analysis method adhering to the above trend. It performs type-checking for an expressive dependent physical type system, TypedC, that extends the type system of C. It focusses on finding memory safety bugs in the original C or binary code, based on an optional file including type definitions in TypedC that may expose hidden invariants and dependencies between memory locations, and relations

Joint research with Matthieu Lemerre, Julien Simonnet and Paul Robert from Univ. Paris-Saclay, CEA, List, Palaiseau, France.

This work has been partially funded by the French National Research Agency project ANR-21-CE48-0011.

between function arguments and results. The type-checking is done semantically, by using abstract interpretation. In addition, we show that our method provides for inference of types specifying complex memory invariants. The method is implemented in the Codex framework and successfully applied to the verification of binary and C programs.

Contents

Producing Shorter Congruence Closure Proofs in a State-of-the-Art SMT
Solver 1
    Bruno Andreotti and Haniel Barbosa

Reachability in Multi-agent Transfer Systems 21
    Nathalie Bertrand, Loïc Hélouët, Engel Lefaucheux, and Luca Paparazzo

Efficiently Verifying Quantum Programs with Few T Gates 44
    Youngchan Cho and Robert Rand

Atomic Gliders and Cellular Automata as Language Generators 58
    Dana Fisman and Noa Izsak

A Hybrid Meta-Learning Framework for Adaptive Safe Controller
Synthesis of Dynamic Systems 82
    Rui Guo, Yang Li, Xiuqing Cao, and Wang Lin

Proof Minimization in Neural Network Verification 99
    Omri Isac, Idan Refaeli, Haoze Wu, Clark Barrett, and Guy Katz

Finding Photonics Circuits via δ-Weakening SMT 125
    Marco Lewis and Benoît Valiron

Forward Symbolic Execution for Trustworthy Automation of Binary Code
Verification 147
    Andreas Lindner, Karl Palmskog, Scott Constable, Mads Dam,
    Roberto Guanciale, and Hamed Nemati

SAT-Based Synthesis of Minimal Deterministic Real-Time Automata
via 3DRTA Representation 173
    Junjie Meng, Jie An, Yong Li, Andrea Turrini, and Miaomiao Zhang

Try-Mopsa: Relational Static Analysis in Your Pocket 197
    Raphaël Monat

Termination Resilience Static Analysis 212
    Naïm Moussaoui Remil and Caterina Urban

Input-Based Three-Valued Abstraction Refinement 237
    Jan Onderka and Stefan Ratschan

Efficient Discovery of Actual Causality in Stochastic Systems 263
Arshia Rafieioskouei, Kenneth Rogale, and Borzoo Bonakdarpour

Verification of Generic VHDL Designs and Their Translation to Rocq 287
Ocan Sankur, Benoît Boyer, and Florian Faissole

Data Race Detection by Digest-Driven Abstract Interpretation 309
Michael Schwarz and Julian Erhard

A Formal Executable Semantics of PROMELA 335
Byoungho Son and Kyungmin Bae

Multi-variable Quantification of BDDs in External Memory using Nested
Sweeping 359
Steffan Christ Sølvsten and Jaco van de Pol

Probabilistic Verification for Modular Network-on-Chip Systems 383
*Nick Waddoups, Jonah Boe, Arnd Hartmanns, Prabal Basu,
Sanghamitra Roy, Koushik Chakraborty, and Zhen Zhang*

Author Index 409

Producing Shorter Congruence Closure Proofs in a State-of-the-Art SMT Solver

Bruno Andreotti[(✉)] and Haniel Barbosa

Universidade Federal de Minas Gerais (UFMG),
Belo Horizonte, Brazil
{bruno.andreotti,hbarbosa}@dcc.ufmg.br

Abstract. An important component of SMT solving is the theory of equality and uninterpreted functions, which is traditionally modelled in solvers via a congruence closure algorithm. Oftentimes, these algorithms are instrumented to provide machine checkable proofs when determining why two terms are equivalent. In a recent work published at FMCAD'22, Flatt et al. presented a modified congruence closure algorithm that could effectively produce demonstrably shorter proofs. This new algorithm relies on computing *redundant* equalities, which are not necessary to prove the equivalence between two terms but can provide shorter proofs. While promising, the modified algorithm was only considered in an equality saturation tool. In this work, we have adapted this algorithm to apply it within an SMT solver, and implemented our approach in the state-of-the-art solver cvc5. We discuss the challenges faced when integrating this algorithm into the backtracking nature of an SMT solver, and how we have addressed them. We evaluate our implementation on a large set of SMT-LIB benchmarks from multiple theories, and demonstrate how this new technique can result in smaller SMT proofs, while having only a moderate impact on runtime performance.

Keywords: SMT solving · Congruence closure · SMT proofs

1 Introduction

Proof-producing SMT solvers have a large potential for increasing the trustworthiness of formal methods applications that rely on solvers to discharge proof obligations. However, proof certificates can be quite large and therefore challenging to be checked quickly, specially in tools with limited performance, such as formally verified checkers or proof assistants. One way to mitigate this issue is to employ solving techniques that can lead to shorter proofs within the same proof calculus, which would be cheaper to check. Flatt et al. [11] recently proposed such a procedure for the congruence closure algorithm [19], a key component of SMT solvers to reason about equality and uninterpreted functions [9,17]. Their approach keeps *redundant* equalities that would normally be discarded, since they are not needed during solving, and takes advantage of them for finding shorter proofs. Their work however was in the context of the equality saturation [25] tool egg [26]. Equality saturation, which has recently seen significant usage in e.g. program optimization, works by constructing a graph that represents all equivalent forms of a program, and selecting an optimized one. Aiming to generate shorter proofs for the congruence closure algorithm as used within SMT solvers, in this paper we adapt the procedure of Flatt et al. for the CDCL($\mathcal{T}$) architecture [20] of modern SMT solvers, and implement it in the state-of-the-art solver cvc5 [4].

Extending the core component of state-of-the-art SMT solvers is notoriously challenging [6], and adapting the algorithm for CDCL($\mathcal{T}$) involves supporting backtracking and orchestrating the congruence closure reasoning with the SAT solver. Since equality saturation tools do not involve backtracking, and in general their congruence closure implementations do not need to interact with external reasoning, these considerations were not relevant in the original implementation in egg. Another difference is positive, however: finding shorter proofs potentially can lead to shorter *conflict clauses*. These clauses are used by SMT solvers to guide the search of the SAT solver, and shorter explanations can prune the search space more aggressively and lead to better performance. This potential advantage is not present in the equality saturation context since it is not performing a backtracking search guided by explanations.

After introducing the necessary background (Sect. 2), we discuss in detail the various challenges specific to the SMT setting and how we tackled them (Sect. 3). We also cover the specific implementation decisions to effectively integrate the algorithm to cvc5 (Sect. 4). Our evaluation of the current implementation (Sect. 5) indicates an encouraging reduction in the size of proofs from the congruence closure algorithm, as well as some significant improvement in runtime for particular families of benchmarks, while also uncovering a number of future directions for improvements.

1.1 Related Work

Producing smaller unsatisfiability proofs has long been a concern in the context of SAT solvers. Fontaine et al. [12] introduced techniques for compressing resolution proofs by optimizing the resolution derivation to more efficiently reuse pivots. Heule et al. [13] proposed producing smaller proofs directly from the SAT solver, by using a more expressive proof system. This new proof system could result in proofs that are not only smaller, but also faster to check. More recently, Reeves et al. [22] leveraged *proof skeletons* to reduce the storage requirements of proofs by only recording a subset of derived clauses, those deemed "important" according to different criteria, and reconstructing the missing clauses at checking-time.

While most of these works could be adapted to work in an SMT solver, they are mainly concerned with the challenge of storing large proofs, which is especially relevant to SAT solving. In a recent work, Otoni et al. [21] showed how theory-specific proof witnessess could be used to compactly certify the execution of specific SMT algorithms. This work encompassed the theories of linear integer and real arithmetic, and equality and uninterpreted functions. For the latter, the authors relied on a compact representation of the equality reasoning, but did not meaningfully modify the proof search in the congruence closure algorithm. To the best of our knowledge, no recent work exists in the direction of producing shorter proofs and smaller explanations from congruence closure reasoning in SMT solvers.

2   Background

2.1   CDCL($\mathcal{T}$) SMT Solvers

The problem of Boolean satisfiability (SAT) consists of determing whether a formula in propositional logic is satisfiable, that is, whether there exists an assignment to the formula's free variables that makes the formula true. It is commonly assumed that the formula is in conjunctive normal form (CNF). The problem of Satisfiability Modulo Theories (SMT) is a generalization of SAT, with a few key differences. Firstly, it operates over many-sorted first-order logic, allowing quantifiers, equality, and arbitrary function symbols; and secondly, the logic is complemented by a set of *theories*, which restrict the possible interpretations of the formula. These theories usually model a real-world domain in mathematics (like the theories of integer and real arithmetic) or in computer science (like the theories of strings and floating point numbers).

Most modern SMT solvers are based on the CDCL($\mathcal{T}$) algorithm [20], a variation of the *Conflict-Driven Clause Learning* (CDCL) algorithm [15] for SAT solving. This algorithm searches for a satisfying assignment by, for each step of the search, selecting an unassigned variable and assigning it a value among $\{\top, \bot\}$. Then, the algorithm computes all other assignemnts that are directly implied by this, in a process called *propagation*. While in SAT solvers only propositional reasoning is used during this step, SMT solvers often also incorporate theory-specific reasoning, in what is referred to as *theory propagation*.

If at any point in the search a clause in the formula is unsatisfied, meaning all its literals are set to $\bot$ by the current variable assignments, we reach what is called a *conflict*. In this case, the algorithm must *backtrack*, undoing a number of assignments until it reaches a previous branching point. The algorithm also performs *conflict analysis* to understand the root cause of the conflict and incorporate that information as a new learned clause, referred to as the *conflict clause*. If a satisfying assignment is found, the CDCL($\mathcal{T}$) algorithm then employs a series of *theory solvers*, to determine whether the assignment is *consistent* with the relevant theories. If it is, the formula is satisfied; otherwise, the theory solver must produce a conflict clause, and the search continues.

2.2   Congruence Closure

Congruence is the property that, given terms $a_1, \ldots, a_n, b_1, \ldots, b_n$ and an arbitrary function f, $a_1 = b_1 \wedge \ldots \wedge a_n = b_n$ implies $f(a_1, \ldots, a_n) = f(b_1, \ldots, b_n)$. For a given set of equalities, we say that the corresponding congruence closure is the minimal equivalence relation that satisfies these equalities as well as the property of congruence.

In the context of SMT solving, solving the theory of equality and uninterpreted functions (EUF) consists in determining whether a conjunction of equalities and disequalities between terms is consistent under the EUF theory. Typically, this is solved by constructing the congruence closure defined by the problem equalities, and verifying that the problem disequalities respect it.

Efficient algorithms for computing congruence closure were first described by Downey et al. [9] and Nelson and Oppen [17], in the context of program verification. The algorithm is based on a *Union-find* data structure [24], where each term is a vertex in a graph (called the *equality graph*, or *e-graph*), and each equivalence class forms a tree whose root is called the class *representative*.

At first, each term is part of an equivalence class containing only itself. The algorithm works by sequentially processing equalities, and merging these equivalence classes when they are found to be equal. When processing an equality $a = b$, if the terms are not already equivalent, the algorithm finds the representatives of a and b, and adds an edge between them (called an *equality edge*), selecting one to be the representative of the new merged class. Then, the algorithm merges all classes that have become equivalent due to congruence (e.g. $f(a)$ and $f(b)$). Edges added in this step are called *congruence edges*, and the terms that caused them to be added (in this case, a and b) are the edges' *justification*.

To avoid having to deal with functions with different numbers of arguments, we will take as a starting assumption that the terms given to the congruence closure algorithm are *currified* [18]. This process, borrowed from functional programming, turns a function application into a series of applications of a special "apply" function symbol, which we will denote f. For example, the term $g(a, b, c)$, when currified, becomes $f(f(f(g, a), b), c)$. From here on, we will assume that all terms have been currified, and are therefore either constants or an application of f over two arguments.

Once the equality graph is built, we can determine if two terms are equivalent by searching for a path between their corresponding vertices. However, for many applications (including SMT solvers), simply determining whether two terms are equivalent is not enough—it is also necessary to produce an *explanation* of their equivalence. Here, an explanation is defined as a subset of the input equalities that is sufficient to ensure the terms are equivalent. In SMT solvers that use the CDCL($\mathcal{T}$) algorithm [20], explanations from congruence closure are crucial, as they are used to construct conflict clauses for the solver. Furthermore, since a smaller conflict clause will prune a larger part of the search space, a small explanation is preferable. However, it is known that finding the smallest possible explanation for the equivalence of two terms is an NP-complete problem [10].

Besides providing an explanation, a congruence closure algorithm may also produce a structured *proof*. If an explanation is simply a set of equalities that justify the equivalence between two terms, a proof is a derivation, using these equalities and the properties of reflexivity, symmetry, transitivity and congruence to demonstrate that the two terms are equivalent. Proof-producing SMT solvers have become increasingly important in the last few years, and are already used in many applications [2,3,5,14,21], including proof automation in interactive theorem provers [23].

An explanation-producing congruence closure algorithm was first described by Nieuwenhuis et al. in [19]. When explaining the equivalence of two terms, this algorithm traverses the path between respective nodes in the equality graph. When traversing an equality edge, the algorithm simply records the input equality associated with that edge. When traversing a congruence edge between the terms $f(a_1, a_2)$ and $f(b_1, b_2)$, the algorithm recursively explains the equivalence of the terms a_1 and b_1, and that of a_2 and b_2; then adds all returned equalities to the explanation. Figure 1 shows the pseudocode for this algorithm. While the algorithm as described can only produce explanations, it can be straightforwardly extended to produce structured proofs.

```
1   function get_explanation(start, end):
2       let explanation = []
3       let lca = find_lowest_common_ancestor(start, end)
4       explanation += explain_along_path(start, lca)
5       explanation += explain_along_path(end, lca)
6       return explanation
7
8   function explain_along_path(lower, upper):
9       let explanation = []
10      let current = lower
11      while current != upper:
12          let edge = current.edge_to_parent()
13          if edge is congruence edge between f(a₁,a₂) and f(b₁,b₂):
14              explanation += get_explanation(a₁, b₁)
15              explanation += get_explanation(a₂, b₂)
16          else:
17              explanation += edge
18          current = current.parent()
19      return explanation
```

Fig. 1. Pseudocode for a classical proof-producing explanation algorithm. The function get_explanation returns a list with the explanation for the equivalence of two terms. Since the equivalence class is represented by a rooted tree, the function works by explaining the path from each of the nodes to their lowest common ancestor.

Importantly, the explanations returned by this algorithm might not be the smallest valid explanations. While there is only one unique path between the two nodes in the tree, if the input equalities were processed in a different order, or if different congruence edges were added, the explanation for the equivalence of these nodes might be different. More generally, the fact that the standard congruence closure algorithm discards redundant equalities means that oftentimes shorter explanations are not found.

2.3   The GREEDY Congruence Closure Algorithm

In an effort to produce shorter explanations from congruence closure, Flatt et al. [11] developed two new congruence closure algorithms, called TREEOPT and GREEDY, that do not discard redundant equalities, and instead use them to provide alternative, possibly shorter paths in the equality graph. Additionally, these new algorithms also compute extra congruence edges each time two classes are merged, contributing to even more alternative paths.

Keeping redundant edges means that the graph for each equivalence class is no longer a tree. Whereas before there was always only one path between two terms in an equivalence class, now the explanation algorithm must have a way to determine which of the possible paths represents the shortest proof. This is where the strategies used by the two algorithms diverge.

The TREEOPT algorithm is a $O(n^5)$ algorithm that finds an optimal proof for a slightly modified metric of proof size. On the other hand, GREEDY is a $O(n \log n)$ heuristic algorithm, that attempts to find small explanations without incurring an increase in complexity when compared to traditional congruence closure algorithms.

Flatt et al. [11] convincingly showed that TREEOPT does not present a significant improvement in proof size when compared to GREEDY, while having a substantial performance overhead. For this reason, we believe that TREEOPT is impractical for use in SMT solvers, and therefore focus our interest on GREEDY.

GREEDY is a greedy congruence closure algorithm that attempts to find a small explanation while keeping the same $O(n \log n)$ complexity as traditional algorithms. To do this, it first computes an estimate of the proof size for each edge in the equality graph. This estimate is obtained by computing the proof size of the edge with the traditional congruence closure algorithm, that is, ignoring redundant equalities.

Once the estimates are calculated, the algorithm simply finds the shortest path between the terms, using the estimates as weights for the edges. The algorithm is parameterized by an integer *fuel*, which, similarly to [11], we set to 10 as a default. When it encounters a congruence edge, if the fuel is greater than 0, the algorithm recurses (and decrements its fuel) to explain the justification of the congruence edge. However, if the fuel is 0, the algorithm explains the justification by simply using the classical congruence closure algorithm, i.e., ignoring redundant equalities.

2.4   Computing Extra Redundant Edges

As mentioned, besides not discarding redundant equalities that are asserted, the new algorithms also compute extra redundant congruence edges everytime two equivalence classes are merged. This is done by finding all terms in the newly merged class that have the same *canonical form*. For an application term $f(a, b)$, we say that its canonical form is the term $f(a', b')$, where t' is the representative of the equivalence class of the term t. For a constant term t, we say that its canonical form is just t. Notably, if two non-identical terms have the same canonical form, it means they are equivalent by congruence, and we may add a congruence edge between them.

```
 1   function get_canonical_form(term):
 2       if term is of the form f(a,b):
 3           let a' = representative_of(a)
 4           let b' = representative_of(b)
 5           return f(a',b')
 6       else:
 7           return term
 8
 9   function compute_extra_edges(eclass):
10       let canonical_map = {}
11       for term in eclass:
12           let canon = get_canonical_form(term)
13           for other in canonical_map[canon]:
14               add_edge(term, other)
15               if number_of_redundant_edges > LIMIT:
16                   return
17           canonical_map[canon] += term
```

Fig. 2. The algorithm for computing extra congruence edges between the two equivalence classes.

Figure 2 shows the algorithm for computing extra congruence edges in an equivalence class. We keep a map from a canonical form term to a list of terms in the class that have that canonical form. When we process a term, we compute its canonical form, and add an edge between it and every other term that shares that canonical form. This process is repeated for every term in the equivalence class, or until an arbitrary limit is reached.

While the original work by [11] implemented the new algorithms in an equality saturation tool, implementing them in an SMT solver requires several changes due to the backtracking nature of the solver and the complex interactions between the congruence closure algorithm and other modules. Over the next sections, we describe in detail the challenges we faced when adapting the GREEDY algorithm, and how we tackled them.

3   Handling Implicit Dependencies in a Congruence Closure Within CDCL($\mathcal{T}$)

Modern SMT solvers work by orchestrating a CDCL SAT solver that finds models for an abstraction of the input formula, with multiple theory solvers that check if these models are consistent. In the EUF theory, the theory solver is also commonly used during *propagation*, that is, even before a full model is found, the EUF solver is queried to ensure that the partial model that is being constructed by the SAT solver is consistent, and also to derive new facts that can be propagated. This results in a back-and-forth interaction between the SAT solver and the theory solver, where the SAT solver adds propagated facts into the congruence closure algorithm, which in turn will derive new literals that are asserted to the SAT solver. Figure 3 shows an example of this interaction between the SAT solver and the congruence closure algorithm during propagation, and the e-graph after this interaction has played out.

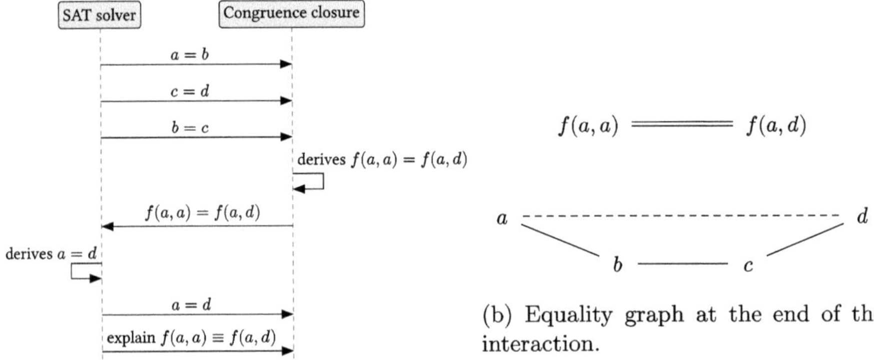

(a) Interaction between SAT solver and congruence closure algorithm.

(b) Equality graph at the end of the interaction.

Fig. 3. An example of a possible interaction between the SAT solver and the congruence closure algorithm, and the e-graph that is constructed after this interaction has played out. Here, we use simple lines to denote equality edges; double lines to denote congruence edges; and dashed lines to denote redundant edges.

In the example, the SAT solver first asserts into the congruence closure the equalities $a = b$, $c = d$ and $b = c$, in that order. When the latter is added, the terms a and d are in the same equivalence class, which means the terms $f(a, a)$ and $f(a, d)$ have become equivalent by congruence. The congruence closure algorithm adds the congruence edge $(f(a, a), f(a, d))$ accordingly. Then, as part of *theory propagation*, the congruence closure algorithm sends the lemma $f(a, a) = f(a, d)$ to the SAT solver. This in turn causes a propagation in the solver, which results in the fact $a = d$ being derived by SAT reasoning. This equality is asserted to the congruence closure algorithm, which results in the redundant edge (a, d) being added.

Now, the edge $a = d$ depends on the equality $f(a, a) = f(a, d)$, even though that dependency is not represented in any way in the equality graph—we call these *implicit dependencies*. When queried for an explanation of $f(a, a) \equiv f(a, d)$, the congruence closure algorithm cannot use the equality $a = d$ in the explanation, as that would result in a circular proof. Instead, the only correct explanation for this equivalence is the one based on the longer path between a and d, namely the equalities $a = b$, $b = c$ and $c = d$.

In the traditional congruence closure algorithm, where no redundant edges are kept, after the path between two terms is first established it will never change. This means that equalities which are implicitly dependent on the equivalence between two terms will never influence the explanation that is returned for those terms' equivalence, preventing this kind of circular proof. In this modified version of the algorithm, however, we need to take special care to ensure this cannot happen.

3.1   Edge Levels

To properly handle implicit dependencies, we need to augment the equality graph with information that encodes these dependency relations. The most precise way of doing this would be to construct the entire implication graph of the SMT solver, and restrict which edges can be used based on that. However, this is complicated to do in practice and might incur a performance cost, since modern SMT solvers rely heavily on lazily computing proofs. For example, most SAT solvers do not store the entire implication graph and instead keep a *trail*, which consists of a list of all assigned literals, in the order that they were assigned. Since a literal can only have been implied by literals that were assigned before it, this trail is a topological ordering of the implication graph.

We follow a similar strategy, and augment the equality graph based on the order in which the equalities are asserted into the congruence closure algorithm. Specifically, we assign to each edge in the e-graph a *level*[1], which is the index in which this edge was added to the graph; as such, these levels also correspond to a topological ordering of the implication graph. We also define the concept of two terms' *merge level*. For two terms in the same equivalence class, we say that their merge level is the level in which the terms were originally determined to be equivalent, i.e., the level of the edge that merged their equivalence classes. When searching for the explanation of the equivalence between two terms, we restrict the search to edges whose level is no greater than the terms' merge level. Effectively, this restricts the search to edges already present in the e-graph when the equivalence of the two terms was first derived.

Figure 4 shows the e-graph from Fig. 3b, but with level information denoted in blue. Consider again querying this e-graph for the explanation of $f(a,a) \equiv f(a,d)$. Their equivalence was first determined when the congruence edge between them was added, so their merge level is the level of that edge, namely 3. Therefore, the redundant edge between a and d, which has level 4, and which implicitly depends on the congruence edge $(f(a,a), f(a,d))$, will not be included in the search for an explanation, preventing a circular proof.

---

[1] Although similar concepts, these levels are distinct from the SAT solver's *decision levels*. In the example in Fig. 3, all of the shown interaction could have taken place during propagation, which is to say, during a single decision level.

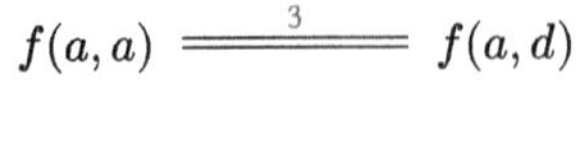

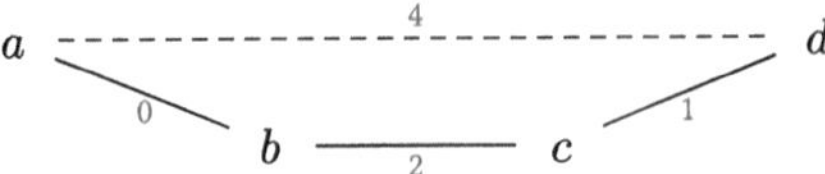

Fig. 4. The e-graph from Fig. 3b, but with the level of each edge shown in blue. (Color figure online)

Figure 5 shows the pseudocode for the GREEDY algorithm, modified to take into account the levels of edges. Now, besides finding the shortest path between the terms according to the computed weights, this shortest-path search must also ignore edges whose level is greater that `max_level`, set to be the merge level of the two terms.

```
1   function get_explanation(start, end, weights):
2       let max_level = get_merged_level(start, end)
3       let path = shortest_path(start, end, weights, max_level)
4       let explanation = []
5       for edge in path:
6           if edge is congruence edge between f(a₁,a₂) and f(b₁,b₂):
7               explanation += get_explanation(a₁, b₁, weights)
8               explanation += get_explanation(a₂, b₂, weights)
9           else:
10              explanation += edge
11      return explanation
```

Fig. 5. Pseudocode for the modified explanation algorithm. Here, we assume `shortest_path` is a function that returns a list of edges representing the shortest path between two nodes, given a set of edge weights, and rejecting all edges whose level is greater than `max_level`.

3.2 Alternative Solutions

Restricting the explanation search with edge levels is an aggressive way of preventing circular proofs, and at first glance, it might seem that other, less restrictive solutions might be possible. In this section, we discuss some alternative solutions that were considered for the implicit dependencies problem, and why ultimately they were not feasible.

Weighing the Edges Based on Proof Size. The GREEDY algorithm already relies on adding weights to the e-graph edges to enable finding smaller proofs. These weights attempt to capture the size of the proof required to justify an edge. Conceivably, it would make sense to make these weights also include the parts of the proof that go beyond congruence closure reasoning. Currently, an equality that is asserted into the congruence closure algorithm would receive the weight 1, even if it implicitly depends on an equivalence, since the weight cannot capture the external reasoning that was done to derive the asserted equality. If instead we could compute an appropriate weight for that edge, which took into account all the reasoning required to derive it, we could ensure that an edge which implicitly depends on an equivalence will have a weight bigger than the proof size for that equivalence. Thus, when explaining this equivalence, the shortest path found would not include this edge.

While this strategy might work in theory, it would require estimating the proof size for each edge, including parts of the proof beyond congruence closure reasoning. In a modern SMT solver, most of these proofs are computed in a lazy fashion [5], and changing that behaviour would greatly impact performance. Therefore, while this technique might work in different contexts, it is not practical for a high-performance SMT solver.

Computing Levels Based on *Provenance*. While constructing the entire implication graph for the SMT solver might be impractical, there are incomplete alternatives that are less restrictive than the arbitrary topological ordering imposed by the edge levels. In particular, when the SAT solver asserts a literal l into the EUF theory solver, it has access to the set of literals that immediately caused that literal to be propagated (in the SAT solver, this would be a clause). This set of literals, which we denote as l's *provenance*, represents the direct ancestors of l in the implication graph. Using this information, we can define the level of a literal l (which will determine the level of edges in the e-graph) to be one more than the maximum level of the literals in its provenance. More formally,

$$\text{level}(l) = 1 + max \{ \text{ level}(p) \mid p \in \text{provenance}(l) \}$$

Contrary to the previous level solution, this would allow multiple edges to share the same level, and in fact each level will correspond to one "layer" of the implication DAG. This would mean that equalities that have the same level have no dependency relation in the implication graph—one could be explained by the other and vice-versa.

While this solution is quite elegant[2], it would still result in explanations that are rejected by the SAT solver as circular. Let l_1 and l_2 be two literals with the same provenance level, and say the EUF solver produces an explanation for l_1 in terms of l_2, but l_2 is situated after l_1 in the SAT solver trail. Since the SAT solver does not store the full implication graph, it cannot ensure that the explanation given is valid, *even if it does respect the dependency relations in the implication graph.* In other words, the SAT solver considers the particular topological ordering of the implication graph that composes the SAT trail as the "source of truth", and will reject explanations that do not follow it, even if they wouldn't lead to circular proofs.

---

[2] It does, however, require changing the API of the EUF theory solver to receive provenance information, and updating all users of that API to provide that information, which is a non-trivial engineering effort.

This invariant sits at the core of modern SAT solvers, and changing it would involve allowing arbitrary *reimplication* of propagated literals. There has been some recent work exploring the benefits of reimplication in SAT solving [8,16], but these techniques are still not well established in SMT solvers. Until that changes, we believe that any correct solution for the implicit dependencies problem will necessarily be as strict as the above solution based on the edge levels.

4   Implementation

We have implemented the modified version of the congruence closure algorithm from the previous section in cvc5 [4], a state-of-the-art SMT solver. The existing theory solver for the theory of equality and uninterpreted functions, called its *equality engine*, implements a version of a proof-producing congruence closure algorithm as seen in Nieuwenhuis et al. [19]. As part of this work, we adapted this existing implementation to not discard redundant equalities, and implemented the GREEDY explanation algorithm, with the adaptations detailed in the previous sections. This required rewriting a large part of the equality engine, with over 900 lines of code changed. In this section, we highlight some of the implementation details that went in to this work.

4.1   Computing the Merge Level of Two Terms

An important part in selecting which edges are allowed in an explanation is determining the merge level of the terms being explained. For a given path between two terms in the e-graph, we said that its *path level* is the maximum of the levels of all edges in path. This corresponds to the level that the path first appeared in the e-graph. Note that the level in which two terms were merged is the level in which any path first appeared between them. Thus, the merge level of two terms is the minimum path level of all the paths between them.

However, we know that when two terms are merged, the path between them with minimal path level will not contain any redundant edge (otherwise that edge would have caused a merge, and would not be redundant). Therefore, we only need to concern ourselves with paths that don't contain redundant edges. There is always only one such path between two terms, which is the path computed by the classical congruence closure algorithms, known as the *tree path*.

All put together, to compute the merge level of two terms we must simply traverse the tree path between them, and record the highest edge level encountered.

4.2   Backtracking-Aware Edge Limits

As mentioned earlier, when computing extra redundant congruence edges, the modified congruence closure algorithm only adds edges up to a limit. This is intended to ensure the algorithm has a linear overhead when compared to traditional congruence closure algorithms, since in theory the extra edges could amount to $O(n^2)$ where n is the number of nodes in the e-graph. In the original implementation [11], this limit was set to $2n$.

When implementing these algorithms in cvc5, however, we found that setting a limit based on the current state of the graph was not sufficient. In particular, in many benchmarks we saw a situation where the SAT solver asserts an equality to the congruence closure algorithm, which caused a large number of redundant edges to be computed; then, the solver backtracks, removing all the redundant edges, as well as the asserted equality; and after that the process repeats, with the solver adding a small number of non-redundant edges followed by computing a large number of redundant edges.

In these benchmarks, we saw that the total number of redundant edges added greatly shadowed the number of non-redundant edges, and there was a substantial slowdown compared to the traditional congruence closure algorithm. To address this, we implemented a smarter limit to the number of computed redundant edges, which takes into account all the edges added throughout the solving process. Now, the total number of redundant edges added cannot exceed twice the number of non-redundant edges.

4.3   Lazy Computation of Weights and Extra Edges

Due to backtracking, it might be the case that edges added to the e-graph will be removed before an explanation using them is computed. Therefore, eagerly recomputing edge weights everytime an edge is added would be inefficient. Instead, we only compute edge weights when searching for an explanation.

Recall that the weight of an edge is based on the proof size for that edge, without using redundant equalities. Therefore, the only edges that affect the weight of a given edge are those that appeared before it. cvc5's equality engine backtracks by removing edges in reverse order, meaning that if an edge was not removed during backtracking, no edge before it was removed either. As such, when backtracking, we don't need to invalidate any computed edge weight, aside from removing the weights of edges that were removed.

We also compute the extra congruence edges as described in Sect. 2.4 in a lazy fashion, only when queried for an explanation. Importantly, we must take care to set the edge level correctly. For a redundant congruence edge between the terms a and b, we can safely set its level to the merge level of a and b. This is because, although we potentially compute this edge much later, it could have conceivably been added as soon as the terms' equivalence classes were merged—it does not have any implicit dependencies.

5   Evaluation

In order to evaluate the effectiveness and performance of our implementation, we tested it against a large set of benchmarks from the SMT-LIB benchmarks library [7], an industry-standard set of benchmarks for SMT solvers. We used the 162,228 SMT problems from a set of 14 logics, notably involving, besides the theory of equality and uninterpreted functions, the theories of strings and arrays[3]. These particular theories were selected because they make heavy use of equality reasoning, and are good testing grounds for the congruence closure algorithm. The results were generated with a cluster equipped with 32 x Intel(R) Xeon(R) CPU E5-2620 v4 @ 2.10 GHz, 256 GiB RAM machines, with one core per solver/benchmark pair, 120 s time limit, and 8 GiB memory limit.

We focus our evaluation on a comparison between the GREEDY and VANILLA algorithms. For each benchmark, we recorded the time taken to solve the benchmark with each of the algorithms, as well as the size of the final proof produced. We also measure directly, for each benchmark, the explanation size returned by all calls to `get_explanation`.

Table 1 shows an overview of our main results. The benchmarks are divided in three groups, based on the logic: the logics that include the theory of arrays, those that include the theory of strings, and those that include neither. For each logic group, the table presents the relative change in runtime, proof size and explanation of the GREEDY algorithm, when using VANILLA as a baseline.

The table shows that the performance of the modified algorithm is heavily theory-dependent. For the theory of strings, the change in explanation size is very minor, but the total proof size is moderately smaller when using GREEDY than with VANILLA. However, the runtime overhead is very large for these logics. On the other hand, for the other two benchmark groups, the explanation size difference is more pronounced, but this difference did not translate into a large difference in the total proof size. The runtime overhead for these groups was significantly smaller, but still substantial.

In the next few sections, we take a more detailed look into each of the evaluated metrics.

5.1 Runtime

Figure 6 presents scatter plots of the runtime of each benchmark, when run with GREEDY and VANILLA. The benchmarks are again divided in three groups,

Table 1. Relative change in runtime, proof size and explanation size of GREEDY when compared to VANILLA, for different groups of logics.

Fragment	Runtime	Proof size	Explanation size
Arrays-based logics	25.96%	−3.26%	−7.97%
Strings-based logics	56.52%	−4.34%	−0.83%
Remaining logics	29.85%	−3.16%	−21.27%
Total	45.59%	−3.33%	−7.91%

---

[3] The specific logics used were ALIA, AUFLIA, AUFLIRA, QF_ALIA, QF_AUFLIA, QF_AX, QF_S, QF_SLIA, QF_UF, QF_UFLIA, QF_UFLRA, UF, UFLIA, and UFLRA.

depending on the logic. Points below the diagonal represent benchmarks in which
GREEDY outperformed VANILLA, and points above the diagonal are the opposite.

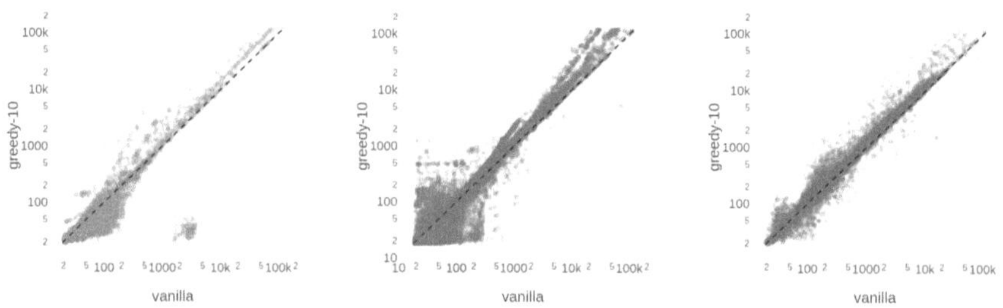

(a) Runtime for arrays-based logics.

(b) Runtime for strings-based logics.

(c) Runtime for remaining logics.

Fig. 6. Scatter plots of the runtime of each benchmark, when run using GREEDY and
VANILLA.

The plot shows that, for most benchmarks, and in particular larger benchmarks, GREEDY presents a significant overhead when compared to VANILLA. In
total, solving all benchmarks with GREEDY took 45.6% more time than with
VANILLA.

Interestingly, there were a number of benchmarks that took significantly less
time to run with GREEDY. In the most extreme case, the plot shows a set of
benchmarks in the array-based logics that take a few seconds to be solved with
VANILLA, but only several milliseconds when using GREEDY. Over all logics,
there were 5,347 benchmarks that were at least twice as fast with GREEDY when
compared to VANILLA (out of 162,228 total benchmarks). We speculate that this
is related to smaller congruence closure explanations resulting in smaller conflict
sets, which in turn results in a more efficient SAT search.

5.2 Proof Size

Figure 7 presents, for each benchmark group, a scatter plot of the final proof size
of each benchmark, when run with GREEDY and VANILLA. Benchmarks where
the proof size was exactly the same with both algorithms are omitted.

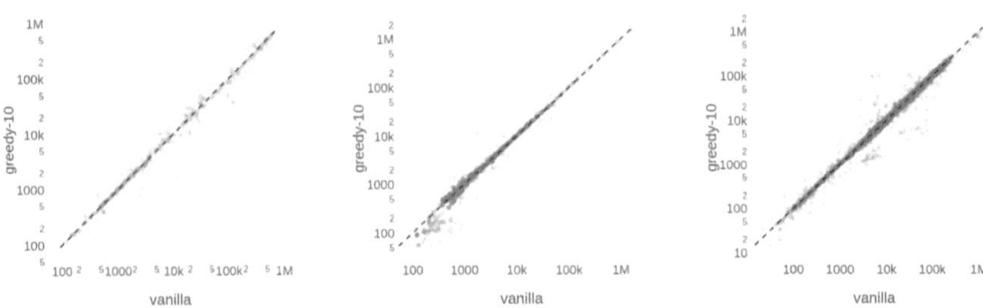

(a) Proof size for arrays-based logics.

(b) Proof size for strings-based logics.

(c) Proof size for remaining logics.

Fig. 7. Scatter plots of the final proof size of each benchmark, when run using Greedy and Vanilla.

For most benchmarks, the final proof size did not change substantially between the two algorithms. However, since this is a measure of the final proof size, which includes reasoning steps besides congruence closure, the impact of the congruence closure algorithm is mitigated. In total, the proof size of all proofs was 3.3% smaller when generated with Greedy, when compared to Vanilla.

5.3 Explanation Size

To more closely analyze the impact of the modified algorithm, we also looked at, for each benchmark, the size of all explanations returned by the congruence closure algorithm. Figure 8 presents scatter plots of the total explanation size of each benchmark, when run with Greedy and Vanilla. The benchmarks were grouped similarly to the previous plots, and benchmarks where the total explanation size was exactly the same with both algorithms are omitted.

This plot more clearly shows that the explanations returned by Greedy are generally smaller than those of Vanilla. However, this is not consistent across

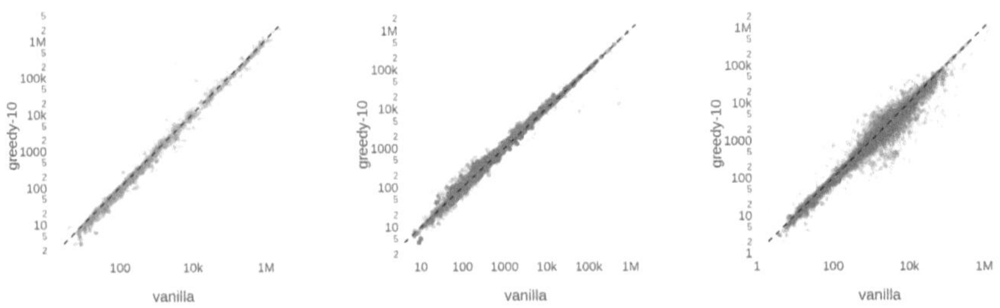

(a) Explanation size for arrays-based logics.

(b) Explanation size for strings-based logics.

(c) Explanation size for remaining logics.

Fig. 8. Scatter plots of the total explanation size of each benchmark, when run using Greedy and Vanilla.

theories. In particular, benchmarks form string-based logics seem to benefit less from the new algorithm, while benchmarks from other logics show a more clear improvement.

Overall, the total explanation size for the GREEDY algorithm was 7.9% smaller that that of VANILLA.

5.4   Comparison with Baseline Implementation from cvc5

As mentioned earlier, implementing the GREEDY algorithm in cvc5 required rewriting a large portion of the equality engine. This involved changing even how the traditional algorithm was implemented, which we refer to as VANILLA. To guarantee our changes did not degrade substantially the performance of cvc5's equality engine, we also compared the new VANILLA implementation with the baseline implementation, before any changes. These results are shown in Fig. 9.

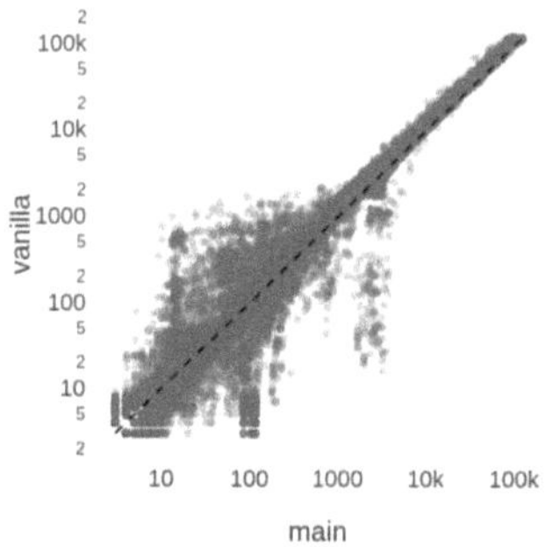

Fig. 9. Scatter plots of the runtime of each benchmark, when run using VANILLA versus the existing implementation in cvc5.

While there is a lot of noise in the results, it is possible to see that there is a small consistent overhead in the implementation of VANILLA when compared to the baseline, especially in larger benchmarks. On average, benchmarks take 8% longer to solve when using VANILLA. While this overhead is not negligible, it is relatively modest, and could be improved with further optimization.

5.5   Other Results

To get a more detailed view into how the redundant edges affect the returned explanations, we also recorded a series of *redundancy* measurements. Over all benchmarks, the number of redundant equalities included in the returned explanations was 3.53% of the total explanation size. Similar to other metrics, this was also theory-dependent: For strings-based logics this ratio was only 0.78%, while for arrays and other logics, it was 5.79% and 6.61%, respectively. We also measured the proportion of all e-graph edges that were redundant. Over all benchmarks, 43.58% of added edges were redundant—this was largely consistent across all logic groups, staying between 41.92% and 47.78%.

6   Conclusion and Future Work

We presented our effort to adapt a new congruence closure algorithm to work in a state-of-the-art SMT solver. We evaluated our implementation on a large, industry-standard set of benchmarks, and obtained mixed but promising results.

Our work shows that congruence closure algorithms that keep redundant equalities can be efficiently implemented in the context of a backtracking SMT solver, with reasonable overhead. Furthermore, we show that these techniques have a measurable and at times substantial impact on explanation and proof size. In particular, for logics involving the theories of arrays and equality and uninterpreted functions, we show that this new algorithm can result in consistently smaller explanations, with a small impact on final proof size. For some benchmarks, we also observed that the new algorithm can result in a drastic reduction in runtime. As for logics involving the theories of strings, the new algorithm did not produce significantly smaller explanations, but it did result in moderately smaller final proofs. However, the runtime overhead was also larger for these benchmarks.

Future opportunities for research may include improving the heuristics or developing new algorithms that make use of redundant edges. For example, the current results seem to show that this technique struggles with very large benchmarks, and maybe a more sophisticated heuristic for limiting the number of redundant edges added might be beneficial.

While our work was initially focused in treating the interaction between the equality engine and the SAT solver, congruence closure reasoning is used by many theories in SMT solving, and future work might explore how to optimize these techniques for the specific interactions between those theories and the congruence closure algorithm. Finally, this work is still somewhat restricted by the specific topological ordering of the implication graph that the SAT solver selects as its trail. One possible avenue for future research is investigating how reimplication, when adapted for an SMT context, might improve this restriction. In conclusion, while the observed experimental results are overall mixed, there is promising evidence that these technique might be useful in some contexts and for some theories.

Data Availability Statement. The source code relevant to this research, as well as the tools and data used for the experimental evaluation are available at [1].

References

1. Andreotti, B., Barbosa, H.: Artifact for "producing shorter congruence closure proofs in a state-of-the-art smt solver" (2025). https://doi.org/10.5281/zenodo.17181344
2. Andreotti, B., Lachnitt, H., Barbosa, H.: Carcara: an efficient proof checker and elaborator for smt proofs in the alethe format. In: Tools and Algorithms for the Construction and Analysis of Systems: 29th International Conference, TACAS 2023, Held as Part of the European Joint Conferences on Theory and Practice

of Software, ETAPS 2023, Paris, France, 22–27 April 2023, Proceedings, Part I, pp. 367–386. Springer-Verlag, Berlin, Heidelberg (2023). https://doi.org/10.1007/978-3-031-30823-9_19

3. Barbosa, H., et al.: Generating and exploiting automated reasoning proof certificates. Commun. ACM **66**(10), 86–95 (2023). https://doi.org/10.1145/3587692

4. Barbosa, H., et al.: cvc5: a versatile and industrial-strength SMT solver. In: TACAS 2022. LNCS, vol. 13243, pp. 415–442. Springer, Cham (2022). https://doi.org/10.1007/978-3-030-99524-9_24

5. Barbosa, H., et al.: Flexible proof production in an industrial-strength smt solver. In: Automated Reasoning: 11th International Joint Conference, IJCAR 2022, Haifa, Israel, 8–10 August 2022, Proceedings, pp. 15–35. Springer, Heidelberg (2022). https://doi.org/10.1007/978-3-031-10769-6_3

6. Barbosa, H., Reynolds, A., Ouraoui, D.E., Tinelli, C., Barrett, C.W.: Extending SMT solvers to higher-order logic. In: Fontaine, P. (ed.) Proceedings of Conference on Automated Deduction (CADE). Lecture Notes in Computer Science, vol. 11716, pp. 35–54. Springer, Heidelberg (2019). https://doi.org/10.1007/978-3-030-29436-6_3

7. Barrett, C., Fontaine, P., Tinelli, C.: The Satisfiability Modulo Theories Library (SMT-LIB) (2016). https://smt-lib.org/

8. Coutelier, R., Fleury, M., Kovács, L.: Lazy reimplication in chronological backtracking. In: Chakraborty, S., Jiang, J.R. (eds.) 27th International Conference on Theory and Applications of Satisfiability Testing, SAT 2024, Pune, India, 21–24 August 2024. LIPIcs, vol. 305, pp. 9:1–9:19. Schloss Dagstuhl - Leibniz-Zentrum für Informatik (2024). https://doi.org/10.4230/LIPICS.SAT.2024.9

9. Downey, P.J., Sethi, R., Tarjan, R.E.: Variations on the common subexpression problem. J. ACM **27**(4), 758–771 (1980). https://doi.org/10.1145/322217.322228

10. Fellner, A., Fontaine, P., Paleo, B.W.: Np-completeness of small conflict set generation for congruence closure. Form. Methods Syst. Des. **51**(3), 533–544 (2017). https://doi.org/10.1007/s10703-017-0283-x

11. Flatt, O., Coward, S., Willsey, M., Tatlock, Z., Panchekha, P.: Small proofs from congruence closure. In: 2022 Formal Methods in Computer-Aided Design (FMCAD), pp. 75–83 (2022). https://doi.org/10.34727/2022/isbn.978-3-85448-053-2_13

12. Fontaine, P., Merz, S., Woltzenlogel Paleo, B.: Compression of propositional resolution proofs via partial regularization. In: Bjørner, N., Sofronie-Stokkermans, V. (eds.) Automated Deduction – CADE-23, pp. 237–251. Springer, Heidelberg (2011). https://doi.org/10.1007/978-3-642-22438-6_19

13. Heule, M.J.H., Kiesl, B., Biere, A.: Short proofs without new variables. In: de Moura, L. (ed.) Automated Deduction – CADE 26, pp. 130–147. Springer, Cham (2017). https://doi.org/10.1007/978-3-319-63046-5_9

14. Hoenicke, J., Schindler, T.: A simple proof format for SMT. In: Déharbe, D., Hyvärinen, A.E.J. (eds.) International Workshop on Satisfiability Modulo Theories (SMT). CEUR Workshop Proceedings, vol. 3185, pp. 54–70. CEUR-WS.org (2022). http://ceur-ws.org/Vol-3185/paper9527.pdf

15. Marques Silva, J., Sakallah, K.: Grasp-a new search algorithm for satisfiability. In: Proceedings of International Conference on Computer Aided Design, pp. 220–227 (1996). https://doi.org/10.1109/ICCAD.1996.569607

16. Nadel, A.: Introducing intel(r) SAT solver. In: Meel, K.S., Strichman, O. (eds.) 25th International Conference on Theory and Applications of Satisfiability Testing, SAT 2022, Haifa, Israel, 2–5 August 2022. LIPIcs, vol. 236, pp 8:1–8:23.

Schloss Dagstuhl - Leibniz-Zentrum für Informatik (2022). https://doi.org/10.4230/LIPICS.SAT.2022.8

17. Nelson, G., Oppen, D.C.: Fast decision procedures based on congruence closure. J. ACM **27**(2), 356–364 (1980). https://doi.org/10.1145/322186.322198

18. Nieuwenhuis, R., Oliveras, A.: Congruence closure with integer offsets. In: Vardi, M.Y., Voronkov, A. (eds.) LPAR 2003. LNCS (LNAI), vol. 2850, pp. 78–90. Springer, Heidelberg (2003). https://doi.org/10.1007/978-3-540-39813-4_5

19. Nieuwenhuis, R., Oliveras, A.: Proof-producing congruence closure. In: Giesl, J. (ed.) Term Rewriting and Applications. pp. 453–468. Springer, Heidelberg (2005). https://doi.org/10.1007/978-3-540-32033-3_33

20. Nieuwenhuis, R., Oliveras, A., Tinelli, C.: Solving sat and sat modulo theories: from an abstract davis–putnam–logemann–loveland procedure to dpll(t). J. ACM **53**(6), 937–977 (2006). https://doi.org/10.1145/1217856.1217859

21. Otoni, R., Blicha, M., Eugster, P., Hyvärinen, A.E.J., Sharygina, N.: Theory-specific proof steps witnessing correctness of SMT executions. In: Design Automation Conference (DAC), pp. 541–546. IEEE (2021). https://doi.org/10.1109/DAC18074.2021.9586272

22. Reeves, J.E., Kiesl-Reiter, B., Heule, M.J.H.: Propositional proof skeletons. In: Sankaranarayanan, S., Sharygina, N. (eds.) Tools and Algorithms for the Construction and Analysis of Systems, pp. 329–347. Springer, Cham (2023). https://doi.org/10.1007/978-3-031-30823-9_17

23. Schurr, H., Fleury, M., Desharnais, M.: Reliable reconstruction of fine-grained proofs in a proof assistant. In: Platzer, A., Sutcliffe, G. (eds.) Proceedings of Conference on Automated Deduction (CADE). Lecture Notes in Computer Science, vol. 12699, pp. 450–467. Springer, Heidelberg (2021). https://doi.org/10.1007/978-3-030-79876-5_26

24. Tarjan, R.E.: Efficiency of a good but not linear set union algorithm. J. ACM **22**(2), 215–225 (1975). https://doi.org/10.1145/321879.321884

25. Tate, R., Stepp, M., Tatlock, Z., Lerner, S.: Equality saturation: a new approach to optimization. SIGPLAN Not. **44**(1), 264–276 (2009). https://doi.org/10.1145/1594834.1480915

26. Willsey, M., Nandi, C., Wang, Y.R., Flatt, O., Tatlock, Z., Panchekha, P.: EGG: fast and extensible equality saturation. Proc. ACM Program. Lang. **5**(POPL) (2021). https://doi.org/10.1145/3434304

Reachability in Multi-agent Transfer Systems

Nathalie Bertrand[1], Loïc Hélouët[1], Engel Lefaucheux[2],
and Luca Paparazzo[1]([envelope])

[1] University of Rennes, Inria, CNRS, Rennes, France
{nathalie.bertrand,loic.helouet,luca.paparazzo}@inria.fr
[2] Université de Lorraine, CNRS, Inria, LORIA, Nancy, France
engel.lefaucheux@inria.fr

Abstract. This paper introduces collaborative reachability games with energy constraints. In the considered arenas, agents can spend or gain energy during moves, or share it with their peers if their current position allows it. We study several variants of energy reachability games where agents move either synchronously or asynchronously, and with/without constraints on energy transfers among peers. We show that these problems have different complexities ranging from NP to EXPSPACE.

Keywords: Multi-Agent Systems · Quantitative verification · VASS · Collaborative games · Planning

1 Introduction

Cooperation of several agents occurs in a variety of applications, such as robotics, traffic control and aviation to name a few. In contrast to adversarial games, in such cooperative settings, the agents collaborate to achieve a common goal. A typical instantiation of this general framework is the multi-agent path finding problem [11], in which one aims at designing a plan to move multiple agents while avoiding collisions to perform a global task. Beyond Boolean objectives such as coverage of an area, or reachability of a position for each agent, introducing quantities in models for multi-agent systems is crucial to represent energy or financial cost. Quantitative settings where multiple agents interact are for instance useful formalisms to find optimal management strategies to control cyber physical systems (CPS) where objectives are not purely Boolean, but also aim at optimizing some measure. A variety of settings of quantitative multi-player games have been proposed in the literature [4–6,12]. Game concepts such as the famous Nash equilibria [20] can then be studied, for instance to efficiently distribute energy in smart grids [5].

This work was supported by the project BisoUS (ANR-22-CE48-0012).

The authors would like to thank an anonymous reviewer for their constructive comments on links between our model and (1-zero-test) VASS.

Y.-F. Chen et al. (Eds.): VMCAI 2026, LNCS 16417, pp. 21–43, 2026.
https://doi.org/10.1007/978-3-032-15700-3_2

In this paper, we introduce a new quantitative multi-agent model, in which agents move on their own local arena and are given a goal, *i.e.*, a particular vertex to reach. Local arenas are equipped with integer weights on edges to represent energy variations. Each agent stores energy and, when moving from a vertex to a consecutive one, gains energy if the weight is positive, or loses energy if the weight is negative. Interestingly, agents may cooperate by sending some or all of their stored energy to other agents. The objective is to design a collaborative plan moving each agent to its target vertex while staying within the energy available in the system, possibly using transfers among peers. We coin this model *multi-agent transfer systems*, or simply *transfer systems*.

Several semantics can be considered for transfer systems: either agents move synchronously or asynchronously. In any case, they can only take an edge if their stored energy is sufficient, as their energy level cannot drop below 0. We consider the natural question of global reachability objectives, where all agents must reach their assigned target simultaneously. Our setting thus shares the objective of multi-agent path finding [11]. There are however several crucial differences: first, in transfer systems, agents move on their respective local arenas rather than on a common space; second, transfer systems are equipped with energy variations and agents must move within energy budget; finally, in transfer systems agents can transfer energy one to another.

Transfer systems form a particularly suited model for modern urban transport networks equipped with regenerative braking systems. In these CPSs, the kinetic energy of a braking vehicle can be converted into electric energy, transferred to the power network and used by other close vehicles. Another possible application is the study of the logistics of complex systems in which resources must be provided at specific locations and times for the success of a mission.

Multi-weighted energy games [10] are close to our transfer systems. In multi-weighted energy games, stored quantities are k-vectors of integers and moves are also labeled by integer vectors of same dimension. Different to our setting, the number of players is fixed to at most two. The objective in these games is to play infinitely while respecting energy bounds on each coordinate: a lower bound or a combination of lower and a weak/strong upper bounds. With a single player, the problem with a lower bound is NP-hard and k-EXPTIME already, and with two players, the complexity is EXPTIME-hard and in k-EXPTIME. These complexity proofs build on results of [3]. One can consider transfer systems with n agents as a reachability question in a multi-weighted game with a single player (representing the coalition of agents) of dimension n, one dimension for each agent that must remain non-negative. For an arbitrary dimension, existence of an infinite run in multiweighted games with lower bounds is EXPSPACE-complete, and becomes PSPACE-complete if integral upper bound are set for each dimension. Notice that this setting has several differences with our questions in transfer systems; one of the main differences is that [10] considers the existence of infinite runs, while the questions addressed in this paper would be encoded as coverability questions. Most importantly, transfer systems are given succinctly by local arenas for each agent, while multi-weighted energy games are monolithic.

As our model deals with transfer of energy, and is close to Petri nets, a natural question is whether reachability in transfer systems is equivalent to a reachability or coverability in transfer Petri nets [8]. Transfer Petri nets extend Petri nets with flow relations that can transfer the whole contents of a place p to another place p' when firing a transition. Our complexity results on transfer systems prove that reachability for transfer systems and coverability/reachability for transfer nets are different questions. Indeed, transfer Petri nets can easily simulate Reset Petri nets a model where reachability is undecidable [1], and coverability is Ackermann-hard [23]. In contrast, our reachablity problems on transfer systems remain decidable in almost all cases, and have at worst complexities in EXPSPACE when decidable. From a modeling perspective, transfers in Petri nets and in transfer systems are quite different: in Petri nets the whole contents of a place is transferred in one step while in our model, an agent can share only a part of its energy.

The semantics of transfer systems can be captured by vector addition systems with states (VASS) [14], or equivalently by Petri nets, and our reachability problems as coverability questions. EXPSPACE-hardness for coverability in VAS was shown by [18], and the matching EXPSPACE upper bound was shown by [21]. A natural question is whether one has to pay the full complexity of VASS to solve our reachability problems in transfer systems. We show in this paper that the answer depends on the chosen characteristics of the model. For instance, reachability for transfer systems with energy transfers always enabled and under asynchronous semantics lies between NP and PSPACE. Also, under asynchronous semantics with arbitrary transfer groups, the complexity lies between PSPACE and EXPSPACE. Finally, if one relaxes synchronicity by allowing agents that lack energy to idle (resulting in the so-called weak synchronous semantics), reachability becomes undecidable.

The rest of the paper is organized as follows. Section 2 presents the model and the notations that will be used throughout the paper. Section 3 details the different possible semantics of the model: asynchronous, strongly synchronous, and weakly synchronous and shows the relations between these semantics. Section 4 studies the complexity of reachability under all semantics when transfer of energy can occur at any time between agents. Section 5 considers reachability for systems with restricted local transfers, that can occur only in some states. Due to space constraints, some proofs are omitted and can be found in an extended version of this paper available at [2].

2   Transfer Systems

Transfer systems are multi-agent systems, in which every agent plays on a local weighted graph, and the communication between agents is limited.

Definition 1. *A local arena* $A = (V, E)$ *is a directed weighted graph where* V *is a finite set of vertices,* $E \subseteq V \times \mathbb{Z} \times V$ *describes the edges of the arena.*

Intuitively, the weight on an edge represents the amount of energy an agent gains (if positive) or loses (if negative) while traversing that edge. Communication between agents is limited to energy transfers, and is formalised by transfer groups that specify conditions on the vertices of the agents to enable transfers.

Definition 2. *Let $n \in \mathbb{N}$, $\mathcal{A} = \{A_1, \ldots, A_n\}$ be a set of local arenas with $A_i = (V_i, E_i)$ for every $i \in [\![1, n]\!]$ (assuming all V_i are disjoint sets) and $\mathcal{T} : \bigcup_{i \in 1..n} V_i \to \mathbb{N}$ is a partial map defining* transfer groups.
$\mathcal{A}$ and $\mathcal{T}$ induce the transfer system *$\mathsf{TS} = \langle \mathcal{A}, \mathcal{T} \rangle$.*

We will say that vertices v, v' belong to the same transfer group if $\mathcal{T}(v) = \mathcal{T}(v')$, and impose that transfer groups are not singletons, are disjoint sets and contain states from at least two arenas. For convenience, we will often define transfer groups as sets of states $T_1, \ldots, T_k$ where $T_i = \{v \mid \mathcal{T}(v) = i\}$. Writing $V = \bigcup_{i \in 1..n} V_i$, the size of a transfer arena is defined as $|\mathsf{TS}| = |V| \cdot (|\mathcal{T}| + 1) + |V|^2 . \log(w_{max})$ where w_{max} is the largest absolute value of a weight appearing in an arena, and $|\mathcal{T}|$ is bounded $|V| \cdot \log(|V|)$. The semantics of a transfer system $\mathsf{TS} = \langle \mathcal{A}, \mathcal{T} \rangle$ is given in terms of a transition system. A *configuration* of TS consists of the current vertex of each agent and their energy level: we write C, C', etc. for a configuration, and $\Gamma = \left(\prod_{i=1}^n V_i \right) \times \mathbb{N}^n$ for the set of all configurations. For a configuration $C = (S, \overrightarrow{e})$, $S \in \prod_{i=1}^n V_i$ is referred to as the *global state* (or simply state) and $\overrightarrow{e}$ as the *energy vector*. When the dimension n is clear from the context, we use $\overrightarrow{0}$ to denote the null energy vector $(0, \ldots, 0) \in \mathbb{N}^n$.

Transitions between configurations are induced by moves of the agents on their local arenas, or energy transfers between agents when permitted by the transfer groups. An agent A_i cannot move along an edge $q_i \xrightarrow{-w} q_i'$ with negative weight $-w$ if its energy level e_i is lower than w. We will say that edge $q_i \xrightarrow{w} q_i'$ is *enabled* if $e_i + w \geq 0$. For move transitions, we distinguish several semantics, depending on whether the agents move simultaneously or not.

Definition 3. *Consider two configurations $C = \langle (q_1, \ldots, q_n), (e_1, \ldots, e_n) \rangle$ and $C' = \langle (q_1', \ldots, q_n'), (e_1', \ldots, e_n') \rangle$.*

move *There is a* move transition *from C to C' if one of the following holds*
  asynchronous *$\exists i \in [\![1, n]\!] : q_i \xrightarrow{w} q_i' \in E_i$, $e_i' = e_i + w \geq 0$ and $\forall j \neq i, (q_j', e_j') = (q_j, e_j)$, corresponding to the single agent A_i moving along an edge of its local arena. This results in an asynchronous* move transition, *and is denoted $C \longrightarrow_m^a C'$.*
  synchronous *$\forall i \in [\![1, n]\!], q_i \xrightarrow{w_i} q_i' \in E_i$ and $e_i' = e_i + w_i \geq 0$, corresponding to all agents moving simultaneously in their respective local arenas. This results in a strongly synchronous* move transition, *denoted $C \longrightarrow_m^s C'$.*
  weakly synchronous *$\forall i \in [\![1, n]\!]$, either $q_i \xrightarrow{w_i} q_i' \in E_i$ and $e_i' = e_i + w_i \geq 0$ or $(q_i', e_i') = (q_i, e_i)$ and $\forall q_i \xrightarrow{w_i} q_i'' \in E_i, e_i + w_i < 0$, corresponding to a synchronous move of all agents that have an enabled edge. This results in a weakly synchronous* move transition, *denoted $C \longrightarrow_m^w C'$.*

transfer *There is a transfer transition from C to C' if $\forall i, q_i = q'_i$, and $\exists i, j \in [\![1, n]\!], \mathcal{T}(q_i) = \mathcal{T}(q_j), e_i + e_j = e'_i + e'_j$ and $\forall k \notin \{i, j\}, e_k = e'_k$, corresponding to a transfer between agents A_i and A_j on vertices of a same transfer group. This transition is denoted $C \longrightarrow_t C'$.*

We use $\longrightarrow^a$ (resp. $\longrightarrow^s$, resp. $\longrightarrow^w$) to denote a transition that is either a transfer or an asynchronous (resp. strongly synchronous, resp. weakly synchronous) move and call it an asynchronous (resp. strongly synchronous, resp. weakly synchronous) transition for short. For instance $\longrightarrow^a = \longrightarrow^a_m \cup \longrightarrow_t$. We also refer to any type of move transition with $\longrightarrow_m$: $\longrightarrow_m = \longrightarrow^a_m \cup \longrightarrow^s_m \cup \longrightarrow^w_m$. Finally, an arbitrary transition is simply denoted $\longrightarrow$. Notice that agents change their local vertex in their arena during moves, and stay on the same vertex during transfers.

As usual, sequences of transitions define runs of the transfer system. A finite/infinite *asynchronous run* (resp. strongly synchronous run, resp. weakly synchronous run) over TS is a finite/infinite sequence of asynchronous (resp. strongly synchronous, resp. weakly synchronous) transitions. We will write $C \rightsquigarrow^a C'$ (resp. $C \rightsquigarrow^s C'$, $C \rightsquigarrow^w C'$) when there exists an asynchronous (resp. synchronous, weakly synchronous) run from C to C'. The set of asynchronous (resp. strongly synchronous, resp. weakly synchronous) runs over TS is denoted $\mathsf{Runs}^a(\mathsf{TS})$ (resp. $\mathsf{Runs}^s(\mathsf{TS})$, resp. $\mathsf{Runs}^w(\mathsf{TS})$). We refer to them as the asynchronous, strongly synchronous and weakly synchronous semantics of the transfer system, respectively, that we sometimes abbreviate into a-semantics, s-semantics and w-semantics.

We observe the following relations between runs of transfer systems under the various semantics. First of all, $\mathsf{Runs}^s(\mathsf{TS}) \subseteq \mathsf{Runs}^w(\mathsf{TS})$. Indeed, by definition, for every two configurations C, C', if $C \longrightarrow^s_m C'$ then $C \longrightarrow^w_m C'$. One can also notice that if $C \longrightarrow^w_m C'$, then there exists a sequence of move transitions $C \longrightarrow^a_m C_1 \longrightarrow^a_m \ldots \longrightarrow^a_m C'$. Thus, a run in the weakly synchronous semantics can be simulated by a run in the asynchronous semantics.

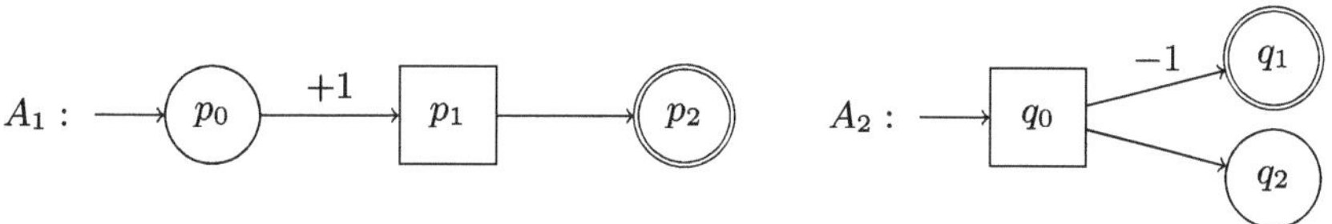

Fig. 1. Transfer system with two agents and a single transfer group with p_1 and q_0: $\mathcal{T} = \{\{p_1, q_0\}\}$. Null weights are omitted.

Example 1. Consider the transfer system with two agents depicted in Fig. 1, where boxed vertices belong to the same transfer group. Let $C_0 = \langle (p_0, q_0), (0, 0) \rangle$ be the initial configuration. Then, there exists an asynchronous run from C_0 to the target state (p_2, q_1), namely $\langle (p_0, q_0), (0, 0) \rangle \longrightarrow^a_m \langle (p_1, q_0), (1, 0) \rangle \longrightarrow_t$

$\langle (p_1, q_0), (0, 1) \rangle \longrightarrow_m^a \langle (p_1, q_1), (0, 0) \rangle \longrightarrow_m^a \langle (p_2, q_1), (0, 0) \rangle$. Yet, there are no strongly nor weakly synchronous runs to the target.

Consider now the transfer system depicted in Fig. 2. Again, there exists an asynchronous execution that reaches the target state : $\langle (p_0, q_0), (0, 0) \rangle \longrightarrow_m^a \langle (p_1, q_0), (0, 0) \rangle \longrightarrow_m^a \langle (p_1, q_0), (1, 0) \rangle \longrightarrow_m^a \langle (p_1, q_1), (1, 1) \rangle \longrightarrow_t \langle (p_1, q_1), (0, 2) \rangle \longrightarrow_m^a \langle (p_1, q_3), (0, 0) \rangle \longrightarrow_m^a \langle (p_2, q_3), (0, 0) \rangle$. Also, since the only transition fireable from $(q_1, 1)$ for A_2' leads to q_2, there is no synchronous run to (p_2, q_3). Finally, even though A_2' has a move transition available from $(q_1, 1)$ leading to q_2, there exists a weakly synchronous run that avoids the sink state q_2 and reaches the global target state: $\langle (p_0, q_0), (0, 0) \rangle \longrightarrow_m^w \langle (p_1, q_1), (0, 1) \rangle \longrightarrow_t \langle (p_1, q_1), (1, 0) \rangle \longrightarrow_m^w \langle (p_1, q_1), (2, 0) \rangle \longrightarrow_t \langle (p_1, q_1), (0, 2) \rangle \longrightarrow_m^w \langle (p_2, q_3), (0, 0) \rangle$. Interestingly, in q_1 it is in A_2''s interest to send energy to A_1', thus not being able to fire any move transition while waiting for A_1' to accumulate enough energy and send it back so that both reach their target.

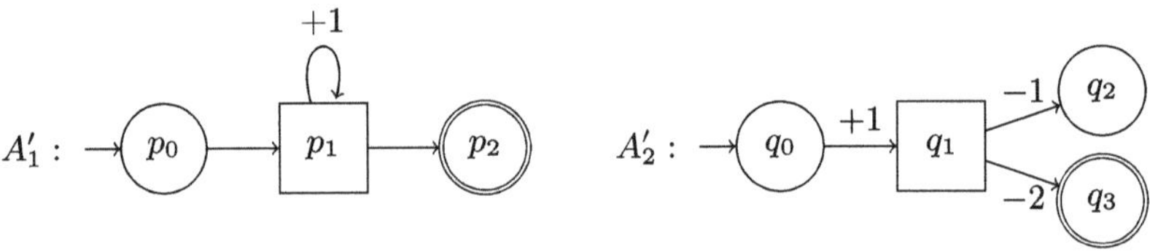

Fig. 2. Transfer system with two agents and a single transfer group with p_1 and q_1: $\mathcal{T} = \{\{p_1, q_1\}\}$. Null weights are omitted.

Reachability in Transfer Systems. The transfer system is equipped with one reachability goal for each agent, given by a set of initial vertices and a set of final vertices. The global objective is then to find a run in which each agent reaches its final vertex, starting from its initial vertex and with initial energy 0. Similarly to vector addition systems with states (VASS), recall that the energy level of agents cannot drop below 0. As agents can reach their final vertices with an arbitrary energy level, we naturally introduce a coverability relation on configurations.

For two configurations $C = \langle S, (e_1, \ldots, e_n) \rangle$ and $C' = \langle S', (e_1', \ldots, e_n') \rangle$, we say C' *covers* C, written $C \triangleleft C'$, if $S = S'$ and $\forall i \in [\![1, n]\!], e_i \le e_i'$. For any global state S we define *the covering* of S as $S{\uparrow} = \{C \mid \langle S, \overrightarrow{0} \rangle \triangleleft C\}$. We are now ready to define the verification problems of interest for transfer systems:

Problem x-REACH

Input: A transfer system TS, an initial state S_0 and a final state S_f

Question: Does there exist $C_f \in S_f{\uparrow}$ and a run $\rho : \langle S_0, \overrightarrow{0} \rangle \rightsquigarrow^x C_f$?

Note that this defines three decision problems, when one varies the semantics (parameter x): asynchronous, strongly synchronous or weakly synchronous.

Moreover, we consider arbitrary transfer groups, as well as the special case of a unique trivial transfer group $T_\top = \bigcup_{i=1}^n V_i$. We later use ℓx-REACH and ux-REACH to highlight the *transfer group type* and respectively denote the variant with arbitrary transfer groups or a unique trivial transfer group.

Example 2. Back to the example of Figs. 1 and 2, $\langle(A_1, A_2), \{\{p_1, q_0\}\}\rangle$ is a positive instance of ℓa-REACH, and $\langle(A_1', A_2'), \{T_\top\}\rangle$ is a positive instance of uw-REACH. but there exists only a single positive run for uw-REACH which is the same as for ℓw-REACH.

In the rest of the paper, we study the complexity of all variants of the tx-REACH problem. The following table summarizes the obtained results:

transfer	semantics		
	asynchronous	strongly synchronous	weakly synchronous
unique group	NP-hard (Th. 3) in PSPACE (Cor. 1)	PSPACE-c. (Th. 4 & Th. 5)	
arbitrary groups	PSPACE-c. (Th. 6 & Th. 7)	PSPACE-hard (Cor. 3) in EXPSPACE (Th. 8)	undecidable (Th. 9)

3 Relationships Between the Different Semantics

A first, immediate, observation is that the unique variants of our decision problem are special cases of the local ones. Indeed, any instance of a ux-REACH with a single trivial transfer group is also an instance of ℓx-REACH. There is thus an immediate polynomial reduction from one to the other:

Proposition 1. *For every semantics* $x \in \{a, s, w\}$, ux-REACH $\preceq_P \ell x$-REACH.

The following theorems relate to the asynchronous, strongly synchronous and weakly synchronous semantics (for a fixed transfer group type):

Theorem 1. *For every transfer group type* $t \in \{u, \ell\}$, ta REACH $\preceq_P ts$ REACH.

Proof. Let $\mathsf{TS} = \langle\{A_1, \ldots, A_n\}, \mathcal{T}\rangle$ be a transfer system, and S_0, S_f initial and final global states. From TS, we build the transfer system TS' in which every vertex of every agent is added a self-loop with weight 0. Formally, $\mathsf{TS}' = \langle\{B_1, \ldots, B_n\}, \mathcal{T}\rangle$ where for every $i \in [\![1, n]\!]$, if $A_i = (V_i, E_i)$ then $B_i = (V_i, F_i)$ with $F_i = E_i \cup \{q \xrightarrow{0} q \mid q \in V_i\}$. We claim that

$$\exists\langle S_0, \overrightarrow{0}\rangle \leadsto^a_{\mathsf{TS}} \langle S_f, \overrightarrow{e}\rangle \in \mathsf{Runs}^a(\mathsf{TS}) \quad \text{iff} \quad \exists\langle S_0, \overrightarrow{0}\rangle \leadsto^s_{\mathsf{TS}'} \langle S_f, \overrightarrow{e}\rangle \in \mathsf{Runs}^s(\mathsf{TS}').$$

Note that the difference between asynchronous and strongly synchronous semantics only lies in move transitions (and do not concern transfer transitions). Intuitively, the 0-self-loops in TS' are used to simulate an asynchronous run over TS by a synchronous run over TS'. Reciprocally, synchronous transitions over TS' can be serialized (and 0-self-loops can be removed) to obtain an asynchronous run over TS. $\qquad\square$

Theorem 2. *For every transfer group type $t \in \{u, \ell\}$, ts-REACH $\preceq_P$ tw-REACH.*

Proof. Let $\mathsf{TS} = \langle \{A_1, \ldots, A_n\}, \mathcal{T} \rangle$ be a transfer system, and S_0, S_f initial and final global states. From TS, we build the transfer system TS' in which each local arena A_i is augmented with a fresh vertex BAD_i and additional edges from every vertex to BAD_i with weight 0. Formally, $\mathsf{TS}' = \langle \{B_1, \ldots, B_n\}, \mathcal{T} \rangle$ where for every $i \in [\![1, n]\!]$, if $A_i = (V_i, E_i)$ then $B_i = (V_i \cup \{\mathrm{BAD}_i\}, F_i)$ with $F_i = E_i \cup \{q \xrightarrow{0} \mathrm{BAD}_i \mid q \in V_i \cup \{\mathrm{BAD}_i\}\}$. We claim that

$$\exists \langle S_0, \overrightarrow{0} \rangle \rightsquigarrow^s_{\mathsf{TS}} \langle S_f, \overrightarrow{e} \rangle \in \mathsf{Runs}^s(\mathsf{TS}) \quad \text{iff} \quad \exists \langle S_0, \overrightarrow{0} \rangle \rightsquigarrow^w_{\mathsf{TS}'} \langle S_f, \overrightarrow{e} \rangle \in \mathsf{Runs}^w(\mathsf{TS}').$$

Note that the difference between strongly and weakly synchronous semantics only lies in move transitions (and do not concern transfer transitions). Since the additional edges have weight 0, every agent always has an enabled edge. This means that for TS', the strongly synchronous runs and weakly synchronous runs coincide. Moreover, every vertex BAD_i is a sink. Hence, a run reaching the global final states cannot visit BAD_i. Finally, by construction, the runs of $\mathsf{Runs}^w(\mathsf{TS}')$ that avoid all vertices BAD_i are exactly the runs of $\mathsf{Runs}^s(\mathsf{TS})$. $\qquad\square$

Remark 1 (Transfer systems and VASS.). In general, reachability in transfer systems can be cast into state-coverability of VASS. The state-space of the VASS is the product of sets of vertices for each agent, thus exponential in the transfer system size. The VASS transitions are induced by transfer transitions and move transitions, and their precise definition depends on the semantics of move transitions. In the case of a unique transfer group, for asynchronous and strongly synchronous semantics, 1-dim VASS even suffice, since intuitively, a unique counter is needed to store the total energy amount shared by the agents. For the weakly synchronous semantics however, it is less obvious how to represent the energy levels of agents with a single counter. Indeed, a global energy level exceeding the energies required to allow one move per agent is not a sufficient condition for all agents to move. For instance, an agent may have the incentive to transfer energy to another agent in order to be temporarilly blocked (see Examples 1 and 2). This suggests that the encoding in 1-dim VASS is not immediate for transfer systems under the weakly synchronous semantics, and that 1 dimension per agent may be needed. Furthermore, up to our knowledge, the state of the art on state-coverability in 1-dim VASS [13, 17] yields worse complexity results than the direct proofs we present in the coming section, since the obtained VASS is exponential in the transfer system size. For arbitrary transfer groups, the situation is even worse since the reduction would be to an exponential size n-dim VASS.

4   Unique Trivial Transfer Group

Let us start with the particular case of a unique and trivial transfer group: $\mathsf{TS} = \langle \{A_1, \ldots, A_n\}, \{T_\top\} \rangle$. In such transfer systems, in every configuration, agents can transfer energy to others, regardless of their respective local vertices.

4.1 Asynchronous Semantics

For the asynchronous semantics, we prove the following complexity lower-bound:

Theorem 3. ua-REACH *is* NP-*hard.*

Proof. We perform a reduction from the SUBSETSUM problem, that we recall now. Given a finite set of integers $\mathcal{S} = \{n_1, \dots, n_m\}$ and a target integer $K \in \mathbb{N}$, the subset sum problem consists in determining whether there exists a subset $I \subseteq [\![1, m]\!]$ such that $\sum_{i \in I} n_i = K$. This problem is known to be NP-complete [15].

From an instance $\mathcal{S} = \langle \{n_1, \dots, n_m\}, K \rangle$ of SUBSETSUM, we build a transfer system $\mathsf{TS} = \langle (A_C, A_1, \dots, A_m), \{T_\top\} \rangle$ together with initial and final states S_0, S_f as represented in Fig. 3, and such that $\mathcal{S}$ is a positive instance of SUBSETSUM iff there exists $C_f \in S_f{\uparrow}$ and $\langle S_0, \overrightarrow{0} \rangle \leadsto^a C_f$ in $\mathrm{Runs}^a(\mathsf{TS})$.

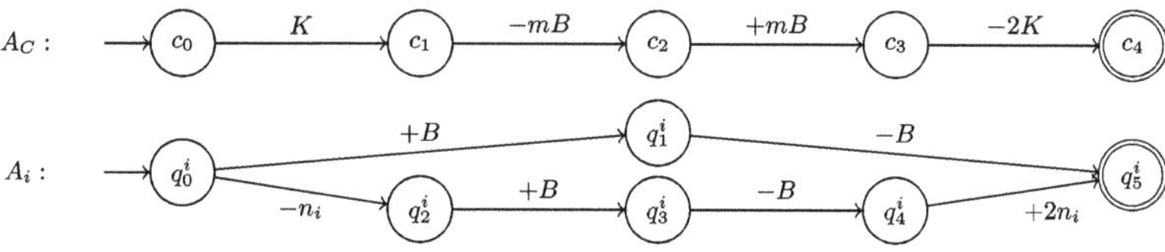

Fig. 3. Transfer system for the NP-hardness of ua-REACH. Incoming arrows point to initial vertices, and doubly circled vertices are final. $B = 2mK$.

Before giving the formal proof, let us explain the intuition of the reduction. The system is formed of one "control" agent A_C, as well as one agent A_i for each integer n_i. The transfer system starts with K energy units (gained by the controller A_C). In a first step, the agents A_i will have to select whether or not the solution set I contains i. This choice corresponds to two branches of the local arena A_i. If they decide $i \in I$, they have to pay n_i, which ensures that the sum of all the values selected by the agents is at most K. After this choice, each A_i agent receives a very large value B. This is a synchronization point: A_C needs all the agents to have received B energy units before it can progress, thus ensuring that every agent made their choice. In the third step, each agent A_i which decided that $i \in I$ receives $2n_i$ before reaching its target. As A_C needs $2K$ to reach its target, this will require that the sum of all the values selected by the agents is at least K. All in all, the sum has to be exactly K. Note that without the synchronization point, some agent A_j could wait for the $2n_i$ to be produced by A_i before making their initial choice.

Let us now formally describe the reduction and establish its correctness. Without loss of generality we assume that for every $i \in [\![1, m]\!]$, $n_i \leq K$, and we set $B = 2mK$. The local arenas are defined by:

- $A_C = (\{c_i \mid i \in [\![0, 4]\!]\}, E)$ with $E = \{c_j \xrightarrow{w_j} c_{j+1} \mid j \in [\![0, 3]\!]\}$, and $w_0 = K, w_1 = -mB, w_2 = mB, w_3 = -2K$.

- for every $i \in [\![1, m]\!]$, $A_i = \left(\{ q_j^i \mid j \in [\![0, 5]\!] \}, E_i \right)$ with :

$$E_i = \left\{ (q_0^i \xrightarrow{w_{01}^i} q_1^i), (q_1^i \xrightarrow{w_{15}^i} q_5^i) \right\}$$
$$\cup \left\{ (q_0^i \xrightarrow{w_{02}^i} q_2^i), (q_2^i \xrightarrow{w_{23}^i} q_3^i), (q_3^i \xrightarrow{w_{34}^i} q_4^i), (q_4^i \xrightarrow{w_{45}^i} q_5^i) \right\};$$
and $w_{01}^i = w_{23}^i = B, w_{15}^i = w_{34}^i = -B, w_{02}^i = -n_i$ and $w_{45}^i = 2n_i$.

Since we consider ua-REACH, the only transfer group is $T_\top$ that contains every vertex of every agent. The initial and target states are $S_0 = (c_0, q_0^1, \cdots q_0^m)$ and $S_f = (c_4, q_5^1, \cdots, q_5^m)$.

We claim that some $C_f \in S_f{\uparrow}$ is reachable from $\langle S_0, \overrightarrow{0} \rangle$ by an asynchronous run if and only if there exists $I \subseteq [\![1, m]\!]$ with $\sum_{i \in I} n_i = K$.

As a first step, let us show that if an agent A_i takes an edge with cost $-B$ before A_C reaches c_2, then the controller cannot reach its target. The maximum amount of energy that can be collected in the system before A_C reaches c_2 is bounded by $mB + K + \sum_{i=1}^{m} 2n_i$. Assume an edge with weight $-B$ is taken, the total energy is now bounded by $(m-1)B + K + \sum_{i=1}^{m} 2n_i$. Due to our choice of B, this value is strictly less than mB and A_C cannot take the edge to c_2.

The previous reasoning implies that every available edge with weight $+B$ must be taken by the agents A_i before A_C can reach c_2. Following this observation, fix a run and assume that every agent took an edge with weight $+B$ but did not take its edge with weight $-B$ yet; assume further that A_C has not reached c_2 yet; finally, without loss of generality, assume that A_C reached c_1. For every $i \in [\![1, m]\!]$, agent A_i is thus either in q_1^i or in q_3^i. Let H be defined as the set of indices i such that A_i is in q_3^i.

Assume that $D = \sum_{i \in H} n_i > K$, then the current total energy is $mB + K - D < mB$, thus A_C cannot reach c_2 and will never reach its target. So $D \leq K$. From this point, as A_C needs to traverse to c_2 not to be blocked, we can assume it goes immediately to c_3. Agent A_C and the agents A_i with $i \notin H$ will no longer gain energy before reaching their target, so we can assume the agents in H act first. They lose an amount of energy of $|H|B$ and gain $2D$. Thus, the total energy in the system is $(m - |H|)B + 2D + (K - D)$. For the remaining agents to reach their target are required exactly $(m - |H|)B + 2K$ energy units. This is only possible if $D \geq K$. As we already showed that $D \leq K$, this means that $D = K$.

This concludes the proof. $\qquad\square$

Note that the hardness proof of Theorem 3 uses acyclic arenas. Moreover, given a transfer system where each local arena is acyclic, the reachability problems under each semantics is in NP. Indeed, the length of a path for each agent from its initial vertex to a final vertex is bounded by the number of vertices. To derive a non-deterministic polynomial time algorithm, one can thus guess for each agent a linear length path, and then check whether they can be combined into a complete run of the transfer system. The latter can be done in polynomial time by checking that at every step the global energy exceeds the one needed for the next transition. Therefore, for acyclic local arenas, ua-REACH is NP-complete.

Towards a complexity upper-bound beyond the acyclic case, we observe that thanks to Theorem 1, ua-REACH reduces in polynomial time to us-REACH. We will state in Theorem 4 that the latter is solvable in polynomial space.

Corollary 1. ua-REACH *is in* PSPACE.

4.2 Strongly and Weakly Synchronous Semantics

Theorem 4. uw-REACH *is in* PSPACE.

Proof. To prove membership in PSPACE, we show a small witness property. Precisely, in the uw-semantics, there exist exponential bounds $B_{\max}$ and $L_{\max}$ such that: if there is a run to the target global state, then there is one (1) of length at most $L_{\max}$ and (2) along which if the energy level of the agents reach $B_{\max}$, then the energy requirements can be ignored for the rest of the run. Further, both bounds are exponential in the size of the input transfer system.

Consider the instance $\mathsf{TS} = \langle \{A_1, \ldots, A_n\}, \{T_\top\}\rangle$ of uw-REACH together with initial and final states S_0 and S_f. Denote by $w_{\max}$ the largest absolute value among the weights of the edges in all A_i's. Suppose agent A_k has $B_{\max} = n \cdot |\mathsf{TS}|^n \cdot w_{\max}$ energy units. It can transfer $K = |\mathsf{TS}|^n \cdot w_{\max}$ energy units to each other agent, and still have energy level K. Now, focusing on states only, not on energy vectors, the length of a cycle-free path in the transfer system is at most $\prod_{i=1}^n |V_i| \leq |\mathsf{TS}|^n$. With K energy units, each agent is hence able to take an acyclic path to its target, losing at most $w_{\max}$ energy units at each step. Hence, from a configuration storing $B_{\max}$ energy units, the energy can be distributed in such a way that each agent reaches its goal assuming it is reachable from its current vertex.

When exploring runs with bounded energy levels, one thus only needs to look for relatively short runs. The number of useful configurations is bounded by $L_{\max} = \prod_{i=1}^n |V_i| \cdot (B_{\max} + 1)$, and each of these is visited at most once in a useful run. Therefore, useful runs are then of length at most $L_{\max}$, a value that is exponential in $|\mathsf{TS}|$.

Let us thus first consider runs with energy levels bounded by B_{max}. The number of configurations such runs visit is bounded by $\prod_{i=1}^n |V_i| \cdot (B_{\max} + 1)$. One can also notice that a run from S_0 to S_f does not need to contain cycles. Hence configurations need only be visited at most once. This induces a bound on the length of relevant runs: $L_{\max} = \prod_{i=1}^n |V_i| \cdot (B_{\max} + 2)$ (notice here that $\prod_{i=1}^n |V_i|$ steps can be required to reach the target when energy level is above $B_{\max}$).

Using this bound on the maximum length of runs to reach the goal, we can design a non-deterministic algorithm that starts from the initial configuration, and explores runs of length at most $L_{\max}$ among configurations that store at most B_{max} energy. The algorithm returns yes if the final state S_f is reached, and fails if the length of the run exceeds L_{max} or if the current configuration is a deadlock. It requires polynomial space to store a configuration and the step-counter. Indeed, a configuration is represented by storing for each agent its vertex and and its energy level. When energy levels are bounded by $B_{\max}$, the space needed to store them is logarithmic in $B_{\max}$. Moreover, the length of runs can be encoded by a counter taking values up to $L_{\max}$, which can be encoded in space logarithmic in $L_{\max}$. Since both bound are exponential in $\mathcal{A}$, polynomial space in $|\mathsf{TS}|$ is sufficient. By Savitch's theorem [22], this non-deterministic polynomial space algorithm proves membership in PSPACE. $\square$

Theorem 5. *us*-REACH *is* PSPACE-*hard.*

Proof. To prove PSPACE-hardness, we reduce the reachability problem for 1-safe Petri nets which is PSPACE-complete. More precisely, w.log, we consider 1-safe Petri nets in which no place is in the postset and the preset of the same transition; reachability is known to be PSPACE-complete for this class [7].

A Petri net is a tuple $\mathcal{N} = \langle P, T; F \rangle$ where $P = \{p_1, \ldots, p_n\}$ is a set of places, $T = \{t_1, \ldots, t_m\}$ is a set of transitions, and $F \subseteq P \times T \cup T \times P$ is a flow relation. A marking of a Petri net is a map $M : P \to \mathbb{N}$ that associates a number of tokens to each place. The preset of a transition is the set of places $^\bullet t = \{p \in P \mid (p, t) \in F\}$ and the postset of t is the set of places $t^\bullet = \{p \in P \mid (t, p) \in F\}$. A transition is firable from marking M if, for every place $p \in {}^\bullet t$, $M(p) > 0$. Firing a transition t from marking M decrements $M(p)$ by 1 for every place of p the preset, and increments $M(p')$ for each p' in its postset. We write $M[t\rangle M'$ when M' is the marking obtained by firing t from M. Given a Petri net $\mathcal{N}$, one can define the set of reachable markings $\mathsf{Reach}(\mathcal{N}, M_0)$ that are reachable from M_0. The net $\mathcal{N}$ is *1-safe* if, for every marking M in $\mathsf{Reach}(\mathcal{N}, M_0)$ and every place p, $M(p) \le 1$. The reachability problem for Petri nets consists in deciding whether a given input marking M belongs to $\mathsf{Reach}(\mathcal{N}, M_0)$.

Let $\mathcal{N} = \langle P, T; F \rangle$ be a 1-safe Petri net. Consider the transfer system with a trivial transfer group $\mathsf{TS} = \langle \{A_C, A_1, \ldots, A_n\}, \{T_\top\} \rangle$ represented in Figs. 4 and 5. This arena is composed of one control agent A_C, and one agent A_i per place $p_i \in P$.

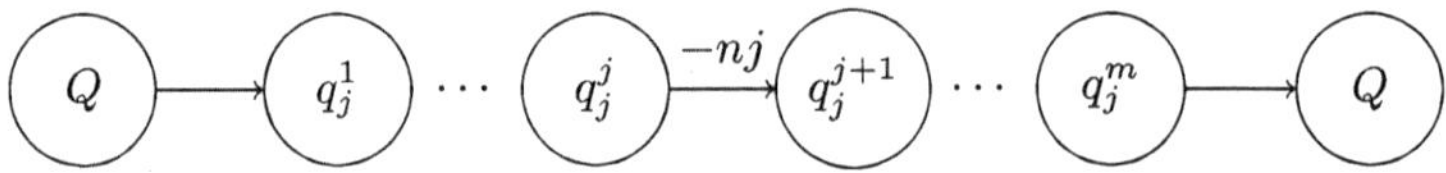

Fig. 4. Local arena of agent A_C. Only vertices relevant for transition t_j are depicted.

Formally, the control agent is $A_C = (V_C, E_C)$ with

$$V_C = \{Q\} \cup \left\{ q_j^{j'} \mid j, j' \in [\![1, m]\!] \right\};$$

$$E_C = \left\{ q_j^{j'} \xrightarrow{w_j^{j'}} q_j^{j'+1} \mid j' \in [\![1, m-1]\!], j \in [\![1, m]\!] \right\}$$

$$\cup \left\{ Q \xrightarrow{0} q_j^1, q_j^m \xrightarrow{w_j^m} Q \mid j \in [\![1, m]\!] \right\}$$

and $\forall j, w_j^j = -n \cdot j$ and all other weights are null.

Intuitively, the controller chooses to fire t_j by moving to vertex q_j and expects to get enough energy in time. We will see that the only way for A_C to pay nj energy units at step j is if other agents have also chosen to fire t_j.

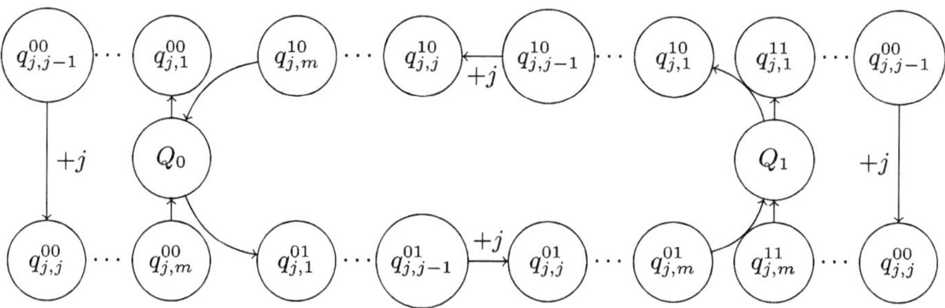

Fig. 5. Local arena of agent A_i. Only edges relevant for transition t_j are depicted.

Then, for every $i \in [\![1, n]\!]$, there is an agent $A_i = (V_i, E_i)$ with $V_i = \{Q_0, Q_1\} \cup \{q_{j,j'}^{AB} \mid j, j' \in [\![1, m]\!]\ A, B \in \{0, 1\}\}$, and

$$E_i = \left\{ q_{j,j'-1}^{AB} \xrightarrow{w_{j,j'}^{AB}} q_{j,j'}^{AB} \mid j' \in [\![2, m]\!], j \in [\![1, m]\!], A, B \in \{0, 1\} \right\}$$

$$\cup \left\{ Q_A \xrightarrow{w_{j,1}^{AB}} q_{j,1}^{AB}, q_{j,m}^{AB} \xrightarrow{0} Q_B \mid j \in [\![1, m]\!], A, B \in \{0, 1\} \right\}$$

and for every transition $t_j \in T$, if $p_i \in {}^\bullet t_j$, then $w_{jj}^{00} = w_{jj}^{01} = w_{jj}^{11} = 0$ and $w_{jj}^{10} = j$; else if $p_i \in t_j^\bullet$, then $w_{jj}^{00} = w_{jj}^{10} = w_{jj}^{11} = 0$ and $w_{jj}^{01} = j$; otherwise $w_{jj}^{10} = w_{jj}^{01} = 0$ and $w_{jj}^{11} = w_{jj}^{00} = j$. All other weights are null. Intuitively, agent A_i at Q_0 represents place p_i having no tokens and agent A_i at Q_1 represents p_i having one token. A_i can provide for the j energy units needed by A_C through simulating the firing of t_j only if A_i is on a path corresponding to the effect of t_j on p_i: if p_i loses one token ($p_i \in {}^\bullet t_j$), the only correct path is through the 10-vertex; if p_i gains one token ($p_i \in t_j^\bullet$), the only correct path is via the 01-vertex; otherwise t_j has no effect on p_i : the correct paths are via the 00- or 11-vertex depending on the current marking.

Obviously, if firing transition t_j from the marking M leads to the marking M', there is a run from $C(M)$ to $C(M')$: For all $p_i \in {}^\bullet t_j$, A_i follows the path 10, for all $p_i \in t_j^\bullet$, A_i follows the path 01 and each other agent A_i follows the path $M(p_i)M(p_i)$. Note that each agent A_i follows the path $M(p_i)M'(p_i)$. With these choices, before states q_{jj} each agent except A_C gains j energy units. On q_{jj} they all transfer that amount to A_C which can leave q_{jj} with exactly enough energy to pay the $-nj$.

Now suppose that there is a run $\rho : C(M) \leadsto_{\mathsf{TS}}^a C'$ with $m + 1$ move transitions. We show that there exists M' such that $C' = C(M')$. If agent A_C follows states q_j. Suppose agents A_i follow states q_{j_i}, they will receive at most j_i energy units through the j_i-th edge of their path. The amount of energy available to A_C before going through its $j+1$-th edge is at most $\sum_{j_i \leq j} j_i$ because if $j_i > j$ this energy has not been gained before the $j+1$-th edge. The only way for $\sum_{j_i \leq j} j_i$ to be greater than or equal to nj is if for all i, $j_i = j$. Thus agents A_i follow q_j states. But because A_i must gain j energy units, it must follow a path that represents a possible behavior of the firing of t_j on place p_i. Thus ρ is exactly as described in the first part of this proof, t_j is enabled by M and C' represents the marking M' resulting from firing t_j in M: $C' = C(M')$.

Finally, this reduction is polynomial : $|A_C| = O(m(m + \log(nm)))$ and $\forall i, |A_i| = O(m(m + \log m))$. $|\mathsf{TS}| = O(nm^2 \log(mn))$. $\qquad\square$

Thanks to Theorem 2, we deduce:

Corollary 2. *us*-REACH *and uw*-REACH *are* PSPACE-*complete*.

5 Arbitrary Transfer Groups

5.1 Asynchronous Semantics

Let us now consider the complexity of ℓa-REACH, i.e., reachability for transfer systems with local transfer groups, and under asynchronous semantics. We show below that this problem is PSPACE-complete. The PSPACE membership is shown by exhibiting an algorithm that requires polynomial space to reach a target configuration. The PSPACE-hardness is proved by a reduction from a reachability problem for safe Petri nets. We give a construction that builds for each safe Petri net, a transfer system of polynomial size w.r.t. the original net, and whose runs simulate that net. We already mentioned that [7] proved PSPACE-completeness of reachability for safe Petri nets. Later, [9] has showed that reachability is NP-complete for free-choice safe Petri nets. However, the encoding shown below applies to any 1-safe Petri net. Let us start with the PSPACE membership.

Theorem 6. ℓa-REACH *is in* PSPACE.

Proof. Consider $\mathsf{TS} = \langle \{A_1, \ldots, A_n\}, \mathcal{T} \rangle$. Call $e_{\max}$ the biggest weight on an edge in TS and $S_{\max}$ the biggest size among sets S_i. From each vertex of its graph, an agent with $B_{\max} = S_{\max} \cdot e_{\max}$ energy units does not need to receive energy from an other agent to reach any other vertex of its graph. In the sequel, we give an upper bound on the useful energy level of an agent, taking into account that it may transfer energy to others to help them achieve their reachability objective. We show that if TS is a positive instance of ℓa-REACH, then there exists a witness execution ρ in which the energy of each agent is bounded by $B_{\max}(2n - 1)$.

Fix $\rho \in \mathsf{Runs}^a(\mathsf{TS})$. We say that A_i *helps* A_j by e energy units along ρ with 0 intermediary if A_i sends at least e energy units to A_j through a transfer transition in ρ and the energy level of A_j is at least e right after the last transfer transition involving this agent in ρ. We say that A_i helps A_j by e energy units along ρ with $k \geq 1$ intermediaries if A_i transfers at least e energy units to another agent A_p that helps A_j along ρ by e energy units with $k-1$ intermediaries. If A_i helps A_j along ρ by e energy units with any number of intermediaries, we say that A_i helps A_j by e energy units for short. This can be thought of as if A_i has ultimately sent e energy units to A_j that it can keep for itself.

We show by induction on k that if an agent starts ρ with $(2k-1)B_{\max}$ energy units, it may help up to k other agents by $B_{\max}$ energy units and have at least $B_{\max}$ energy units after its last transfer transition. The case $k = 0$ is immediate. Suppose the property holds until $k \geq 0$. If agent A_i has $(2k + 1)B_{\max}$ energy units, it may try to meet an other agent A_j at cost at most $B_{\max}$ leaving at least $2kB_{\max}$ energy units. If A_i sends $(2k' - 1)B_{\max}$ energy units to A_j for some $k' \in [\![0, k]\!]$, then A_j may help up to k' other agents by $B_{\max}$ and A_i may help up to $k - k' - 1$ other agents by $B_{\max}$. Note that A_i helps all the up to k' agents that A_j helps (with an additional intermediary) and because A_j has at least $B_{\max}$ energy units after its last transfer transition, A_i also helps A_j which adds up to a total of k agents helped. As a consequence, an agent never needs to have more than $(2n - 1)B_{\max}$ energy units since helping every other agents by $B_{\max}$ while still having that much afterwards is enough to reach the final state.

The same way as we showed it in the proof of Theorem 4, we have now an exponential bound in $O(n|\mathsf{TS}|2^{|\mathsf{TS}|})$ on useful energy levels which results in a polynomial bound in $O(|\mathsf{TS}|\log(n|\mathsf{TS}|))$ on the space needed to store a useful configuration and an exponential bound $L_{\max} \in O(n|\mathsf{TS}|2^{|\mathsf{TS}|})$ on the length of useful runs. In the end, there exists an $\mathsf{NPSPACE}$ algorithm that explores runs of size at most $L_{\max}$ and either fails if the final state S_f is not reached in $L_{\max}$ steps (the current length of the run can be stored in space $O(|\mathsf{TS}|\log(n|\mathsf{TS}|))$) or if a deadlock is reached, and succeeds otherwise. By Savitch's theorem, we get that ℓa-REACH is in PSPACE. $\qquad\square$

Theorem 7. ℓa-REACH *is* PSPACE-*hard.*

Proof (sketch). We encode a reachability problem for safe Petri nets in a ℓa-REACH problem with a transfer system whose size is linear in the size of the considered net. Let $\mathcal{N} = (P, T; F)$ be a safe Petri net with initial marking M_0. We build a transfer system composed of $n+1$ agents, $\mathsf{TS}_{\mathcal{N}} = \langle \{A_C, A_1, \ldots, A_n\}, \mathcal{T} \rangle$ simulating the behavior of $\mathcal{N}$. We do not give the whole construction here, and refer to [2] for details. The first agent A_C is a controller that initiates the simulation of a transitions firing. Agents of the form A_i encode the contents of place p_i through their states, and simulate the effect of a transition firing via sequences of moves. We distinguish in particular two states $A_{i,0}$ and $A_{i,1}$, used to encode $M(p_i) = 0$ and $M(p_i) = 1$ in order to represent the marking M. Then we set an ordering on places, and ensure that when the controller agent chooses a particular transition t, all place agents choose the transition they simulate, but have to wait for energy from their predecessor to progress in this simulation. If two agents choose different transitions, the system deadlocks. Upon agreement on the chosen transition to simulate, the last agent A_n eventually sends back energy to the controller, acknowledging the fact that all places are engaged in the simulation of the same transition from the same marking. The controller then launches another round among place agents (still by transferring energy) who successively update their state to encode the effect of t on their place contents before acknowledging all changes to the controller. For instance, if transition t consumes a token from place p_i and $M(p_i) = 1$, then agent A_i will start its interactions from state $A_{i,1}$ and will end the simulation of t's firing in state $A_{i,0}$. If a transition t is chosen, p_i is in the preset of t, but A_i started from state $A_{i,0}$, then choosing to simulate t will send A_i to a deadlock state, and will prevent reaching global states that encode markings. During this simulation process, if a place agent does not transfer energy to its successor and moves to its next state, then the system necessarily deadlocks or can return to the situation where the transfer was missed. When an agent keeps energy for its future moves, it can only repeat the choice of a new transition to simulate, hence canceling its previous choice. The only way to simulate properly a Petri net transition is if all agents choose the same transition and transfer their energy to their successor. Other choices lead either to deadlocks or livelocks in configurations that do not encode markings.

Configurations of $\mathsf{TS}_\mathcal{N}$ of the form $\langle (q_1^C, A_{1,b_1}, \cdot, A_{n,b_n}), \overrightarrow{0} \rangle$ will be called *stable* configurations (and other configurations *unstable*). Stable configurations represent an encoding of a marking of $\mathcal{N}$. A first step in the proof is to show that, for a pair of markings M, M' of $\mathcal{N}$, such that $M[t\rangle M'$, there exists a run from $C(M) = \langle (q_1^C, A_{1,M(p_1)}, \ldots, A_{n,M(p_n)}), \overrightarrow{0} \rangle$ to $C(M')$ in $\mathsf{TS}_\mathcal{N}$. The shape of such runs is depicted in Fig. 6. In this figure, agents steps are organized as local sequences, red dashed arrows depict energy transfers, and vertices that belong to the same transfer group have identical shape and color. This first part of the proof shows that a transfer system can simulate a safe Petri net, as for $M[t\rangle M'$, there exists a "canonical" run $\rho_{M[t\rangle M'}$ from stable configuration $C(M)$ to stable configuration $C(M')$ that does not visit any other stable configuration.

It then remains to show that $\mathsf{TS}_\mathcal{N}$ does not allow the reachability of stable configurations that are not encodings of reachable markings. To this extent, we look at the transition system composed of possible configurations and moves of the transfer system and highlight its properties. One can show that the runs encoding firing of a transition t of a safe net follow a particular pattern: Agents choose their transition and guess their predecessor's bit. When all agents agree on a common transition, one unit of energy flows from A_c to A_1, $A_2 \ldots A_n$ and then back to A_c. A second round then starts, with two units of energy transferred successively to A_c to A_1, $A_2 \ldots A_n$. If two agents did incompatible choices of simulated transition or control bit, then a deadlock (without reaching any stable configuration) is unavoidable. Similarly, if an agent does not transfer all its energy to its successor , then the system either deadlocks, or enters an infinite sequence of moves that can be only exited by sending back the faulty agent to the state from which the wrong choice was performed, still without visiting a stable configuration.

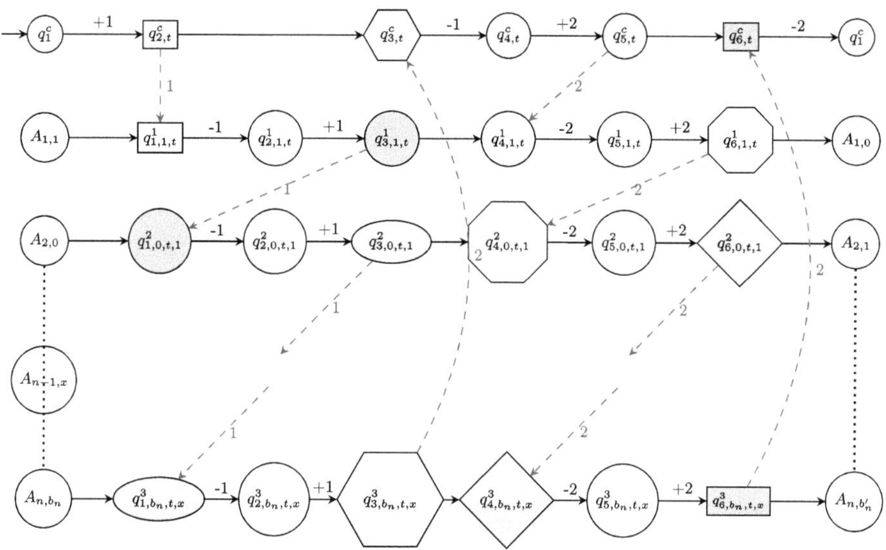

Fig. 6. Simulating a transition t moving a token from p_1 to p_2

We can hence conclude that a marking M of the safe Petri net $\mathcal{N}$ is reachable from marking M_0 if and only if the stable configuration $C(M)$ is reachable from the configuration $C(M_0)$ in $\mathsf{TS}_{\mathcal{N}}$. Hence, ℓa-REACH is PSPACE-hard. $\qquad\square$

5.2 Strongly Synchronous Semantics
Theorem 8. ℓs-REACH *is in* EXPSPACE.

Proof. The exponential space algorithm we exhibit to show membership in EXPSPACE involves the construction of a VASS $\mathcal{V}$ of exponential size yet polynomial dimension. Rackoff's backward algorithm for VASS coverability requires $2^{O(d)} \cdot \log|\mathcal{V}|$ space [16] where d is the dimension of $\mathcal{V}$. This results in an overall exponential space algorithm for ℓs-REACH.

The above-mentioned VASS $\mathcal{V}$ is built as follows. Given an instance $\mathsf{TS} = \langle \{A_1, \cdots, A_n\}, \mathcal{T} \rangle$ with $\forall i, A_i = (V_i, E_i)$, we define $\mathcal{V} = \langle \mathcal{S}, \delta \rangle$ such that:

- $\mathcal{S} = \prod_{i=1}^{n} V_i$
- $\forall (q_1 \xrightarrow{w_1} q_1', \ldots, q_n \xrightarrow{w_n} q_n') \in \prod_{i=1}^{n} E_i, (q_1, \ldots, q_n) \xrightarrow{w_1, \ldots, w_n} (q_1', \ldots, q_n') \in \delta$
- $\forall \overrightarrow{q} = (q_1, \ldots, q_n) \in \mathcal{S}, \forall q_i \neq q_j, \left[\exists T \in \mathcal{T} \mid q_i, q_j \in T \implies \overrightarrow{q} \xrightarrow{\overrightarrow{w_{ij}}} \overrightarrow{q} \in \delta \right]$

where $\overrightarrow{w_{ij}}$ is the vector with only zeros except $+1$ at index i and -1 at index j.

With this construction, whether there exists a run $\rho_{\mathsf{TS}} : \langle S, \overrightarrow{e} \rangle \rightsquigarrow^{s}_{\mathsf{TS}} \langle S', \overrightarrow{e'} \rangle$

is equivalent to whether there exists a run $\rho_{\mathcal{V}} : \langle S, \overrightarrow{e} \rangle \rightsquigarrow_{\mathcal{V}} \langle S', \overrightarrow{e'} \rangle$.

We show the direct implication by induction on the length of ρ_{TS}:

- If ρ_{TS} is empty, $\langle S, \overrightarrow{e} \rangle = \langle S', \overrightarrow{e'} \rangle$ and the property holds.

- Let ρ_{TS} be of length $l+1$. Let $\langle S^{-1}, \overrightarrow{e^{-1}} \rangle$ be the penultimate configuration of $\rho_{\mathcal{A}}$. By induction, there exists $\rho_{\mathcal{V}} : \langle S, \overrightarrow{e} \rangle \leadsto_{\mathcal{V}} \langle S^{-1}, \overrightarrow{e^{-1}} \rangle$. If the last transition of ρ_{TS} is a move transition, for all i there is a transition $q_i^{-1} \xrightarrow{w_i} q_i' \in E_i$ such that all energy levels $e_i' = e_i^{-1} + w_i$ are non-negative. $\rho_{\mathcal{V}}$ may reach $\langle S', \overrightarrow{e'} \rangle$ with the transition $S^{-1} \xrightarrow{w_1,\dots,w_n} S'$ induced by these transitions. Otherwise, there are some $i, j \in [\![1, n]\!]$ and some $T \in \mathcal{T}$ with $q_i', q_j' \in T$, $e_i^{-1} + e_j^{-1} = e_i' + e_j'$ and $\forall k \notin \{i,j\}$, $e_k^{-1} = e_k'$. The transfer from coordinate i to coordinate j may be decomposed using the transition $S' \xrightarrow{\overrightarrow{w_{ij}}} S'$ since q_i' and q_j' share the same group T. In all cases, $\langle S', \overrightarrow{e'} \rangle$ is reachable.

Conversely, by induction on the length of $\rho_{\mathcal{V}}$:

- The case of $\rho_{\mathcal{V}}$ empty is immediate.

- Let $\rho_{\mathcal{V}}$ be of length $l+1$. Let $\langle S^{-1}, \overrightarrow{e^{-1}} \rangle$ be the penultimate configuration of $\rho_{\mathcal{V}}$. By induction, there exists $\rho_{\mathsf{TS}} : \langle S, \overrightarrow{e} \rangle \leadsto_{\mathsf{TS}}^{s} \langle S^{-1}, \overrightarrow{e^{-1}} \rangle$. The last transition of $\rho_{\mathcal{V}}$ has two possible forms. Either there is $(q_1 \xrightarrow{w_1} q_1', \dots, q_n \xrightarrow{w_n} q_n') \in \prod_{i=1}^{n} E_i$ such that this last transition is $(q_1, \dots, q_n) \xrightarrow{w_1,\dots,w_n} (q_1', \dots, q_n')$, in which case ρ_{TS} can be extended by the move transition induced by these edges to reach S'. Note that because the coordinates are maintained non-negative in $\mathcal{V}$, so will the energy levels. Or, there are some coordinates $i, j \in [\![1, n]\!]$ such that the last transition of $\rho_{\mathcal{V}}$ is $S' \xrightarrow{\overrightarrow{w_{ij}}} S'$ with some transfer group $T \in \mathcal{T}$ such that $q_i', q_j' \in T$. In that case, a transfer transition from $\rho_{\mathcal{A}}$ that sends 1 energy unit from A_i to A_j reaches $\langle S', \overrightarrow{e'} \rangle$.

According to the previous result, we conclude that if an instance is positive for ℓs-REACH then its joined insatnce for VASS-cover is also positive. Furthermore, $\mathcal{V}$ is of size $O(|\mathsf{TS}|^{2n})$ since there are $O(|\mathsf{TS}|^n)$ states, $O(n^2)$ loops on each state and $O((|\mathsf{TS}|^n)^2)$ other transitions.

$\square$

For the complexity lower-bound, recall that ℓa-REACH is PSPACE-hard (Theorem 7) and conclude with Theorem 1 that:

Corollary 3. *ℓs-REACH is PSPACE-hard.*

5.3   Weakly Synchronous Semantics

Perhaps surprisingly, the relaxation of agents synchronization from strong to weak synchronous semantics, *i.e.* the fact that agents with no enabled edges may not move simulaneously with other agents, leads to undecidability. The main reason is the following. Both the asynchronous and (strongly) synchronous semantics enjoy a monotonicity property: higher energy levels can only enable more transitions and allow to reach more configurations. This monotonicity does not hold under the w-semantics. Indeed, it can be profitable for an agent to reach a vertex with a low energy level, so that it is allowed to "wait" for other agents to move and later gain energy through a transfer. Intuitively, this behaviour allows one to test whether an agent has energy left, thus encoding a zero test.

Theorem 9. ℓw-REACH *is undecidable.*

Proof. We give a reduction from the termination problem of Minsky machines, which is known to be undecidable [19]. Let us start with a quick recall on Minsky machines. A Minsky machine $\mathcal{M}$ is described by two counters x and y, as well as a sequence of commands $l_0, \ldots, l_m$ where l_0 is the starting command, l_m ends the run of the system, and l_0 to l_{m-1} are among the following three types:

- increment counter $c \in \{x, y\}$, move to the next command;
- decrement counter $c \in \{x, y\}$, move to the next command[1];
- if counter $c \in \{x, y\}$ is equal to 0, move to command l_k, otherwise move to command l_j.

In summary, the machine goes through a list of commands starting with l_0, incrementing, decrementing counters, or testing whether a counter is equal to 0 in order to select the new command to jump to—and it terminates whenever it reaches l_m. The *termination problem* for Minsky machines consists in deciding, given a machine $\mathcal{M}$, whether $\mathcal{M}$ terminates.

In our reduction, we use two agents with local arenas A_x and A_y, storing as energy level the current value of each counter, one control agent A_C that encodes the control flow of the Minsky machine, and one additional agent A_i for each command l_i to simulate the effect of l_i when it is activated by A_C. Sink vertices BAD are used to punish agents that reach a vertex with an energy level that differs from the one that is expected in the Minsky machine simulation.

We now detail how to encode a decrement: assume command l_i decrements counter $z \in \{x, y\}$. Only three agents take part in the simulation of l_i: the control agent A_C, agent A_z associated to counter z and agent A_i dedicated to l_i.

Formally, we define the transfer system associated with command line l_i as $\mathsf{TS}_i = \langle \{A_z, A_C, A_i\}, \mathcal{T}_i \rangle$, depicted in Fig. 7, with:

- $A_z = (\{q_0^z\}, \emptyset)$;

- $A_i = (V_i, E_i)$, where $V_i = \{q_j^i \mid j = 0 \ldots 4\}$; $E_i = \{q_j^i \xrightarrow{w_j^i} q_k^i \mid j \in [\![0, 4]\!] \wedge (k \equiv j + 1 \ (\mathrm{mod}\ 5))\}$; with $w_0^i = -1, w_1^i = 1, w_2^i = -3, w_3^i = 4$ and $w_4^i = 0$.

- $A_C = (V_i^C, E_i^C)$, where $V_i^C = \{l_j^i \mid j \in [\![0, 8]\!]\} \cup \{\text{BAD}, l_0^{i+1}\}$ and $E_i^C = \{l_j^i \xrightarrow{w^i} l_r^i \mid j = 0 \ldots 7 \wedge r = j + 1\} \cup \{l_8^i \xrightarrow{w^8} l_0^{i+1}, l_4^i \xrightarrow{0} \text{BAD}, l_8^i \xrightarrow{0} \text{BAD}\}$; and $w^0 = 1, w^4 = -2, w^5 = 3, w^8 = -4$ and other weights are null.

---

[1] Note that the machine must be designed so that decrement can only occur when the counter value is positive.

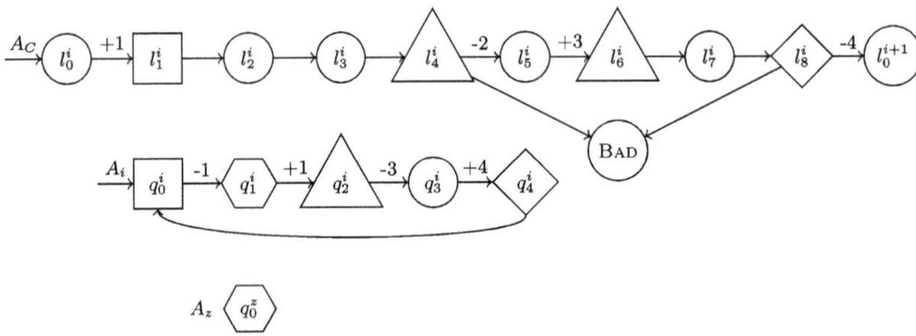

Fig. 7. Encoding a decrement. To the exception of circles, states with the same shape belong to the same transfer group.

- $\mathcal{T}_i$ contains the groups $\{l_1^i, q_0^i\}$, $\{q_0^z, q_1^i\}$, $\{l_4^i, l_6^i, q_2^i\}$ and $\{l_8^i, q_4^i\}$.

Recall that this addresses a single command line l_i. To encode a complete Minsky machine $\mathcal{M}$ with k instructions, we assemble the arenas for all command lines into the transfer system $\mathsf{TS}_{\mathcal{M}} = \langle (A_C, A_x, A_y, A_1, \ldots A_k), \bigcup \mathcal{T}_i \rangle$ A_C, in which A_x and A_y are the unions of sets of vertices and edges of the arenas of every command line. In particular, the vertex l_0^{i+1} is shared with the local arena associated with the command line l_{i+1}.

Let us show that if the agents A_C, A_i and A_z start in $\langle (l_0^i, q_0^i, q_0^z), (0, 0, n) \rangle$ with $n > 0$, then the only way to avoid BAD leads them to the configuration $\langle (l_0^{i+1}, q_0^i, q_0^z), (0, 0, n - 1) \rangle$. Hence, executing the command line l_i will indeed decrement the energy of A_z by 1.

In the first step, only A_C can move, reaching l_1^i with 1 unit of energy. There it can either transfer this energy to A_i or keep it. If it chooses not to transfer, it remains the only agent able to move, and when reaching l_4^i, it will have 1 unit of energy, forcing it to go to BAD. Assume thus that A_C tranfers its energy unit to A_i. Both then move synchronously to l_2^i and q_1^i. In q_1^i, A_i can interact with A_z. Let m be the energy level of of A_i at that point. Agents A_C and A_i then reach l_3^i and q_2^i with energy levels 0 and $m + 1$. If $m + 1 \geq 3$, then both agents move to l_4^i and q_3^i and in the next step, A_C is forced to go to BAD. In order for A_C to avoid its BAD vertex, it must be the case that $m + 1 < 3$. In this case, A_C progresses to l_4^i while A_i remains in q_2^i, because the only available edge consumes 3 units of energy. With this move, A_C can receive energy from A_i, as their current vertices belong to the same transfer group. Now A_C needs 2 energy units to avoid going to BAD, which requires $m + 1 \geq 2$. Hence $m + 1 = 2$ which implies that during their interaction, A_z transferred 1 energy unit to A_i, leaving A_z with $n - 1$ units of energy. After transferring 2 energy units to A_C, A_i remains stuck in q_2^i while A_C progresses to l_5^i and then l_6^i with 3 energy units. Again, A_C can choose to move on its own without transferring energy to A_i, but it will eventually reach vertex l_8^i, and lacking 4 energy units will be forced to go to BAD. If A_C transfers 3 energy units to A_i they both move to l_8^i and q_4^i where A_i can then transfer to A_C the 4 energy units required to avoid BAD. This then leads to a configuration where A_C is in state l_0^{i+1} with no energy left, A_i is back in vertex q_0^i also with no energy left, and A_z is in vertex q_0^z with an energy level $n-1$. In this construction, transferring other quantities of energy always leads to deadlock configurations.

The encoding of the increment is similar. The zero test however is even more involved, allowing the agents to remain stuck in some vertices and wait for the other agents iff the counter value is 0. The constructions for these two operations are provided in details in [2]. Altogether, the three constructions ensure that in order to avoid the BAD vertices, the agents must correctly implement the command of the Minsky machine.

To complete the reduction, the reachability objective for the transfer system is defined as follows. The last command l_m of the Minsky machine, is represented by a single vertex l_0^m which is the target vertex of the control agent. The targets of other agents are the set of vertices l_0^m, q_0^i for $i \in [\![0, m-1]\!]$ and q_0^z for $z \in \{x, y\}$. As the agents must avoid the BAD vertices, the above constructions ensure that the target state is covered in the transfer system if and only if the Minsky machine terminates. $\square$

6 Conclusion

This paper introduced and studied a cooperative game model, in which agents move on local weighted arenas, and can help each other by transferring energy to their peers. We considered a global reachability question, *i.e.*, whether it is possible to reach a system configuration where each agent is in their goal vertex, while always keeping all energy levels non-negative. While transfer systems can be easily encoded as vector addition systems with states, and our reachability problems as a coverability question, we showed that the energy transfer feature induces a complexity drop, with complexities ranging from PSPACE to EXPSPACE. For asynchronous and strongly synchronous semantics, we exploited a form of monotonicity and a small witness property. However, monotonicity does not hold under weak synchronous semantics, leading to undecidability.

An obvious future work is to close the complexity gaps for ua-REACH and ℓs-REACH. The similarities between transfer systems and subclasses of VASS, in particular 1-dim VASS for ua-REACH might help solving this issue. Considering extensions of transfer systems is another interesting research direction, for instance with features that have been considered for Petri nets while maintaining decidability such as transfers or resets. Finally, beyond the purely cooperative question we tackled here, it would also be interesting to consider alternative problems in which the agents have conflicting objectives.

References

1. Araki, T., Kasami, T.: Some decision problems related to the reachability problem for Petri nets. Theoret. Comput. Sci. **3**(1), 85–104 (1976)

2. Bertrand, N., Hélouët, L., Lefaucheux, E., Paparazzo, L.: Reachability in multi-agent transfer systems (extended version). Technical report, HAL Inria https://inria.hal.science/hal-05366409 (2025)

3. Brázdil, T., Jančar, P., Kučera, A.: Reachability games on extended vector addition systems with states. In: Abramsky, S., Gavoille, C., Kirchner, C., Meyer auf der Heide, F., Spirakis, P.G. (eds.) ICALP 2010. LNCS, vol. 6199, pp. 478–489. Springer, Heidelberg (2010). https://doi.org/10.1007/978-3-642-14162-1_40

4. Brihaye, T., Bruyére, V., Goeminne, A., Raskin, J.F., Van den Bogaard, M.: The complexity of subgame perfect equilibria in quantitative reachability games. In: Proceedings of the 30th International Conference on Concurrency Theory (CONCUR'19), volume 140 of LIPIcs, pp. 13:1–13:16. Schloss Dagstuhl - Leibniz-Zentrum für Informatik (2019)

5. Brihaye, T., Dhar, A.K., Geeraerts, G., Haddad, A., Monmege, B.: Efficient energy distribution in a smart grid using multi-player games. In: Proceedings of Cassting Workshop on Games for the Synthesis of Complex Systems and 3rd International Workshop on Synthesis of Complex Parameters (Cassting/SynCoP'16), volume 220 of EPTCS, pp. 1–12, (2016)

6. Bulling, N., Goranko, V.: Combining quantitative and qualitative reasoning in concurrent multi-player games. Auton. Agent. Multi-Agent Syst. **36**(1), 2 (2022)

7. Cheng, A., Esparza, J., Palsberg, J.: Complexity results for 1-safe nets. Theor. Comput. Sci. **147**(1&2), 117–136 (1995)

8. Ciardo, G.: Petri Nets with Marking-Dependent Arc Cardinality: Properties and Analysis, pp. 179–198. Springer (1994)

9. Esparza, J.: Reachability in live and safe free-choice Petri nets is NP-complete. Theoret. Comput. Sci. **198**(1–2), 211–224 (1998)

10. Fahrenberg, U., Juhl, L., Larsen, K.G., Srba, J.: Energy Games in Multiweighted Automata. In: Cerone, A., Pihlajasaari, P. (eds.) ICTAC 2011. LNCS, vol. 6916, pp. 95–115. Springer, Heidelberg (2011). https://doi.org/10.1007/978-3-642-23283-1_9

11. Felner, A., et al.: Search-based optimal solvers for the multi-agent pathfinding problem: Summary and challenges. In: Proceedings of the 10th International Symposium on Combinatorial Search (SOCS'17), pp. 29–37. AAAI Press (2017)

12. Gutierrez, J., Murano, A., Perelli, G., Rubin, S., Steeples, T., Wooldridge, M.J.: Equilibria for games with combined qualitative and quantitative objectives. Acta Informatica **58**(6), 585–610 (2021)

13. Haase, C., Kreutzer, S., Ouaknine, J., Worrell, J.: Reachability in succinct and parametric one-counter automata. In: Bravetti, M., Zavattaro, G. (eds.) CONCUR 2009. LNCS, vol. 5710, pp. 369–383. Springer, Heidelberg (2009). https://doi.org/10.1007/978-3-642-04081-8_25

14. Hopcroft, J.E., Pansiot, J.-J.: On the reachability problem for 5-dimensional vector addition systems. Theoret. Comput. Sci. **8**, 135–159 (1979)

15. Karp, R.M.: Reducibility Among Combinatorial Problems, pp. 85–103. Springer (1972)

16. Künnemann, M., Mazowiecki, F., Schütze, L., Sinclair-Banks, H., Wegrzycki, K.: Coverability in VASS revisited: Improving Rackoff's bound to obtain conditional optimality. In: Proceeding of the 50th International Colloquium on Automata, Languages, and Programming (ICALP'23), volume 261 of LIPIcs, pages 131:1–131:20. Schloss Dagstuhl - Leibniz-Zentrum für Informatik (2023)

17. Leroux, J.: In: Petri net reachability problem (invited talk). Schloss Dagstuhl - Leibniz-Zentrum f"ur Informatik (2019)

18. Lipton, R.: The Reachability Problem Requires Exponential Space. Yale University (1976). Technical report
19. Minsky, M.L.: Computation: Finite and Infinite Machines. Prentice-Hall Inc, Upper Saddle River, NJ, USA (1967)
20. Nash, J.F.: Equilibrium points in n-person games. Proc. Natl. Acad. Sci. **36**(1), 48–49 (1950)
21. Rackoff, C.: The covering and boundedness problems for vector addition systems. Theoret. Comput. Sci. **6**, 223–231 (1978)
22. Savitch, W.J.: Relationships between nondeterministic and deterministic tape complexities. J. Comput. Syst. Sci. **4**(2), 177–192 (1970)
23. Schnoebelen, P.: Revisiting ackermann-hardness for lossy counter machines and reset petri nets. In: Hliněný, P., Kučera, A. (eds.) MFCS 2010. LNCS, vol. 6281, pp. 616–628. Springer, Heidelberg (2010). https://doi.org/10.1007/978-3-642-15155-2_54

Efficiently Verifying Quantum Programs
with Few T Gates

Youngchan Cho(✉) and Robert Rand

University of Chicago, Chicago, USA
{youngchan,rand}@uchicago.edu

Abstract. We mechanize a lightweight quantum logic in the Rocq proof
assistant to verify Clifford+T quantum programs, particularly those with
low T counts. Our tool is lightweight and automation-centric: Hoare
triple validation composes simple rules, reduces all obligations to syntactic well-formedness side-conditions, and efficiently discharges those
conditions. We demonstrate our tool using several case studies: a standard low-T Toffoli decomposition, graph-state generators, and the 7-
qubit Steane encoder. A small empirical study on graph-state families
shows near-linear growth in the number of Clifford gates and a slightly
super-quadratic trend in the number of qubits, consistent with implementation overheads from list-based tensor representations, while repeated
T gates on a single qubit exhibit an exponential blow-up due to additive
branching. The results indicate that a rule-driven, syntax-directed approach suffices to verify low-T quantum circuits while keeping the trusted
core and user-facing proofs simple.

Keywords: hoare logic · quantum computing · proof assistants · Rocq

1 Introduction

Quantum computing poses unique challenges for program verification: measurement is inherently probabilistic, the state space grows exponentially with the
number of qubits, and realistic toolchains must reason modularly about practical quantum circuits. As hardware scales, the demand for efficient verification
scales accordingly.

One influential line of work adapts Hoare logic to the quantum setting. Following Ying's Floyd–Hoare Logic for Quantum Programs [19], a number of systems have explored proof rules for quantum programs with different surface languages (circuits vs. while-programs) and different forms of assertions [17,18,21].
Among mechanized efforts, there are implementations in Isabelle/HOL for quantum Hoare logic [9], while CoqQ [20] verifies a quantum while-language in
Rocq [15].

This paper develops a complementary syntax-directed path: Hoare triple verification is purely syntactic and rule driven, while the base logic is formally verified. We formalize a variant of Hoare–Heisenberg logic [14] in the Rocq proof

Y.-F. Chen et al. (Eds.): VMCAI 2026, LNCS 16417, pp. 44–57, 2026.
https://doi.org/10.1007/978-3-032-15700-3_3

assistant, building on the QuantumLib library [16]. The logic takes an operator-level view rooted in the stabilizer formalism [5] and is tailored to Clifford+T programs. Our implementation is intentionally lightweight and automation-first: Hoare triple validation composes a core set of rules for H, S, CNOT, and T together with a small set of structural rules; all remaining obligations are reduced to well-formedness side-conditions that are discharged syntactically.

We demonstrate the tool on three representative classes of programs: a standard low-T Toffoli decomposition, graph-state generators, and the 7-qubit Steane encoder. Our benchmark on graph-state families exhibits near-linear growth in the number of gates and a slightly super-quadratic trend in the number of qubits, which we attribute to list-based data structures in the current implementation; raw data points are included in the artifact.

2   Quantum Computation

We begin with a short primer on quantum computation, discussing quantum states and operators, the Pauli and Clifford groups, and the Heisenberg representation of quantum computing.

2.1   Qubits and Quantum States

An n-qubit state is a unit vector in the complex Hilbert space $\mathbb{C}^{2^n}$. We use Dirac notation in which a *ket* $|\psi\rangle$ denotes a column vector and its bra is the complex conjugate transpose $\langle\psi| = |\psi\rangle^\dagger$. Computational basis states are given in the form $|x_1 \cdots x_n\rangle = |x_1\rangle \otimes \cdots \otimes |x_n\rangle$ (for $x_i \in \{0,1\}$) where $|0\rangle = \left(\begin{smallmatrix}1\\0\end{smallmatrix}\right)$ and $|1\rangle = \left(\begin{smallmatrix}0\\1\end{smallmatrix}\right)$. Any state can be written as a superposition $|\psi\rangle = \sum_{x \in \{0,1\}^n} \alpha_x |x\rangle$, i.e., a linear combination of basis states. The complex amplitudes satisfy $\sum_x |\alpha_x|^2 = 1$ and induce measurement probabilities $\Pr[x] = |\alpha_x|^2$. We use $|+\rangle$ and $|-\rangle$ to represent the states $\frac{1}{\sqrt{2}}(|0\rangle + |1\rangle)$ and $\frac{1}{\sqrt{2}}(|0\rangle - |1\rangle)$, each of which has a 50% chance of being measured as 0 or 1.

2.2   Unitary and Hermitian Operators

A quantum circuit is a sequence of quantum gates, and each gate corresponds to a unitary matrix on $\mathbb{C}^{2^n}$ (where n is the number of qubits). A matrix U is *unitary* if $U^\dagger U = I = UU^\dagger$. If U is unitary and $Uv = \lambda v$, then the eigenvalue satisfies $|\lambda| = 1$. Because unitary matrices preserve inner products ($\langle U\psi, U\phi\rangle = \langle\psi, \phi\rangle$), they preserve norms; a unitary applied to a unit vector is a unit vector. This guarantees that the total probability over all computational-basis outcomes remains 1.

A matrix H is *Hermitian* if $H^\dagger = H$. Hermitian matrices have real eigenvalues, so Hermitian operators are used to model measurable physical quantities. If a matrix is both unitary and Hermitian, then $H^2 = I$ and its eigenvalues are ± 1. Conjugation by a unitary preserves both unitarity and Hermiticity: If Q is unitary (resp. Hermitian) and U is unitary, then $UQU^\dagger$ is unitary (resp. Hermitian).

2.3   The Pauli and Clifford Groups

The Pauli matrices are

$$I = \begin{pmatrix} 1 & 0 \\ 0 & 1 \end{pmatrix}, \quad X = \begin{pmatrix} 0 & 1 \\ 1 & 0 \end{pmatrix}, \quad Y = \begin{pmatrix} 0 & -i \\ i & 0 \end{pmatrix}, \quad Z = \begin{pmatrix} 1 & 0 \\ 0 & -1 \end{pmatrix}.$$

The matrices X, Y, Z are *unitary, Hermitian*, and *trace-zero* (the sum of diagonal entries is 0). Since X, Y, Z are unitary Hermitian, all their eigenvalues are ± 1; and because $\operatorname{tr}(X) = \operatorname{tr}(Y) = \operatorname{tr}(Z) = 0$, the spectral decomposition implies that the number of $+1$ and -1 eigenvalues is equal: $\operatorname{tr}(X) = \sum_k \lambda_k = 0$ with each $\lambda_k \in \{\pm 1\}$ forces $|\{\lambda_k \mid \lambda_k = +1\}| = |\{\lambda_k \mid \lambda_k = -1\}|$, and the same for Y, Z.

The Pauli matrices multiplied by phases $\pm 1, \pm i$ form a group $\mathbf{P}$ under matrix multiplication, and the n-qubit Pauli group is the set of all *Pauli strings*

$$\mathbf{P}_n = \{ P_1 \otimes \cdots \otimes P_n \mid P_j \in \mathbf{P} \},$$

again under matrix multiplication. Since $\operatorname{tr}(A \otimes B) = \operatorname{tr}(A)\operatorname{tr}(B)$, a Pauli string is trace-zero whenever at least one element in the string is not cI.

The n-qubit Clifford group $\mathbf{Cl}_n$ consists of the unitaries that normalize $\mathbf{P}_n$:

$$\mathbf{Cl}_n = \{ U \in \mathbb{C}^{2^n \times 2^n} \mid U^\dagger U = I, \; U\mathbf{P}_n U^\dagger = \mathbf{P}_n \}.$$

We use the standard generators of the Clifford group—Hadamard (H), Phase (S), and Controlled-NOT (CNOT)—with matrix forms

$$H = \tfrac{1}{\sqrt{2}} \begin{pmatrix} 1 & 1 \\ 1 & -1 \end{pmatrix}, \quad S = \begin{pmatrix} 1 & 0 \\ 0 & i \end{pmatrix}, \quad \text{CNOT} = \begin{pmatrix} 1 & 0 & 0 & 0 \\ 0 & 1 & 0 & 0 \\ 0 & 0 & 0 & 1 \\ 0 & 0 & 1 & 0 \end{pmatrix}.$$

Gates acting on specific qubits are padded with identities using tensors. For a 1- or 2-qubit operator U, we write $U\, i$ (respectively $U\, i\, j$) for the operator on $\mathbb{C}^{2^n}$ that acts as U on qubit i (respectively qubits i, j) and as the identity on all other qubits; we take CNOT $i\, j$ to mean "control i, target j". For example, in a 3-qubit circuit (tensor order $1 \otimes 2 \otimes 3$), $H\, 2 = I \otimes H \otimes I$ and CNOT $3\, 1 = I \otimes I \otimes |0\rangle\langle 0| + X \otimes I \otimes |1\rangle\langle 1|$. When the register size matches the gate's arity, we will write S and CNOT for $S\, 1$ and CNOT $1\, 2$, respectively.

Adding the non-Clifford gate $T = \begin{pmatrix} 1 & 0 \\ 0 & e^{i\pi/4} \end{pmatrix}$ to the Clifford generators yields a universal gate set: any unitary on $\mathbb{C}^{2^n}$ can be approximated to arbitrary precision using Clifford+T. This is the complete set of gates used in this paper.

2.4   Heisenberg Representation

An n-qubit *stabilizer state* is a quantum state that is completely determined by listing n independent, commuting Pauli strings (excluding $-I^{\otimes n}$) that have eigenvalue $+1$ on the state. For stabilizer states, rather than printing the full state vector, we describe a state by such Pauli strings.

Called the *Heisenberg representation* by Gottesman [5], the evolution of a stabilizer state under a unitary matrix U can be tracked by transforming its stabilizers through group conjugation by U (i.e. mapping each stabilizer s to $UsU^\dagger$). In the Heisenberg representation a circuit with unitary U acts on operators by conjugation, $A \mapsto UAU^\dagger$. Conjugation similarly describes the evolution of $+1$ eigenvectors: If $UAU^\dagger = B$ and a vector v satisfies $Av = v$, then Uv satisfies $B(Uv) = Uv$. Hence, a stabilizer description evolves by transforming its Pauli strings P_i into $UP_iU^\dagger$. Conjugation by Clifford gates maps Pauli strings to Pauli strings, so we can track evolution by updating Pauli strings rather than simulating a 2^n-dimensional state vector (the Gottesman–Knill Theorem [5]).

For example, if v is a 2-qubit stabilizer state of the Pauli strings $X \otimes I$ and $I \otimes X$, then $v = |{+}{+}\rangle$. Suppose we want to compute $(H \otimes I)v$ without directly calculating $(H \otimes I)|{+}{+}\rangle$. Then, using the conjugation equation $HXH^\dagger = Z$, we see that $H \otimes I$ maps $X \otimes I$ to $Z \otimes I$ and $I \otimes X$ to itself. The Pauli strings $Z \otimes I$ and $I \otimes X$ describe the output state $(H \otimes I)v = |0{+}\rangle$.

This observation forms the basis for Hoare-Heisenberg logic.

3   Hoare-Heisenberg Logic

Guided by the observation above, Hoare-Heisenberg logic [14] defines a triple

$$\{A\}\ U\ \{B\} \qquad \text{to mean} \qquad \forall v \in E(A),\ Uv \in E(B)$$

where $E(A)$ is the set of $+1$ eigenvectors of A. That is, U takes states satisfying A to states satisfying B. In the implementation, we also use an equivalent interpretation at the operator level:

$$UAU^\dagger = B.$$

We will refer to these as the $+1$-*eigenvector semantics* and the *Heisenberg semantics*, respectively. The two semantics capture the same intent, but the Heisenberg semantics is often convenient in proofs because it reduces obligations to equalities between matrices. For well-formed atomic predicates (Sect. 4.1), the two semantics coincide; a mechanized proof of this equivalence is included in the accompanying Rocq development.

3.1   Syntax

The grammar of our implementation of Hoare-Heisenberg logic has four syntactic levels:

$$G ::= I \mid X \mid Y \mid Z$$
$$T ::= c\,(G \otimes \cdots \otimes G)$$
$$A ::= T + \cdots + T$$
$$P ::= A \mid A \cap A \mid T_{[n]}$$

We separate *tensor* predicates (T) from *additive* predicates (A) so that well-formedness (Hermitian, unitary, trace-zero, and matching dimensions) can be checked modularly at the T level and conservatively lifted to A (see Sect. 4.1).

Addition of T level terms in the A level is implemented as lists, so addition does not compute. At the predicate level (P), we include conjunction $(A \cap A)$ and separability $(T_{[n]})$ of predicates where the $[n]$ in separability denotes the list of qubit positions that are separable from the rest of the system; these predicates are called *non-atomic*.

We first give the core rules; each follows from an equality $UAU^\dagger = B$.

$$\{X\} \; H \; \{Z\} \qquad \{X \otimes I\} \; CNOT \; \{X \otimes X\} \qquad \{X\} \; T \; \{\tfrac{1}{\sqrt{2}}(X + Y)\}$$

$$\{Z\} \; H \; \{X\} \qquad \{I \otimes X\} \; CNOT \; \{I \otimes X\} \qquad \{Y\} \; T \; \{\tfrac{1}{\sqrt{2}}(Y - X)\}$$

$$\{X\} \; S \; \{Y\} \qquad \{Z \otimes I\} \; CNOT \; \{Z \otimes I\} \qquad \{Z\} \; T \; \{Z\}$$

$$\{Z\} \; S \; \{Z\} \qquad \{I \otimes Z\} \; CNOT \; \{Z \otimes Z\}$$

In our notation, an unindexed gate symbol denotes the gate in its canonical placement on the first qubit(s), stated in the minimal-arity setting. For example, in the core rules above, CNOT means CNOT 1 2. To act on other qubits or within larger registers, we use tensor lifting rules that apply the gates to the relevant position in the Pauli string: A simple instance of such a rule says that if $\{P\} \; U \; \{Q\}$ then $\{A_1 \otimes P \otimes A_3\} \; U \; 2 \; \{A_1 \otimes Q \otimes A_3\}$.

Note that we did not include rules where Y is a precondition. Instead, using the equality $Y = iXZ$ and the following rule

$$\frac{\{T_1\} \; g \; \{T_1'\} \qquad \{T_2\} \; g \; \{T_2'\} \qquad acom(T_1, T_2) \qquad acom(T_1', T_2')}{\{iT_1 T_2\} \; g \; \{iT_1' T_2'\}} \; \text{MUL-ACOM}$$

(where *acom* says that the operators anticommute) we can derive the behavior on Y for any gate in Clifford gate. For example,

$$\frac{\{X\} \; H \; \{Z\} \qquad \{Z\} \; H \; \{X\} \qquad acom(X, Z) \qquad acom(Z, X)}{\{Y\} \; H \; \{-Y\}}$$

For commuting operators such as $X \otimes I$ and $I \otimes Z$, we have a similar rule that simply multiplies the predicates together without the i factor.

Common Clifford gates like X, Y, Z, and CZ (controlled-Z) are also excluded from the core rules. Instead, we can define them in terms of H, S and $CNOT$: for instance, $CZ \; 1 \; 2 = H \; 2; CNOT \; 1 \; 2; H \; 2$. We can then use the sequential composition rule

$$\frac{\{A\} \; g_1 \; \{B\} \qquad \{B\} \; g_2 \; \{C\}}{\{A\} \; g_1 \; ; \; g_2 \; \{C\}} \; \text{SEQ}$$

to derive CZ's behavior. For instance

$$\frac{\dfrac{\{I \otimes Z\} \; H \; 2 \; \{I \otimes X\} \qquad \{I \otimes X\} \; CNOT \; 1 \; 2 \; \{I \otimes X\}}{\{I \otimes Z\} \; H \; 2 \; ; \; CNOT \; 1 \; 2 \; \{I \otimes X\}} \; \text{SEQ} \qquad \{I \otimes X\} \; H \; 2 \; \{I \otimes Z\}}{\{I \otimes Z\} \; H \; 2 \; ; \; CNOT \; 1 \; 2 \; ; \; H \; 2 \; \{I \otimes Z\}} \; \text{SEQ}$$

When we want to specify the $+1$ eigenspace as an intersection of multiple predicates, we can use the intersection rule

$$\frac{\{A_1\}\ g\ \{B_1\} \qquad \cdots \qquad \{A_k\}\ g\ \{B_k\}}{\{\bigcap\{A_i\}\}\ g\ \{\bigcap\{B_i\}\}}\ \text{CAP}$$

For example,

$$\frac{\{X \otimes I\}\ H\,1\ \{Z \otimes I\} \qquad \{I \otimes X\}\ H\,1\ \{I \otimes X\}}{\{(X \otimes I) \cap (I \otimes X)\}\ H\,1\ \{(Z \otimes I) \cap (I \otimes X)\}}$$

shows that $H1$ takes $|{+}{+}\rangle$, a common eigenstate of both $X \otimes I$ and $I \otimes X$, to $|0{+}\rangle$, a common eigenstate of both $Z \otimes I$ and $I \otimes X$.

In the presence of T gates, we often need to reason about a sum of predicates. In these cases, we use the linear combination rule

$$\frac{\{A_1\}\ g\ \{A_1'\} \qquad \cdots \qquad \{A_n\}\ g\ \{A_n'\}}{\{c_1 A_1 + \cdots + c_n A_n\}\ g\ \{c_1 A_1' + \cdots + c_n A_n'\}}\ \text{LINCOM}$$

where $c_1, \ldots, c_n \in \mathbb{C}$. For example,

$$\frac{\{X\}\ T\ \{\frac{1}{\sqrt{2}}(X + Y)\} \qquad \{Y\}\ T\ \{\frac{1}{\sqrt{2}}(Y - X)\}}{\{\frac{1}{\sqrt{2}}(X + Y)\}\ T\ \{\frac{1}{\sqrt{2}}(\frac{1}{\sqrt{2}}(X + Y) + \frac{1}{\sqrt{2}}(Y - X))\}}$$

Note that this is not exactly the outcome we would like: The consequent is equal to Y, which we would expect given that $T; T = S$. We can derive this using the consequence rule:

$$\frac{A' \Rightarrow A \quad \{A\}\ g\ \{B\} \quad B \Rightarrow B'}{\{A'\}\ g\ \{B'\}}\ \text{CONS}$$
.

$A \Rightarrow B$ means that every vector satisfying A also satisfies B (i.e., $E(A) \subseteq E(B)$). This is the part of the logic where the user can appeal to the underlying semantics. However, in our automation, we limit our application of the consequence rule to simple syntactic implications, like $A \cap B \Rightarrow B \cap A$.

Our most interesting implication rules concern separability. Separability rules allow us to state judgments of the form $T_1 \cap \cdots \cap T_n \Rightarrow (T_{i_1} \cap \cdots \cap T_{i_k})_{L_1} \cap \cdots \cap (T_{j_1} \cap \cdots \cap T_{j_l})_{L_n}$ where the L_is are lists of qubit positions that are separable from the other system. For example, $(X \otimes X \otimes I) \cap (Y \otimes Y \otimes I) \cap (I \otimes I \otimes Z) \Rightarrow ((X \otimes X) \cap (Y \otimes Y))_{[1,2]} \cap Z_{[3]}$, in which the first two qubits are separably from the third. Our logic can be lifted to separable systems in the expected manner.

4　Implementation

We implement the Hoare-Heisenberg logic in Rocq [15] on top of the QuantumLib library [16], extending it with subspace reasoning (dimension theorem) needed by our predicates. Automation does not manipulate $(+1)$-eigenvectors or $UAU^\dagger$ directly: the tactics apply the core and composition rules, while well-formedness side-conditions are discharged syntactically. The equivalence of the two semantics for atomic predicates holds under the well-formedness conditions below.

4.1　Well-Formedness

Atomic predicates should denote matrices that are **Hermitian, unitary**, and **trace-zero** with the **expected dimension**. We enforce this via syntactic well-formedness restrictions:

- **Tensors:** $T = c\,(G_1 \otimes \cdots \otimes G_n)$ with $c = \pm 1$ and at least one $G_i \neq I$. Since $\mathrm{tr}(G) = 0$ for non-identity Pauli and $\mathrm{tr}(A \otimes B) = \mathrm{tr}(A)\mathrm{tr}(B)$, this ensures the predicate trace-zero; $c = \pm 1$ ensures Hermiticity and unitarity. The dimensions are embedded in the tensor predicates using dependent types.
- **Additive predicates:** $A = \frac{1}{\sqrt{2}}(B + C)$ where B, C are well-formed and pairwise anticommuting (including nested predicates).

We check syntactic well-formedness for all atomic predicates that appear *above the line* in all inference rules except SEQ and CAP. This guarantees that all proven triples will contain only valid (Hermitian, unitary and trace-zero) predicates that characterize quantum states. The only exception is the LINCOMB rule, where the user can provide incorrect coefficients to derive an invalid postcondition from an invalid precondition. To avoid this, deductions should begin with stabilizer (tensor) predicates.

4.2　Automation

Verifying Triples. Our main automation tactic, `validate`, proves Hoare triples by forward reasoning. Given $\{P\}\ g_1; \cdots ; g_n\ \{Q\}$, it repeatedly applies SEQ to reduce the goal to single–gate triples $\{P\}\ g_1\ \{P_1\}$, $\{P_1\}\ g_2\ \{P_2\}$, $\cdots$, where each P_i is initially an existential variable (or "evar"). We apply the CAP rule as needed to reduce each P_i to atomic predicates, and use forward reasoning to instantiate each P_{i+1}, and so on. When atomic predicates are additive, the forward reasoning may invoke LINCOMB, which can introduce additional additive terms. After the forward reasoning yields P_n, the tactic discharges $P_n \Rightarrow Q$ via simple permutation checks; the syntactic checker handles all well-formedness obligations.

Inferring Postconditions. The reader may note that `validate` infers its own postcondition P_n before attempting to unify it with the given postcondition. This allows us to automatically generate postconditions using the tactic `solvePlaceholder`, which solves triples of the form $\exists Q,\ \{P\}\, g_1; \cdots ; g_n\, \{Q\}$. To be precise, it proves a triple of the form $\{P\}\ g\ \{P_n\}$, allowing the user to unify Q with P_n or revise their lemma to say $\{P\}\, g_1; \cdots ; g_n\, \{P_n\}$. We use `solvePlaceholder` for several of our benchmarks (Sect. 5.3).

5   Case Studies and Benchmarks

We now evaluate the tool on three representative classes of programs. Unless stated otherwise, all proofs use `validate`.

5.1   Toffoli Gates

A Toffoli gate is a doubly controlled NOT and a standard stress test for Clifford+T verification. We verify the usual low-T decomposition by validating the three triples

$$\{Z \otimes I \otimes I\}\ \mathsf{Toffoli}\ \{Z \otimes I \otimes I\},$$
$$\{I \otimes Z \otimes I\}\ \mathsf{Toffoli}\ \{I \otimes Z \otimes I\},$$
$$\{I \otimes I \otimes X\}\ \mathsf{Toffoli}\ \{I \otimes I \otimes X\}.$$

The first two examples show that the Toffoli gate preserves the state when one of the input qubits is zero. The last example illustrates that performing a bit flip operation on the target $|+\rangle = \frac{1}{\sqrt{2}}(|0\rangle + |1\rangle)$ state does not change the target state. We also validate triples with the remaining preconditions $X \otimes I \otimes I$, $I \otimes X \otimes I$, and $I \otimes I \otimes Z$, whose postconditions expand to sixteen additive summands, illustrating how multiple T gates grow additive predicates. The circuit listing and Rocq definitions are provided in the artifact.

5.2   Graph States

Graph states [7] form a widely used, highly entangled family with applications in measurement-based computation and quantum information [6]. For a graph $G = (V, E)$, let $K_v^G = (X\ v) \cdot \prod_{\{v,w\}\in E}(Z\ w)$ be the standard stabilizer at v; that is, for a fixed qubit position v, we apply the X operator to qubit v and for each undirected edge $\{v, w\}$ in the graph G, we apply Z to w. These generators are independent and commute [7]. Writing $|G\rangle$ for the common $+1$-eigenvector of $\{K_v^G\}_{v\in V}$, our specification matches the usual preparation procedure from $|+\rangle^{\otimes |V|}$:

$$\{X \otimes I \otimes \cdots \otimes I \cap \cdots \cap I \otimes \cdots \otimes I \otimes X\} \left(\prod_{\{v,w\}\in E} \mathrm{CZ}\ v\ w \right) \left\{ \bigcap_{v\in V} K_v^G \right\}.$$

Our `validate` tactic solves these goals immediately.

5.3 Benchmarks

Setup 1 (Graph State Variations). We benchmark Hoare triple validation on graph state generators using the `validate` tactic. Specifications are written with CZ edges, but in the implementation each CZ $i\ j$ is expanded to $H\ j$; CNOT $i\ j$; $H\ j$ before validation, so we report m as the number of *primitive* gates and have $m = 3\,|E|$ where E is the chosen edge set. For the *qubit sweep* (left plot in Fig. 1) we fix E to the 10 edges of the complete graph K_5 and embed that pattern into an n–qubit register (unused qubits are isolated), hence $m = 3\binom{5}{2} = 30$ for all n. For the *gate sweep* (right plot in Fig. 1) we fix $n = 10$ and take $E = E(K_k)$ on the first $k \in \{5, 6, 7, 8, 9, 10\}$ qubits, so $m = 3\binom{k}{2} \in \{30, 45, 63, 84, 108, 135\}$.

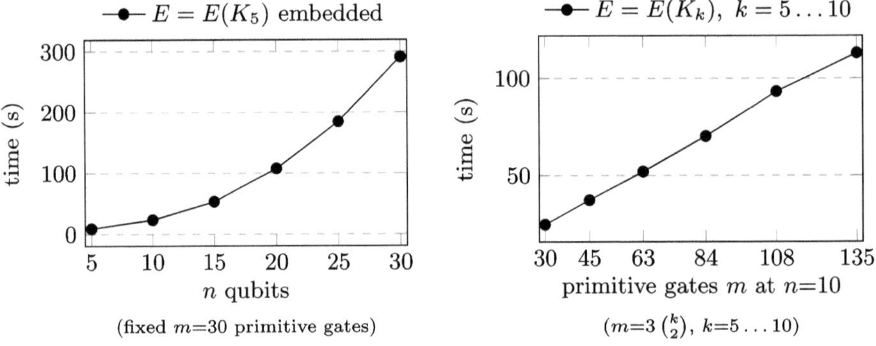

Fig. 1. Hoare triple verification wall-clock time on graph–state generators. Left: scaling in n with a fixed primitive-gate(H, S, CNOT, T) budget $m = 30$ from the complete graph K_5 ($|E| = 10$). Right: scaling in the number of primitive gates m at fixed $n = 10$ by growing the edge set from K_5 to K_{10} ($m = 3\binom{k}{2}$).

Results 1. Runtime grows nearly linearly in the primitive–gate count m—about 0.83 s per primitive gate at n=10 (i.e., ~ 2.5 s per CZ)—and increases more steeply with n; for $n \geq 10$, time divided by $n^2 \log n$ is roughly constant (Fig. 1). Conceptually, the logic entails $O(mn)$ rule applications, and our list-based tensor representation lifts the practical baseline to $O(mn^2)$; empirically the n–dependence is slightly heavier and is well captured by a slowly varying extra factor, giving an overall fit close to $O(mn^2 \log n)$ on our ranges.

Setup 2 (GHZ Diagonal Sweep). A (generalized) GHZ state is $\frac{1}{\sqrt{2}}|0\rangle^{\otimes n} + \frac{1}{\sqrt{2}}|1\rangle^{\otimes n}$. We validate GHZ circuits that use H followed by a chain of CNOTs (H 1; CNOT 1 2; ... CNOT (n-1) n) using the `solvePlaceholder` tactic. Here, each circuit uses $m = n$ *primitive* gates (one H and $n-1$ CNOTs), so this suite probes a diagonal slice of the (m, n) plane. We measure wall-clock time to close the triple with precondition $\bigcap_{i=1}^{n} Z\ i$ and vary $n = m \in \{5, 10, 15, 20, 25, 30\}$.

Results 2. Runtime grows as a smooth cubic-log curve in n and is well explained by the model $T(n) \propto n^3 \log n$ (which matches the general $O(mn^2 \log n)$ trend under $m{=}n$); see Fig. 2.

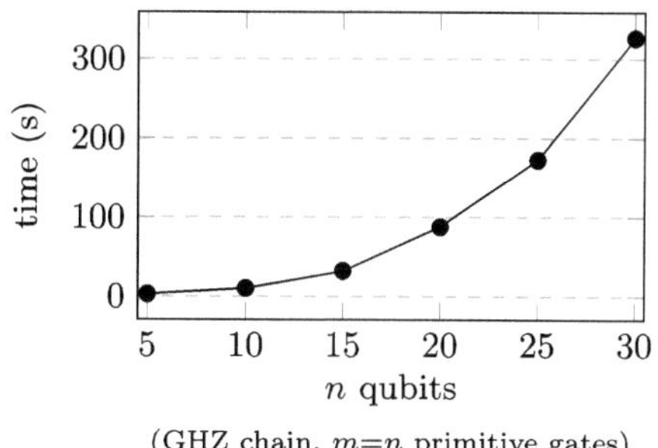

Fig. 2. Hoare triple verification wall-clock time for the n-qubit GHZ preparation (chain). Each point uses $m{=}n$ primitive gates (one H and $n{-}1$ CNOTs).

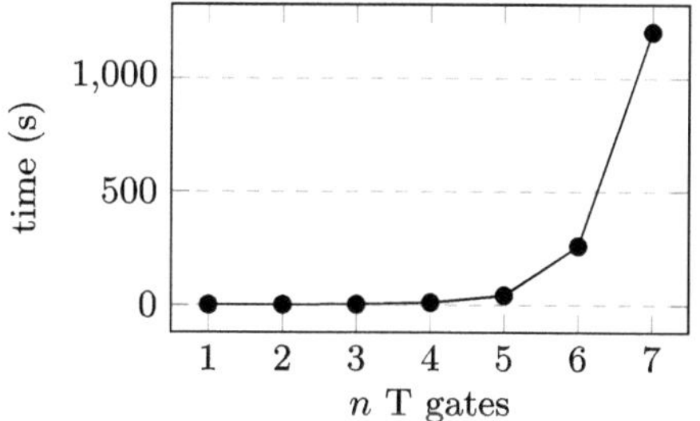

Fig. 3. T gates on a single qubit. Wall-clock time for $H; (T)^n$ with $n = 1 \dots 7$.

Setup 3 (T Gates on a Single Qubit). We isolate additive branching on one qubit using the `solvePlaceholder` tactic. Starting from $|0\rangle$ we apply one H to expose an X-stabilizer and then apply n T gates to the same qubit: That is, the precondition predicate P_{init} and the circuits Ts_n depending on n are

$$\mathsf{Ts}_n := H; (T)^n, \qquad P_{\mathrm{init}} = Z.$$

The intended algebra would keep the predicate in a small, cycling form, since repeated T gates rotate X within the XY plane. After n applications one obtains (up to global phase) $X \mapsto \cos\left(\frac{n\pi}{4}\right) X + \sin\left(\frac{n\pi}{4}\right) Y$, so the predicate should cycle with period 8. However, since `validate` does *not* merge or cancels terms, each application of T doubles the number of summands in the synthesized postcondition; after n applications the list has 2^n terms. We measure wall-clock time to validate the triple for $n \in \{1, 2, 3, 4, 5, 6, 7\}$.

Results 3. Runtime rises rapidly with n (seconds: 0.305, 0.869, 2.469, 10.841, 42.227, 260.36, 1204.312). The plot of Fig. 3 shows raw time. The runtime thus shows an exponential blow-up, as we would expect from a circuit with many T gates (and no cancellation).

5.4   Error Correction

A central application domain is stabilizer codes. For the $[7, 1]$ Steane code [13], let $g_1, \ldots, g_6$ be the standard independent commuting generators and let $\overline{X}, \overline{Z}$ denote the logical Paulis [11, Chs. 10.4.2, 10.5.1]. We confirm that the encoder maps computational and Hadamard bases to logical codewords stabilized by these operators by validating

$$\{Z\,1 \cap Z\,2 \cap \cdots \cap Z\,7\} \ \mathsf{Steane7} \ \{g_1 \cap \cdots \cap g_6 \cap \overline{Z}\},$$

$$\{X\,1 \cap Z\,2 \cap \cdots \cap Z\,7\} \ \mathsf{Steane7} \ \{g_1 \cap \cdots \cap g_6 \cap \overline{X}\}.$$

In the proof scripts, we solve the existential postcondition produced by SEQ and then check equivalence using a canonical ordering of intersections; an admissible rule allows us to replace a conjunctive postcondition B by its equivalent canonical form inspired by the row-echelon form of matrices [14, Section 3]. We separate these steps for performance. We include a similar validation of the $[9, 1]$ Shor code [12] in the artifact.

6   Discussion

Our benchmarks show that rule-driven, syntax-directed validation scales predictably on Clifford circuits such as graph states and GHZ preparations. While the logic entails $O(mn)$ rule applications, our list-based tensor representation adds overhead, and observed runtimes fit a model close to $O(mn^2 \log n)$. Even so, the growth remains polynomial and confirms that the approach is lightweight and practical for low-T pipelines. We hope to address this in future iterations of the tool by using arrays or binary trees to represent lists of tensors, further reducing the time complexity of verification.

By contrast, the single-qubit T-gate experiment shows exponential additive blow-up. We can expect this to reliably be a limiting factor for our logic (the Gottesman-Knill theorem [4] applies only to Clifford circuits) but not of this particular example: Repeated T applications should cycle through a fixed set of stabilizers with period 8, bounding the branching factor. Future work includes addressing this exponential blow-up by introducing a tactic that simplifies predicates during the validation procedure: $\frac{1}{\sqrt{2}}(X + Y) + \frac{1}{\sqrt{2}}(Y - X))$ should become simply Y. Reflection-based tactics inspired by Rocq's `ring_simplify` should make this straightforward.

We can also revise the framework to reduce well-formedness checks. Revising our triples so that $\{P\}\ g\ \{Q\}$ also states that "if P is well-formed, so is Q" will push all the well-formedness checks to the leaves of our derivations, where they should be trivial to check.

The current implementation is missing rules for measurement from the Hoare-Heisenberg logic [14]. Measurement will allow us to reason disjunctively about the different branches of a measurement, allowing us to reason effectively about quantum error correction.

7   Related Work

A number of quantum Hoare logics emerged from Ying's QHL [19] which has been extended to encompass richer language features and proof principles [17, 18, 21]. These systems vary along two axes that matter for tooling: (i) the program model (circuit fragments versus quantum while-programs), and (ii) the shape of assertions (from positive operators to algebraic combinations of Paulis). Notably, with the exception of Huang et al. [8] (discussed below), none of these logics feature a syntactically-defined assertion language.

Mechanized instances include an Isabelle/HOL development of quantum Hoare logic [9] and the formalization of a similar logic within Rocq in CoqQ [20]. The Isabelle/HOL development treats square matrices P where both P and $I - P$ are positive operators as predicates, whereas CoqQ allows arbitrary linear operators as predicates while acknowledging that these *should* be Hermitian. By contrast, our work targets the *circuit* model and restricts atomic predicates so that they are Hermitian, unitary, and trace-zero. These choices are aligned with stabilizer-style reasoning and friendly to syntactic well-formedness checks. QECV [18] explores universal expressiveness of stabilizers by allowing arithmetic combinations of Pauli strings; but there is no available implementation, let alone an automated checker. Finally, Huang et al. [8], concurrently developed a Hoare logic for error-correction, inspired by Hoare-Heisenberg logic and QECV, which contains both a Rocq formalization and a Python verifier integrated with SMT-solvers. While the Python tool is powerful, it is not verified and cannot be integrated with other formal verification efforts, and the authors did not explore proof automation in Rocq.

Outside of quantum Hoare logic [19], other circuit-level verification frameworks include Feynman, QBricks, and AutoQ. Feynman [1] uses the *path-sum* formalism, based on Feynman's path integrals, to efficiently verify concrete quantum programs. QBricks generalizes this approach to parametric circuits and is implemented as a DSL inside the Why3 deductive platform [2], with its own Hybrid Quantum Hoare Logic. AutoQ [3] encodes pre- and postconditions as level-synchronized tree automata and uses the Z3 [10] SMT solver for checking entailment.

8   Conclusion and Future Work

We presented a Rocq formalization of Hoare–Heisenberg logic and rule-driven tactics for verifying Clifford-T circuits. Hoare triple validation applies a small set of rules and reduces obligations to syntactic well-formedness checks, keeping the trusted core and user-facing proofs small. Case studies of Toffoli gates, graph and GHZ states, and the 7-qubit Steane encoder, along with benchmarking, indicate that this lightweight approach is practical for circuits with low T counts. We discuss the strengths and weaknesses of our tool, and how it can be extended to efficiently reason about a broader class of quantum programs.

In our experimental development, there are two main improvements. One is a function, `compute_PC`, that computes the postcondition given the precondition and the circuit by forward reasoning of the Hoare logic. The correctness of this function is proved using the core and composition rules: If A is a well-formed predicate and c is a circuit, the triple $\{A\}\, c\, \{\texttt{compute\_PC}(A)\}$ is valid. This allows us to reduce the number of syntactic well-formedness checks to only once in the precondition. The other improvement is the reflective calculation of scalars. From the observation that only scalars of the form $\pm(\frac{1}{\sqrt{2}})^k$ for some natural number k appear in additive predicates due to well-formedness, we reflect this in a separate data structure to facilitate the calculation of powers of $1/\sqrt{2}$ which is unwieldy to handle as a general complex number. We structured Rocq's type classes so that the parsing of $1/\sqrt{2}$ into its reflected data structure is automatically done by Rocq's type class inference. This reflective calculation changes the T-gate benchmark to be very efficient because additive terms with the same Pauli strings but different scalars are computationally combined.

Acknowledgments. Thanks to William Spencer for his help in making the tool more efficient. This material is based upon work supported by the Air Force Office of Scientific Research under Grant No. FA95502310406.

Data Availability Statement. An environment with the tools and data used for the experimental evaluation in this study is available at https://doi.org/10.5281/zenodo.17180231.

References

1. Amy, M.: Towards large-scale functional verification of universal quantum circuits. Electron. Proc. Theor. Comput. Sci. **287**, 1–21 (2019). https://doi.org/10.4204/eptcs.287.1
2. Bobot, F., Filliâtre, J.C., Marché, C., Paskevich, A.: Why3: Shepherd your herd of provers. In: Boogie 2011: First International Workshop on Intermediate Verification Languages (2012)
3. Chen, Y., Chung, K., Hsieh, M., Huang, W., Lengál, O., Lin, J., Tsai, W.: AutoQ 2.0: From verification of quantum circuits to verification of quantum programs. In: Gurfinkel, A., Heule, M. (eds.) Tools and Algorithms for the Construction and Analysis of Systems - 31st International Conference, TACAS 2025, Held as Part of the International Joint Conferences on Theory and Practice of Software, ETAPS 2025, Hamilton, ON, Canada, 3–8 May 2025, Proceedings, Part III. Lecture Notes in Computer Science, vol. 15698, pp. 87–108. Springer, Heidelberg (2025). https://doi.org/10.1007/978-3-031-90660-2_5

4. Gottesman, D.: The heisenberg representation of quantum computers. arXiv preprint quant-ph/9807006 (1998)

5. Gottesman, D.: The heisenberg representation of quantum computers. In: Group22: Proceedings of the XXII International Colloquium on Group Theoretical Methods in Physics. pp. 32–43. No. LA-UR-98-2848, International Press (1998)

6. Hein, M., Dür, W., Eisert, J., Raussendorf, R., den Nest, M.V., Briegel, H.J.: Entanglement in graph states and its applications (2006). https://arxiv.org/abs/quant-ph/0602096

7. Hein, M., Eisert, J., Briegel, H.J.: Multiparty entanglement in graph states. Phys. Rev. A **69**(6) (2004). https://doi.org/10.1103/physreva.69.062311

8. Huang, Q., Zhou, L., Fang, W., Zhao, M., Ying, M.: Efficient formal verification of quantum error correcting programs. Proc. ACM Program. Lang. **9**(PLDI), 190 (2025). https://doi.org/10.1145/3729293

9. Liu, J., et al.: Quantum hoare logic. Archive of Formal Proofs (2019). https://isa-afp.org/entries/QHLProver.html, Formal proof development

10. de Moura, L., Bjørner, N.: Z3: an efficient SMT solver. In: Ramakrishnan, C.R., Rehof, J. (eds.) TACAS 2008. LNCS, vol. 4963, pp. 337–340. Springer, Heidelberg (2008). https://doi.org/10.1007/978-3-540-78800-3_24

11. Nielsen, M.A., Chuang, I.L.: Quantum Computation and Quantum Information, 10th Anniversary Cambridge University Press, Cambridge (2010). https://doi.org/10.1017/CBO9780511976667

12. Shor, P.W.: Scheme for reducing decoherence in quantum computer memory. Phys. Rev. A **52**, R2493–R2496 (1995). https://doi.org/10.1103/PhysRevA.52.R2493. https://link.aps.org/doi/10.1103/PhysRevA.52.R2493

13. Steane, A.: Multiple-particle interference and quantum error correction. Proc. R. Soc. Lond. Ser. A: Math. Phys. Eng. Sci. **452**(1954), 2551–2577 (1996). https://doi.org/10.1098/rspa.1996.0136

14. Sundaram, A., Rand, R., Singhal, K., Lackey, B.: Hoare meets heisenberg: a lightweight logic for quantum programs (2025). https://arxiv.org/abs/2101.08939

15. Team, T.R.: The Rocq proof assistant. https://rocq-prover.org/. Accessed 14 Mar 2025

16. The INQWIRE Developers: INQWIRE QuantumLib (2022). https://github.com/inQWIRE/QuantumLib

17. Unruh, D.: Quantum hoare logic with ghost variables. In: Proceedings of the 34th Annual ACM/IEEE Symposium on Logic in Computer Science, LICS '19, pp. 1–13. IEEE Computer Society (2019). https://doi.org/10.1109/LICS.2019.8785779

18. Wu, A., Li, G., Zhang, H., Guerreschi, G.G., Xie, Y., Ding, Y.: Qecv: quantum error correction verification (2021)

19. Ying, M.: Floyd-hoare Logic for Quantum Programs. ACM Trans. Program. Lang. Syst. **33**(6), 19 (2012). https://doi.org/10.1145/2049706.2049708

20. Zhou, L., Barthe, G., Strub, P.Y., Liu, J., Ying, M.: Coqq: foundational verification of quantum programs. Proc. ACM Program. Lang. **7**(POPL), 833–865 (2023)

21. Zhou, L., Yu, N., Ying, M.: An applied quantum Hoare logic. In: Proceedings of the 40th ACM SIGPLAN Conference on Programming Language Design and Implementation, PLDI '19, pp. 1149–1162. ACM (2019). https://doi.org/10.1145/3314221.3314584. https://opus.lib.uts.edu.au/bitstream/10453/140615/2/3314221.3314584.pdf

Atomic Gliders and Cellular Automata as Language Generators

Dana Fisman[1] and Noa Izsak[1,2](✉)

[1] Ben Gurion University, Beer-Sheva, Israel
[2] CISPA Helmholtz Center for Information Security, Saarbrücken, Germany
`noa.izsak@cispa.de`

Abstract. Cellular automata (CA) are well-studied models of decentralized parallel computation, known for their ability to exhibit complex global behavior from simple local rules. While their dynamics have been widely explored through simulations, a formal treatment of CA as genuine *language generators* remains underdeveloped. We formalize CA-expressible languages as sets of finite words obtained by projecting the non-quiescent segments of configurations reachable by one-dimensional, deterministic, synchronous CA over bi-infinite grids. These languages are defined with respect to sets of initial configurations specified by a regular language as in *regular model checking*. To capture structured dynamics, we propose a *glider-based generative semantics* for CA. Inspired by the classical notion of gliders, we define a glider as a one-cell entity carrying a symbol in a certain velocity under well defined interaction semantics. We show that despite the regularity of the initial configurations and the locality of the transition rules, the resulting languages can exhibit non-regular and even non-context-free structure. This positions regular-initialized CA languages as a surprisingly rich computational model, with potential applications in the formal analysis of linearly ordered MAS.

Keywords: Cellular automata · Glider-based systems · Symbolic dynamics · Regular Model checking · Beyond regularity

1 Introduction

Multi-agent systems (MAS) arranged in a linear topology, consisting of identical agents (or processes) that operate in parallel, arise naturally in many applications. In these systems, agents are typically organized in a line or a ring. The agents evolve in synchronous rounds; at each step, all agents update their state simultaneously, each according to the state of its finite neighborhood (including itself). Analyzing and verifying such systems is a central challenge in formal methods. For a single finite-state transition system, classical verification techniques apply directly since the system has an explicit finite-state space. In contrast, in a multi-agent system, all agents implement the same protocol, P.

N. Izsak—This author was partially supported by ISF grant 2507/21.

Executing P with n indistinguishable agents yields a finite system P^n. The resulting MAS induces an infinite family $\{P^n\}_{n \in \mathbb{N}}$, referred to as a *parametrized system*, since the number of agents is unbounded. Research in this area spans verification (e.g., [2,4,7,15,29]), synthesis [30,34,42] and learning [19,22,45].

In their seminal work [33], Kesten et al. proposed the use of *rich assertional languages* to symbolically capture the unbounded nature of the system families induced by such systems. This idea gave rise to the method now known as *regular model checking* (RMC), which has since developed into a well-established verification technique [3,8–10,22,31,59].

In RMC, *sets of configurations* are represented by regular languages over a finite alphabet, while *transition relations* are expressed as regular relations, typically implemented by finite-state transducers. Using regular languages, one can succinctly describe the set of initial configurations for systems of all sizes. The alphabet specifies the local state-space of a process, and a word of length n encodes the states of n processes arranged along a line or a ring, with position i corresponding to process i. For example, the regular language 10^* compactly denotes infinitely many systems, one for each $n \in \mathbb{N}$, where the leftmost process is in state 1 (signifying that it holds a token), while all processes to its right are in state 0 (not holding the token).

In each synchronous step, *all* processes update their states simultaneously, based on their current state and the states of finitely many neighbors (e.g., their immediate left and right). In the case of the *token-passing protocol* [16,43], the relation specifies that if a process does not currently hold a token but its left neighbor does, then in the next step the process acquires the token while the neighbor relinquishes it. The RMC framework has proven to be highly effective for automatic verification of *parameterized systems*, supporting the analysis of a wide range of protocols [1,3,9,17,48].

The RMC framework assumes that: the initial configurations, the transition relation, and the set of bad states are all described by regular languages. Only limited work has explored the use of non-regular specifications [20,21]. Moreover, there is often an implicit assumption that starting from a regular initial configuration and repeatedly applying a regular transition relation necessarily yield a regular language. As the following simple *two-captains protocol* illustrates, this assumption does not always hold.

Imagine a group of children forming two opposing teams during recess. At first, there are only two captains: one for *Team-A* (abbreviated as **a**) on the left and one for *Team-B* (**b**) on the right: 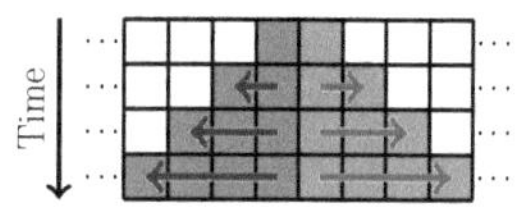. In each round, both captains simultaneously recruit a new player to join their side. The captain of Team-A extends the line to the left by adding an **a**, while the captain of Team-B extends the line to the right by adding a **b**. That is: ▬▨◪▬. Which in turn becomes: ▬▰▨◪▬. As rounds proceed, the two teams grow in perfect synchrony.

$$\mathsf{ab} \mapsto \mathsf{aabb} \mapsto \mathsf{aaabbb} \mapsto \cdots \mapsto \overbrace{\mathsf{a}\cdots\mathsf{a}}^{n}\overbrace{\mathsf{b}\cdots\mathsf{b}}^{n} \mapsto \cdots$$

After m rounds, the playground is exactly $\mathsf{a}^m\mathsf{b}^m$. This simple example illustrates how a trivial initial configuration (of two captains) evolves, under a uniform growth rule, into the classical non-regular language $\{a^n b^n \mid n \in \mathbb{N}\}$.

In this paper, we study the expressiveness and limitations of systems that start from a regular set of initial configurations and evolve according to a deterministic transition relation, determined by the states of processes within a fixed-radius neighborhood r. Such a transition can be viewed as a function from Σ^{2r+1} to Σ, and can naturally be modeled by a regular transducer. We consider our evolution of the processes over a bi-infinite tape, that is, a sequence indexed by the integers $\mathbb{Z}$, rather than a one-sided infinite tape indexed by the naturals $\mathbb{N}$. This bi-infinite representation is often preferable, as it facilities reasoning about systems whose evolution extends in both directions. The two-captains protocol already illustrates why this perspective is natural.

This bi-infinite view, together with the local update rule (each process updating its state based on its own state and the states of its neighbors within radius r), aligns closely with the classical framework of one-dimensional cellular automata (CA). The correspondence is natural; as in both settings, a collection of homogeneous processes (or cells, in CA terminology) evolves synchronously under a uniform local transition rule.

Cellular automata (CA) have long served as canonical models of distributed computation and emergent behavior [6, 39, 49, 56]. First, they were popularized by Conway [13] and later systemically explored by Wolfram [57]. CA have become emblematic of how simple local rules can give rise to rich, complex, and sometimes unpredictable global behavior. In this tradition, CA are studied not only as mathematical abstractions of parallel systems, but also as laboratories for exploring the emergence of complexity from homogeneous local behaviors. Their appeal stems precisely from this dual role as both minimalist models of computation and vivid illustrations of collective phenomena.

In parallel with these explorations, a line of work connected CA directly to formal language theory. Smith's seminal work [52] established a correspondence between one-dimensional CA and linearly bounded automata, situating CA within the Chomsky hierarchy and clarifying their recognition capabilities. Subsequent contributions by Hurd [28] and Nordahl [50], refined this perspective, showing that a one-dimensional CA defined over finite grids *preserves language complexity under evolution*. These studies demonstrated that CA, when viewed as recognizers, maintain well-defined connections to classical automata-theoretic complexity. However, as in much of the CA literature, they treated the grid as bounded and focused on acceptance, stabilization, or halting conditions. The potential of CA as *language generators* on an unbounded grid and an unbounded time evolution has remained comparatively underexplored. Specifically, the evolution of a CA from a fixed initial configuration over time naturally yields sequences of global states, which may be viewed as the language of reachable states.

In this work, we investigate the *expressive power* of the language generated by a CA starting from regular initial configurations. Specifically, we ask which families of words are *generated* by a CA as it evolves on an unbounded grid? We refer to this perspective as the *generative view* of CA languages. This shifts the role of CA from the common focus on *simulating* behaviors or *recognizing languages* [32,41,54] to *generating* languages. A viewpoint that, while natural from a formal-methods perspective, has remained comparatively underexplored in the CA literature. To avoid misunderstanding, we emphasize that prior work on CA, although seldom using the term *generation*, has been considered under specific constraints, such as bounded evolutions (in time or space) [11,23,44], and analysis of *limit set*, which requires infinite recurrence [24,40,47,57]. These contributions highlight important aspects of reachability in restricted settings. In contrast, our framework adopts a *symbolic* view, admits arbitrary alphabet and radii, no evolution bounds are imposed, and recurrence is not required. Thus, this allows us to position CA as genuine *language generator*. At the same time, the generative view offers a symbolic framework for analyzing the global dynamics of such systems and opens the door to future applications of formal verification in this domain.

The *two-captains protocol* also sets the stage for the central notion of this paper—*glider mechanisms*. In the classical literature on CA, the term glider refers to a recurring multi-cell pattern that travels the grid, most famously in Conway's Game of Life [13] and later in Cook's universality construction [14]. Such gliders typically cycle through a sequence of shapes while advancing in the grid, returning to their initial form after several steps, a kind of "wheel of life" whose periodicity underlies their evolution.

Our usage is inspired by these traditions, but adopts a more structured perspective. Here, a glider is not a composite pattern but a *single cell*, endowed with the semantics of the automaton's local rules: its symbol, its velocity, i.e., speed and direction (left or right), and its interaction behavior. In this sense, the cycle sustaining a Conway-style glider is compressed to size one: the state of the glider remains fixed as it propagates. On the other hand, since our automata may have arbitrary radius, a glider's velocity need not be limited to one cell per step, but may reflect any displacement determined by the local rule.

As in the classical setting, our glider persists indefinitely in the absence of interaction, but may be terminated when colliding with other gliders. The key difference is that persistence, velocity, and interaction behavior are all dictated directly by the CA local rule function, rather than being emergent from multi-cell dynamics. This makes gliders in our setting *symbolic building blocks*; their semantics are modular and explicit, even when the underlying CA has a large interaction radii. Seen through this lens, the captains of the *two-captain protocol* are instances of gliders: we have two gliders of state a one of velocity of -1 and the second of velocity 0, that way, in every iteration, Team-A "expands" one step to the left (the -1 velocity) while maintaining all existing locations (the 0 velocity). Similarly, for state b we have two gliders as well, one of velocity 0 and the second of velocity $+1$, which allows the expansion of Team-B towards the right. These structured gliders provide the building blocks for the generative semantics developed in the reminder of the paper, where we show how these gliders yield a modular explanation of emergent pattern and enable constructive proofs of expressive power in a system where population growth is essential.

Expressiveness Results. In exploring the generative power of CA, we set out to find languages that push the expressive boundary. We began with a simple non-regular example, showing that $a^n b^n$ arises naturally for local growth rules. From there, we traveled alongside variations of the one-counter automata languages, a class that strictly contains the regular languages and reaches into the context-free. Here, we construct CA generating $a^n b a^n$, $a^n b^m$ for $n \geq m$, demonstrating how these patterns can be captured in one-dimensional growth. Pushing further, we exhibited even context-sensitive behavior; as language $a^n b^n c^n$ can be generated in this framework. More generally, we show that languages of the form

$$L = \{w_1^{e_1(n)} \dots w_m^{e_m(n)} \mid n \in \mathbb{N}\}$$

are CA-expressible (and, in particular, glider-expressible), for any $m \in \mathbb{N}$, any words $w_1 \dots w_m$, and any positive linear expressions $e_i(n)$. We further note that glider-expressible languages form a strict subclass of CA-expressible languages.

Contributions. Beyond these expressiveness results, the paper makes the following two contributions.

– *Generative framework.* We introduce a formulation of cellular automata as genuine *language generators.*
– *Glider mechanisms.* We define gliders as atomic carriers specified by a symbol, velocity, and interaction semantics derived from the CA rule.

Combining the two, we show how gliders provides modular explanations for CA-generated languages and enable constructive proofs of expressive power.

To the best of our knowledge, this is the first formal analysis of CA-expressible languages over unbounded, bi-infinite grids with arbitrary finite-size alphabets and radii. In contrast to earlier work on bounded grids or finite-time evolutions [28,50], which proved that language complexity is preserved, we show that even regular initializations (which sit at the bottom of the Chomsky hierarchy) can lead to languages of strictly higher complexity. This work aims to highlight an underexplored connection between local dynamics and global formal properties.

Paper Structure. The remainder of the paper is organized as follows. Section 2 provides preliminaries. Section 3 introduces the cellular automata (CA) framework and its generative point of view. Section 4 provides a walkthrough one counter automata languages, which leads us to Sect. 5 where we define gliders mechanism and establishes the strict separation between CA-expressible and glider-expressible. Section 6 develops our expressiveness results from illustrative cases to general families. Then we conclude our discussion at Sect. 7.

2   Preliminaries

We begin by fixing key notational conventions and reviewing background from formal language theory. We then formally define the cellular automaton (CA) model considered in this work. These conventions are provided to avoid ambiguity in the technical development that follows.

Conventions. Let $\mathbb{N}$ be the set of positive natural numbers, and let $\mathbb{N}_0 = \mathbb{N} \cup \{0\}$ be the set of non-negative integers. Let $\mathbb{Z}$ denote the set of all integers, with $\mathbb{Z}^+ = \mathbb{N}$ and $\mathbb{Z}^-$ representing the positive and negative integers, resp. For any $n, m \in \mathbb{Z}$ where $n \leq m$, we write $[n, m]$ to denote the interval $\{n, n+1, ..., m\}$, and we abbreviate $[1, m]$ as $[m]$. Given a domain $D = [x, y] \subseteq \mathbb{Z}$ we use $D+s$ for the *domain shift* of D by $s \in \mathbb{Z}$, i.e., for the interval $[x+s, y+s]$. When convenient, we use the notation $(-\infty, k]$ and $[k, \infty)$ for $k \in \mathbb{Z}$ in the classical manner, e.g., $\mathbb{N}$ can be seen as $[1, \infty)$.

Words. An *alphabet* Σ is a finite, non-empty set of elements called *letters*. A finite *word* w over Σ is an assignment of Σ to consecutive locations. It could be considered as a function $w : [n] \to \Sigma$ for some $n \in \mathbb{N}_0$. That is, $w = x_1 ... x_n$ where $x_i = w[i]$ is from Σ. Its *length* is denoted by $|w| = n$. The empty word is denoted by ε and $|\varepsilon| = 0$. The set of all finite words over Σ is Σ^*, and $\Sigma^+ = \Sigma^* \setminus \{\varepsilon\}$.

Infinite words are defined over infinite domains. Words over domain $\mathbb{Z}$ are referred to as *bi-ω words* (or *bi-infinite*). When the domain is $(-\infty, k]$ (resp. $[k, \infty)$) for some $k \in \mathbb{Z}$ the words are referred to as *left-ω* (resp. *right-ω*) words.

Infix. Let $\mathbb{D}, \mathbb{D}' \subseteq \mathbb{Z}$ be two integer intervals. Let $w : \mathbb{D} \to \Sigma$ and $w' : \mathbb{D}' \to \Sigma$ be two (finite or infinite) words. We say that w' is an *infix* of w if there exists $s \in \mathbb{Z}$ such that $\mathbb{D}' + s \subseteq \mathbb{D}$ and $w'[i+s] = w[i]$ for every $i \in \mathbb{D}'$. We write $w' \sqsubseteq w$ to denote that w' is an *infix* of w.

Concatenation. Given two finite words $x, y \in \Sigma^*$, their *concatenation* is denoted $x \cdot y$ (or simply xy). For $k \in \mathbb{N}$, we write x^k for the k-times right concatenation of x with itself. Similarly, x^ω, $^\omega x$, and $^\omega x^\omega$ for an infinite concatenation of x to itself to the right, left, or both sides, respectively.

The concatenation of a finite or left-ω (resp. right-ω) word to the left (resp. right) of a right-ω (resp. left-ω) word is defined naturally.

Languages. A *language* over Σ, is a set of words over Σ. We use $\Sigma^*, \Sigma^\omega, {}^\omega\Sigma$ and ${}^\omega\Sigma^\omega$ to denote the sets of all finite, right-ω, left-ω and bi-ω words, resp. Subsets of Σ^*, are *finitary language.* The subsets of Σ^ω and ${}^\omega\Sigma$ are referred to as *infinitary languages*, consisting of right- and left-ω words, resp. A subset of ${}^\omega\Sigma^\omega$ is a *bi-infinitary language.* For any $W \subseteq \Sigma^*$ and $k \in \mathbb{N}$, we write $W^k = \{w_1 \cdot w_2 \cdots w_k \mid \forall i \in [k]. \ w_i \in W\}$ for the set of all k concatenations of words from W. The sets W^ω, $^\omega W$, and $^\omega W^\omega$ are defined analogously for infinite repetition.

2.1 Cellular Automata (CA)

In the following we formally introduce and define our models of interest, *cellular automata.* These are dynamical systems, exhibiting a variety of organized and complex behaviors over time [12, 25, 35–38]. We restrict our attention to one-dimensional cellular automata (1-CA) defined over bi-infinite grids, with arbitrary finite alphabets and neighborhood radii. We henceforth refer to them simply as CA, omitting the dimensional qualifier.

Definition 1 (Cellular Automata (CA)). *A CA $\mathcal{A} = (\Sigma, r, f)$ consists of an alphabet Σ of states, a radius $r \in \mathbb{N}_0$, and a local rule $f : \Sigma^{2r+1} \to \Sigma$.*

Let $\mathcal{A} = (\Sigma, r, f)$ be a CA, we use $\mathcal{A}$ to define the following definitions.

Configurations. A *configuration* c of $\mathcal{A}$, is an assignment of Σ elements to each cell indexed by $\mathbb{Z}$. A configuration is viewed as a bi-ω word, thus $c \in {}^\omega\Sigma^\omega$. Throughout this paper, we denote this *set of bi-infinite words by* $\Sigma^{\mathbb{Z}}$, which is the standard notation in the CA literature.

Given a configuration $c \in \Sigma^{\mathbb{Z}}$ and $i, j \in \mathbb{Z}$ with $i < j$, we write $c[i]$ (or $\langle i \rangle_c$) to denote the cell state at position i. Similarly, $c[i, j]$ or $\langle i, j \rangle_c$ denotes the infix of the configuration from index i to j, inclusive. Hence, $\langle i, j \rangle_c \in \Sigma^{j-i+1}$.

Neighborhood. Each cell $i \in \mathbb{Z}$ of $\mathcal{A}$, updates its state according to the local rule function (also referred to as the *rule function*), which acts according to the current states of a fixed set of adjacent cells centered in cell i; this is the *neighborhood* of i. For a *radius* r, the state update depends on the $2r+1$ consecutive cells centered at the current position. Formally, given a configuration c the *neighborhood* of cell i is defined as $\langle i - r, i + r \rangle_c$.

Quiescent States and Configuration Support. Given $\mathcal{A}$, a state $s \in \Sigma$ is termed a *quiescent state* if $f(s^{2r+1}) = s$, that is, the rule function f maps a neighborhood of s's to state s. We follow the common assumption that there is a unique quiescent state, and denote it by $\bot$. Given a configuration c, the *support* of c is the set of cells not in state $\bot$, denoted $\mathrm{supp}_\bot(c) = \{i \in \mathbb{Z} \mid c[i] \neq \bot\}$.

Finite configurations are those with only finitely many non-quiescent states. Formally, a configuration c is *finite* if and only if its support; $\mathrm{supp}_\bot(c)$, is a finite set. We denote the set of all finite configurations by $C_{\mathcal{F}}$. Note that $C_{\mathcal{F}} \subseteq \Sigma^{\mathbb{Z}}$. Let $c \in C_{\mathcal{F}}$, we define the *active interval* of c as the smallest integer interval containing its support, formally defined as:

$$\mathrm{int}(c) := [\min(\mathrm{supp}_\bot(c)), \max(\mathrm{supp}_\bot(c))] \tag{1}$$

Let $c, c' \in C_{\mathcal{F}}$ be two finite configurations, and let $\mathrm{int}(c), \mathrm{int}(c')$ be their intervals. If there exists $s \in \mathbb{Z}$ such that $\mathrm{int}(c)+s = \mathrm{int}(c')$, and $c[\mathrm{int}(c) + s] = c'[\mathrm{int}(c')]$, then the two configurations are considered *equivalent* under a domain shift. Consequently, *width* of c, denoted $\mathrm{width}(c)$, is defined as:

$$\mathrm{width}(c) := \max(\mathrm{supp}_\bot(c)) - \min(\mathrm{supp}_\bot(c)) + 1 \tag{2}$$

Orbits. The local rule function f of $\mathcal{A}$ is extended to the *global rule function*, denoted $G : \Sigma^{\mathbb{Z}} \to \Sigma^{\mathbb{Z}}$, where f is applied to all cells simultaneously. Starting from an initial configuration $c_0 \in \Sigma^{\mathbb{Z}}$, applying the global function once, we get the next configuration, that is, $c_1 = G(c_0)$ and generally, $c_n = G(c_{n-1}) = G^n(c_0)$ for every $n \in \mathbb{N}$. The *orbit* of a configuration c, denoted $\mathcal{O}(c) = \{G^n(c) \mid n \in \mathbb{N}_0\}$, is the infinite configuration sequence $c_0, c_1, c_2, \ldots$ where $c_0 = c$ and $c_n = G^n(c)$ for every $n \in \mathbb{N}$. Given a set of configurations $C \subseteq \Sigma^{\mathbb{Z}}$, its orbit is denoted $\mathcal{O}(C) = \bigcup_{c \in C} \mathcal{O}(c)$. Note that an orbit of any $c \in \Sigma^{\mathbb{Z}}$ is a bi-infinite language.

3   CA Languages – Generative Perspective

We now formalize what it means for a CA to *generate a language*, viewing its reachable configurations as a formal object of study. This framework allows us to examine how parameters such as alphabet size and neighborhood radius influence expressiveness and to relate the generated languages to established classes in formal language theory.

Padding and Finite Configurations. First we define a $\perp$-padding operator on words, then we lift it to languages. The function, $\mathsf{pad}_\perp : \Sigma^* \to C_\mathcal{F}$, takes a finite word $w \in \Sigma^*$ and returns $^\omega\!\perp w\perp^\omega$. The lifting of the function to languages is as expected $\mathsf{pad}_\perp : \mathcal{P}(\Sigma^*) \to \mathcal{P}(C_\mathcal{F})$ where $\mathsf{pad}_\perp(L) := \{\mathsf{pad}_\perp(w) \mid w \in L\}$.

Given a finitary language F, we define the set of finite configurations induce by F to be $\mathsf{pad}_\perp(F)$, as explained above. In the sequel, we consider CA languages derived from initial configurations $I = \mathsf{pad}_\perp(F)$ for a *regular language* F. Regular languages are in particular finitary languages.

CA Language. Given $\mathcal{A} = (\Sigma, r, f)$ and $I \subseteq C_\mathcal{F}$. The language of $\mathcal{A}$ is defined wrt the set of initial configurations I and is denoted $\mathcal{L}(\mathcal{A}, I) = \{w \mid {}^\omega\!\perp w\perp^\omega \in \mathcal{O}(I)\}$. Given $F \subseteq \Sigma^*$, we abbreviate $\mathcal{L}(\mathcal{A}, \mathsf{pad}_\perp(F))$ as $\mathcal{L}(\mathcal{A}, F)$.

Remark 2. Although our definitions allow general initial sets $I \subseteq C_\mathcal{F}$, we focus on the structured case $I = \mathsf{pad}_\perp(F)$, for a regular language $F \subseteq \Sigma^*$. Regular initial sets are standard in formal verification, as many natural systems admit such representations [9]. Despite this syntactic constraint, we show that such cases can still yield CA-expressible languages of considerable complexity.

$\mathbb{CA}$ *and CA-Expressible.* We use $\mathbb{CA}_{r,\Sigma}$ to denote the class of 1-CA with radius r over Σ, respectively. $\mathbb{CA}$ denotes the class of all 1-CA (for any r and Σ). A language $L \subseteq \Sigma^*$ is $\mathbb{CA}_{r,\Sigma}$-expressible (thus, CA-expressible in general), if there exists $\mathcal{A} \in \mathbb{CA}_{r,\Sigma}$ and $I \subseteq C_\mathcal{F}$ such that $L = \mathcal{L}(\mathcal{A}, I)$. In which case we write $L \in \mathbb{CA}_{r,\Sigma}$. When $I = \mathsf{pad}_\perp(F)$ and F is a regular language, we often write $\mathcal{L}(\mathcal{A}, I) \in \mathbb{CA}_{r,\Sigma}^{\mathsf{REG}}$ to emphasize the regular origin.

3.1   Feasible Neighborhoods

Let $\mathcal{A} \in \mathbb{CA}_{r,\Sigma}$ be a fixed CA. By the above definitions, not every neighborhood $w \in \Sigma^{2r+1}$, where $w = x_{-r}\ldots x_0 x_1 \ldots x_r$, needs to appear along orbits starting from an item of I. We say that a neighborhood $w \in \Sigma^{2r+1}$ is *feasible* for $\mathcal{A}$ under I if there exists $c \in \mathcal{O}(I)$ satisfying $w \sqsubset c$; otherwise w is *infeasible*. Accordingly, we define the set of *feasible neighborhood* of $\mathcal{A}|_I$ (read $\mathcal{A}$ under I).

Definition 3 (Feasible Neighborhoods). *For $\mathcal{A} \in \mathbb{CA}_{r,\Sigma}$ and $I \subseteq C_\mathcal{F}$, the set of feasible neighborhoods is $FN_{\mathcal{A},I} = \{w \in \Sigma^{2r+1} \mid \exists c \in \mathcal{O}(I) \text{ with } w \sqsubset c\}$.*

Example 4. Let $\mathcal{A} \in \mathbb{CA}_{1,\{a,\perp\}}$, where f is defined by:

$$f(\perp\perp\perp) = \perp \quad f(\perp\perp a) = a \quad f(\perp a \perp) = \perp \quad f(\perp aa) = a$$
$$f(a\perp\perp) = \perp \quad f(a\perp a) = a \quad f(aa\perp) = a \quad f(aaa) = a$$

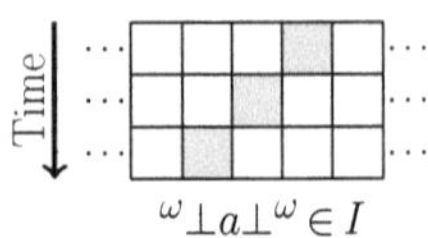

Let $I = \mathsf{pad}_\perp(\{a\})$. Thus, only $\perp\perp\perp$, $\perp\perp a$, $\perp a \perp$, and $a\perp\perp$ are feasible neighborhoods (*FNs*) for $\mathcal{A}|_I$. Consider the illustration to the right, of applying the rule function on I, where $\square = \perp$ and $\blacksquare = a$. Therefore, it suffices to define f as a partial function, on the *FN*, namely:

$$f(\perp\perp\perp) = \perp \quad f(\perp\perp a) = a \quad f(\perp a \perp) = \perp \quad f(a\perp\perp) = \perp$$

Infeasible neighborhoods (those outside $FN_{\mathcal{A},I}$) can be ignored in the analysis. Therefore, henceforth we define f as a partial function with domain $FN_{\mathcal{A},I}$.

Given a $\mathcal{A} \in \mathbb{CA}_{r,\Sigma}$, the local rule f assigns to each neighborhood $w \in \Sigma^{2r+1}$ a value from Σ. It is often more convenient to ask for each $\sigma \in \Sigma$ which (feasible) neighborhoods are assigned to it. Consider Ex.4, its local rule can be succinctly given by a table consisting of only feasible neighborhoods. Each row corresponds to a letter $\sigma \in \Sigma$ and lists all the *FNs* for which $f(w) = \sigma$.

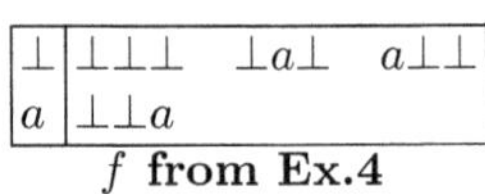

$\perp$	$\perp\perp\perp$	$\perp a \perp$	$a\perp\perp$
a	$\perp\perp a$		

f from Ex.4

3.2 Observations and Implications

We begin with elementary observations on the CA and its generated language, which underpin the developments that follow. The full proof appears in the extended version [18, App. A].

Given a language $L \in \Sigma^*$, let $w_0, w_1, w_2, \ldots$ be a list of all words in L, arranged in a non-decreasing order by their length, (i.e., $|w_i| \le |w_{i+1}|$ for any $i \in \mathbb{N}_0$). Let $a_1, a_2, \ldots$ be the sequence of size differences of words in this sequence, i.e., $a_i = |w_i| - |w_{i-1}|$ which we term the *differences sequence*. We say that the difference sequence of L is *bounded* if there is a $k \in \mathbb{N}$ such that $a_i \le k$ for all $i \in \mathbb{N}$. If the difference sequence is bounded, we define the minimal such k as the *difference-bound* of L. We proceed with some observations regarding $\mathrm{int}(c)$ and $\mathrm{width}(c)$ defined in Eqs. (1) and (2).

Observation 5. Let $\mathcal{A} \in \mathbb{CA}_{r,\Sigma}$ and let $c \in C_\mathcal{F}$ be a finite configuration. Assume $\mathrm{int}(c) = [m,n]$ and $\mathrm{int}(G(c)) = [i,j]$. Then $i \ge m-r$ and $j \le n+r$.

Corollary 6. Let $\mathcal{A} \in \mathbb{CA}_{r,\Sigma}$. Then there exists a natural number $k \le 2 \cdot r$ such that each finite configuration $c \in C_\mathcal{F}$ of $\mathcal{A}$ satisfies $\mathrm{width}(G(c)) - \mathrm{width}(c) \le k$.

Corollary 6 can be used to show that a language L is not $\mathbb{CA}_{r,\Sigma}$-expressible for certain r's. That is, let $L = \mathcal{L}(\mathcal{A}, \{c\})$ for some $\mathcal{A} \in \mathbb{CA}_{r,\Sigma}$ and $c \in C_\mathcal{F}$. If L is k-difference bounded, then it is necessarily not CA-expressible with $r \le \lfloor \frac{k-1}{2} \rfloor$.

4   OCA Separation Witnesses via CA

In this section, we examine the expressive capabilities of regular-initialized CA. First note that every regular language, $L \in \Sigma^*$, is trivially $\mathbb{CA}_{0,\Sigma}$-expressible, by setting the initial configuration to $I = \mathsf{pad}_\perp(L)$ and using the identity rule function: $\forall \sigma \in \Sigma. \ f(\sigma) = \sigma$. That is, $\mathbb{CA}_{r,\Sigma}^{\mathtt{REG}} \supseteq \mathbb{REG}$ where $\mathbb{REG}$ stands for the class of regular languages. Next, we use one-counter automata (OCA) and their known subclasses as a framework for comparison.

One-counter automata (OCA) are a fundamental model of infinite-state systems. They are a special case of pushdown automata (PDA) where the stack is restricted to a single symbol. As a result, the stack behaves as a single unbounded natural counter, which can be incremented, decremented (only if positive), and tested for zero. While seemingly simple, OCA recognizes a rich class of languages that extends the regular languages and captures basic counting. This makes OCA a natural tool for probing the expressive scope of our model. By positioning regular-initialized CA generated languages against increasingly expressive subclasses beyond the regular class, we demonstrate a non-trivial increase in language expressiveness.

OCA Subclasses. Although one-counter automata (OCA) are inherently non-deterministic, their *deterministic* variant DOCA-a strict subclass of OCA-has been widely studied.[1] Within DOCA, further syntactic restrictions yield *real-time* variant ROCA. Finally, restricting ROCA yields a *visible* model VOCA. Despite being the most restricted type, even VOCA strictly subsumes the regular languages. Full formal definitions of OCA and their relevant subclasses are provided in the extended version [18, App. B].

OCA Languages. Given an OCA $\mathcal{A}$, the language it recognizes is denoted by $\mathcal{L}(\mathcal{A})$. The class of languages recognized by OCA is denoted by $\mathbb{OCL}$. Similarly, $\mathbb{DOCL}$, $\mathbb{ROCL}$, and $\mathbb{VOCL}$ are defined based on DOCA, ROCA and VOCA.

These subclasses form a hierarchy of increasing restrictions and decreasing expressive power, see [5,26,27,53] for formal definitions and proofs.

Theorem 7 (OCL hierarchy [53]). $\mathbb{REG} \subsetneq \mathbb{VOCL} \subsetneq \mathbb{ROCL} \subsetneq \mathbb{DOCL} \subsetneq \mathbb{OCL}$

Theorem 7 establishes a strict hierarchy. For each consecutive pair of classes, one can show a CA-expressible language that separates them, as we next show.

Proposition 8. *For every pair of consecutive classes $\mathbb{C}$ and $\mathbb{C}'$ in Theorem 7 there exists $L \in \mathbb{C}' \setminus \mathbb{C}$ that is expressible by a regular initialized CA, i.e., $L \in \mathbb{CA}_{r,\Sigma}^{REG}$.*

Claim 9. *There exists $L \in \mathbb{VOCL} \setminus \mathbb{REG}$ such that $\mathbb{CA}_{r,\Sigma}^{REG}$.*

---

[1] Their study is also motivated by the fact that many decision problems are undecidable for general OCAs and PDAs [46,51,55].

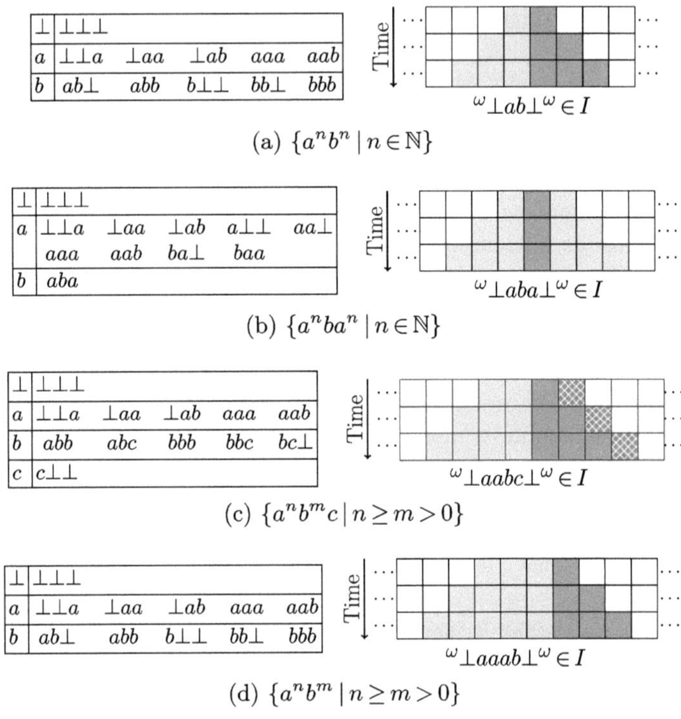

(a) $\{a^n b^n \mid n \in \mathbb{N}\}$

(b) $\{a^n b a^n \mid n \in \mathbb{N}\}$

(c) $\{a^n b^m c \mid n \geq m > 0\}$

(d) $\{a^n b^m \mid n \geq m > 0\}$

Fig. 1. Legend: $\bot = \square$, $a = \square$, $b = \blacksquare$, $c = \boxtimes$

Proof. The language $L = \{a^n b^n \mid n \in \mathbb{N}\} \in \mathrm{VOCL} \setminus \mathrm{REG}$. Let $\mathcal{A} \in \mathbb{CA}_{1,\{a,b,\bot\}}$ and $I = \mathsf{pad}_\bot(\{ab\})$. Figure 1a shows f such that $\mathcal{L}(\mathcal{A}, I) = L$. Thus, $L \in \mathbb{CA}^{\mathrm{REG}}_{1,\{a,b,\bot\}}$.

Claim 10. *There exists $L \in \mathrm{ROCL} \setminus \mathrm{VOCL}$ such that $\mathbb{CA}^{REG}_{r,\Sigma}$.*

Proof. The language $L = \{a^n b a^n \mid n \in \mathbb{N}\} \in \mathrm{ROCL} \setminus \mathrm{VOCL}$. Let $\mathcal{A} \in \mathbb{CA}_{1,\{a,b,\bot\}}$ and $I = \mathsf{pad}_\bot(\{aba\})$. Figure 1b shows f s.t. $\mathcal{L}(\mathcal{A}, I) = L$. Thus, $L \in \mathbb{CA}^{\mathrm{REG}}_{1,\{a,b,\bot\}}$.

Claim 11. *There exists $L \in \mathrm{DOCL} \setminus \mathrm{ROCL}$ such that $\mathbb{CA}^{REG}_{r,\Sigma}$.*

Proof. $L = \{a^n b^m c \mid n \geq m > 0\} \in \mathrm{DOCL} \setminus \mathrm{ROCL}$. Let $\mathcal{A} \in \mathbb{CA}_{1,\{a,b,c,\bot\}}$ and $I = \mathsf{pad}_\bot(\{a^* abc\})$. Figure 1c shows f such that $\mathcal{L}(\mathcal{A}, I) = L$. Thus, $L \in \mathbb{CA}^{\mathrm{REG}}_{1,\{a,b,c,\bot\}}$

Claim 12. *There exists $L \in \mathrm{OCL} \setminus \mathrm{DOCL}$ such that $\mathbb{CA}^{REG}_{r,\Sigma}$.*

Proof. The language $L = \{a^n b^m \mid n \geq m > 0\} \in \mathrm{OCL} \setminus \mathrm{DOCL}$. Let $\mathcal{A} \in \mathbb{CA}_{1,\{a,b,\bot\}}$ and $I = \mathsf{pad}_\bot(\{a^* ab\})$. Figure 1d shows f s.t. $\mathcal{L}(\mathcal{A}, I) = L$. Thus, $L \in \mathbb{CA}^{\mathrm{REG}}_{1,\{a,b,\bot\}}$.

To show that cellular automata can express languages beyond the OCA hierarchy, we introduced the notion of gliders, defined next.

5   Gliders Perspective

In this section, we define *gliders* within regular-initialized CA and formalize their interactions and semantic role. Gliders offer a symbolic structure that reflects the emergent behavior of CA dynamics, especially when the rule function is viewed as being shaped by dominant or persistent symbolic patterns. Recall from the introduction that our definition of *gliders* differ from the conventional one. Specifically, we regard a glider as a one-cell (*atomic*) entity moving at a fixed velocity, as we formally define below.

5.1   Formal Definition of Gliders

Given a neighborhood $w \in \Sigma^{2r+1}$, we index it as $w = x_{-r} \ldots x_{-1} x_0 x_1 \ldots x_r$.

Notation. For any $\sigma \in \Sigma$ and $i \in [-r, r]$, we define:

$$\mathcal{N}_i^\sigma := \{w \in \Sigma^{2r+1} \mid x_{-i} = \sigma\}$$

That is, $\mathcal{N}_i^\sigma$ is the set of neighborhoods where the symbol σ appears in the $-i$ position.[2] The parameters Σ and r are those of the CA $\mathcal{A} \in \mathbb{CA}_{r,\Sigma}$ currently in context, and are omitted from the notation for brevity.

This shorthand will be used extensively in what follows.

A glider $\mathbf{g}(\sigma, i)$ is a pair (σ, i), where σ denotes its *value* and $i \in \mathbb{Z}$ denotes its *velocity*. The velocity i determines both its *speed*, given by $|i|$, and its *direction*, given by $\mathbf{sign}(i)$. Positive values indicate motion to the right, negative values indicate motion to the left, and $i = 0$ corresponds to no horizontal movement, i.e., the object is static. The formal definition follows.

Definition 13 (Glider). *Let $\mathcal{A} \in \mathbb{CA}_{r,\Sigma}$. For $i \in [-r, r]$ and $\sigma \in \Sigma$, we say that $\mathbf{g}(\sigma, i)$ is a glider in $\mathcal{A}$ if there exists $w \in \mathcal{N}_i^\sigma$ such that $f(w) = \sigma$.*

Thus, given $\mathcal{A} \in \mathbb{CA}_{r,\Sigma}$, the gliders of $\mathcal{A}$ are subsumed in the following set: $\{\mathbf{g}(\sigma, i) \mid i \in [-r, r], \ \sigma \in \Sigma\}$. Consequently, the total number of distinct gliders admitted by $\mathcal{A} \in \mathbb{CA}_{r,\Sigma}$ is at most $|\Sigma| \cdot (2r+1)$.

Next, we define various types of gliders, the interactions that may arise between them, and the resulting effects on the system, together with behaviors that may or may not be exhibited.

Definition 14 (Persistent Glider). *Let $\mathbf{g}(\sigma, i)$ be a glider in $\mathcal{A} \in \mathbb{CA}_{r,\Sigma}$. We say that $\mathbf{g}(\sigma, i)$ is persistent if $f(w) = \sigma$ for <u>every</u> $w \in \mathcal{N}_i^\sigma$. We say that $\mathbf{g}(\sigma, i)$ is non-persistent if $f(w) = \sigma$ for <u>some but not all</u> of $w \in \mathcal{N}_i^\sigma$.*

Note that, if $\mathbf{g}(\sigma, i)$ is *persistent* in $\mathcal{A}$ then any configuration $c \in \Sigma^{\mathbb{Z}}$ satisfies that $\langle z \rangle_c = \sigma$ implies $\langle z+i \rangle_{G(c)} = \sigma$ for any $z \in \mathbb{Z}$.

---

[2] This corresponds to the classical *cylinder set* notation, typically denoted $[\sigma]_{-i}$.

Spreading State. Let $m, n \in \mathbb{Z}$ for $m < n$, $\sigma \in \Sigma$, and $[m, n] \subseteq [-r, r]$. Assume that $\mathcal{A} \in \mathbb{CA}_{r,\Sigma}$ implements a set of *persistent gliders*, $\{\mathsf{g}(\sigma, i) \mid i \in [m, n]\}$, i.e., all *persistent gliders* of value σ and velocity i for any $i \in [m, n]$. Then we say that $\mathcal{A}$ has a σ-*spreading state*. We consider the following types of σ-spreading state:

i. $m = 0$: Then $\mathcal{A}$ has a *right*-spreading state (at speed n) (Fig. 2a).
ii. $n = 0$: Then $\mathcal{A}$ has a *left*-spreading state (at speed $|m|$) (Fig. 2c).
iii. $0 \in (m, n)$: Then $\mathcal{A}$ has a *two-way* spreading-state (Fig. 2b).
iv. $0 \notin [m, n]$: Then $\mathcal{A}$ has a *shift*-spreading state (Fig. 2d).

Intuitively, a σ-spreading state enables a prediction of the orbit behavior of a $c \in \Sigma^{\mathbb{Z}}$, based solely on the fact that $\langle i \rangle_c = \sigma$. E.g., having a two-way σ-spreading state (resp. right/left spreading), means that eventually every cell of c in positions $\mathbb{Z}$ (resp. every cell in position $z \geq i$ or $z \leq i$) would be equal to σ.

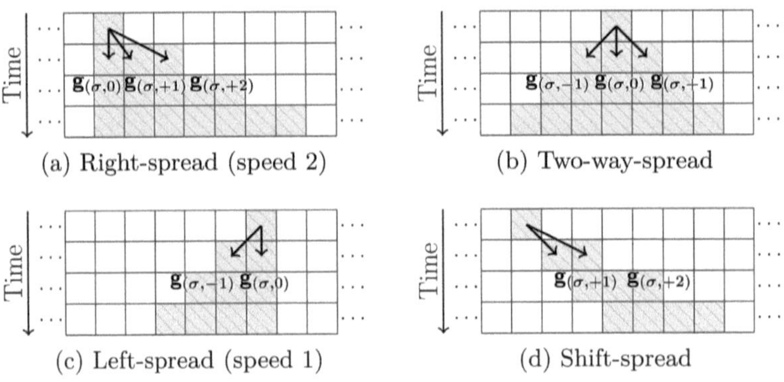

(a) Right-spread (speed 2)   (b) Two-way-spread

(c) Left-spread (speed 1)   (d) Shift-spread

Fig. 2. Examples of spreading state behavior

Coexistence of Persistent Gliders. Not all persistent gliders can coexist. Since each glider is defined by a value and velocity, four pairwise cases arise. Let $\mathcal{A} \in \mathbb{CA}_{r,\Sigma}$ and let $\mathsf{g}(\sigma, i), \mathsf{g}(\tau, j)$ be two persistent gliders in $\mathcal{A}$.

Same Velocity. If $i = j$, the gliders never collide. If, in addition, $\sigma = \tau$, then they are identical. If $\sigma \neq \tau$, then $\mathcal{N}_i^{\sigma} \cap \mathcal{N}_i^{\tau} = \emptyset$, as every cell admits a single value, so no conflict arises. (See Figs. 3a, 3b)

Different Velocities, Same Value. If $i \neq j$ and $\sigma = \tau$, then the gliders may *collide*, but agree on the resulting value. That is, for any $w \in \mathcal{N}_i^{\sigma} \cup \mathcal{N}_j^{\sigma}$, the rule function satisfies both by assigning $f(w) = \sigma$. Hence, no conflict arises. (Fig. 3c)

Different Velocities, Different Values. If $i \neq j$ and $\sigma \neq \tau$, then $\mathcal{N}_i^{\sigma} \cap \mathcal{N}_j^{\tau} \neq \emptyset$. Any $w \in \mathcal{N}_i^{\sigma} \cap \mathcal{N}_j^{\tau}$ must satisfy both $f(w) = \sigma$ and $f(w) = \tau$. However, $\sigma \neq \tau$. Since a single cell cannot hold multiple values at once, this leads to contradiction. Thus such persistent gliders cannot coexists. (Fig. 3d)

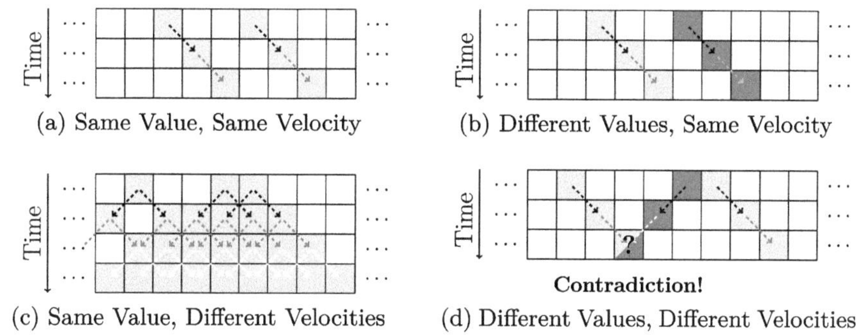

(a) Same Value, Same Velocity

(b) Different Values, Same Velocity

(c) Same Value, Different Velocities

Contradiction!

(d) Different Values, Different Velocities

Fig. 3. Coexistence of Persistent Gliders

Corollary 15 (Coexistence of Persistent Gliders). *If* $\mathrm{g}(\sigma, i)$ *and* $\mathrm{g}(\tau, j)$ *are both persistent gliders in* $\mathcal{A}$*, then either* $i = j$ *or* $\sigma = \tau$*.*

Thus, the set G of persistent gliders must consist of gliders that either share the same velocities or carry the same value.

Observation 16. Two limiting cases illustrate the system-wide effect of persistent gliders. Let $\mathcal{A} \in \mathbb{CA}_{r,\Sigma}$, $i \in [-r, r]$, and consider the set $G_i = \{\mathrm{g}(\sigma, i) \mid \sigma \in \Sigma\}$ of persistent gliders.

i. If $\mathcal{A}$ implements all gliders in G_i, then any $c \in \Sigma^{\mathbb{Z}}$ shifts uniformly by i. That is, for any $z \in \mathbb{Z}$, it holds that $G(c)[z] = c[z-i]$. Hence, $\mathcal{O}(I) = I$ and $\mathcal{L}(\mathcal{A}, F) = F$ for any $I \subseteq \Sigma^{\mathbb{Z}}$ and $F \subseteq \Sigma^*$.
ii. If $\mathcal{A}$ implements all gliders in $G_i \setminus \{\mathrm{g}(\bot, i)\}$, then for any $w \in (\Sigma \setminus \{\bot\})^*$ and every infix $w' \sqsubseteq w$, it holds that $w' \sqsubseteq c'$ of every $c' \in \mathcal{O}(\mathsf{pad}_\bot(\{w\}))$.

These cases show how implemented persistent gliders constrain global behavior-ranging from uniform shifts to infix preservation.

Recall that a CA language is defined with respect to a set $I \subseteq C_{\mathcal{F}}$ of initial configurations. Furthermore, in this setting, the local rule function, f, may be partial, defined only for the feasible neighborhoods, namely, $FN_{\mathcal{A},I}$. Accordingly, we define the notion of an *I-persistent glider* as follows.

Definition 17 (*I*-Persistent Gliders). *Let* $\mathcal{A} \in \mathbb{CA}_{r,\Sigma}$ *and* $I \subseteq C_{\mathcal{F}}$*. A glider* $\mathrm{g}(\sigma, i)$ *is* I*-persistent in* $\mathcal{A}$ *if for all* $w \in \mathcal{N}_i^{\sigma} \cap FN_{\mathcal{A},I}$*, we have* $f(w) = \sigma$*.*

Note that I-persistence generalizes persistence, as every persistent-glider is I-persistent for any I, but not vice versa. The following example illustrates the concept of being persistent with respect to a particular initial configuration I.

Example 18 (I-Persistent Gliders). Let $\mathcal{A} \in \mathbb{CA}_{1,\{a,b,\bot\}}$ and f defined as follows.

$\bot$	$\bot\bot\bot$	$a\bot b$	$aa\bot$	aba	$b\bot b$	
a	$\bot\bot a$	$\bot a\bot$	$\bot aa$	$\bot ab$	$\bot b\bot$	
	$\bot ba$	aaa	aab	$ba\bot$	bba	
b	$\bot\bot b$	$\bot bb$	$a\bot\bot$	$a\bot a$	$ab\bot$	abb
	$b\bot\bot$	$b\bot a$	baa	bab	$bb\bot$	bbb

Let $I = \mathsf{pad}_\bot(\{ab\})$, so f is defined as follows (corresponds to Claim 9).

$\bot$	$\bot\bot\bot$				
a	$\bot\bot a$	$\bot aa$	$\bot ab$	aaa	aab
b	$ab\bot$	abb	$b\bot\bot$	$bb\bot$	bbb

Note that while $\mathsf{g}(a,-1)$, $\mathsf{g}(a,0)$, $\mathsf{g}(b,0)$ and $\mathsf{g}(b,+1)$ are not persistent in $\mathcal{A}$ owing to $f(a\bot a) = b$, $f(aa\bot) = \bot$, $f(\bot ba) = a$, $f(b\bot b) = \bot$ resp. They are I-persistent.

Remark 19. Note that Corollary 15 considers all possible neighborhoods. Under a particular initial configuration I, there can be two I-persistent gliders $\mathsf{g}(\sigma,i), \mathsf{g}(\tau,j)$, such that both $i \neq j$ and $\sigma \neq \tau$. If this is the case, this is because the contradicting environment is not feasible under I.

5.2 Gliders Derived by a CA Under I

We can now reason not only about individual gliders, but also about their interactions under a fixed initial configuration set. Given $\mathcal{A} \in \mathbb{CA}_{r,\Sigma}$ and $I \subseteq C_\mathcal{F}$, we denote the set of gliders derived from $\mathcal{A}|_I$ by $\boldsymbol{G}_{\mathcal{A}|_I}$. By Definition 13, this is the set of gliders $\mathsf{g}(\sigma,i)$ for which there exists a $w \in FN_{\mathcal{A},I}$ that is in $\mathcal{N}_i^\sigma$, namely for which $x_{-i} = \sigma$. Some of them might be I-persistent, while others might not.

The Derived Partial Order. Conflicting gliders may overlap in their neighborhood constraints w.r.t. $\mathcal{A}$ and $I \subseteq C_\mathcal{F}$. To compare such cases, we define a *dominance relation*. Let $\mathsf{g}(\sigma,i), \mathsf{g}(\tau,j) \in \boldsymbol{G}_{\mathcal{A}|_I}$ with $\sigma \neq \tau$ and $i \neq j$. We write

$$\mathsf{g}(\sigma,i) >_{\mathcal{A}|_I} \mathsf{g}(\tau,j) \quad \text{iff} \quad f(w) = \sigma \text{ for all } w \in \mathcal{N}_i^\sigma \cap \mathcal{N}_j^\tau \cap FN_{\mathcal{A},I}$$

When the context is clear, we simply write $\mathsf{g}(\sigma,i) > \mathsf{g}(\tau,j)$. Intuitively, $\mathsf{g}(\sigma,i)$ dominates $\mathsf{g}(\tau,j)$ when the local rule f assigns σ to every FN that overlaps. Thus, a dominating glider consistently overrides a weaker one in every overlapping context. For example, if every $w \in \mathcal{N}_i^x \cap \mathcal{N}_j^y$ satisfies $f(w) = x$, then $\mathsf{g}(x,i) > \mathsf{g}(y,j)$.

Maximality. We say that a glider g is *maximal* in a set of gliders $\boldsymbol{G}$ if no $\mathsf{g}' \in \boldsymbol{G}$ satisfies $\mathsf{g}' > \mathsf{g}$. In this sense, *maximality* aligns with I-*persistence*: a glider that is not dominated by any other conflicting one behaves stably under the neighborhoods that are derived by $\boldsymbol{G}$, while non-maximal gliders may be suppressed in evolution or reconstruction.

Example 20 (Maximality under a strict partial order). Let $\boldsymbol{G} = \{\mathsf{g}_1, \mathsf{g}_2, \mathsf{g}_3, \mathsf{g}_4\}$ be a glider set with the following ordering relations: $\mathsf{g}_1 > \mathsf{g}_2$, $\mathsf{g}_2 > \mathsf{g}_3$, $\mathsf{g}_1 > \mathsf{g}_3$ and $\mathsf{g}_4 > \mathsf{g}_3$. Note that the strict partial order, $>$, allows glider sets to have more than one maximal element. E.g., the maximal gliders of $\boldsymbol{G}$ are g_1 and g_4.

Sound Derivation. We have defined the set of gliders G and the strict partial order $>$ derived from a CA $\mathcal{A}$ and an initial configuration I. If the derivation is sound, as formally defined next, it would be possible to reconstruct the CA from $(G, >)$. Formally, given $\mathcal{A} \in \mathbb{CA}_{r,\Sigma}$ and $I \subseteq C_{\mathcal{F}}$. Let G and $>$ be the set of gliders and their order, derived from $\mathcal{A}|_I$ as defined above. We say that the *derivation is sound* if the following conditions are satisfied.

   i. The graph induced by $>$ on G is acyclic,
  ii. Every $w \in FN_{\mathcal{A},I}$ has some $g(\sigma, i) \in G$ with $w \in \mathcal{N}_i^\sigma$ s.t. $f(w) = \sigma$, and
 iii. For every $w \in FN_{\mathcal{A},I}$ if $G_w \subseteq G$ is the set of gliders $\{g(\sigma, i) \mid w \in \mathcal{N}_i^\sigma\}$, and $g(\tau, j)$ is maximal in G_w then $f(w) = \tau$.

The first condition establishes that the dominance relation is well defined. The second condition makes sure that every feasible neighborhood is defined according to a glider, and that this glider is in G. Lastly, the third condition states that if the evolution of a certain feasible neighborhood w intersects with more than one glider of G and can presumably be defined by any such glider, then the one that actually determines the evolution is the most dominant among these.

We now consider the inverse direction: Given a finite set of gliders G equipped with a partial order $>$, can we reconstruct a CA whose local rule is derived from $(G, >)$? That is, does there exist a CA that this glider system *induces*?

5.3   Glider-Based Rule Reconstruction

Given a set of gliders G ordered by $>$, we construct a (possibly partial) rule function consistent with their dominance structure.

Example 21. Let $G = \{g(a, 0), g(b, 1)\}$ over $\Sigma = \{a, b\}$ with $r = 1$, and assume $g(b, 1) > g(a, 0)$. Then, for any $w \in \Sigma^3$, the rule induced by G satisfies: if $x_{-1} = b$, then $f(w) = b$; otherwise, if $x_0 = a$, then $f(w) = a$.

We briefly describe the procedure, GTOR, which derives a partial rule function f from an ordered glider system $(G, >)$. Each glider $g(\sigma, i) \in G$ constrains f to map certain neighborhoods in $\mathcal{N}_i^\sigma$ to σ, unless a more dominant glider overrides it. The procedure utilizes simple subroutines that process gliders in dominance order and incrementally assign values to f.

Maximal Glider (MAX). The procedure $\text{MAX}(G, >)$ returns a pair (σ, i) such that $g(\sigma, i) \in G$ is a *maximal* glider in G with respect to $>$.

Rule Extension (EXTEND). Given a value σ, a position i, and a partial rule function $f' : \Sigma^{2r+1} \rightharpoonup \Sigma$, the EXTEND procedure extends f' by assigning σ to every neighborhood $w \in \mathcal{N}_i^\sigma$ for which $f'(w)$ is undefined. Formally, we define $f' := \text{EXTEND}(f', \sigma, i)$, where:

$$f'(w) := \begin{cases} f'(w) & \text{if } w \in \mathbf{dom}(f') \\ \sigma & \text{if } w \in \mathcal{N}_i^\sigma \setminus \mathbf{dom}(f') \end{cases}$$

Glider-to-Rule (GtoR) Reconstruction Algorithm

We define $\text{GtoR}(G,>,\Sigma,r)$ to construct a partial rule $f':\Sigma^{2r+1} \rightharpoonup \Sigma$ by iteratively applying Extend to maximal gliders, following any linearization of $>$, until all of G is processed.[3]

Algorithm 1. $\text{GtoR}(G,>,\Sigma,r)$

1: $f' \leftarrow \emptyset$
2: **while** $G \neq \emptyset$ **do**
3:     $(\sigma,i) \leftarrow \text{Max}(G)$
4:      $G \leftarrow G \setminus \{\text{g}(\sigma,i)\}$
5:      $f' \leftarrow \text{Extends}(f',\sigma,i)$
6: **return** f'

We now show that if a CA derives a sound glider system $(G,>)$, then its language can be reconstructed from this set.

Lemma 22 (Soundness of Glider Reconstruction). *Let $\mathcal{A} \in \mathbb{CA}_{r,\Sigma}$ and $I \subseteq C_{\mathcal{F}}$. Suppose that $(G,>)$ is the glider system derived from $\mathcal{A}|_I$. If the derivation is sound, then the partial rule $f' = \text{GtoR}(G,>,\Sigma,r)$ satisfies $f'(w) = f(w)$ for all $w \in FN_{\mathcal{A},I}$.*

Proof. Let $(G,>)$ be the glider system *derived* from $\mathcal{A}|_I$. Assume towards contradiction that although the derivation is sound, there exists a $w \in FN_{\mathcal{A},I}$ for which the partial rule $f' = \text{GtoR}(G,>,\Sigma,r)$ returns $f'(w) \neq f(w)$. By GtoR construction, the only gliders that can define the assignment of w's rule are from $\{\text{g}(\sigma,i) \mid w \in \mathcal{N}_i^\sigma\}$, denoted G_w in the sound-derivation definition. Procedure GtoR chooses a maximal glider among them. By item (iii) of the sound derivation, this maximal glider, $\text{g}(\tau,j)$, imposes $f(w) = \tau$. Thus, $f'(w)$ would also assign it τ, which contradicts $f'(w) \neq f(w)$. Therefore, for every sound derivation, the partial rule function satisfies $f'(w) = f(w)$ for every $w \in FN_{\mathcal{A},I}$. $\square$

Remark 23. Note that in general, there are multiple linearizations that GtoR could follow and that different linearizations, l' and l'', yield different rule functions, f' and f'' resp. However, given that $(G,>)$ was derived from $\mathcal{A} \in \mathbb{CA}_{r,\Sigma}$ and I as per Sec.5.2, both will agree with f on any feasible neighborhood with respect to $\mathcal{A}|_I$. That is, $f(w) = f'(w) = f''(w)$ for every $w \in FN_{\mathcal{A},I}$.

Definition 24 (Glider Expressible). *Let $\mathcal{A} \in \mathbb{CA}_{r,\Sigma}$ and $I \subseteq C_{\mathcal{F}}$. We term language $\mathcal{L}(\mathcal{A},I)$ glider-expressible if the glider system derived from $\mathcal{A}|_I$ is sound.*

We denote the class of such languages by $\mathbb{CA}_{r,\Sigma}^{\text{Glider}}$. The following inclusion is a direct consequence of Lem.22.

Corollary 25. $\mathbb{CA}_{r,\Sigma}^{\text{Glider}} \subseteq \mathbb{CA}_{r,\Sigma}$

That is, every glider-expressible language is also CA-expressible. However, the converse does not hold in general; there exist CA-expressible languages that are not glider-expressible. An explicit counterexample is provided in Example 27. It follows from the following useful observation.

Observation 26. Let $\mathcal{A} \in \mathbb{CA}_{r,\Sigma}$, $I \subseteq C_{\mathcal{F}}$, and $L = \mathcal{L}(\mathcal{A},I)$. If there exists a $w \in FN_{\mathcal{A},I}$ such that $f(w) = \sigma$ while $\sigma \notin \{x_i \mid i \in [-r,r]\}$, then $L \notin \mathbb{CA}_{r,\Sigma}^{\text{Glider}}$.

---

[3] A linearization of a strict partial order is a total order that respects the partial order: if $\text{g} > \text{g}'$, then g precedes g' in the sequence.

This phenomenon occurs in the following example.

Example 27 (Rule 90 [58]). Let $\mathcal{A} \in \mathbb{CA}_{1,\{0,1\}}$ where $f(xyz) = x \oplus z$, for every $x, y, z \in \{0, 1\}$ and $\oplus$ is the XOR operation. Note that the assignment $f(111) = 0$, is not derivable from any glider $\mathbf{g}(\sigma, i)$ with $i \in \{-1, 0, 1\}$. Hence, this CA (aka Wolfram's Elementary-CA *Rule 90*) is not glider-expressible.

Corollary 28. $\mathbb{CA}_{r,\Sigma}^{\text{Glider}} \subsetneq \mathbb{CA}_{r,\Sigma}$.

6 $\mathbb{CA}_{r,\Sigma}^{\text{Glider}}$ Can Count

We turn to show that $\mathbb{CA}_{r,\Sigma}^{\text{Glider}}$ (therefore $\mathbb{CA}_{r,\Sigma}$) can generate languages beyond the OCA hierarchy. A classic example is the language $\{a^n b^n c^n \mid n \in \mathbb{N}\}$, which is known to be context-sensitive but not context-free, whereas all OCA-recognizable languages are context-free. Figure 4 presents a local rule function that generates this language for $I = \mathsf{pad}_\perp(\{abc\})$, along with an illustration of its evolution. In this section, we explore a powerful generalization of this result, given in Theorem 35. We progress towards it gradually, generalizing one aspect at a time.

Our first result shows that CA can count any given number of letters with different multiplications in their powers.

Proposition 29. *Let* $L = \{a_1^{k_1 \cdot n} a_2^{k_2 \cdot n} \ldots a_m^{k_m \cdot n} \mid n \in \mathbb{N}\}$ *where* $a_i \in \Sigma$ *and* $k_i \in \mathbb{N}$ *for every* $i \in [m]$. *If* $\sum_{i=1}^{m} k_i \leq 2r$, *then* L *is* $\mathbb{CA}_{r,\Sigma}^{REG}$*-expressible.*

We can assume without loss of generalization that $m = 2r$, by padding $\perp$ if needed and rewriting $a_i^{k_i \cdot n}$ as $a_i^n \cdots a_i^n$ (with k_i occurrences). By relabeling the first r symbols as $a_{-r}, \ldots, a_{-1}$ and the last r as $a_1, \ldots a_r$, the result becomes:

Proposition 30 (Proposition 29 Reformulated). *Let* $r \in \mathbb{N}$ *and* L_r *be the language* $\{a_{-r}^n \ldots a_{-1}^n a_1^n \ldots a_r^n \mid n \in \mathbb{N}\}$ *over* $\Sigma = \{\perp\} \cup \{a_{-i}, a_i \mid i \in [r]\}$. *Then* L_r *is* glider-expressible, *that is,* $L_r \in \mathbb{CA}_{r,\Sigma}^{\text{Glider}}$, *and thus is also* $\mathbb{CA}_{r,\Sigma}^{REG}$*-expressible.*

Before providing the proof, we illustrate the construction in the concrete case of $r = 2$ in Example 31, where the symbols a_{-2}, a_{-1}, a_1, a_2 are renamed a, b, c, d respectively for readability. Note that the base case of $r = 1$ is given in Claim 9.

Example 31. Let $r = 2$, $L = \{a^n b^n c^n d^n \mid n \in \mathbb{N}\}$, and $F = \{abcd\}$. This corresponds to L_2 with $a_{-2} = a$, $a_{-1} = b$, $a_1 = c$, $a_2 = d$. The CA uses eight gliders, two for each symbol; they propagate the respective symbol at the appropriate velocities. The inner letters (b and c) need to shift less than the outer letters (a and d). As a result, b and c override locations previously occupied by a and d resp. Formally, lower absolute index gliders ($|i|$, for a_i), dominate the higher ones when overlapping, ensuring nested growth. A detailed illustration of f's behavior is given in the extended version [18, App. C].

Proof (sketch). We construct a CA $\mathcal{A} \in \mathbb{CA}_{r,\Sigma}^{\text{REG}}$ that generates L_r from an initial configuration $I = \text{pad}_\perp(\{a_{-r}\ldots a_{-1}a_1\ldots a_r\})$. Each symbol $i \in [-r,r]\setminus\{0\}$ is assigned to two gliders $\boldsymbol{G}_i = \{\text{g}(a_i,i), \text{g}(a_i, i - \text{sign}(i))\}$ with a velocity relative to i. The set of induced gliders $\boldsymbol{G} = \bigcup_{i \in [r]} \boldsymbol{G}_i \cup \boldsymbol{G}_{-i}$, is associated with a strict partial order $>$, where; $\text{g}(a_i,x) > \text{g}(a_j,y)$ iff $\text{sign}(i) = \text{sign}(j)$, $|i| < |j|$, $a_i \neq a_j$, and $x \neq y$. As gliders with the same symbol or the same velocity are incomparable under $>$. Accordingly, the maximal gliders are those of the symbols a_{-1} and a_1. That is, the gliders of $\boldsymbol{G}_1$ and $\boldsymbol{G}_{-1}$ are maximal in $\boldsymbol{G}$; other gliders are suppressed when dominated. The local rule function of the CA induced by applying $\textsc{GtoR}$ to $(\boldsymbol{G}, >)$, determines the assignment of a neighborhood respecting the dominance relation induced by the gliders. This ensures deterministic growth of the nested pattern $c_n = {}^\omega\perp a_{-r}^{n+1}\ldots a_{-1}^{n+1} a_1^{n+1}\ldots a_r^{n+1} \perp^\omega$, such that $\mathcal{L}(\mathcal{A}, I) = L_r$. The full proof is given in the extended version [18, App. D]. $\qquad\square$

$\perp$	$\perp\perp\perp\perp\perp$	$\perp\perp\perp\perp a$					
a	$\perp\perp\perp aa$	$\perp\perp\perp ab$	$\perp\perp aaa$	$\perp\perp aab$	$\perp\perp abc$	$\perp aaaa$	$\perp aaab$
	$\perp aabb$	$aaaaa$	$aaaab$	$aaabb$			
b	$\perp abc\perp$	$aabbb$	$aabbc$	$abbbb$	$abbbc$	$abbcc$	$abc\perp\perp$
	$bbbbb$	$bbbbc$	$bbbcc$	$bbcc\perp$	$bbccc$		
c	$bc\perp\perp\perp$	$bcc\perp\perp$	$bccc\perp$	$bcccc$	$c\perp\perp\perp\perp$	$cc\perp\perp\perp$	$ccc\perp\perp$
	$cccc\perp$	$ccccc$					

Illustration of $I = {}^\omega\perp abc\perp^\omega$

Fig. 4. $\{a^n b^n c^n \mid n \in \mathbb{N}\}$ Legend: $\perp = \square$, $a = \square$, $b = \blacksquare$, $c = \boxtimes$

The construction above demonstrates that CA which are *glider-expressible*, can produce arbitrarily segment-aligned languages with precise symbolic spacing and coordination. Crucially, the mechanism used to generate L_r (i.e., the induced $(\boldsymbol{G}, >)$), does not depend on the specific value of each symbol. Rather, it depends on the *structure* and positioning of the segments, as well as on their bounded total width. We now generalize this idea to arbitrary repeated blocks.

Theorem 32. *Let $m \in \mathbb{N}$, $w_1, \ldots, w_m \in \Sigma^+$ and $r = \lceil \frac{1}{2}\sum_{i=1}^{m} |w_i| \rceil$. If the language $L = \{w_1^n w_2^n \ldots w_m^n \mid n \in \mathbb{N}\}$ is in $\mathbb{CA}_{r,\Sigma}$, then, in particular, $L \in \mathbb{CA}_{r,\Sigma}^{\text{Glider}}$.*

To prove this, we build on the idea that gliders can encode both *repetition* and *shifting*. Intuitively, the construction uses gliders to repeatedly emit new copies of each w_i, in the correct order and alignment, from a compact initial configuration. The following lemma provides the base case and will be used as a building block.

Lemma 33 (Shift & Concat). *For any $w \in \Sigma^+$ and direction $d \in \{\text{left}, \text{right}\}$, there exists $\mathcal{A} \in \mathbb{CA}_{r,\Sigma}^{\text{Glider}}$ that repeatedly shifts the existing copies of w by $s \in \mathbb{Z}$ and appends a new copy to the d side, resulting in the language $\{w^n \mid n \in \mathbb{N}\}$.*

The idea is to use persistent gliders with aligned shift velocities to replicate w incrementally. Figure 5 illustrates a word $w = \blacksquare\square\square\square\boxtimes$ that is shifted one step to the right, (i.e., $s = 1$) and is concatenated to the *right* of the shifted block. This results in the glider set $G = \{\mathbf{g}(\sigma, +1), \mathbf{g}(\sigma, +5) \mid \sigma \in \{\square, \blacksquare, \boxtimes\}\}$. To avoid clutter, we only annotated the gliders of $\blacksquare$ and $\boxtimes$ in the first step. The proof of the lemma is given in [18, App. E]. We now extend to two block patterns $L = \{w_1^n w_2^n \mid n \in \mathbb{N}\}$, which introduce asymmetry and spatial coordination. When, without loss of generalization $|w_1| > |w_2|$, naive repetition breaks alignment; we overcome this by letting the boundary between the blocks "move dynamically" during growth.

Lemma 34 (Two-block case). *Let $w_1, w_2 \in \Sigma^+$, and define $r = \lceil \frac{|w_1| + |w_2|}{2} \rceil$. Then $L = \{w_1^n w_2^n \mid n \in \mathbb{N}\} \in \mathbb{CA}_{r,\Sigma}^{\text{Glider}}$.*

The gliders that are associated with w_1 and w_2 interleave and overwrite part of each other's space, maintaining consistent symbolic alignment through carefully assigned shifts and dominance. See [18, App. F] for the full proof.

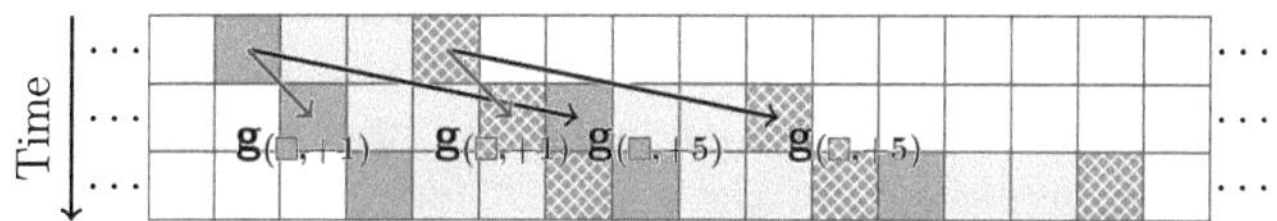

Fig. 5. $\blacksquare\square\square\square\boxtimes$, $s = +1$, $d = \text{right}$

In the general case, for arbitrary many words of arbitrary lengths, no global shift suffices. Each w_i is defined with its own *shift* and glider family. A key insight is that if L is $\mathbb{CA}_{r,\Sigma}$-expressible, then its derived gliders and partial order are sound. Thus, the CA can be reconstructed from the gliders. Moreover, this construction naturally extends to languages where the powers differ per block. Formally,

Theorem 35. *Let $m \in \mathbb{N}$, $w_1, \ldots, w_m \in \Sigma^+$, $r = \lceil \frac{1}{2} \sum_{i=1}^{m} a_i \cdot |w_i| \rceil$ and $e_i(n) = a_i \cdot n + b_i$ be positive linear expressions, i.e., $a_i, b_i \in \mathbb{N}_0$ and $a_i > 0$ for any $i \in [m]$. Consider $L = \{w_1^{e_1(n)} \ldots w_m^{e_m(n)} \mid n \in \mathbb{N}\}$. If $L \in \mathbb{CA}_{r,\Sigma}$, then $L \in \mathbb{CA}_{r,\Sigma}^{\text{Glider}}$.*

The proof of Theorem 35, including glider resolution, is provided in the extended version [18, App. H]. A specific example, $(aba)^{2n}c^{n+1}$, is given in [18, App. I].

7   Discussion

We set out to explore the symbolic expressiveness of deterministic, one dimensional cellular automata (CA) through a new lens–the *generative perspective*

To this end, we introduced a glider-based framework, where global behavior is induced by finitely many local symbolic patterns ("gliders") equipped with a strict partial order that specifies their dominance relation upon collision. This dominance order formalizes which glider prevails in each interaction, ensuring that the resulting dynamics are compositional and well-defined. The framework naturally led us to define the class of *glider-expressible languages*–those CA languages generated by sound glider systems associated with a CA in $\mathbb{CA}_{r,\Sigma}$, i.e., finitely many gliders interacting under such a dominance relation.

We showed that, despite their syntactic constraints, the class of *glider expressible CA*, and therefore CA expressible in general, captures complex and interesting behaviors examples include: synchronized multi-counter patterns such as $a^n b^n c^n d^n$, which illustrate fair division of unbounded resources, and asymmetric growth patterns such as $(aba)^{2n} c^{n+1}$, which reflect skewed resource consumption. These examples highlights that the glider generative model retains sufficient structural richness to encode nontrivial, long-range dependencies.

From a verification standpoint, this generative perspective provides a direct language-theoretic handle. Let $(G, >)$ be a sound glider system derived from $\mathcal{A} \in \mathbb{CA}_{r,\Sigma}$ and an initial regular set $I = \mathsf{pad}_\perp(L) \subseteq C_{\mathcal{F}}$, for $L \in \mathbb{REG}$, and let $B = \mathsf{pad}_\perp(\mathcal{B})$ denote the regular specification of bad configurations, where $\mathcal{B} \in \mathbb{REG}$. The safety question "$\exists t : G^t(c_0) \in B$ for some $c_0 \in I$?" Can be reduced to checking whether $\mathcal{L}(\mathcal{A}, I) \cap \mathcal{B} \neq \emptyset$. If the generated language is regular, classical forward or backward regular model checking applies directly [9]. When $\mathcal{L}(\mathcal{A}, I)$ is a *context-free language* (CFL), the problem is still decidable [21]: CFLs are closed under intersection with regular languages, and the emptiness problem is decidable, enabling workflows in the spirit of forward regular model checking. However, for systems whose generated language is not context-free (as in our multi-counter examples), this reduction no longer applies. This motivates the need for new verification techniques capable of exploiting the structured generative semantics that our framework exposes.

Overall, the glider-based viewpoint provides an interpretable and compositional bridge between symbolic dynamics and language-theoretic verification. By exposing the generative mechanisms behind CA evolution, it supports reasoning about system that are simultaneously unbounded and highly structured, and highlights the potential of CA under this generative perspective; as analyzable models for emergent computations.

Acknowledgments. Noa Izsak carried out this work in part as a member of the Saarbrücken Graduate School of Computer Science.

References

1. Abdulla, P.A.: Regular model checking. Int. J. Softw. Tools Technol. Transfer **14**(2), 109–118 (2012)
2. Abdulla, P.A., Bouajjani, A., Jonsson, B., Nilsson, M.: Handling global conditions in parametrized system verification. In: Halbwachs, N., Peled, D. (eds.) CAV 1999.

LNCS, vol. 1633, pp. 134–145. Springer, Heidelberg (1999). https://doi.org/10.1007/3-540-48683-6_14

3. Abdulla, P.A., Jonsson, B., Nilsson, M., Saksena, M.: A survey of regular model checking. In: Gardner, P., Yoshida, N. (eds.) CONCUR 2004. LNCS, vol. 3170, pp. 35–48. Springer, Heidelberg (2004). https://doi.org/10.1007/978-3-540-28644-8_3

4. Abdulla, P.A., Sistla, A.P., Talupur, M.: Model checking parameterized systems. In: Handbook of Model Checking, pp. 685–725. Springer, Cham (2018). https://doi.org/10.1007/978-3-319-10575-8_21

5. Alur, R., Madhusudan, P.: Visibly pushdown languages. In: Proceedings of STOC, pp. 202–211. Association for Computing Machinery (2004)

6. Bandini, S., Mauri, G., Serra, R.: Cellular automata: from a theoretical parallel computational model to its application to complex systems. Parallel Comput. **27**(5), 539–553 (2001)

7. Bloem, R., et al.: Decidability in parameterized verification. SIGACT News **47**(2), 53–64 (2016)

8. Boigelot, B., Godefroid, P.: Symbolic verification of communication protocols with infinite state spaces using QDDs. FMCAD **14**(3), 237–255 (1999)

9. Bouajjani, A., Jonsson, B., Nilsson, M., Touili, T.: Regular model checking. In: Emerson, E.A., Sistla, A.P. (eds.) CAV 2000. LNCS, vol. 1855, pp. 403–418. Springer, Heidelberg (2000). https://doi.org/10.1007/10722167_31

10. Bouajjani, A., Habermehl, P., Vojnar, T.: Abstract regular model checking. In: Alur, R., Peled, D.A. (eds.) CAV 2004. LNCS, vol. 3114, pp. 372–386. Springer, Heidelberg (2004). https://doi.org/10.1007/978-3-540-27813-9_29

11. Byg, J., Jørgensen, K.Y.: Regular Model Checking and Verification of Cellular Automata. Master thesis, Aalborg University, Department of Computer Science (2008)

12. Béal, M.-P., Perrin, D.: Symbolic dynamics and finite automata. In: Rozenberg, G., Salomaa, A. (eds.) Handbook of Formal Languages, pp. 463–506. Springer, Heidelberg (1997). https://doi.org/10.1007/978-3-662-07675-0_10

13. Conway, J.: The game of life. Sci. Am. **223**(4), 4 (1970)

14. Cook, M.: Universality in elementary cellular automata. Complex Syst. **15**(1), 1–40 (2004)

15. Emerson, E.A., Kahlon, V.: Reducing model checking of the many to the few. In: McAllester, D. (ed.) CADE 2000. LNCS (LNAI), vol. 1831, pp. 236–254. Springer, Heidelberg (2000). https://doi.org/10.1007/10721959_19

16. Emerson, E.A., Namjoshi, K.S.: Reasoning about rings. In: Proceedings of ACM SIGPLAN-SIGACT Symposium, pp. 85–94 (1995)

17. Esparza, J., Raskin, M., Welzel-Mohr, C.: Regular model checking upside-down: an invariant-based approach. arXiv preprint arXiv:2205.03060 (2022)

18. Fisman, D., Izsak, N.: Atomic Gliders and CA as Language Generators (Extended Version) (2025). https://arxiv.org/abs/2511.12656

19. Fisman, D., Izsak, N., Jacobs, S.: Learning broadcast protocols. In: Proceedings of the AAAI Conference on Artificial Intelligence, vol. 11, pp. 12016–12023 (2024)

20. Fisman, D., Kupferman, O., Lustig, Y.: On verifying fault tolerance of distributed protocols. In: Proceedings of TACAS, pp. 315–331 (2008)

21. Fisman, D., Pnueli, A.: Beyond regular model checking. In: FST TCS 2001: Foundations of Software Technology and Theoretical Computer Science, 21st Conference, Bangalore, India, 13–15 December 2001, Proceedings, pp. 156–170 (2001)

22. Fribourg, L., Olsén, H.: Reachability sets of parameterized rings as regular languages. Electron. Notes Theor. Comput. Sci. **9**, 40 (1997)

23. Gershenson, J.A.: Model Checking Omega Cellular Automata. Master's thesis, Carnegie Mellon University, School of Computer Science, Pittsburgh, PA (2010)
24. Guillon, P.: Automates cellulaires : dynamiques, simulations, traces. Ph.D. thesis, Université Paris-Est (2008). https://theses.hal.science/tel-00432058
25. Hedlund, G.A.: Endomorphisms and automorphisms of the shift dynamical system. Math. Syst. Theory **3**, 320–375 (1969)
26. Herbst, T.: On a subclass of context-free groups. RAIRO-Theor. Inf. Appl. **25**(3), 255–272 (1991)
27. Hopcroft, J.E., Motwani, R., Ullman, J.D.: Introduction to automata theory, languages, and computation. ACM SIGACT News **32**(1), 60–65 (2001)
28. Hurd, L.P.: Formal language characterization of cellular automaton limit sets. Complex Syst. **1** (1987)
29. Jaber, N., Jacobs, S., Wagner, C., Kulkarni, M., Samanta, R.: Parameterized verification of systems with global synchronization and guards. In: Lahiri, S.K., Wang, C. (eds.) CAV 2020. LNCS, vol. 12224, pp. 299–323. Springer, Cham (2020). https://doi.org/10.1007/978-3-030-53288-8_15
30. Jacobs, S., Bloem, R.: Parameterized synthesis. LMCS **10**(1) (2014)
31. Jonsson, B., Nilsson, M.: Transitive closures of regular relations for verifying infinite-state systems. In: Graf, S., Schwartzbach, M. (eds.) TACAS 2000. LNCS, vol. 1785, pp. 220–235. Springer, Heidelberg (2000). https://doi.org/10.1007/3-540-46419-0_16
32. Kari, J.: Theory of cellular automata: a survey. Theor. Comput. Sci. **334**(1), 3–33 (2005)
33. Resten, Y., Maler, O., Marcus, M., Pnueli, A., Shahar, E.: Symbolic model checking with rich assertional languages. In: Grumberg, O. (ed.) CAV 1997. LNCS, vol. 1254, pp. 424–435. Springer, Heidelberg (1997). https://doi.org/10.1007/3-540-63166-6_41
34. Khalimov, A., Jacobs, S., Bloem, R.: Towards efficient parameterized synthesis. In: VMCAI, pp. 108–127. Springer, Heidelberg (2013). DOI: https://doi.org/10.1007/978-3-642-35873-9_9
35. Kitchens, P.: One-sided, two-sided and countable state markov shifts. Symbolic Dynamics. Universitext, Springer (1998)
36. Kurka, P.: Languages, equicontinuity and attractors in cellular automata. Ergodic Theory Dynam. Syst. **17**(2), 417–433 (1997)
37. Kurka, P.: Zero-dimensional dynamical systems, formal languages, and universality. Theory Comput. Syst. **32**(4), 423–433 (1999)
38. Kurka, P.: Topological and symbolic dynamics. SMF France (2003)
39. Kutrib, M.: Cellular Automata and Language Theory, pp. 513–542. Springer, Heidelberg (2018)
40. Kutrib, M., Malcher, A.: String generation by cellular automata. Complex Syst. **30**(2), 111–132 (2021)
41. Kutrib, M., Worsch, T.: Self-verifying cellular automata. J. Cell. Autom. **15**(3), 223–242 (2020)
42. Lazic, M., Konnov, I., Widder, J., Bloem, R.: Synthesis of distributed algorithms with parameterized threshold guards. In: OPODIS 2017, pp. 32–1 (2018)
43. Le Lann, G.: Distributed systems-towards a formal approach. In: IFIP Congress, vol. 7, pp. 155–160 (1977)
44. Lu, G., Wang, Y.: Research on model checking cellular automata. Int. J. Digital Content Technol. Appl. **5**(11), 44–51 (2011)

45. Ma, H., Goel, A., Jeannin, J.B., Kapritsos, M., Kasikci, B., Sakallah, K.A.: I4: incremental inference of inductive invariants for verification of distributed protocols. In: SOSP, pp. 370–384 (2019)
46. Mathew, P., Penelle, V., Sreejith, A.: Learning deterministic one-counter automata in polynomial time. arXiv preprint arXiv:2503.04525 (2025)
47. Mazurkiewicz, A.: Concurrent program schemes and their interpretations. DAIMI Rep. Ser. **6**(78) (1977)
48. Neider, D., Jansen, N.: Regular model checking using solver technologies and automata learning. In: Proceedings of RWTH Aachen University (2013)
49. Neumann, J.V.: Theory of self-reproducing automata. IEEE **5**(1), 3–14 (1966)
50. Nordahl, M.G.: Formal languages and finite cellular automata. Comput. Syst. (1989)
51. Roos, R.S.: Deciding equivalence of deterministic one-counter automata in polynomial time with applications to learning. The Pennsylvania State University (1988)
52. Smith, A.: Cellular automata and formal languages. In: FOCS, pp. 216–224 (1970)
53. Staquet, G.: Active Learning of Automata with Resources. Ph.D. thesis, University of Antwerp (2024)
54. Sutner, K.: Model checking one-dimensional cellular automata. J. Cell. Autom. **4**(3), 213–224 (2009)
55. Valiant, L.G., Paterson, M.S.: Deterministic one-counter automata. Computer Syst. Sci. **10**(3), 340–350 (1975)
56. Wolfram, S.: Statistical mechanics of cellular automata. RMP **55**(3), 601 (1983)
57. Wolfram, S.: Computation theory of cellular automata. Commun. Math. Phys. **96**(1), 15–57 (1984)
58. Wolfram, S.: A New Kind of Science. Wolfram Media (2002)
59. Wolper, P., Boigelot, B.: Verifying systems with infinite but regular state spaces. In: Hu, A.J., Vardi, M.Y. (eds.) CAV 1998. LNCS, vol. 1427, pp. 88–97. Springer, Heidelberg (1998). https://doi.org/10.1007/BFb0028736

A Hybrid Meta-Learning Framework for Adaptive Safe Controller Synthesis of Dynamic Systems

Rui Guo[1], Yang Li[1], Xiuqing Cao[1], and Wang Lin[1,2]($\boxtimes$)

[1] School of Computer Science and Technology, Zhejiang Sci-Tech University, Hangzhou 310018, China
[2] Zhejiang Key Laboratory of Reliability Technology for Mechanical and Electrical Product, Zhejiang Sci-Tech University, Hangzhou 310018, China
`linwang@zstu.edu.cn`

Abstract. Safety-critical cyber physical systems usually require controllers that not only meet performance goal but also guarantee safety. However, the existing neural controller synthesis methods require a large number of data samples and gradient steps, which makes the real-time training infeasible when system dynamics vary over time. To address the challenge, we propose HyML-ASCS, a hybrid meta-learning framework for adaptive safety controller synthesis. HyML-ASCS combines model-based meta-learning, to generate task-specific embeddings that capture the patterns of safety controller synthesis tasks, with gradient-based meta-learning, to efficiently adapt to new synthesis tasks. This approach speeds up convergence, reduces the number of iterations required, and improves synthesis success rates, making HyML-ASCS a scalable and efficient solution for real-time control synthesis in dynamic environments. We evaluate HyML-ASCS on several benchmarks, demonstrating that it outperforms existing state-of-the-art methods in terms of synthesis efficiency, success rate, and scalability.

Keywords: Formal synthesis · safety verification · barrier certificates · meta-learning

1 Introduction

Cyber physical systems (CPSs), such as self-driving cars, unmanned aerial vehicles, and advanced robotics, represent a rapidly evolving domain at the intersection of computational and physical systems. These systems rely on software to perform critical decision-making tasks without human intervention, enabling unprecedented levels of autonomy [3]. For safety-critical CPSs, such as those used

This work was supported in part by the National Natural Science Foundation of China under Grant 62272416 and 62132014, in part by the Zhejiang Provincial Natural Science Foundation of China under Grant LD24F020006, and in part by the Fundamental Research Funds of Zhejiang Sci-Tech University under Grant 24232090-Y.

in aerospace, industrial automation, and intelligent transportation, controllers must not only achieve desired performance objectives but also guarantee that the system avoids unsafe or undesirable states under all conditions [4,5]. Thus, safety controller synthesis is one of the most fundamental problems in ensuring the reliability of CPSs.

Barrier certificate synthesis is a key technique for verifying the safety of controllers synthesized for CPSs. A barrier certificate is a function used to prove the safety of a dynamical system by ensuring that all trajectories of the system starting from a given initial set remain within a safe set and do not reach unsafe regions. Recent advancements in deep learning have provided new opportunities for addressing the challenges of controller synthesis in dynamical systems. Neural networks are well-suited for complex control tasks due to their ability to approximate nonlinear functions, making them a promising candidate for simultaneously synthesizing controllers and barrier certificates to meet both safety and performance requirements. Zhao et al. [18] proposed a supervised learning method to synthesize controllers and barrier certificates for nonlinear systems using neural networks. Deshmukh et al. [35] introduced a deep reinforcement learning-based approach with a verification-in-the-loop algorithm, using barrier certificates to ensure the safety of controllers.

While the neural network based methods have shown great potential in synthesizing controllers, they face significant challenges in real-world applications. Neural networks heavily rely on training data and often behave unpredictably when operating in environments outside their training distribution. This lack of robustness poses serious risks for safety-critical CPSs, where unanticipated scenarios are inevitable, and unsafe behavior leads to catastrophic consequences. Furthermore, when the actual dynamics of a system deviate from the system used for training, due to parameter variations, environmental changes, or estimation errors, the learned controllers may fail to ensure the safety of the actual system. Retraining safety controllers to address these discrepancies is often impractical in real-world settings due to high computational costs and the need to quickly synthesize the safety controllers, especially when data samples are limited. Therefore, developing safety controllers that can efficiently adapt to the real-world system with the dynamics may vary over time is crucial.

Meta-learning offers a promising solution to these challenges. One widely used meta-learning approach is Model-Agnostic Meta-Learning (MAML) [8], a gradient-based method designed to learn model parameters that can be efficiently fine-tuned for new tasks. Jena et al. [27] has proposed integrating MAML with neural Lyapunov functions to address parametric uncertainties in dynamical systems, enabling adaptive stability certification and improved performance in computing Regions of Attraction for nonlinear systems. Another research [34] employs a safe meta reinforcement learning framework to synthesize safety controllers for nonlinear systems, combining meta-learning for controller pretraining with formal safety verification via polynomial optimization. However, MAML relies on a single model initialization, which poses challenges for safe controller synthesis and barrier certificate generation. These synthesis tasks often involve dynamic systems with varying dynamics and safety constraints, requiring task-specific controllers and barrier certificates. A single initialization fails to generalize across systems with parametric shifts or dynamic variations, making it hard to ensure safety guarantees. Furthermore, MAML's lack of leveraging shared task structures limits its ability to quickly adapt to new tasks with minimal data and computation. When real-world system dynamics differ from the offline model used during training, MAML requires extensive fine-tuning with additional data, which is impractical in scenarios where data is limited and rapid adaptation is critical.

In contrast to MAML, model-based meta-learning methods excel at extracting and encoding shared structures across tasks through task-specific embeddings [7]. To address the limitations of MAML based method, we propose a hybrid meta learning framework for adaptive safety controller synthesis, called HyML-ASCS, which integrates model-based and gradient-based meta learning approaches. Our HyML-ASCS framework consists of two primary phases: meta-learning and task adaptation. In the meta-learning phase, the model-based meta-learner generates task-specific embeddings from trajectory data, to capture the patterns of shared safety constraints and system dynamics across synthesis tasks, and the gradient-based meta-learner then fine-tunes the network parameters for each specific synthesis task by minimizing a task-specific loss function. In the task adaptation phase, the learned meta parameters rapidly adapt to new synthesis tasks, by fine-tuning the controller and barrier certificate networks with the task-specific embeddings.

In summary, the main contributions of the paper are as follows:

- Our HyML-ASCS framework uses task-specific embeddings to identify the patterns of safety controller synthesis tasks, ensuring adaptive synthesis of controllers and barrier certificates under system dynamics varying.
- Our HyML-ASCS framework combines model-based task embeddings with gradient-based updates, enabling efficient synthesis of tailored controllers and barrier certificates with a few data and gradient update steps.
- We evaluate HyML-ASCS on a wide range of benchmarks, and demonstrate its superiority over state-of-the-art methods, such as MAML based method and FOSSIL 2.0. HyML-ASCS achieves faster synthesis task adaptation, reduced computational cost, and robust safety guarantees across all scenarios.

2   Related Work

Verifying the safety of dynamical systems is crucial in the design and deployment of CPSs, with barrier certificates being a prominent approach. Early work on safety verification using barrier certificates, such as [10,11], provides a framework for ensuring safety over an infinite time horizon without explicitly computing reachable sets [12,13]. Recent advances integrate learning-based methods to enhance scalability, using reinforcement learning (RL) and supervised learning for synthesizing safety certificates, including control barrier functions (CBFs) [1,16,17]. While these methods improve scalability, they still face challenges in ensuring robustness when system dynamics change significantly.

Meta-learning, or "learning to learn," provides a framework for enabling agents or machine learning models to leverage past experience across multiple tasks to improve learning efficiency and adaptability on new tasks [19, 22]. Meta-learning has been explored through various approaches in recent studies [20, 21, 23, 24]. Existing meta-learning methods can be broadly categorized into gradient-based and model-based approaches. Gradient-based methods, such as MAML [8], aim to learn task-agnostic initialization parameters that can be quickly fine-tuned to new tasks with a few gradient updates. MAML has been widely recognized for its flexibility and successfully applied in various domains, including computer vision [7], natural language processing [25], reinforcement learning [24], and control systems [26]. A recent work combines MAML with nonlinear Lyapunov functions (NLFs) to enable adaptive stability certification in dynamical systems under parametric uncertainty [27]. Model-based methods focus on learning task-specific embeddings or representations that enable rapid adaptation to new tasks [28, 29]. For example, [30] proposed Context Adaptation for Meta-Learning (CAML), which learns task-specific context parameters in the inner loop while updating task-independent model parameters in the outer loop, enabling rapid adaptation to new tasks. Recent works have explored hybrid meta-learning methods that combine model-based task representation with gradient-based adaptation to leverage the strengths of both approaches [9, 31].

3   Preliminaries

3.1   Safety in Controlled Continuous Dynamical Systems

A *Controlled Continuous Dynamical System (CCDS)* is defined as a quadruple $\Gamma = (\mathbf{f}, \mathbf{X}_D, \mathbf{X}_I, \mathbf{X}_U)$, where:

- $\mathbf{f} : \mathbb{R}^{n+m} \rightarrow \mathbb{R}^n$ is a continuous vector field defining the system dynamics, governed by:

$$\dot{\mathbf{x}} = \mathbf{f}(\mathbf{x}, \mathbf{u}), \quad \text{where } \mathbf{u} = \mathbf{g}(\mathbf{x}), \tag{1}$$

  where $\dot{\mathbf{x}}$ denotes the derivative of the state $\mathbf{x}$ with respect to t, $\mathbf{u} \in \mathbb{R}^m$ represents the control inputs, and $\mathbf{g} : \mathbb{R}^n \rightarrow \mathbb{R}^m$ denotes the feedback controller function.
- $\mathbf{X}_D \subseteq \mathbb{R}^n$ denotes the system domain.
- $\mathbf{X}_I \subseteq \mathbf{X}_D$ represents the set of permissible initial states.
- $\mathbf{X}_U \subseteq \mathbf{X}_D$ specifies the set of unsafe states that must be avoided.

Both $\mathbf{f}$ and $\mathbf{g}$ are assumed to be locally Lipschitz continuous. Specifically, for any initial state $\mathbf{x}_0 \in \mathbf{X}_I$, there exists a time interval $[0, T]$ and a unique trajectory $\mathbf{x}(t, \mathbf{x}_0)$ satisfying the differential Eq. (1). We formally define the safety and the safe controller synthesis of a CCDS as follows:

Definition 1 (Safety). *A CCDS $\Gamma = (\mathbf{f}, \mathbf{X}_D, \mathbf{X}_I, \mathbf{X}_U)$ is safe if, for each initial state $\mathbf{x}_0 \in \mathbf{X}_I$, there exists no time $t \geq 0$ such that the system trajectory $\mathbf{x}(\mathbf{t}, \mathbf{x}_0)$ enters the unsafe region $\mathbf{X}_U$. Formally, $\forall \mathbf{x}_0 \in \mathbf{X}_I, \forall t \geq 0, \mathbf{x}(\mathbf{t}, \mathbf{x}_0) \notin \mathbf{X}_U$.*

Definition 2 (Safe Controller Synthesis). *Given a CCDS $\Gamma = (\mathbf{f}, \mathbf{X}_D, \mathbf{X}_I, \mathbf{X}_U)$, design a locally Lipschitz feedback controller $\mathbf{g}$ such that the closed-loop system Γ is safe.*

To formally guarantee the safety of CCDSs, control barrier certificates (CBCs) are introduced [1]. A barrier certificate is a real-valued function that acts as a "barrier" between the reachable states and the unsafe region. Given a CCDS Γ, if such a function exists, it ensures that trajectories starting from the initial set $\mathbf{X}_I$ never enter the unsafe region $\mathbf{X}_U$. The following theorem guarantees the safety of a given CCDS.

Theorem 1 [10]. *Given a CCDS $\Gamma = (\mathbf{f}, \mathbf{X}_D, \mathbf{X}_I, \mathbf{X}_U)$, if there exists a real-valued function $B : \mathbf{X}_D \to \mathbb{R}$ satisfying the following conditions:*

$$
\begin{aligned}
B(\mathbf{x}) \leq 0 \quad &\forall \mathbf{x} \in \mathbf{X}_I, \\
B(\mathbf{x}) > 0 \quad &\forall \mathbf{x} \in \mathbf{X}_U, \\
\mathcal{L}_\mathbf{f} B(\mathbf{x}) < 0 \quad &\forall \mathbf{x} \in \mathbf{X}_D \ s.t. \ B(\mathbf{x}) = 0,
\end{aligned}
\tag{2}
$$

where $\mathcal{L}_\mathbf{f} B(\mathbf{x}) = \sum_{i=1}^{n} \frac{\partial B}{\partial x_i} \cdot \mathbf{f}_i(\mathbf{x}, \mathbf{u})$, then $B(\mathbf{x})$ is a barrier certificate, and the safety of the system is guaranteed.

3.2   Model-Agnostic Meta-Learning

Model-Agnostic Meta-Learning (MAML) is a gradient-based meta-learning method designed to learn model initializations that can adapt to new tasks with minimal updates [8]. MAML is particularly appealing in the context of safety-critical controller synthesis, as it enables rapid adaptation to varying system dynamics, which is crucial when dealing with parameter uncertainties or changing environments. For controlled dynamical systems, each task T_i corresponds to a specific system $\Gamma = (\mathbf{f}_{\theta_i}, \mathbf{X}_D, \mathbf{X}_I, \mathbf{X}_U)$, parameterized by θ_i.

The meta-learner maintains a set of meta-parameters (ϕ_B, ϕ_g), where ϕ_B parameterizes the barrier certificate network and ϕ_g parameterizes the controller network.

During the inner loop, these meta-parameters are adapted to a given task T_i using its training data $D_{\text{tr}}^{T_i}$ by applying one or more steps of gradient descent. The adapted parameters (ϕ_B', ϕ_g') represent task-specific solutions that account for the dynamics and constraints of T_i.

The outer loop then optimizes the meta-parameters (ϕ_B, ϕ_g) by minimizing the expected loss on the test data of tasks sampled from a distribution $\rho(\mathcal{T})$:

$$
\min_{\phi_B, \phi_g} \ \mathbb{E}_{T_i \sim \rho(\mathcal{T})} \left[L_B^{T_i} \left(D_{\text{te}}^{T_i}; \phi_B', \phi_g' \right) \right],
\tag{3}
$$

where $L_B^{T_i}$ denotes the task-specific evaluation loss. This formulation ensures that the meta-parameters are optimized such that, after adaptation, the model performs well on unseen tasks.

In summary, the MAML framework consists of an inner loop that adapts the meta-parameters (ϕ_B, ϕ_g) to task-specific parameters (ϕ'_B, ϕ'_g) using limited training data, and an outer loop that updates the meta-parameters to minimize the meta-loss across tasks. This dual-loop structure enables task-specific adaptation while simultaneously learning an initialization that generalizes effectively to unseen tasks.

3.3 Problem Statement

We formally define the problem of safety controller synthesis studied in this paper.

Problem 1. Given CCDSs Γ_i, $i = 1, 2, \cdots, N$, with neural safety controller $\mathbf{g}_i(\mathbf{x})$ guaranteed via neural control barrier certificate $B_i(\mathbf{x})$. Consider a new system Γ, the goal is to synthesize a safe controller for the system using the data from the verified CCDSs Γ_i, $i = 1, 2, \cdots, N$.

In general, the neural network based methods train neural safety controllers via optimizing a barrier loss that captures the barrier certificate constraints. However, the methods require a large number of data samples and gradient update steps, which makes the real-time training for a new system infeasible. We aim to address the above problem via meta learning approach, with a small number of training samples and gradient steps.

4   The Hybrid Meta-Learning Framework

In this section, we present a hybrid meta-learning framework for synthesizing safe controllers of dynamical systems. Our framework operates in two distinct phases: the meta-training phase and the task adaptation phase. The workflow of the proposed framework is illustrated in Fig. 1.

- **Meta Learning**: During the meta-training phase, we learn meta-controllers and meta-barrier certificates through the combination of model-based and gradient-based meta-learning. It uses a model-based meta-learner to generate task-specific embeddings and a gradient-based meta-learner to adapt network parameters for each synthesis task.
- **Task Adpatation**: During the task adaptation, the meta-learned controller and barrier certificate, parameterized by ϕ_B, ϕ_g, are fine-tuned via gradient-based optimization using the entirety of the new synthesis task's dataset. This full-data utilization enables rapid convergence to a task-adapted solution, bypassing the iterative meta-learning steps and facilitating the efficient generation of controllers and barrier certificates tailored to the target synthesis task's dynamics and safety specifications.

This two-phase approach enables the synthesis of safe controllers and barrier certificates that are both highly generalizable and adaptable to new tasks with minimal data and computation.

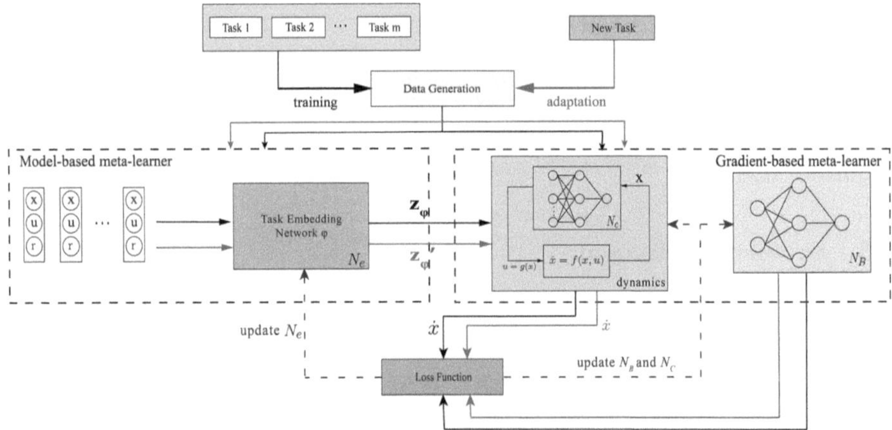

Fig. 1. The framework of HyML-ASCS to synthesize safe controller.

4.1 Training Process via Meta-Learning

In meta learning phase, model-based meta-learner aims to identify the patterns of sample safety controller synthesis tasks through analyzing task-related trajectories, and gradient-based meta-learner uses the patterns to fine-tune network parameters efficiently through inner and outer loop optimization.

Model-Based Meta-Learning. Model-based meta-learning extracts task embeddings from trajectory sequences, for identifying the patterns of safety controller synthesis tasks. It is implemented by an LSTM network that takes trajectory sequences as input and outputs a task embedding.

We first generate the trajectory sequences, which consist of state-action-reward pairs, required for the model-based learner. With a state $\mathbf{x}_0$ from the initial set D_I, we iteratively compute the next state over a time horizon T. Concretely, for the state $\mathbf{x}_t$, the controller $\mathbf{g}$ maps it to a control signal $\mathbf{u_t}$, and computes the next state $\mathbf{x}_{t+1}$ according to the system dynamics. Moreover, the reward r_t is computed based on the Euclidean distance between the current state $\mathbf{x}_t$ and the closest point on the unsafe set boundary:

$$r_t = \begin{cases} \|\mathbf{x}_t - \mathbf{x}_{\mathrm{unsafe}}\|, & \text{if } \mathbf{x}_t \notin X_U, \\ 0, & \text{otherwise.} \end{cases}$$

Repeating this process for T steps, we obtain a trajectory sequence:

$$S = \{\mathbf{s}_t = [\mathbf{x}_t, \mathbf{u}_t, r_t]\}_{t=0}^{T-1}.$$

Next, with M different initial states $\{\mathbf{x}_{0,1}, \cdots, \mathbf{x}_{0,M}\}$ from D_I, we obtain the corresponding trajectory sequences $\{S_1, \cdots, S_M\}$.

The LSTM network, denoted as N_e, processes these sequences and generates a task embedding $\mathbf{z}_{\phi_e}$. In our implementation, $\mathbf{z}_{\phi_e}$ is obtained from the final hidden state of the LSTM, capturing the identified patterns of each task. This embedding is subsequently provided as an additional input to the gradient-based meta-learning module. Note that the parameters ϕ_e of the task embedding network N_e will be updated in the gradient-based meta-learning phase.

Gradient-Based Meta-Learning. The gradient-based meta-learning aims to provide the network parameters of the meta controllers and the meta CBCs, based on the system states and the task embedding $\mathbf{z}_{\phi_e}$ extracted earlier. Specifically, we directly combine the task embedding with the system state to form the input for these networks, enabling the networks to generalize across different synthesis tasks. Similar to the MAML based method, the process consists of two stages: the inner loop, where the networks are fine-tuned for each synthesis task, and the outer loop, where the initial parameters of the networks are updated to improve generalization across synthesis tasks.

We first generate the necessary datasets for training and testing in the inner and outer loops. For the i-th CCDS Γ_i, we sample data from the domain X_D^i, the initial set X_I^i, and the unsafe area X_U^i. A grid-based sampling method is employed to discretize these regions into uniform grids, filtering out points that do not satisfy the region constraints. The resulting datasets D_D^i, D_I^i, and D_U^i correspond to the domain, the initial set, and the unsafe set, respectively. These datasets are divided into the training (D_{tr}^i, 60%) and testing (D_{te}^i, 40%) subsets.

To formally connect our framework with barrier certificates, recall that a barrier certificate $B : \mathbf{X}_D \to \mathbb{R}$ is a function satisfying the conditions in Theorem 1, which ensure system safety. In our approach, we parameterize this function by introducing both the system state $\mathbf{x}$ and the task embedding $\mathbf{z}_{\phi_e}$ as inputs, i.e. $B(\mathbf{x}, \mathbf{z}_{\phi_e})$.

In the inner loop, for each CCDS, the network is fine-tuned by minimizing the task-specific loss using the training data D_{tr}^i. This loss function plays a key role in adapting the model to the specific synthesis task's dynamics and safety constraints, which is formalized as follows:

$$L_B = \sum_{\mathbf{x} \in D_I} \mathrm{ReLU}(B(\mathbf{x}, \mathbf{z}_{\phi_e}) + \varepsilon_1)$$
$$+ \sum_{\mathbf{x} \in D_U} \mathrm{ReLU}(-B(\mathbf{x}, \mathbf{z}_{\phi_e}) + \varepsilon_2)$$
$$+ \sum_{\mathbf{x} \in D_D} \mathrm{ReLU}(\mathcal{L}_{\mathbf{f}} B(\mathbf{x}, \mathbf{z}_{\phi_e}) + \varepsilon_3), \tag{4}$$

where $\varepsilon_1, \varepsilon_2, \varepsilon_3$ are non-negative tolerances for improving numerical stability during training.

This formulation explicitly conditions the loss function and subsequent updates on both the system state and the task embedding, ensuring that the learned CBCs and controllers adapt to different synthesis tasks.

Starting from the parameters provided by the outer loop, which represent the initial parameters of the barrier certificate network, the controller network, and the task embedding network, the networks are updated through gradient descent:

$$\phi_i' = \phi - \alpha \nabla_\phi L_B(D_{\mathrm{tr}}^i, \phi),$$

where ϕ_i' are the updated parameters for the i-th synthesis task, and α is the learning rate. The optimization continues until the loss L_B converges to 0, ensuring the networks are well-adapted to the specific synthesis task.

In the outer loop, the adapted parameters ϕ_i' are evaluated on the test dataset of the i-th synthesis task D_{te}^i to compute the current synthesis task's test loss:

$$L_{\mathrm{te}}^i = L_B(D_{\mathrm{te}}^i, \phi_i'),$$

The meta-loss is then computed by averaging the test losses over all synthesis tasks:

$$L_{\mathrm{meta}} = \frac{1}{N} \sum_{i=1}^{N} L_{\mathrm{te}}^i,$$

where N is the number of synthesis tasks. The initial parameters of the barrier certificate network, the controller network, and the task embedding network, $\phi = (\phi_B, \phi_g, \phi_e)$, are updated based on the gradient of the meta-loss:

$$\phi \leftarrow \phi - \beta \nabla_\phi L_{\mathrm{meta}},$$

where β is the meta-learning rate.

At the conclusion of the meta-training process, the framework produces meta-learned parameters (ϕ_B, ϕ_g, ϕ_e) that enable the system to efficiently adapt to new synthesis tasks.

Algorithm 1 outlines the joint optimization procedure of the model-based and gradient-based meta-learners. This approach combines the model-based learner's advantage of identifying the patterns of synthesis tasks with the gradient-based meta-learner's ability to consistently improve performance through a small number of gradient descent steps.

4.2   New Task Adaptation and Verification

During adaptation on a new safety controller synthesis task T_{new}, a dataset D_{new} is generated following the training procedure. Unlike meta-training, this dataset is used in its entirety for rapid adaptation, bypassing the support/query set partitioning. The task embedding network N_e then extracts a task-specific embedding $\mathbf{z}_\phi$, which captures the distinct system's dynamics and safety constraints. Given the meta-learned parameters (ϕ_B, ϕ_g) and the task embedding $\mathbf{z}_\phi$, the barrier certificate and the controller are synthesized through fine-tuning. Specifically, both networks are optimized to minimize the task-specific loss L_B:

$$\phi' = \phi - \alpha \nabla_\phi L_B(D_{new}, \phi),$$

Algorithm 1. HyML-ASCS training process

Input: Task distribution $p(\mathcal{T})$, learning rates α, β, threshold ϵ
Output: trained controller parameters ϕ_g, trained barrier certificate parameters ϕ_B, trained task embedding parameters ϕ_e.

1: Initialize meta-parameters ϕ_B, ϕ_g, ϕ_e.
2: **for** each training step **do**
3:     Sample synthesis tasks $\{\mathcal{T}_i\} \sim p(\mathcal{T})$.
4:     **for** each synthesis task $\mathcal{T}_i$ **do**
5:         Sample data D^i; split into D^i_{tr} and D^i_{te}.
6:         Initialize $\phi'_B, \phi'_g, \phi'_e$ to ϕ_B, ϕ_g, ϕ_e.
7:         **while** $L > \epsilon$ **do**
8:             Sample a batch of trajectories τ
9:             Generate task embedding z_ϕ according to τ
10:             Evaluate $\nabla_{\phi_B,\phi_g,\phi_e} L_B(D^i_{tr}; \phi'_B, \phi'_g, z_\phi)$
11:             Update parameters $\phi'_B, \phi'_g, \phi'_e$ using gradient descent.
12:         **end while**
13:         Compute task loss L^i_{te}.
14:     **end for**
15:     Evaluate $\nabla_{\phi_B,\phi_g,\phi_e} L_{meta}$
16:     Update meta-parameters ϕ_B, ϕ_g, ϕ_e via gradient descent.
17: **end for**

---

where ϕ includes the parameters of both the barrier certificate and controller networks, and α is the learning rate. As the loss converges to a minimum, the learned networks can serve as candidate control barrier certificate and controller that satisfy the safety conditions in Theorem 1 over the sampled data regions.

However, these candidates do not provide formal safety guarantees. Therefore, we perform formal verification using a counterexample-guided verification process using dReal [2]. To translate the barrier conditions into SMT formula, we negate each condition in (2) as:

$$\mathbf{x} \in X_I \wedge B(\mathbf{x}) > 0,$$
$$\mathbf{x} \in X_U \wedge B(\mathbf{x}) \leq 0, \tag{5}$$
$$\mathbf{x} \in X_D \wedge \mathcal{L}_{\mathbf{f}} B(\mathbf{x}) \geq 0 \wedge B(\mathbf{x}) = 0,$$

The solver checks for the existence of counterexamples that satisfy any of the negated conditions. If dReal returns UNSAT (unsatisfiable) for (5), it means that no counterexamples exist, and the candidate B and g are formally verified as safe in the state space. Conversely, it does not necessarily imply that the system is unsafe; rather, it indicates that the current candidates fails to satisfy the sufficient conditions for formal verification. In this case, dReal provides specific counterexample state that violates the safety conditions. New data points are then sampled around the counterexample and added to the dataset for further refinement of both networks. This verification and refinement process is repeated iteratively until dReal returns "UNSAT", ensuring that the synthesized controller and barrier certificate globally satisfy the safety requirements. In practice, this process converges within a few iterations, highlighting the efficiency of our approach.

5   Experiments

We have implemented a tool named *Meta-Controller* based on PyTorch platform to evaluate the performance of our proposed method on dynamic systems of varying dimensions. Specifically, we conduct experiments on 2D, 3D, and 4D dynamic systems, where a separate meta-model was trained for each dimension. Each meta-model is only responsible for synthesis task adaptation within its corresponding system dimension. We compare the performance of our method with the MAML based approach and FOSSIL 2.0 [2], a counterexample-guided tool.

5.1   Implementation Details

In our experiments, each system dimension (2D, 3D, 4D) has a set of representative dynamic systems selected for meta-model training. Each synthesis task corresponds to a CCDS with varying initial conditions and, in some cases, varying system parameters (e.g., control gains, coefficients a_1, a_2).

Table 1. Performance comparison of benchmarks in terms of success rates and computation times.

Benchmark	n_x	NN	FOSSIL 2.0		MAML		HyML-ASCS	
			round	*time(s)*	*round*	*time(s)*	*round*	*time(s)*
2D-C1 [35]	2	2-10-10-1	6	32.09	1	16.30	1	**14.90**
2D-C2 [36]	2	2-10-10-1	6	13.77	2	13.44	1	**9.48**
2D-C3 [37]	2	2-10-10-1	1	17.93	1	8.32	1	**5.65**
2D-C4 [37]	2	2-10-10-1	1	12.66	1	3.41	1	**1.77**
2D-C5 [32]	2	2-10-10-1	6	7.57	1	8.49	1	**3.14**
3D-C1 [35]	3	3-64-1	8	203.04	1	103.87	1	**18.69**
3D-C2 [35]	3	3-64-1	2	41.66	2	38.02	1	**26.88**
3D-C3 [38]	3	3-64-1	5	24.51	1	14.13	1	**7.34**
3D-C4 [33]	3	3-64-1	/	×	1	3.64	1	**3.34**
3D-C5 [10]	3	3-64-1	1	18.74	1	13.58	1	**13.03**
4D-C1 [37]	4	4-32-1	1	581.45	1	1555.73	1	**55.88**
4D-C2 [39]	4	4-32-1	7	435.27	1	1734.70	1	**3.23**
4D-C3 [39]	4	4-32-1	2	627.92	/	×	4	**541.10**
4D-C4 [37]	4	4-32-1	1	518.47	6	356.89	1	**219.51**
4D-C5 [37]	4	4-32-1	1	600.50	6	483.34	4	**54.74**

The experiments were divided into training and testing phases. In the training phase, we selected representative dynamic systems within each dimension

(2D, 3D, and 4D) to train the corresponding meta-model. These representative systems were chosen to cover a variety of dynamics and parameter regimes in each dimension, ensuring that the meta-model learns generalizable patterns across tasks. The meta-model learns to synthesize controllers and barrier certificates by capturing commonalities and differences among synthesis tasks in that dimension. As shown in Table 1, we evaluated our trained meta-models on 15 new systems, all of which were selected from previously published benchmarks, as indicated by the references in the table. These test tasks differ from the training systems in both initial conditions and selected system parameters, providing a robust assessment of the meta-controller's generalization ability. This setup ensures that each test task represents a new adaptation challenge, rather than simply repeating training tasks, and demonstrates the effectiveness of the approach across multiple system dimensions. Note that the computation times reported in Table 1 include both the new task adaptation phase and the formal verification process using dReal.

For 2D systems, both the barrier function network and controller network consist of two hidden layers, each with 10 neurons. For 3D systems, the network has one hidden layer with 64 neurons, and for 4D systems, the network has one hidden layer with 32 neurons. All networks use the Sigmoid activation function. We employ AdamW as the optimizer, with an inner-loop learning rate $\alpha = 0.001$, an outer-loop meta-learning rate $\beta = 0.01$ and a total of 50 meta-training iterations. In addition, the task embedding network N_e is implemented as a single-layer LSTM with hidden size 16. The final hidden state is mapped via a fully connected layer to an 8-dimensional task embedding. The same optimizer (AdamW) and learning rates ($\alpha = 0.001$, $\beta = 0.01$) are applied to update N_e jointly with the controller and barrier networks during meta-training. The parameters $\varepsilon_1, \varepsilon_2, \varepsilon_3$ in loss function (4) are set to 0, and the convergence threshold $\epsilon = 10^{-16}$. For formal safety verification, we use dReal solver to ensure that the generated controller and barrier certificate satisfy safety constraints. All experiments are conducted on Ubuntu 22.04, with AMD Ryzen 5 5600G CPU and 16GB memory.

5.2   Experiments Results

We evaluated the effectiveness of the proposed tool by synthesizing controllers for various dynamic systems and verifying their safety. To clearly illustrate the performance, we select two representative examples for detailed discussion, while the remaining results are summarized in a table to compare our method with baseline approaches.

Example 1 (Pendulum [32]). The controlled dynamic system is as follows:

$$\begin{bmatrix} \dot{x}_1 \\ \dot{x}_2 \end{bmatrix} \begin{bmatrix} x_2 \\ \frac{3g\sin(x_1)}{2l} + \frac{3u}{ml^2} \end{bmatrix},$$

where u is the control input, m = 1 and l = 1 denote the pendulum mass and length respectively, g = 9.8 is the gravitational acceleration and

- the domain is $X_D = \{\mathbf{x} \in \mathbb{R}^2 \mid -\pi \leq x_1 \leq \pi, -8 \leq x_2 \leq 8\}$,
- the initial region is $X_I = \{\mathbf{x} \in \mathbb{R}^2 \mid 0 \leq x_1 \leq \pi, -3 \leq x_2 \leq 3\}$,
- and the unsafe region is $X_U = \{\mathbf{x} \in \mathbb{R}^2 \mid -\pi \leq x_1 \leq -\frac{3}{5}\pi, -8 \leq x_2 \leq 0\}$.

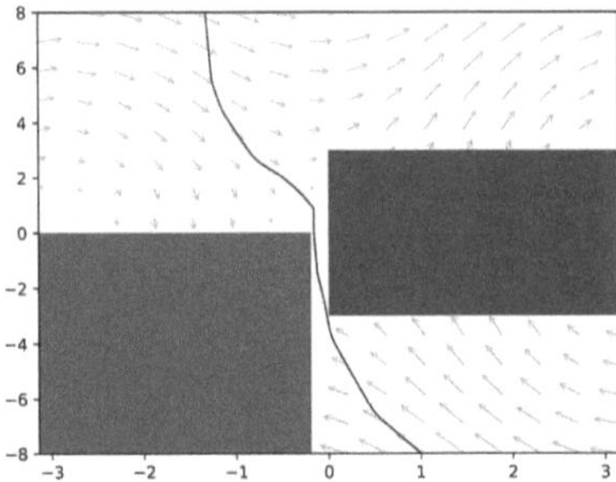

Fig. 2. Adapted and verified NN controller and barrier certificate for Example 1.

Both the barrier certificate and controller networks adopt [2,10,10,1] structure. Compared to the MAML (8.49 s) and FOSSIL 2.0 (7.57 s), our Meta-Controller demonstrates a significant improvement in efficiency by synthesizing the correct controller and barrier certificate in just 3.14 s, reducing computation time by up to 63.0% faster than MAML and 58.5% faster than FOSSIL 2.0, as illustrated in Fig. 2. In the figure, the green region represents the safe state space, while the red region denotes the unsafe state space and the blue line indicates the zero-level set of the barrier function $B(\mathbf{x}) = 0$.

Example 2 (Ball [33]). The controlled dynamic system is as follows:

$$\begin{bmatrix} \dot{x}_1 \\ \dot{x}_2 \\ \dot{x}_3 \end{bmatrix} = \begin{bmatrix} -x_1{}^3 + a_1 x_1 (x_3)^3 \\ -x_2 - (x_1)^2 x_2 \\ a_2 (x_1)^2 x_3 + u \end{bmatrix},$$

where u denotes the control input, and

- the domain is $X_D = \{\mathbf{x} \in \mathbb{R}^3 \mid -3.0 \leq x_1 \leq 0.0, -3.0 \leq x_2 \leq 0.0, -3.0 \leq x_3 \leq 0.0\}$,
- the initial region is $X_I = \{\mathbf{x} \in \mathbb{R}^3 \mid -1.5 \leq x_1 \leq 0.5, -1.5 \leq x_2 \leq 0.5, -2.5 \leq x_3 \leq -1.5\}$,
- and the unsafe region is $X_U = \{\mathbf{x} \in \mathbb{R}^3 \mid (x_1 + 2.5)^2 + (x_2 + 2.5)^2 + (x_3 + 0.5)^2 \leq 0.25\}$.

We utilized [3,64,1] network architecture for the barrier certificate and the controller networks, our tool synthesized correct and verified controller in just 3.34 s, which is faster than the MAML (3.64 s), as illustrated in Fig. 3. In this figure, the yellow surface represents the zero level set of learned barrier certificate, seperating the red sphere (X_U) and the green cuboid (X_I). Note that FOSSIL 2.0 failed to synthesize a safe controller.

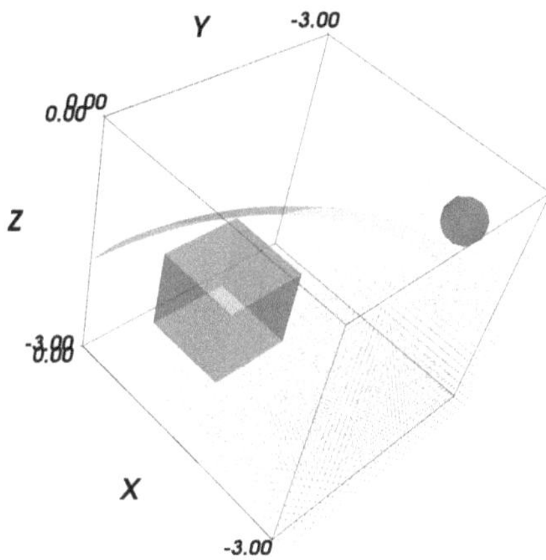

Fig. 3. Adapted and verified NN controller and barrier certificate for Example 2.

Table 1 provides a more comprehensive comparison of our method, the MAML based approach, and FOSSIL 2.0. Through fifteen benchmarks, it demonstrates that our method can achieve higher synthesis efficiency and success rates for safe controller synthesis with fewer training samples and gradient steps.

In terms of synthesis success rate, HyML-ASCS successfully synthesized safety controllers for all systems, while FOSSIL 2.0 failed to synthesize a safety controller for 3D-C4, and MAML failed to do so for 4D-C3. This is because HyML-ASCS, through model-based meta-learning, captures the shared patterns of system dynamics and safety constraints across different synthesis tasks, generating task-specific embeddings. This enables the model to adapt more quickly to new tasks, thereby improving the success rate of synthesis.

With respect to synthesis efficiency, HyML-ASCS shows significant improvements over the other two tools. The average number of iterations for HyML-ASCS is 1.4, compared to 1.8 for MAML and 3.4 for FOSSIL 2.0. This is because, in FOSSIL 2.0, the synthesis rate of the safety controller is affected by the initial parameters of the neural network. The use of random initialization in training leads to slower convergence. In contrast, HyML-ASCS generates task-specific embeddings through model-based meta-learning, enabling the gradient-based meta-learning to complete the synthesis with fewer data and fewer gradient update steps. As a result, our method requires fewer iterations and shorter synthesis time.

HyML-ASCS demonstrates stronger scalability compared to the other two tools. In four-dimensional systems, the average time for HyML-ASCS to synthesize the safety controller is 174.892 s, achieving a 68% efficiency improvement over FOSSIL 2.0 and an 83% improvement over MAML. Moreover, MAML failed to generate a safety controller for 4D-C3 within the specified time. This is attributed to the model's ability to recognize shared patterns early in the meta-learning phase, accelerating convergence. Additionally, the task-specific embeddings and gradient-based updates allow the framework to rapidly adapt to complex, high-dimensional systems, enabling it to efficiently handle large-scale tasks.

6   Conclusion

This paper proposes a hybrid meta learning framework for adaptive safety controller synthesis (HyML-ASCS), to address the challenges of synthesizing adaptive and safe controllers for nonlinear dynamical systems. Many safety-critical systems require controllers that can adapt to changing system parameters, but existing methods often fail to maintain performance under such conditions, limiting their practical applications. HyML-ASCS combines task-specific embeddings from model-based methods with rapid gradient-based updates from gradient-based methods, achieving a balance between efficient task adaptation and computational cost. Moreover, HyML-ASCS is designed to support fast adaptation to dynamic environments, addressing the limitations of traditional methods in scenarios with varying system parameters. Experimental results on several benchmark examples demonstrate that HyML-ASCS excels in adapting to changing system parameters and significantly outperforms state-of-the-art methods. Future work will focus on extending HyML-ASCS to higher-dimensional systems and developing strategies to reduce computational costs, enhancing its efficiency for complex and resource-constrained applications.

References

1. Ames, A.D., Coogan, S., Egerstedt, M., Notomista, G., Sreenath, K., Tabuada, P.: Control barrier functions: theory and applications. In: 18th European Control Conference (ECC), pp. 3420–3431. IEEE (2019)
2. Edwards, A., Peruffo, A., Abate, A.: Fossil 2.0: formal certificate synthesis for the verification and control of dynamical models. In: Proceedings of 27th ACM International Conference on Hybrid Systems: Computation and Control, pp. 1–10 (2024)
3. Tuncali, C.E., Kapinski, J., Ito, H., Deshmukh, J.V.: Reasoning about safety of learning-enabled components in autonomous cyber-physical systems. In: Proceedings of 55th Annual Design Automation Conference, pp. 1–6. ACM (2018)
4. Chinelato, C.I.G., Neves, G.P.D., Angélico, B.A.: Safe control of a reaction wheel pendulum using control barrier function. IEEE Access **8**, 160315–160324 (2020)
5. Squires, E., Pierpaoli, P., Egerstedt, M.: Constructive barrier certificates with applications to fixed-wing aircraft collision avoidance. In: IEEE Conference on Control Technology and Applications (CCTA), pp. 1656–1661. IEEE (2018)
6. Hospedales, T., Antoniou, A., Micaelli, P., Storkey, A.: Meta-learning in neural networks: a survey. IEEE Trans. Pattern Anal. Mach. Intell. **44**(9), 5149–5169 (2021)
7. Achille, A., et al.: Task2vec: task embedding for meta-learning. In: Proceedings of IEEE/CVF International Conference on Computer Vision, pp. 6430–6439 (2019)
8. Finn, C., Abbeel, P., Levine, S.: Model-agnostic meta-learning for fast adaptation of deep networks. In: International Conference on Machine Learning, pp. 1126–1135. PMLR (2017)

9. Xu, Z., Chen, X., Cao, L.: Fast task adaptation based on the combination of model-based and gradient-based meta learning. IEEE Trans. Cybern. **52**(6), 5209–5218 (2020)
10. Prajna, S., Jadbabaie, A., Pappas, G.J.: A framework for worst-case and stochastic safety verification using barrier certificates. IEEE Trans. Autom. Control **52**(8), 1415–1428 (2007)
11. Prajna, S., Jadbabaie, A.: Safety verification of hybrid systems using barrier certificates. In: Alur, R., Pappas, G.J. (eds.) HSCC 2004. LNCS, vol. 2993, pp. 477–492. Springer, Heidelberg (2004). https://doi.org/10.1007/978-3-540-24743-2_32
12. Sun, X., Khedr, H., Shoukry, Y.: Formal verification of neural network controlled autonomous systems. In: Proceedings of 22nd ACM International Conference on Hybrid Systems: Computation and Control, pp. 147–156 (2019)
13. Ivanov, R., Carpenter, T.J., Weimer, J., Alur, R., Pappas, G.J., Lee, I.: Verifying the safety of autonomous systems with neural network controllers. ACM Trans. Embed. Comput. Syst. **20**(1), 1–26 (2020)
14. Saveriano, M., Lee, D.: Learning barrier functions for constrained motion planning with dynamical systems. In: IEEE/RSJ International Conference on Intelligent Robots and Systems (IROS), pp. 112–119. IEEE (2019)
15. Srinivasan, M., Dabholkar, A., Coogan, S., Vela, P.A.: Synthesis of control barrier functions using a supervised machine learning approach. In: IEEE/RSJ International Conference on Intelligent Robots and Systems (IROS), pp. 7139–7145. IEEE (2020)
16. Taylor, A., Singletary, A., Yue, Y., Ames, A.: Learning for safety-critical control with control barrier functions. In: Learning for Dynamics and Control, pp. 708–717. PMLR (2020)
17. Wang, L., Theodorou, E.A., Egerstedt, M.: Safe learning of quadrotor dynamics using barrier certificates. In: IEEE International Conference on Robotics and Automation (ICRA), pp. 2460–2465. IEEE (2018)
18. Zhao, H., Zeng, X., Chen, T., Liu, Z., Woodcock, J.: Learning safe neural network controllers with barrier certificates. Formal Aspects Comput. **33**, 437–455 (2020)
19. Younger, A.S., Hochreiter, S., Conwell, P.R.: Meta-learning with backpropagation. In: International Joint Conference on Neural Networks (IJCNN), vol. 3. IEEE (2001)
20. Vuorio, R., Sun, S.H., Hu, H., Lim, J.J.: Toward multimodal model-agnostic meta-learning. arXiv preprint arXiv:1812.07172 (2018)
21. Nichol, A.: On first-order meta-learning algorithms. arXiv preprint arXiv:1803.02999 (2018)
22. Hochreiter, S., Younger, A.S., Conwell, P.R.: Learning to learn using gradient descent. In: Dorffner, G., Bischof, H., Hornik, K. (eds.) ICANN 2001. LNCS, vol. 2130, pp. 87–94. Springer, Heidelberg (2001). https://doi.org/10.1007/3-540-44668-0_13
23. Kim, T., Yoon, J., Dia, O., Kim, S., Bengio, Y., Ahn, S.: Bayesian model-agnostic meta-learning. arXiv preprint arXiv:1806.03836 (2018)
24. Nagabandi, A., et al.: Learning to adapt in dynamic, real-world environments through meta-reinforcement learning. arXiv preprint arXiv:1803.11347 (2018)
25. Huang, P.S., Wang, C., Singh, R., Yih, W., He, X.: Natural language to structured query generation via meta-learning. arXiv preprint arXiv:1803.02400 (2018)
26. Shi, G., Azizzadenesheli, K., O'Connell, M., Chung, S.J., Yue, Y.: Meta-adaptive nonlinear control: theory and algorithms. Adv. Neural. Inf. Process. Syst. **34**, 10013 10025 (2021)

27. Jena, A., Kalathil, D., Xie, L.: Meta-learning-based adaptive stability certificates for dynamical systems. In: Proceedings of AAAI Conference on Artificial Intelligence, vol. 38, no. 11, pp. 12801–12809 (2024)
28. Santoro, A., Bartunov, S., Botvinick, M., Wierstra, D., Lillicrap, T.: Meta-learning with memory-augmented neural networks. In: International Conference on Machine Learning, pp. 1842–1850. PMLR (2016)
29. Mishra, N., Rohaninejad, M., Chen, X., Abbeel, P.: A simple neural attentive meta-learner. arXiv preprint arXiv:1707.03141 (2017)
30. Zintgraf, L., Shiarli, K., Kurin, V., Hofmann, K., Whiteson, S.: Fast context adaptation via meta-learning. In: International Conference on Machine Learning, pp. 7693–7702. PMLR (2019)
31. Park, J., Berto, F., Jamgochian, A., Kochenderfer, M.J., Park, J.: Meta-sysid: a meta-learning approach for simultaneous identification and prediction. arXiv preprint arXiv:2206.00694 (2022)
32. Brockman, G.: OpenAI Gym. arXiv preprint arXiv:1606.01540 (2016)
33. Wang, Y., et al.: Joint differentiable optimization and verification for certified reinforcement learning. In: Proceedings of 14th ACM/IEEE International Conference on Cyber-Physical Systems, pp. 132–141 (2023)
34. Zhao, H., Zeng, X., Qi, N., Yang, Z., Zeng, Z.: Safe DNN-type controller synthesis for nonlinear systems via meta reinforcement learning. In: 60th ACM/IEEE Design Automation Conference (DAC), pp. 1–6. IEEE (2023). https://doi.org/10.1109/DAC56929.2023.10247837
35. Deshmukh, J.V., Kapinski, J.P., Yamaguchi, T., Prokhorov, D.: Learning deep neural network controllers for dynamical systems with safety guarantees. In: IEEE/ACM International Conference on Computer-Aided Design (ICCAD), pp. 1–7. IEEE (2019)
36. Zhu, H., Xiong, Z., Magill, S., Jagannathan, S.: An inductive synthesis framework for verifiable reinforcement learning. In: Proceedings of 40th ACM SIGPLAN Conference on Programming Language Design and Implementation, pp. 686–701 (2019)
37. Chesi, G.: Computing output feedback controllers to enlarge the domain of attraction in polynomial systems. IEEE Trans. Autom. Control **49**(10), 1846–1853 (2004)
38. Clavière, A., Asselin, E., Garion, C., Pagetti, C.: Safety verification of neural network controlled systems. In: IEEE/IFIP International Conference on Dependable Systems and Networks Workshops (DSN-W), pp. 47–54. IEEE (2021)
39. Jarvis-Wloszek, Z.W.: Lyapunov based analysis and controller synthesis for polynomial systems using sum-of-squares optimization. PhD Thesis, University of California, Berkeley (2003)

Proof Minimization in Neural Network Verification

Omri Isac[1]([✉]), Idan Refaeli[1], Haoze Wu[2], Clark Barrett[3], and Guy Katz[1]

[1] The Hebrew University of Jerusalem, Jerusalem, Israel
omri.isac@mail.huji.ac.il
[2] Amherst College, Amherst, USA
[3] Stanford University, Stanford, USA

Abstract. The widespread adoption of deep neural networks (DNNs) requires efficient techniques for verifying their safety. DNN verifiers are complex tools, which might contain bugs that could compromise their soundness and undermine the reliability of the verification process. This concern can be mitigated using *proofs*: artifacts that are checkable by an external and reliable proof checker, and which attest to the correctness of the verification process. However, such proofs tend to be extremely large, limiting their use in many scenarios. In this work, we address this problem by minimizing proofs of unsatisfiability produced by DNN verifiers. We present algorithms that remove facts which were learned during the verification process, but which are unnecessary for the proof itself. Conceptually, our method analyzes the dependencies among facts used to deduce UNSAT, and removes facts that did not contribute. We then further minimize the proof by eliminating remaining unnecessary dependencies, using two alternative procedures. We implemented our algorithms on top of a proof producing DNN verifier, and evaluated them across several benchmarks. Our results show that our best-performing algorithm reduces proof size by 37%–82% and proof checking time by 30%–88%, while introducing a runtime overhead of 7%–20% to the verification process itself.

1 Introduction

Deep neural networks (DNNs) have made great strides in recent years, becoming the state-of-the-art solution for a variety of tasks in medicine [55], autonomous driving [12], natural language processing [53], and many other domains. However, DNNs lack structure that humans can readily interpret, rendering them opaque and potentially jeopardizing their trustworthiness [19]. One prominent example is the sensitivity of DNNs to small input perturbations, which can lead to significant and undesirable changes to the networks' outputs. This sensitivity can be exploited maliciously [64], possibly leading to catastrophic results.

The pervasiveness of DNNs, combined with their potential vulnerability, has made DNN verification a growing research field within the verification community [2,5,6,15,16,31,33,37,51,54,58,62,66,68,72]. Modern DNN verifiers

Y.-F. Chen et al. (Eds.): VMCAI 2026, LNCS 16417, pp. 99–124, 2026.
https://doi.org/10.1007/978-3-032-15700-3_6

typically use techniques such as SMT solving [1,8,44,52], abstract interpretation [31,33,51,62,69], LP solving [65], and combinations thereof. Many such verifiers are available [15,16], and some tools have even been applied to industrial case studies [3,30,48]. Given a DNN and a property over its input and outputs, DNN verifiers will typically attempt to either find an input to the network that violates the property or conclude that so such input exists.

Despite this success, one significant issue that the DNN verification community faces is related to the *reliability* of verifiers. One concern is that even state-of-the-art verifiers may contain implementation bugs. A second concern is that verifiers typically use floating-point arithmetic, which is crucial for scalability but could lead to rounding errors and numerical instability issues. These issues might be exploited to compromise a verifier's soundness [40,74], thus undermining user trust in these tools.

Due to their complexity, verifying the correctness of DNN verifiers themselves is typically infeasible. Instead, DNN verifiers can produce *proofs* for their verification results, as is commonly done in SAT [36] and SMT [9]. These proofs, which serve as witnesses of correctness of the verification result, constitute a mathematical object and can be checked by an external, trusted *proof checker* [25,26]. When a violation of the property is discovered, the returned counter-example may serve as a proof of correctness. However, proving the absence of a violation is less straightforward, due to the NP-hardness of the problem [39,44,57].

So far, only a handful of attempts have been made to enable DNN verifiers to produce proofs [38]. These approaches typically include eager bookkeeping of many intermediate lemmas deduced during verification [38]. Therefore, the resulting proofs tend to be extremely large, increasing the memory consumption of the verifier and harming the performance of proof checkers [25]. These problems limit the applicability of proof-producing verifiers, and call for further analysis to minimize proof sizes.

In this work, we address this problem by minimizing the proof objects constructed using the framework of Isac et al. [38]. Our proof reduction method involves two main steps. First, we apply a *dependency analysis* procedure that recursively examines the dependencies of each lemma on previously learned lemmas. This helps to identify and remove lemmas that were eagerly learned and stored during the verification process, but are not actually required in the final proof. Second, we apply two alternatives for *dependency minimization* procedures, which further refine the proof by removing unnecessary lemma dependencies, keeping only a minimal subset of dependencies for each lemma. This results in additional lemmas being classified as unused, leading to a smaller proof.

After the termination of the dependency analysis and minimization, the unused lemmas can be safely removed from the proof, without needing to be checked. Furthermore, our method allows lemma deletion on-the-fly, thus avoiding accumulation of unused lemmas and minimizing the memory consumption during the verification process. This, in turn, increases the scalability of proof producing DNN verifiers.

We evaluated our method across multiple benchmarks, using [38] as a baseline. We measured the effectiveness of our algorithms along two axes: the size of the resulting proof, and the time overhead for applying our algorithms. The results for our best-performing algorithm indicate reduction of proof size of 37%–82% on average and of proof checking time by 30%–88% on average. The reduction incurs a overhead of 7%–20% to the verifier's runtime, where for some benchmarks the average verification time was slightly improved, possibly due to the use of more efficient data structures.

Our contributions are summarized as follows: (i) an algorithm for analyzing the dependencies of each lemma learned within proof objects, removing unused lemmas; (ii) two algorithms for further detecting a minimal subset of dependencies; and (iii) a demonstration of a substantial reduction of the proof size over several benchmarks, while adding a reasonable overhead in performance speed.

The rest of this paper is organized as follows: In Sect. 2 we provide the necessary background on DNNs and their verification, and on proof production. In Sect. 3 we explain our method to reduce the proof size. Section 4 is dedicated to evaluation of our method over several benchmarks, with respect to both proof size and verification speed. In Sect. 5 we discuss related work, and lastly, we conclude and outline ideas for future research in Sect. 6.

2   Background

2.1   Deep Neural Networks and Their Verification

Deep Neural Networks (DNNs) [32]. Formally, a DNN $\mathcal{N} : \mathbb{R}^n \to \mathbb{R}^m$ is a sequence of k layers $L_0, ..., L_{k-1}$ where each layer L_i consists of $s_i \in \mathbb{N}$ nodes $v_i^1, ..., v_i^{s_i}$ and s_i biases $b_i^j \in \mathbb{Q}$. Each directed edge in the DNN is of the form (v_{i-1}^l, v_i^j) and is labeled with a weight $w_{i,j,l} \in \mathbb{Q}$. The assignment to the nodes in the input layer is defined by $v_0^j = x_j$, where $\overline{x} \in \mathbb{R}^n$ is the input vector. The assignment for the jth node in the $1 \leq i < k - 1$ layer is computed as

$$v_i^j = f_i \left(\sum_{l=1}^{s_{i-1}} w_{i,j,l} \cdot v_{i-1}^l + b_i^j \right)$$ for some activation function $f_i : \mathbb{R} \to \mathbb{R}$. Neurons in the output layer are computed similarly, where f_{k-1} is the identity function.

Onc of thc most common activation functions is the *rectified linear unit* (ReLU), defined as $\text{ReLU}(x) := \max(x, 0)$. The function has two linear phases: if the input is non-negative, the functions is *active*; otherwise, the function is *inactive*. For simplicity, we focus on DNNs with ReLU activations, though our work can be extended to support any piecewise-linear activation (e.g., *maxpool*).

Example 1. *A simple DNN with four layers appears in Fig. 1. For simplicity, all biases are set to 0 and are omitted. For input $\langle 2, 1 \rangle$, the node in the second layer evaluates to $ReLU(2 \cdot 1 + 1 \cdot (-1)) = ReLU(1) = 1$; the nodes in the third layer evaluate to $ReLU(1 \cdot (-2)) = 0$ and $ReLU(1 \cdot 1) = 1$; and the node in the output layer evaluates to $0 + 1 \cdot 2 = 2$.*

The DNN Verification Problem. Consider a DNN $\mathcal{N} : \mathbb{R}^n \to \mathbb{R}^m$ where $\overline{x} \in \mathbb{R}^n, \overline{y} \in \mathbb{R}^m$ denote the network's inputs and outputs. The *DNN verification problem* is the problem of deciding whether there exist $(\overline{x}, \overline{y}) \in \mathbb{R}^{n+m}$ such that

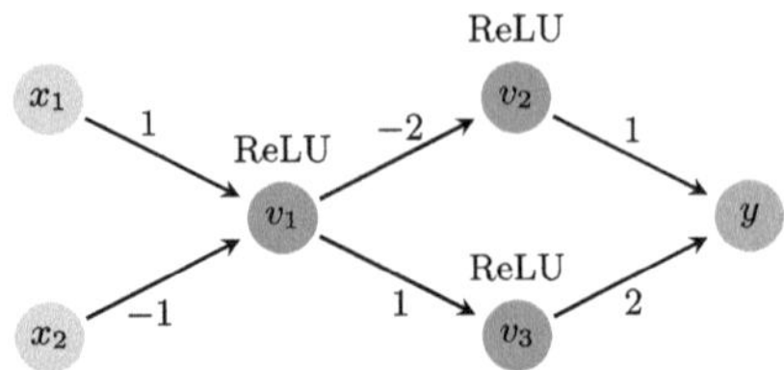

Fig. 1. A toy DNN with ReLU activations.

$(\mathcal{N}(\overline{x}) = \overline{y}) \wedge \varphi_I(\overline{x}) \wedge \varphi_O(\overline{y})$ for some property $\varphi_I \wedge \varphi_O$. If such $\overline{x}, \overline{y}$ exist, we say the problem is satisfiable, denoted **SAT**. Otherwise, we say that it is unsatisfiable, denoted **UNSAT**. In this work, we consider *quantifier-free linear properties*, i.e., formulas without quantifiers whose atoms are linear equations and inequalities.

Example 2. *Consider the DNN from Fig. 1, the input property*

$$\varphi_I(\overline{x}) := (1 \leq x_1) \wedge (1 \leq x_2) \wedge (x_1 \leq 2) \wedge (x_2 \leq 2)$$

and the output property

$$\varphi_O(y) := (y \leq -1).$$

*This instance is **UNSAT**, as the output is a sum of two non-negative components with positive weights.*

Linear Programming (LP) [24] **and DNN Verification.** In LP we seek an assignment to a set of variables that satisfies a set of linear constraints, while maximizing a given objective function. As is common in the context of DNN verification, we assume here that the objective function is constant and ignore it. Formally, let $V = [x_1, \ldots, x_n]^{\mathsf{T}} \in \mathbb{R}^n$ denote variables, $A \in M_{m \times n}(\mathbb{R})$ denote a constraint matrix and $l, u \in (\mathbb{R} \cup \{\pm\infty\})^n$ denote vectors of bounds. The LP problem is to decide the satisfiability of $(A \cdot V = 0) \wedge (l \leq V \leq u)$. Throughout the paper, we use $l(x_i)$ and $u(x_i)$, to refer to the lower and upper bounds (respectively) of $x_i \in V$, and denote $l \leq u$ to indicate that for each element i, $l(x_i) \leq u(x_i)$.

A DNN verification query can be encoded as a tuple $Q = \langle V, A, l, u, R \rangle$, where $\langle V, A, l, u \rangle$ constitutes a linear program [24,44] that represents the property and the affine transformations within $\mathcal{N}$, and where R is the set of ReLU activation constraints of the form $f_i = \mathrm{ReLU}(b_i)$, for variables $b_i, f_i \in V$. In addition, a fresh auxiliary variable a_i is added for each ReLU constraint, representing the non-negative difference of its output and input. The variable is added to V, with an equation $a_i = f_i - b_i$ added to A, and the bounds $l(a_i) = 0$, $u(a_i) = u(f_i) - l(b_i)$ added to l, u respectively.

Example 3. *We demonstrate this encoding with the example presented in Fig. 1 and the property from Example 2. Let x_1, x_2 denote the network's inputs and y denote its output, and let b_i, f_i ($i \in \{1, 2, 3\}$) denote the input and output of* *neuron v_i, respectively. The affine transformations of the network are reduced to the equations:*

$$x_1 - x_2 - b_1 = 0$$
$$2f_1 + b_2 = 0$$
$$f_1 - b_3 = 0$$
$$f_2 + 2f_3 - y = 0$$

accompanied by $f_i - b_i - a_i = 0$ for $i \in \{1, 2, 3\}$. The property is encoded using the inequalities: $(1 \leq x_1), (1 \leq x_2), (x_1 \leq 2), (x_2 \leq 2), (y \leq -1)$. We also add the inequalities $(0 \leq f_i)$ (for $i \in \{1, 2, 3\}$) to express the non-negativity of ReLUs, and $(0 \leq a_i)$ as described. The aforementioned constraints form the LP part of the query; and its piecewise-linear portion is comprised of the constraints $f_i = ReLU(b_i)$ for $i \in \{1, 2, 3\}$.

Using this encoding, the verification query can be dispatched through a series of invocations of an LP solver, combined with *case splitting* to handle the piecewise-linear constraints [44]. Schematically, the DNN verifier begins by invoking the LP solver over the linear part of the query. An UNSAT result of the LP solver implies the overall query is UNSAT. Otherwise, the LP solver finds an assignment that satisfies the linear part of the query. If this assignment satisfies the piecewise-linear part of the query, the DNN verifier may conclude SAT. If neither case applies, the DNN verifier splits the query into two subqueries, by deciding the phase of a single ReLU constraint $f_i = ReLU(b_i)$. One subquery is augmented with the bounds $b_i \geq 0$ and $a_i \leq 0$ corresponding to the active phase; the other is augmented with $b_i \leq 0$ and $f_i \leq 0$, corresponding to the inactive phase. Then, the LP solver is invoked again over each subquery, and the process is repeated. Note that the introduction of auxiliary variables is intended to reduce case splitting to bound updates, without needing to add new equations; empirically, this is known to improve performance [70].

This splitting approach creates an abstract tree structure, where each node represents a case split. If the LP solver deduces UNSAT for each leaf of the tree, then the original query is UNSAT as well. If not, then due to the completeness of LP solvers, it finds a satisfying assignment for one of the subqueries.

The search tree may be exponentially large in the number of piecewise linear constraints (i.e., the number of neurons). Therefore, it is common to use case splitting rarely, and apply algorithms to deduce tighter bounds of variables. These algorithms may conclude in advance that some neurons' phases are fixed, i.e., they are always active or inactive, avoiding the need to split on them [44].

2.2   Proof Production for DNN Verification

When DNN verification is regarded as a satisfiability problem, a satisfying assignment is an efficient proof of SAT: one can simply run it through the network and observe that it satisfies the given property. The UNSAT case is more complex, and requires producing a *proof certificate*. The first method for proof production in DNN verification performed via LP solving and case splitting was introduced in [38]. These proof certificates consist of a *proof tree* that replicates the search tree created by the verifier, and is constructed during the verification process. Recall that for an UNSAT query, all leaves of the search tree represent UNSAT subqueries. Thus, each internal node of the proof tree represents a case split performed by the verifier, and each leaf corresponds to a subquery for which the verifier has deduced UNSAT. Therefore a *proof checker* for these proofs is required to traverse the tree, check its structural correctness (i.e., all splits are correct), and certify the unsatisfiability of each leaf.

In each leaf of the search tree, the verifier was able to conclude UNSAT using solely the available linear constraints. Thus, a proof for the leaf's unsatisfiability can be described as a vector, according to the Farkas lemma [21], which combines the linear constraints to derive an evident, easily-checkable, contradiction [38]. More formally:

Theorem 1. (Farkas Lemma Variant (From [38]**)).**   *Observe the constraints $A \cdot V = \bar{0}$ and $l \leq V \leq u$, where $A \in M_{m \times n}(\mathbb{R})$ and $l, V, u \in \mathbb{R}^n$. The constraints are UNSAT if and only if $\exists w \in \mathbb{R}^m$ such that for $w^{\mathsf{T}} \cdot A \cdot V := \sum_{i=1}^{n} c_i \cdot x_i$, we have that $\sum_{c_i > 0} c_i \cdot u(x_i) + \sum_{c_i < 0} c_i \cdot l(x_i) < 0$ whereas $w^{\mathsf{T}} \cdot \bar{0} = 0$. Thus, w is a proof of the constraints' unsatisfiability, and can be constructed during LP solving.*

Example 4. *We demonstrate this theorem with our example. Consider the query based on the DNN in Fig. 1 and the property in Example 2. Assume, for simplicity, that all lower and upper bounds not explicitly stated are -1 or 1, respectively. Further assume that both v_2, v_3 are inactive (i.e. $f_2 = f_3 = 0$ and $b_2, b_3 \leq 0$). The linear part of the query yields the LP instance:*

$$
A = \begin{bmatrix}
1 & -1 & -1 & 0 & 0 & 0 & 0 & 0 & 0 & 0 & 0 & 0 \\
0 & 0 & 0 & 1 & 0 & 2 & 0 & 0 & 0 & 0 & 0 & 0 \\
0 & 0 & 0 & 0 & -1 & 1 & 0 & 0 & 0 & 0 & 0 & 0 \\
0 & 0 & 0 & 0 & 0 & 0 & 1 & 2 & 0 & 0 & 0 & -1 \\
0 & 0 & -1 & 0 & 0 & 1 & 0 & 0 & -1 & 0 & 0 & 0 \\
0 & 0 & 0 & -1 & 0 & 0 & 1 & 0 & 0 & -1 & 0 & 0 \\
0 & 0 & 0 & 0 & -1 & 0 & 0 & 1 & 0 & 0 & -1 & 0
\end{bmatrix}
$$

$$
\begin{aligned}
u &= \begin{bmatrix} 2 & 2 & 1 & 0 & 0 & 1 & 0 & 0 & 2 & 1 & 1 & -1 \end{bmatrix}^{\mathsf{T}} \\
V &= \begin{bmatrix} x_1 & x_2 & b_1 & b_2 & b_3 & f_1 & f_2 & f_3 & a_1 & a_2 & a_3 & y \end{bmatrix}^{\mathsf{T}} \\
l &= \begin{bmatrix} 1 & 1 & -1 & -1 & -1 & 0 & 0 & 0 & 0 & 0 & 0 & -1 \end{bmatrix}^{\mathsf{T}}
\end{aligned}
$$

Note that in A the first four rows corresponds to the affine transformations of the DNN, and the latter three rows are introduced with the auxiliary variables. Consider the vector

$$
w = \begin{bmatrix} 0 & 0 & -1 & -2 & 0 & 0 & 0 \end{bmatrix}^{\mathsf{T}}.
$$

The product $w^{\mathsf{T}} \cdot A \cdot V$ yields the equation $-f_1 + b_3 - 2f_2 - 4f_3 + 2y = 0$. The left hand side of the equation is at most: $-l(f_1) + u(b_3) - 2 \cdot l(f_2) - 4 \cdot l(f_3) + 2 \cdot u(y) = -2 < 0$. Therefore, according to Theorem 1, w is a proof of UNSAT. Note that there may be multiple proofs of this result. An overall proof of UNSAT for the query consists of a proof tree, where each leaf has a proof vector proving UNSAT for the corresponding subquery.

Bound Tightening Lemmas. DNN verifiers often employ bound tightening procedures that deduce bounds using the non-linear activations, called *bound tightening lemmas*. Each bound tightening lemma consists of a ground bound and a learned bound, which replaces a bound currently in l or u. As opposed to deduction of bounds using linear equations, proving such lemmas generally cannot be captured by the Farkas vector described above, and require separate proofs [38]. For example, given $f = \mathrm{ReLU}(b)$ and present bounds $f \leq 5$ and $b \leq 7$, it is possible to deduce that $b \leq 5$. The lemma can be proven using a Farkas vector that proves $f \leq 5$, and another proof rule that captures this form of bound derivation.

Example 5. *Consider the query Q in Example 4. We can use the equation $b_2 = -2f_1$, which is equivalent to $2f_1 + b_2 = 0$, to deduce that $u(b_2) = -2 \cdot l(f_1) = 0$. Then, based on the ReLU constraint, we deduce that the neuron v_2 is inactive, i.e., that $u(f_2) = 0$. A lemma with a ground bound $u(b_2) = 0$, learned bound $u(f_2) = 0$ and a proof vector for generating $b_2 = -2f_1$ (the vector $\begin{bmatrix} 0 & -1 & 0 & 0 & 0 & 0 & 0 \end{bmatrix}^{\mathsf{T}}$) is learned at the root of the proof tree, without any case splits. Similarly, we can deduce that v_3 is inactive. Observe that $f_3 = \frac{1}{2}y - \frac{1}{2}f_2$, and thus $u(f_3) = \frac{1}{2}u(y) - \frac{1}{2}l(f_2) = -0.5$. Then a lemma with a ground bound $u(f_3) = -0.5$, learned bound $u(b_3) = 0$ and a proof vector $\begin{bmatrix} 0 & 0 & 0 & -0.5 & 0 & 0 & 0 \end{bmatrix}^{\mathsf{T}}$ is learned. Combined with Example 4, we get a proof of UNSAT for Q that consists of a single node with two lemmas, as illustrated in Fig. 2.*

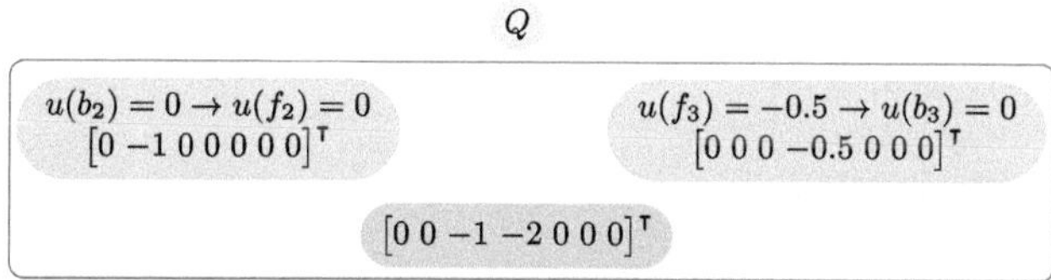

Fig. 2. A proof tree example with a single node (orange). Given the query Q, two lemmas (blue) are learned before deriving UNSAT (red). (Color figure online)

To avoid ambiguity in the definition of the bound vectors l, u, unless stated otherwise, l, u denote the tightmost bound vectors, while l_{input}, u_{input} denote the original bound vectors, given as inputs.

3   Proof Dependency Analysis and Minimization

In this section, we elaborate on our algorithms for proof minimization. Recall that each lemma and each leaf within the proof includes a proof vector; and that all these vectors have the same dimension, since the number of rows in A is constant. As proving UNSAT of all leaves in the proof tree is necessary for proving UNSAT for the whole query, proof minimization is achieved by reducing

106   O. Isac et al.

the number of lemmas used within the proof. The minimization algorithms are applied upon deducing UNSAT on each proof tree leaf, thus enabling on-the-fly minimization and preventing the accumulation of unnecessary lemmas. We begin by describing our algorithm for dependency analysis in Sect. 3.1 and two algorithms for dependency minimization in Sect. 3.2. We conclude in Sect. 3.3, by introducing an additional minimization technique, which can be used in both algorithms, by explaining the procedure for removing the unused lemmas and by introducing engineering optimizations to improve performance speed.

Algorithm 1. *proofDeps()*: Construct the dependency list of a proof vector

---

Input: A DNN verification subquery $Q = \langle V, A, l, u, R \rangle$, a list of learned lemmas Lem, and a proof of UNSAT w

Output: A list of lemmas sufficient to deduce UNSAT

    // Let $\sum_{i=1}^{n} c_i \cdot x_i$ denote the linear combination $w^\mathsf{T} \cdot A \cdot V$

1: $Deps \leftarrow \emptyset$ ▷ An empty list of lemmas
2: **for** $i \in [n]$ **do**
3:     **if** $c_i > 0$ and $u(x_i)$ is learned by $\ell_i^u \in Lem$ **then**
4:        $Deps \leftarrow Deps \cup \ell_i^u$
5:     **else if** $c_i < 0$ and $l(x_i)$ is learned by $\ell_i^l \in Lem$ **then**
6:        $Deps \leftarrow Deps \cup \ell_i^l$
7:     **end if**
8: **end for**
9: **for** $lem \in Deps$ **do**   ▷ Repeat recursively
10:     $lem.includeInProof \leftarrow true$
11:     $l', u' \leftarrow$ bounds with ID at most $lem.getID()$
12:     $w' \leftarrow lem.getProof()$
13:     $Deps' \leftarrow proofDeps(\langle V, A', l', u', R \rangle, Lem, w')$
14:     $Deps \leftarrow Deps \cup Deps'$
15: **end for**
16: **return** $Deps$

3.1 Analyzing Proof Dependencies

We now describe our algorithm for minimizing proof-trees by dependency analysis. Given a proof vector—either a proof of unsatisfiability for a search-tree leaf, or a proof for a bound tightening lemma—our analysis identifies which bound tightening lemmas are involved in its proof checking process. Our approach for analyzing the dependencies of a given proof vector w appears as Algorithm 1.

Conceptually, Algorithm 1 reconstructs the linear combination $w^\mathsf{T} \cdot A \cdot V$, denoted as $\sum_{i=1}^{n} c_i \cdot x_i$. Then, as the proof only uses the bounds $l(x_i)$ for variables with a negative coefficients, and $u(x_i)$ for variables with a positive coefficient, their corresponding lemmas form the *dependency list* for w. The dependency list provides a sufficient set of lemmas for proving the unsatisfiability of w—and this set can potentially be minimized. Then, the algorithm further analyses the dependencies of the discovered lemmas, only after we know they are used in the proof. Recall that each lemma has its own proof vector, and thus this analysis is carried out through a recursive call on each lemma's proof.

To expedite the algorithm, we store for every bound $l(x_i), u(x_i)$, the lemma or split that was used to deduce it. In addition, to avoid cycles and ensure the completeness of our algorithms, each lemma is given a unique, chronological ID. Whenever a lemma with ID k is analyzed, only lemmas with ID $l < k$ are considered in the analysis.

Example 6. *Recall that in Example 4 and Example 5, after learning two lemmas we are able to derive* **UNSAT** *from a proof vector* $w = \begin{bmatrix} 0 & 0 & -1 & -2 & 0 & 0 \end{bmatrix}^\mathsf{T}$. *Using this vector generates the equation* $-f_1 + b_3 - 2f_2 - 4f_3 + 2y = 0$, *which uses the bounds* $l(f_1), u(b_3), l(f_2), l(f_3), u(y)$ *to prove* **UNSAT**. *As* $u(b_3)$ *is deduced using one of the lemmas in Example 5, this lemma forms the dependency list of* w. *Then, the algorithm recursively generates the dependency list of the proof vector of the lemma in a similar manner. As shown in Example 5, that lemma is not dependent on* $u(f_2)$. *Thus, the lemma for deducing* $u(f_2)$ *is not actually used in the proof of* **UNSAT**, *and can then be removed from the proof tree. An illustration of this example appears in Fig. 3.*

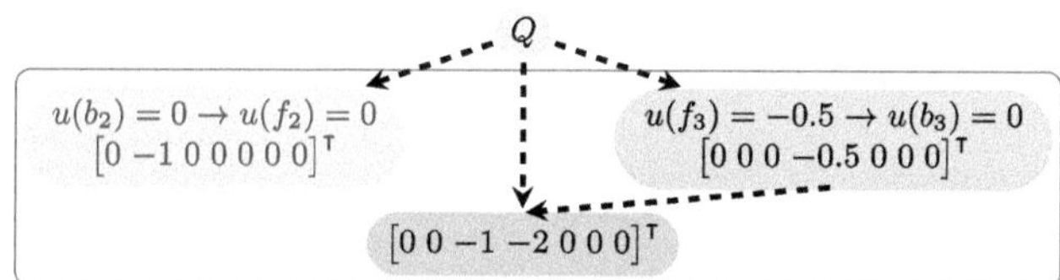

Fig. 3. Using Algorithm 1 over the proof vector used to derive **UNSAT** (red) shows that a single lemma (blue) is used during the proof process, and the other (gray) is not. All proof vectors use information in the root query Q (orange).(Color figure online)

3.2   Minimizing the Number of Proof Dependencies

We now elaborate on our two alternatives of algorithms for minimizing the dependency list of a proof vector. By doing so, we aim to increase the number of lemmas that do not appear in any dependency list, i.e., are unused. Such lemmas can then be omitted from the proof, further reducing its overall size.

Let w be a proof vector proving the unsatisfiability of a search-tree leaf. Our algorithm focuses on minimizing the number of bound tightening lemmas that participate in the refutation. Recall that a refutation is checked by computing the upper bound of $w^\mathsf{T} \cdot A \cdot V = \sum\limits_{i=1}^{n} c_i \cdot x_i$, i.e.,

$$\Delta := \sum_{c_i > 0} c_i \cdot u(x_i) + \sum_{c_i < 0} c_i \cdot l(x_i)$$

and ensuring it is negative. The idea of our algorithm lies in the fact that Δ can be very far from zero. In this case, we can relax some of the bounds derived from the participating bound lemmas, without altering the sign of Δ, and thus maintain the proof's correctness. Analyzing the dependencies of a proof vector $\hat{w}$, used to prove a bound tightening lemmas, is performed similarly. In this case, Δ represents the difference between the bound it is used to prove and the actual bound achieved with $\hat{w}^\mathsf{T} \cdot A \cdot V$.

Proof Minimization. The first algorithm for proof minimization is greedy, and removes the maximal number of unnecessary dependencies for each proof vector separately. To do so, we first define for each lemma its *contribution to* w, which can be seen as its effect on Δ. More formally, given the input bound vectors l_{input}, u_{input}, and the bound vectors l, u with bounds derived by a set of bound lemmas and case splits. For each $c_j > 0$, the contribution of $u(x_j)$ to $w^\mathsf{T} \cdot A \cdot V$ is defined as:

$$cont(u_j, w) := c_j \cdot (u(x_j) - u_{input}(x_j))$$

For each $c_j < 0$, the contribution of $l(x_j)$ to $w^\mathsf{T} \cdot A \cdot V$ is:

$$cont(l_j, w) := c_j \cdot (l_{input}(x_j) - l(x_j))$$

Note that in all cases the contribution is non-positive, as $l_{input} \leq l \leq u \leq u_{input}$.

These definitions naturally give rise to Algorithm 2, which sorts the various lemmas according to their contributions to the proof, from small to large. Those lemmas with only a small contribution could potentially be omitted from the proof, without altering the sign of Δ. Attempting to remove lemmas in this order ensures the removal of the maximum number of lemmas from the dependency list, and consequently the minimality of the output, as we prove in Theorem 2. Algorithm 2 can be used within Algorithm 1, before any recursive application of the algorithm. By that, Algorithm 2 reduces the time overhead of Algorithm 1, by reducing the number of the latter's recursive calls.

Theorem 2. *Given $\langle V, A, l, u, R \rangle$, a proof vector w, a list of its lemma dependencies D, and the input bound vectors l_{input}, u_{input}, Algorithm 2 returns a dependency subset of Deps that is minimal in size.*

Proof. Assume towards contradiction that there exists D', which constitutes a dependency list of w, with $|D'| < |D|$. First, observe that if $D' \subsetneq D$, then there is some lemma $\ell \in D \setminus D'$ that can be removed from D when considered in line 6, this contradicts the definition of the algorithm.

Therefore, $D \setminus D'$ and $D' \setminus D$ are not empty. Let $\ell \in D \setminus D'$, with the minimal contribution δ, and let $\ell' \in D' \setminus D$ with the minimal contribution δ'. Since Algorithm 2 is greedy, we must have that $\delta' \leq \delta$, as otherwise the algorithm would have removed ℓ from D before removing ℓ'. This means that we can replace ℓ' with ℓ in D' without increasing the overall contribution removed from *Deps*. We can repeat the process until saturation, i.e., when $D' \subsetneq D$, which leads to a contradiction. $\qquad\square$

Algorithm 2. *proofMin()*: Minimize the dependency list of a proof vector

Input: A DNN verification subquery $Q = \langle V, A, l, u, R \rangle$, a proof vector w, a list of its lemma dependencies $Deps$, and the input bound vectors l_{input}, u_{input}

Output: A minimial dependency list of w

// Let $\sum_{i=1}^{n} c_i \cdot x_i$ denote the linear combination $w^{\mathsf{T}} \cdot A \cdot V$

1: $\Delta \leftarrow \sum_{c_i > 0} c_i \cdot u(x_i) + \sum_{c_i < 0} c_i \cdot l(x_i)$
2: $Deps \leftarrow Deps.sortByContribution()$
3: $\delta \leftarrow 0$
4: **for** $dep \in Deps$ **do**
5:      $\delta{+}{=} |cont(dep.getBoundType(), w)|$ $\triangleright$ Use $A, V, l, u, l_{input}, u_{input}$
6:      **if** $\delta \leq |\Delta|$ **then**
7:          $Deps.remove(dep)$
8:      **else return** $Deps$
9:      **end if**
10: **end for**
11: **return** $\emptyset$

Example 7 *In Example 6 we saw that the proof vector $w = \begin{bmatrix} 0 & 0 & -1 & -2 & 0 & 0 \end{bmatrix}^{\mathsf{T}}$, presented in Example 4, is dependent on only one of the two lemmas learned as in Example 5. In addition, the analysis in Example 4 shows that $\Delta = -2$, and that the proof vector w is dependent on the lemma used to deduce $u(b_3) = 0$, whose input bound is $u(b_3) = 1$. Therefore, the contribution of the lemma is $1 \cdot (0 - 1) = -1$, indicating this lemma can be removed from the dependency list as well. Observe that indeed, given the input bounds, we have:*

$$- l(f_1) + u_{input}(b_3) - 2 \cdot l(f_2) - 4 \cdot l(f_3) + 2 \cdot u(y) = -1 < 0$$

Together with Example 6, we conclude that both lemmas from Example 5 can be removed from the proof tree. An illustration for this example appears in Fig. 4.

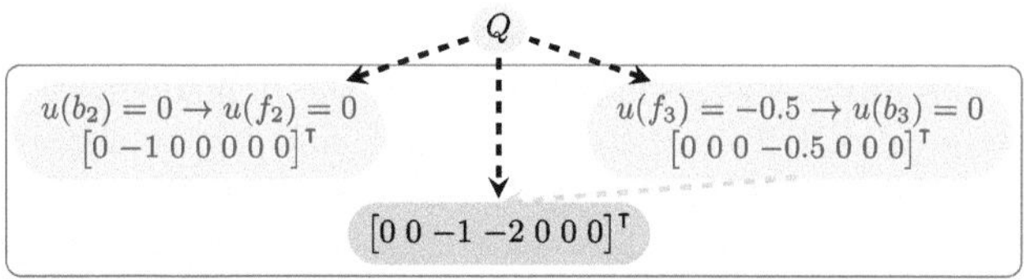

Fig. 4. Using Algorithm 2 over the proof vector used to derive UNSAT (red) shows that both lemmas (gray) are unnecessary for proving UNSAT.

Global Heuristic for Proof Minimization. Although Algorithm 2 guarantees achieving the minimum number of dependencies for each proof vector, this does not necessarily ensure achieving the global minimum across the entire proof tree. To incorporate global considerations, we adjust the order in which lemmas are considered for removal, prioritizing those with more dependencies and whose elimination is likely to trigger the removal of additional downstream lemmas.

Specifically, before applying the minimization algorithm, we recursively apply the dependency analysis procedure to each lemma and compute its total number of dependencies, including those discovered through the recursive applications of the minimization procedure. The lemmas are then sorted in decreasing order by their overall dependency count (i.e., lemmas with more dependencies are considered for removal first). In case of ties, we fall back on the previous contribution score. We apply the analysis until saturation, meaning no further reduction of dependencies is possible. This ensures minimality of the subset of dependencies, though it does not guarantee the resulting subset is smallest. This can occur if Algorithm 3 diverges significantly from Algorithm 2, for example, when lemmas with high dependency counts contribute heavily. In such cases, the former algorithm may remove these lemmas first, while the latter will instead remove many more lemmas with a lower dependency count. To maintain the applicability of the algorithm during proof generation, we apply it each time an UNSAT leaf is detected, considering all lemmas learned globally up to that point.

This intuition is formalized in Algorithm 3, and it opens up possibility to explore additional heuristics for prioritizing lemmas for removal, e.g., by considering different combinations of each lemma's number of dependencies and its contribution.

Note that introducing global considerations for proof minimization involves dependency analysis for many lemmas, and thus might incur a relatively large time overhead.

3.3   Additional Improvements and Lemma Deletion

Further Reduction of Proof Size. Both minimization algorithms presented above consider the dependencies for each proof vector separately. However, it is possible that minimizing dependencies at the proof-vector level leads to redundant dependencies at the proof-tree level. Consider the case where two proof vectors w_1, w_2 depend on a lemma ℓ, and that the dependencies of w_1 are minimized before those of w_2. If the dependency minimization of w_1 indicates that ℓ is retained in the proof, then removing it from the dependencies of w_2 will not lead to the removal of ℓ from the overall proof, and will unnecessarily increase the contributions counter δ Algorithm 2 (line 5). Therefore, a straightforward improvement to both minimization algorithms is to consider removing dependencies only from lemmas that are not currently used in the proof. This approach allows us to skip lemmas that are already essential and would be retained regardless.

Algorithm 3. *globProofMin()*: Minimize the dependency list of a proof vector

Input: A DNN verification subquery $Q = \langle V, A, l, u, R \rangle$, a proof vector w, a list of its lemma dependencies $Deps$, the input bound vectors l_{input}, u_{input}, and the overall list of learned lemmas Lem
Output: A minimial dependency list of w

$$// \text{ Let } \sum_{i=1}^{n} c_i \cdot x_i \text{ denote the linear combination } w^{\mathsf{T}} \cdot A \cdot V$$

1: $\Delta \leftarrow \sum_{c_i > 0} c_i \cdot u(x_i) + \sum_{c_i < 0} c_i \cdot l(x_i)$
2: $Deps \leftarrow Deps.sortByDepsLength(Q, Deps, Lem)$
3: $\delta \leftarrow 0$
4: $saturation \leftarrow false$
5: **while** $!saturation$ **do**
6:     $saturation \leftarrow true$
7:     **for** $dep \in Deps$ **do**
8:         $\delta + = |cont(dep.getBoundType(), w)|$ $\triangleright$ Use $A, V, l, u, l_{input}, u_{input}$
9:         **if** $\delta \leq |\Delta|$ **then**
10:             $Deps.remove(dep)$
11:             $saturation \leftarrow false$
12:         **else**
13:             $\delta - = |cont(dep.getBoundType(), w)|$ $\triangleright$ Revert
14:         **end if**
15:     **end for**
16: **end while**
17: **return** $Deps$

    sortByDepsLength($Q, Deps, Lem$):
1: **for** $\ell \in Deps$ **do**
2:     **if** $!\ell.wasAnalyzed$ **then**
3:         $\ell.score \leftarrow proofDeps(Q, Lem, \ell.getProof()).size()$
4:         $\ell.tieBreaker \leftarrow \ell.getContribution()$
5:         $\ell.wasAnalyzed \leftarrow true$
6:     **end if**
7: **end for**
8: **return** $Deps.sortByScore()$

Example 8 *Consider a lemma ℓ_1 with $\Delta_1 = -1$. Suppose that proof analysis of Algorithm 1 determines that ℓ_1 is dependent on ℓ_2, ℓ_3 with respective contributions of -0.5 and -0.6. Suppose also that other lemmas depend on ℓ_2, so it is retained in the proof, but none depend on l_3 yet. In this case, Algorithm 2 would first remove the dependency of ℓ_1 in ℓ_2 and then stop as $|-0.5| + |-0.6| > |-1|$. Therefore, this application of Algorithm 2 will keep both ℓ_2 and ℓ_3 in the proof. Alternatively, by ignoring ℓ_2 as it is already used in the proof, and removing the dependency on ℓ_3 instead, we may reduce the overall number of lemmas required for the proof.*

Avoiding Duplicate Computations. Recall that Algorithm 1 recursively iterates through lemmas and analyzes their dependencies. Since multiple lemmas can depend on the same lemma, it is useful to avoid redundant or repeated analysis. To address this, we add a flag to each lemma indicating whether it has already been analyzed. This ensures that each lemma is analyzed at most once.

Unused Lemma Deletion. To reduce the memory consumption of the DNN verifier during the verification process, we propose to remove unused lemmas as early as possible. Since each lemma is applicable only in the search state S where it was deduced and in all descendant states of S, it becomes irrelevant once the sub-tree rooted at S has been fully explored. At that point, any lemma in S that was not involved in any dependency analysis can be safely deleted.

Supporting Additional Constraints. Although so far we focused on DNNs with ReLU constraints, our minimization algorithms can be applied to additional kinds of constraints in a straightforward manner. Therefore, assuming the underlying verifier supports proof production for additional kinds of piecewise-linear constraints, the minimization algorithms are readily applicable.

4   Implementation and Evaluation

To assess the effectiveness of our proposed proof minimization techniques, we conducted an empirical evaluation across six benchmark suites in neural network verification (ordered alphabetically): (i) the ACAS-XU drone collision avoidance networks [41]; (ii) the CERSYVE benchmark [71] for verification of neural safety certificate in control systems; (iii) a subset of the CORA benchmark [15,47], where disjuncts from the original queries where separated into different queries; (iv) MAXLEAKY, a fresh benchmark that includes simple queries over a newly-trained DNN employing *Maxpool* and *LeakyRelu* [27] activation functions; (v) a robotic navigation system benchmark [4]; and, (vi) the SAFENLP benchmark [17,18]. These benchmarks were selected due to their relevance to safety-critical applications their diversity in size and activation functions, and their established use in previous work or at the DNN verification competition (VNNCOMP) [15,16]. Notably, we included all VNNCOMP benchmarks for which the proof-producing version of Marabou reasonably scales.

4.1   Experimental Setup and Implementation

Experiments were conducted on machines running Debian 12 Linux, each on a single CPU. We divided the benchmarks into two sets according to network sizes, with ACAS-XU and CORA consisting of larger DNNs, and the remaining benchmarks consisting of smaller DNNs. Memory and time configurations depended on category: for the larger benchamrks we used 16GB RAM and a 5-hour TIMEOUT, and for the smaller ones—2GB RAM and a 2.5-hour TIMEOUT.

Our implementation is based on the proof-producing version of the Marabou DNN verifier [38,46,70], into which we integrated our proof minimization algorithms and which we used as a baseline. The key metrics evaluated include (i) *Proof size:* the number of proof vectors (for both lemmas and leaves) in the produced proofs of UNSAT queries; (ii) *Total Runtime:* the total runtime of the verification process, for all queries; and (iii) *Total Proof-Checking time:* the total proof-checking time for the UNSAT queries, using the native Marabou proof checker. Note that a partial proof object is constructed during verification in satisfiable queries as well (before concluding SAT); and thus, measuring verification time for all queries offers a more comprehensive comparison.

4.2 Evaluation

We begin by presenting a summary of the experimental results in Table 1, which reports the average proof size generated during proof production, the average total verification time, and the average total proof-checking time, for each benchmark and proof minimization technique. In this table, averaging was conducted with respect to queries solved by all methods. Furthermore, we provide a visual summary in Fig. 5, which compares the proof sizes of all UNSAT queries, and overall runtime across all queries. We evaluated three variants of our method: (i) *Dependency Analysis* (**Dep. Analysis**), as described in Algorithm 1; (ii) *Dependency Minimization* (**Dep. Min.**), which includes the implementation of Algorithm 2 on top of Algorithm 1; and (iii) *Global Dependency Minimization* (**Dep. Min. Glob.**), which replaces Algorithm 2 with Algorithm 3. For a deeper study of the effect on the total runtime, we consider an additional running configuration, **Min. Baseline**, which includes all data structures used for implementing our algorithms as exemplified in Sect. 3.1, but without actually executing Algorithm 1. By running this version, we are able to separate the running time effect of the changes in data structures, from the effect of the analysis and minimization algorithms themselves. All methods were implemented with the optimizations detailed in Sect. 3.3.

The results demonstrate a clear reduction in average proof size, achieving a reduction rate of 37%–82%. The incurred verification time overhead is 5%–24% for **Dep. Min.** and 7%–20% for **Dep. Min. Glob.**. In addition, the results indicate a reduction in proof checking time of up to 88%. We observe that this reduction is not necessarily correlated with the reduction rate of proof sizes; and we hypothesize that this stems from the fact that checking time of each lemma is dependent on its sparsity, and that our approach retains sparser lemmas in the proof tree. Comparing the minimization algorithms, the results show that Algorithm 1 is responsible for most of the removed lemmas, and both Algorithm 2 and Algorithm 3 improve it further with generally similar impact, except for the CORA benchmarks. However, both algorithms incur a small overhead comparing to using Algorithm 1 alone. Analyzing **Min. Baseline** indicates that changes in data structures contributed to a minor improvement in both verification time and proof-checking time, for most benchmarks. This has moderated the time overhead of the minimization algorithms. As expected, bookeeping more information slightly increased the number of overall MEMOUTs of **Min. Baseline** compared to **Baseline**. When applying the minimization algorithms however, the number of MEMOUTs decreases significantly, due to the reduction of proof size (see subsections below).

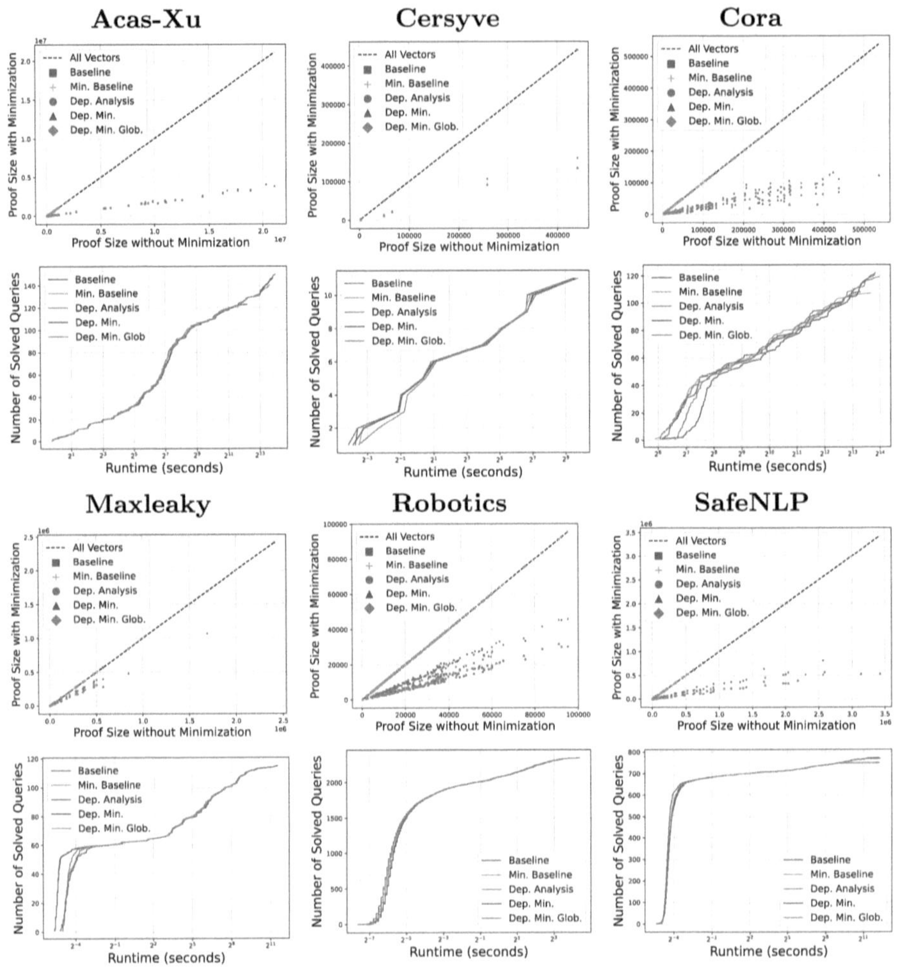

Fig. 5. Proof size and runtime comparison across all benchmarks and methods.

In the subsequent subsections, we provide a more detailed analysis of the results for each benchmark suite.

Acas-Xu Benchmark. The ACAS-XU suite consists of fully connected feedforward networks trained to support drone collision avoidance decision-making. We ran experiments on a set of 45 networks, each comprised of 6 hidden layers, each of which contained 50 neurons with ReLU activation functions. For each network we verified 4 different specifications [44]. This results in a total of 180 queries. The comparison in Table 1 was conducted on 82 UNSAT and 39 SAT queries solved by all methods. Notably, a single query was either solved with a failure during proof certification, or terminated due to memory out. This result is consistent

Table 1. Average proof size (number of leaves and lemmas), verification time for all queries and proof-checking time for UNSAT queries, using Marabou across benchmarks. Lower values indicate more compact proofs. Each non-baseline entry includes both reported numbers, and percentage of change with respect to baseline. Best results among our three algorithms are marked in **bold**.

Benchmark	Avg. Proof Size [±% of Baseline]			
	Avg. Verification Time (Sec.) [±% of Baseline]			
	Avg. Checking Time (Sec.) [±% of Baseline]			
	Baseline[38] Min.	Baseline	Dep. Analysis	Dep. Min.	Dep. Min. Glob.
Acas-Xu	302021.14	302021.14 0%	59231.07 -80.4%	**52890.17** **-82.5%**	53144.84 -82.4%
	265.98	269.42 +1.2%	299.96 +12.8%	294.58 +10.7%	**288.42** **+8.4%**
	66.37	46.58 -29.8%	8.67 -86.9%	**7.78** **-88.3%**	8.15 -87.7%
cersyve	163592.4	163592.4 0%	60764.6 -62.9%	**51283.6** **-68.6%**	51492.6 -68.5%
	86.99	84.84 -2.5%	96.90 +11.4%	**93.02** **+6.9%**	104.57 +20.2%
	13.77	13.66 -0.8%	3.14 -77.2%	**2.62** **-81%**	2.77 -79.9%
Cora	49202.05	49202.05 0%	15199.58 -69.1%	**8819.75** **-82.1%**	11549.88 -76.5%
	1277.88	1231.5 -3.6%	**1262.87** **-1.2%**	1583.02 +23.9%	1405.65 +10%
	1633.47	1572.66 -3.7%	455.88 -72.1%	**266.17** **-83.7%**	365.32 -77.6%
MaxLeaky	55342.88	55342.88 0%	42224.96 -23.7%	34852.52 -37%	**34847.12** **-37%**
	60.85	56.34 -7.4%	**59.01** **-3%**	59.7 -1.9%	60.01 -1.4%
	2.67	2.63 -1.5%	2.07 -22.5%	**1.89** **-29.2%**	**1.89** **-29.2%**
Robotics	25549.65	25549.65 0%	11981.7 -53.1%	**8016.64** **-68.6%**	8026.99 -68.6%
	0.55	0.53 -3.6%	**0.57** **+3.6%**	0.58 +5.4%	0.59 +7.3%
	0.32	0.31 -3.1%	0.15 -53.1%	**0.11** **-65.6%**	**0.11** **-65.6%**
SafeNLP	55066.71	55066.71 0%	17135.91 -68.9%	10118.98 -81.6%	**10062.95** **-81.7%**
	14.69	14.27 -2.8%	**14.73** **+0.3%**	14.94 +1.7%	15.88 +8.1%
	5.89	7.78 +32%	1.59 -73%	**0.96** **-83.7%**	0.97 -83.5%

with the original evaluation of the ACAS-XU benchmark in [38], and we hypothesize that for this particular verification instance, the proof-producing Marabou is numerically unstable.

To assess robustness under resource constraints, we also report the number of queries that were successfully solved within the 5-hour timeout and 16GB memory limit. Table 2 summarizes the number of solved queries for each method.

As shown, all three minimization methods outperform the baseline, solving more queries by avoiding memory blowup during proof construction.

Table 2. Number of Acas-Xu SAT queries solved, UNSAT queries solved, timed out, and failed due to memory exhaustion, out of 180 queries.

Method	#SAT	#UNSAT	#Timeout	#Memout
Baseline	41	82	0	57
Min. Baseline	39	82	0	59
Dep. Analysis	44	107	29	0
Dep. Min.	44	104	21	11
Dep. Min. Glob.	44	96	20	20

cersyve Benchmark. The CERSYVE suite consists of fully connected feedforward networks serving as neural certificates for control systems. We ran experiments on a set of 12 networks, with 65 to 198 ReLU neurons. For each network we verified a single property, as in the VNNCOMP 2025 benchmark [67]. This results in a total of 12 queries, out of which 6 are UNSAT. For this benchmark, all methods successfully verified 11 queries, whereas a single UNSAT query fails due to memory restrictions when tested on all methods.

Cora Benchmark. The CORA benchmark consists of local adversarial robustness queries for image classifiers trained on several datasets. For our experiments, we used a subset of these queries on a feedforward DNN with seven layers of 250 ReLU neurons each, trained on the MNIST dataset. To extend the benchmark and improve scalability, each query was further divided into nine sub-queries, each corresponding to a single possible misclassification option. In total, this yields 189 queries. The comparison in Table 1 was conducted on 104 UNSAT queries solved by all methods.

In Table 3, we report the number of instances verified under a time limit of 5 h and a memory limit of 16GB. As previously observed, the analysis and min-

Table 3. Number of CORA SAT queries solved, UNSAT queries solved, timed out, and failed due to memory exhaustion, out of 189 queries.

Method	#SAT	#UNSAT	#Timeout	#Memout
Baseline	1	107	81	0
Min. Baseline	1	106	81	1
Dep. Analysis	1	118	70	0
Dep. Min.	0	122	67	0
Dep. Min. Glob.	1	119	69	0

imization algorithms increase the total number of solved queries, this time primarily due to improvements in overall running time. Since the reduction in proof-checking time is more significant than the reduction in verification time (as shown in Table 1), we conclude that the overall improvement is mainly attributable to the reduced time of proof-checking.

MaxLeaky Benchmark. The MAXLEAKY benchmark includes a single DNN trained to classify wine types based on chemical composition [23]. To test our approach over a DNN which employs various activation functions, we trained a fresh DNN with 64 ReLU, 32 LeakyReLU, and 16 Maxpool neurons, to classify the dataset. We then generated a single adversarial robustness query for 133 train and 44 test data points. This results in a total of 177 queries. The comparison in Table 1 was conducted on 99 UNSAT and 12 SAT queries solved by all methods.

Table 4 includes the number of instances verified under the limitation of 2.5-hour time and 2GB memory. The results indicate a minor reduction in the number of solved queries when introducing the minimization algorithms. This result is unique, and is likely a result of the relatively moderate reduction of proof size (see Table 1), combined with the small overhead introduced when keeping additional information for each lemma (see Sect. 3.1).

Table 4. Number of MAXLEAKY SAT queries solved, UNSAT queries solved, timed out, and failed due to memory exhaustion, out of 177 queries.

Method	#SAT	#UNSAT	#Timeout	#Memout
Baseline	12	163	1	1
Min. Baseline	12	160	0	5
Dep. Analysis	12	160	1	4
Dep. Min.	12	162	0	3
Dep. Min. Glob.	12	160	0	5

Robotics Navigation Benchmark. The robotic navigation benchmark [4] involves DNN controllers for obstacle avoidance. We tested our approach on 2340 queries from this benchmark, where 2126 of them are SAT queries, and 214 are UNSAT. Each query includes a DNN with the ReLU activation only. The DNN is comprised of an input layer with 9 neurons, two hidden layers with 16 neurons each, and an output layer of 3 neurons. For this benchmark, all methods were able to verify all queries and produce proofs in the time and memory limitations.

SafeNLP Benchmark. The SAFENLP benchmark [17,18] examines robustness to word- and sentence-level perturbations in the embedding space of several sentences over two fully-connected DNNs. Both DNNs have the same architecture, which is comprised of an input layer of 30 nodes, a single hidden layer of 128 nodes with ReLU activation functions, and an output layer of 2 nodes. The benchmark consists of 1080 queries in total. The comparison in Table 1 was conducted on 149 UNSAT and 599 SAT queries solved by all methods.

Table 5 reports the number of queries successfully verified within the resource limits (2.5 h and 2GB RAM). Here again, our minimization methods demonstrate improved performance, reducing memory exhaustion and increasing the number of successful verification queries compared to the baseline.

Table 5. Number of safeNLP SAT queries solved, UNSAT queries solved, timed out, and failed due to memory exhaustion, out of 1080 queries.

Method	#SAT	#UNSAT	#Timeout	#Memout
Baseline	599	196	13	272
Min. Baseline	599	194	7	280
Dep. Analysis	599	214	36	231
Dep. Min.	599	219	54	208
Dep. Min. Glob.	599	215	42	224

5   Related Work

Producing proofs as witnesses of correctness is a common practice in both SMT and SAT solving. In both communities, the proof size tends to be significant, limiting their applicability [7,9,36,63]. Our work builds on concepts inspired by *lazy proofs* [45], where proof objects are constructed gradually, using only necessary lemmas. In DNN verification, proof production has been studied only recently, and in a handful of works [38,61]. The authors of [61] consider a symbolic approach for increasing the reliability of DNN verifiers, which reduces the problem to an additional SMT query. Our approach is then not directly applicable to [61].

The study of dependencies analysis of UNSAT results has been explored in various contexts [14,35,42,43,49] often using the name *UNSAT core*. To the best of our knowledge, ours is the first to investigate this in the context of DNN verification. In DNN verification, previous studies have attempted to analyze dependencies between the phases of neurons as part of the DNN verification process [13]. As fixing neuron phases could be captured by proof lemmas, it would be interesting to study the synergies between that approach and ours.

In linear programming, the well-established concept of *Irreducible Infeasible Systems* [20] is defined as a minimal set of constraints that maintain unsatisfiablity. The method to detect IISs from [20] has been used in DNN verifiers [29], and is implemented in the Gurobi linear optimizer [34] often used in DNN verifiers. Although IIS detection algorithms typically rely on repeated LP solver invocations, which can be computationally expensive, it presents a compelling direction for future exploration in this work's context.

6   Conclusion and Future Work

In this work, we developed methods to reduce the size of UNSAT proofs generated by DNN verifiers, as presented in [38]. These proofs can be checked by a trusted, external proof checker in order to improve reliability of the DNN verifier. Importantly, DNN verifiers currently assume ideal mathematical implementation of the DNN itself, and modeling any specific DNN implementations remains a challenge for the DNN verification community in general [22]. Our goal here is to improve memory consumption in proof-producing verifiers and reduce proof checking time, thereby enhancing the practicality of reliable DNN verification.

To this end, we designed an algorithm that analyzes, for each lemma in the proof, its dependencies on previously derived lemmas. This enables us to eliminate lemmas that are eagerly learned but are ultimately irrelevant for proving UNSAT. In addition, we designed an algorithm to minimize the number of such dependencies, with two variants. All algorithms have been implemented on top of the proof-producing version of the Marabou DNN verifier [70].

Our results show that Algorithm 1 reduces proof size by an order of magnitude across multiple benchmarks. The dependency minimization algorithms further reduces proof size, though more moderately. When comparing the two variants of the minimization algorithm, both versions achieve similar results on almost all benchmarks. Moreover, the performance overhead introduced by our methods is reasonable, especially when considering improvements in proof checking time. Notably, the time overhead from the dependency analysis is reduced due to the use of some alternative data structures.

Future Work. Moving forward, we intend to continue this work along several paths. First, we aim to explore different variants of proof minimization. In particular, our current algorithms consider a fixed proof vector w, which we could try to alter in order to reduce the number of dependencies even further. Additionally, it would be interesting to explore additional alternatives to the minimization algorithm, attempting to remove dependencies according to other heuristics, e.g., by their stage of deduction. Furthermore, it would also be interesting to consider the sparsity of the proof vectors and prioritize removing dense vectors, as they (heuristically) depend on many other bound lemmas.

Second, we aim to extend the proof checker as in [25], to support checking proofs of bound tightening lemmas. We will then evaluate the reduction of proof-checking time for this version of the checker as well.

A third direction for future research is extending our work to support additional subroutines commonly used in DNN verifiers, particularly abstract interpretation. This integration is inherently more complex, as it requires proving linear relaxations of non-linear constraints. However, once this support is established, we believe our proof minimization approach and technique could also be useful in this context.

Fourth, our work's analysis method paves the way for a *conflict analysis* method, as a part of *Conflict-Driven Clause Learning* [10,59,60] scheme, which is commonly used in SAT and SMT solving [8,11]. Conceptually, in search-based solving, CDCL identifies subspaces of the search space which are "similar" to subspaces already traversed, and which were shown not to contain any satisfying assignment. This similarity guarantees that the new subspaces also do not contain any such assignment, and they can be safely skipped by the verifier. A core component in CDCL is the generation of *conflict clauses*, a representation of an UNSAT subspaces, that guide the solver not to search in other subspaces that are also guaranteed to be UNSAT. Naturally, the algorithm for deriving conflict clauses is a key to a successful CDCL framework. Our work can serve as a basis for the derivation of potentially useful conflict clauses for DNN verifiers—leading to a significant improvement in their performance. Although several DNN verifiers employ CDCL or similar algorithms [28,29,50,73], to the best of our knowledge, none of these use UNSAT proofs for deriving conflict clauses.

Acknowledgments. The work of Isac, Refaeli and Katz was supported by the Binational Science Foundation (grant numbers 2020250 and 2021769), the Israeli Science Foundation (grant number 558/24) and the European Union (ERC, VeriDeL, 101112713). Views and opinions expressed are however those of the author(s) only and do not necessarily reflect those of the European Union or the European Research Council Executive Agency. Neither the European Union nor the granting authority can be held responsible for them.

The work of Barrett was supported in part by the Binational Science Foundation (grant number 2020250), the National Science Foundation (grant numbers 1814369 and 2211505), and the Stanford Center for AI Safety.

Data Availability Statement. The implementation and scripts used for the experiments in Sect. 4 are publicly available [56].

References

1. Abraham, E., Kremer, G.: SMT solving for arithmetic theories: theory and tool support. In: Proceedings of the 19th International Symposium on Symbolic and Numeric Algorithms for Scientific Computing (SYNASC), pp. 1–8 (2017)
2. Akintunde, M., Kevorchian, A., Lomuscio, A., Pirovano, E.: Verification of RNN-based neural agent-environment systems. In: Proceedings of the 33rd AAAI Conference on Artificial Intelligence (AAAI), pp. 197–210 (2019)
3. Amir, G., Freund, Z., Katz, G., Mandelbaum, E., Refaeli, I.: veriFIRE: verifying an industrial, learning-based wildfire detection system. In: Proceedings of the 25th International Symposium on Formal Methods (FM), pp. 648–656 (2023)
4. Amir, G., et al.: Verifying learning-based robotic navigation systems. In: Proceedings of the 29th International Conference on Tools and Algorithms for the Construction and Analysis of Systems (TACAS), pp. 607–627 (2023)
5. Avni, G., Bloem, R., Chatterjee, K., Henzinger, T., Konighofer, B., Pranger, S.: Run-time optimization for learned controllers through quantitative games. In: Proceedings of the 31st International Conference on Computer Aided Verification (CAV), pp. 630–649 (2019)
6. Baluta, T., Shen, S., Shinde, S., Meel, K., Saxena, P.: Quantitative verification of neural networks and its security applications. In: Proceedings of the 26th ACM Conference on Computer and Communication Security (CCS) (2019)

7. Barbosa, H., et al.: Generating and Exploiting Automated Reasoning Proof Certificates. Commun. ACM **66**(10), 86–95 (2023)

8. Barrett, C., Tinelli, C.: Satisfiability Modulo Theories. In: Clarke, E., Henzinger, T., Veith, H., Bloem, R. (eds.) Handbook of Model Checking, pp. 305–343. Springer International Publishing (2018)

9. Barrett, C., de Moura, L., Fontaine, P.: Proofs in satisfiability modulo theories. All about Proofs, Proofs for All **55**(1), 23–44 (2015)

10. Bayardo Jr, R., Schrag, R.: Using CSP look-back techniques to solve real-world SAT Instances. In: Proceedings of the 14th National Conference on Artificial Intelligence (AAAI), pp. 203–208 (1997)

11. Biere, A., Faller, T., Fazekas, K., Fleury, M., Froleyks, N., Pollitt, F.: CaDiCaL 2.0. In: Proceedings of the 36th International Conference on Computer Aided Verification (CAV), pp. 133–152 (2024)

12. Bojarski, M., et al.: End to end learning for self-driving cars (2016). technical Report. http://arxiv.org/abs/1604.07316

13. Botoeva, E., Kouvaros, P., Kronqvist, J., Lomuscio, A., Misener, R.: Efficient verification of relu-based neural networks via dependency analysis. In: Proceedings of the 34th AAAI Conference on Artificial Intelligence (AAAI), pp. 3291–3299 (2020)

14. Bradley, A., Manna, Z.: Property-directed incremental invariant generation. Formal Aspects Comput. **20**, 379–405 (2008)

15. Brix, C., Bak, S., Johnson, T., Wu, H.: The Fifth International Verification of Neural Networks Competition (VNN-COMP 2024): Summary and Results (2024). technical Report. http://arxiv.org/abs/2412.19985

16. Brix, C., Müller, M., Bak, S., Johnson, T., Liu, C.: First three years of the international verification of neural networks competition (VNN-COMP). Int. J. Softw. Tools Technol. Transfer 1–11 (2023)

17. Casadio, M., et al.: ANTONIO: towards a systematic method of generating NLP benchmarks for verification. In: Proc. 6th Workshop on Formal Methods for ML-Enabled Autonomous Systems (FoMLAS), vol. 16, pp. 59–70 (2023)

18. Casadio, M., et al.: NLP verification: towards a general methodology for certifying robustness. Europ. J. Appl. Math. 1–58 (2025)

19. Cheng, C.H., et al.: Neural networks for safety-critical applications — challenges, experiments and perspectives. In: Proceedings of the International Conference on Design, Automation and Test in Europe (DATE), pp. 1005–1006. IEEE (2018)

20. Chinneck, J., Dravnieks, E.: Locating minimal infeasible constraint sets in linear programs. ORSA J. Comput. **3**(2), 157–168 (1991)

21. Chvátal, V.: Linear Programming. W. H, Freeman and Company (1983)

22. Cordeiro, L.C., et al.: Neural Network Verification is a Programming Language Challenge. In: Proceedings of the 34th European Symposium on Programming (ESOP), pp. 206–235 (2025)

23. Cortez, P., Cerdeira, A., Almeida, F., Matos, T., Reis, J.: Modeling Wine Preferences by Data Mining from Physicochemical Properties. Decis. Support Syst. **47**(4), 547–553 (2009)

24. Dantzig, G.: Linear Programming and Extensions. Princeton University Press (1963)

25. Desmartin, R., Isac, O., Komendantskaya, E., Stark, K., Passmore, G., Katz, G.: a certified proof checker for deep neural network verification in imandra. In: Proceedings of the 16th International Conference on Interactive Theorem Proving (ITP), pp. 1–21 (2025)

26. Desmartin, R., Isac, O., Passmore, G., Stark, K., Komendantskaya, E., Katz, G.: Towards a certified proof checker for deep neural network verification. In: Proceedings of the 33rd International Symposium on Logic-Based Program Synthesis and Transformation (LOPSTR), pp. 198–209 (2023)
27. Dubey, S., Singh, S., Chaudhuri, B.: Activation functions in deep learning: a comprehensive survey and benchmark. Neurocomputing (2022)
28. Duong, H., Nguyen, T., Dwyer, M.: A DPLL(T) Framework for Verifying Deep Neural Networks (2023), technical Report. http://arxiv.org/abs/2307.10266
29. Ehlers, R.: Formal verification of piece-wise linear feed-forward neural networks. In: Proceedings of the 15th International Symposium on Automated Technology for Verification and Analysis (ATVA), pp. 269–286 (2017)
30. Elboher, Y., et al.: Robustness assessment of a runway object classifier for safe aircraft taxiing. In: Proceedings of the 43rd International Digital Avionics Systems Conference (DASC) (2024)
31. Gehr, T., Mirman, M., Drachsler-Cohen, D., Tsankov, E., Chaudhuri, S., Vechev, M.: AI2: safety and robustness certification of neural networks with abstract interpretation. In: Proceedings of the 39th IEEE Symposium on Security and Privacy (S&P) (2018)
32. Goodfellow, I., Bengio, Y., Courville, A.: Deep Learning. MIT Press (2016)
33. Goubault, E., Palumby, S., Putot, S., Rustenholz, L., Sankaranarayanan, S.: Static analysis of ReLU neural networks with tropical polyhedra. In: Proceedings of the 28th International Static Analysis Symposium (SAS), pp. 166–190 (2021)
34. The Gurobi Optimizer. https://www.gurobi.com/
35. Guthmann, O., Strichman, O., Trostanetski, A.: Minimal unsatisfiable core extraction for SMT. In: In Proceedings 16th International Conference on Formal Methods in Computer-Aided Design (FMCAD), pp. 57–64 (2016)
36. Heule, M.J., Biere, A.: Proofs for satisfiability problems. All about Proofs, Proofs for all **55**(1), 1–22 (2015)
37. Huang, X., Kwiatkowska, M., Wang, S., Wu, M.: Safety verification of deep neural networks. In: Proceedings of the 29th International Conference on Computer Aided Verification (CAV), pp. 3–29 (2017)
38. Isac, O., Barrett, C., Zhang, M., Katz, G.: Neural network verification with proof production. In: Proceedings of the 22nd International Conference on Formal Methods in Computer-Aided Design (FMCAD), pp. 38–48 (2022)
39. Isac, O., Zohar, Y., Barrett, C., Katz, G.: DNN verification, reachability, and the exponential function problem. In: Proceedings of the 34th International Conference on Concurrency Theory (CONCUR) (2023)
40. Jia, K., Rinard, M.: Exploiting Verified Neural Networks via Floating Point Numerical Error. In: Proc. 28th Int. Static Analysis Symposium (SAS). pp. 191–205 (2021)
41. Julian, K., Kochenderfer, M., Owen, M.: Deep Neural Network Compression for Aircraft Collision Avoidance Systems. J. Guid. Control. Dyn. **42**(3), 598–608 (2019)
42. Junker, U.: Quickxplain: conflict detection for arbitrary constraint propagation algorithms. In: Proceedings of the Workshop on Modelling and Solving Problems with Constraints (2001)
43. Junker, U.: Quickxplain: preferred explanations and relaxations for over-constrained problems. In: Proceedings of the 19th National Conference on Artificial Intelligence (AAAI), pp. 167–172 (2004)
44. Katz, G., Barrett, C., Dill, D., Julian, K., Kochenderfer, M.: Reluplex: a Calculus for Reasoning about Deep Neural Networks. Formal Methods in System Design (FMSD) (2021)

45. Katz, G., Barrett, C., Tinelli, C., Reynolds, A., Hadarean, L.: Lazy Proofs for DPLL(T)-based SMT solvers. In: Proceedings of the 16th International Conference on Formal Methods in Computer-Aided Design (FMCAD), pp. 93–100 (2016)
46. Katz, G., et al.: The marabou framework for verification and analysis of deep neural networks. In: Proceedings of the 31st International Conference on Computer Aided Verification (CAV), pp. 443–452 (2019)
47. Koller, L., Ladner, T., Althoff, M.: Set-Based Training for Neural Network Verification (2024), technical Report. http://arxiv.org/abs/2401.14961
48. Kouvaros, P., Leofante, F., Edwards, B., Chung, C., Margineantu, D., Lomuscio, A.: Verification of semantic key point detection for aircraft pose estimation. In: Proceedings of the 20th International Conference on Principles of Knowledge Representation and Reasoning (KR), pp. 757–762 (2023)
49. Liffiton, M.H., Sakallah, K.A.: Algorithms for computing minimal unsatisfiable subsets of constraints. J. Autom. Reason. **40**, 1–33 (2008)
50. Liu, Z., Yang, P., Zhang, L., Huang, X.: DeepCDCL: A CDCL-based neural network verification framework. In: Proceedings of the 18th International Symposium on Theoretical Aspects of Software Engineering (TASE), pp. 343–355 (2024)
51. Lyu, Z., Ko, C.Y., Kong, Z., Wong, N., Lin, D., Daniel, L.: Fastened crown: tightened neural network robustness certificates. In: Proceedings of the 34th AAAI Conference on Artificial Intelligence (AAAI), pp. 5037–5044 (2020)
52. de Moura, L., Bjørner, N.: Satisfiability modulo theories: introduction and applications. Commun. ACM **54**(9), 69–77 (2011)
53. OpenAI: ChatGPT. https://chatgpt.com
54. Pulina, L., Tacchella, A.: An Abstraction-Refinement Approach to Verification of Artificial Neural Networks. In: Proceedings of the 22nd International Conference on Computer Aided Verification (CAV), pp. 243–257 (2010)
55. Rajpurkar, P., Chen, E., Banerjee, O., Topol, E.J.: AI in health and medicine. Nat. Med. **28**(1), 31–38 (2022)
56. Refaeli, I., Isac, O.: Proof Minimization in Neural Network Verification (Artifact) (2025). https://doi.org/10.5281/ZENODO.17169557, https://zenodo.org/doi/10.5281/zenodo.17169557
57. Sälzer, M., Lange, M.: Reachability Is NP-complete even for the simplest neural networks. In: Proceedings of the 15th International Conference on Reachability Problems (RP), pp. 149–164 (2021)
58. Sankaranarayanan, S., Dutta, S., Mover, S.: Reaching out towards fully verified autonomous systems. In: Proceedings of the 13th International Conference on Reachability Problems (RP), pp. 22–32 (2019)
59. Silva, J., Sakallah, K.: GRASP-a new search algorithm for satisfiability. In: Proceedings of the 15th International Conference on Computer Aided Design (ICCAD), pp. 220–227 (1996)
60. Silva, J.M., Sakallah, K.A.: GRASP: a search algorithm for propositional satisfiability. IEEE Trans. Comput. **48**(5), 506–521 (1999)
61. Singh, A., Sarita, Y.C., Mendis, C., Singh, G.: Automated verification of soundness of DNN Certifiers. Proc. ACM Programm. Lang. **9** (2025)
62. Singh, G., Gehr, T., Puschel, M., Vechev, M.: An abstract domain for certifying neural networks. In: Proceedings of the 46th ACM SIGPLAN Symposium on Principles of Programming Languages (POPL) (2019)
63. Sörensson, N., Biere, A.: Minimizing learned clauses. In: Proceedings of the 12th International Conference on Theory and Applications of Satisfiability Testing (SAT), pp. 237–243. Springer (2009)

64. Szegedy, C., et al.: Intriguing Properties of Neural Networks (2013), technical Report. http://arxiv.org/abs/1312.6199
65. Tjeng, V., Xiao, K., Tedrake, R.: Evaluating Robustness of Neural Networks with Mixed Integer Programming (2017), technical Report. http://arxiv.org/abs/1711.07356
66. Tran, H.D., Bak, S., Xiang, W., Johnson, T.: Verification of deep convolutional neural networks using imagestars. In: Proceedings of the 32nd International Conference on Computer Aided Verification (CAV), pp. 18–42 (2020)
67. The VNNCOMP 2025 Benchmarks Repository. https://github.com/VNN-COMP/vnncomp2025_benchmarks (2025)
68. Wang, S., Pei, K., Whitehouse, J., Yang, J., Jana, S.: Formal security analysis of neural networks using symbolic intervals. In: Proceedings of the 27th USENIX Security Symposium, pp. 1599–1614 (2018)
69. Wang, S., et al.: Beta-crown: efficient bound propagation with per-neuron split constraints for neural network robustness verification. In: Proceedings of the 35th Conference on Neural Information Processing Systems (NeurIPS (2021)
70. Wu, H., et al.: Marabou 2.0: a versatile formal analyzer of neural networks. In: Proceedings of the 36th International Conference on Computer Aided Verification (CAV) (2024)
71. Yang, Y., Hu, H., Wei, T., Li, S.E., Liu, C.: Scalable synthesis of formally verified neural value function for Hamilton-Jacobi reachability analysis. J. Artif. Intell. Res. **83** (2025)
72. Zhang, H., Shinn, M., Gupta, A., Gurfinkel, A., Le, N., Narodytska, N.: Verification of recurrent neural networks for cognitive tasks via reachability analysis. In: Proceedings of the 24th European Conference on Artificial Intelligence (ECAI), pp. 1690–1697 (2020)
73. Zhou, D., Brix, C., Hanasusanto, G.A., Zhang, H.: Scalable neural network verification with branch-and-bound inferred cutting planes. In: Proceedings of the 38th Conferenc on Neural Information Processing Systems (NeurIPS) (2024)
74. Zombori, D., Bánhelyi, B., Csendes, T., Megyeri, I., Jelasity, M.: Fooling a complete neural network verifier. In: Proceedings of the 9th International Conference on Learning Representations (ICLR) (2021)

Finding Photonics Circuits
via δ-Weakening SMT

Marco Lewis[(✉)] and Benoît Valiron

Université Paris-Saclay, CNRS, CentraleSupélec,
ENS Paris-Saclay, Inria, Laboratoire Méthodes
Formelles, 91190 Gif-sur-Yvette, France
`macro-john.lewis@inria.fr`

Abstract. For quantum computers based on photonics, one main problem is the synthesis of a photonic circuit that emulates quantum computing gates. The problem requires using photonic components to build a circuit that act like a quantum computing gate with some probability of success. This involves not only finding a circuit that can correctly act like a quantum gate, but also optimizing the probability of success. Whilst many approaches have been given in the past and applied to specific gates, they often lack ease of reusability. We present a tool that uses dReal, a δ-weakening SMT solver, to find such photonic circuits, optimize the likelihood of occurring, and provide some guarantee that the result is optimal. We demonstrate the usage of our tool by recreating known results in the literature, extending upon them, and presenting new results for Givens rotation gates.

1 Introduction

The field of quantum computing has been making strides in recent years due to developments of different quantum computers. These devices are developed using a variety of different physical techniques, such as ion trap [8], photonic devices [23], topological qubits [17,20], *etc.* Many of the techniques directly implement quantum computing gates from theory onto the physical devices. However, some techniques require quantum gates to be implemented using basic building blocks. Thus, the problem arises of how to synthesize a circuit of the basic building blocks that implements the desired quantum gate.

This is the case for quantum computers implemented using photonics, or linear optics. The usage of linear optics to perform quantum computation was first introduced by Knill et al. [22]. Linear optics consists of wires that photons can be sent down and uses a variety of components (beam splitters, phase shifters, emitters, detectors, *etc.*.) to implement unitary quantum gates. The main difficulties of representing gates on a linear optical device is that there are many ways to represent the gate, and each representation has a probability to act as the desired gate and a probability that it acts in some other way. Thus, whilst it is important to find a representation, it is also important to find the representation that is optimal, *i.e.*, has the highest probability of performing the desired operation.

In the past, various techniques have been used to find linear optical circuits to represent a quantum gate. These include using computational tools to solve and simplify expressions [21], building a circuit [27,29], and through different decomposition schemes [2,10,25,30]. With these past techniques, it is usually the case that they return a circuit that implements the desired gate, but it does not guarantee optimality.

The usage of SMT and SAT solvers to perform forms of synthesis [7] or optimization for quantum computing has been investigated previously. Many recent works investigate the reduction in the depth of a quantum circuit [26,32, 34], reducing the number of specific gates in a circuit (such as $CNOT$ gates) [9, 31,33], and the depth of certain gates [18]. Additionally, whilst δ-weakening SMT solvers have been used to investigate program correctness in quantum programs [6], they have not been used for circuit synthesis. Our usage of (δ-weakening) SMT solvers to investigate the synthesis of photonics circuits is a new line of research that has not been investigated before.

In this work, we present a search technique based on SMT solvers [5] to synthesize correct and optimal circuits for quantum gates. We introduce the synthesis problem for linear optics and present the theory behind the developed search technique. Our search method is primarily based on using a δ-weakening SMT solver [14,15] to find an approximation of a circuit (although the search can be adapted to standard SMT solvers). This approximation is refined to an exact representation that can be implemented on a linear optics device. These techniques have been implemented, and we are able to check against known results in the literature and find new results. Notably, our technique is automated in that one can input a quantum computing gate and the setup of the linear optics circuit, and the search is automatically performed. This makes the technique generalisable; other techniques often require some work to apply it to different setups.

2   Background

We provide a comprehensive background to quantum computing and linear optics in Sects. 2.1 and 2.2 respectively. A full introduction to linear optics and its application to quantum computing can be found in [23]. For a summary of the problem we are trying to solve, see Sect. 2.3.

2.1   Quantum Computing

In quantum computing, states are normally described by a complex vector and the base unit of information is the qubit. A qubit resides in the unit circle of $\mathbb{C}^2$ and consists of the computational basis states $|0\rangle = (1,0)^\top$ and $|1\rangle = (0,1)^\top$. Thus, a valid quantum state for a qubit can be written as $|\phi\rangle = \alpha_0 |0\rangle + \alpha_1 |1\rangle$, where $\alpha_i \in \mathbb{C}$ and $|\alpha_0|^2 + |\alpha_1|^2 = 1$.

Qubits can be combined together using the tensor product. Thus, an n-qubit system resides in the unit circle of $\mathbb{C}^{2^n}$ and a state can be written as $|\phi\rangle = \sum_{i=0}^{2^n-1} \alpha_i |i\rangle$, where $|i\rangle = |b_{n-1}\rangle \otimes |b_{n-2}\rangle \otimes \cdots \otimes |b_0\rangle$ and the binary

(a) Beam splitter   (b) Phase shifter

Fig. 1. Linear optical components.

representation of i is $b_{n-1}b_{n-2}\ldots b_0$. Additionally, $\alpha_i \in \mathbb{C}$ and $\sum_{i=0}^{2^n-1}|\alpha_i|^2 = 1$. Quantum systems can also be combined by tensor product; the global state of an n-qubit and m-qubit system can be written as $|\phi\rangle \otimes |\psi\rangle \in \mathbb{C}^{2^n+2^m}$, where $|\phi\rangle \in \mathbb{C}^{2^n}$ and $|\psi\rangle \in \mathbb{C}^{2^m}$.

The standard operation on an n-qubit system is the unitary operation, $U \in \mathbb{C}^{2^n} \times \mathbb{C}^{2^n}$. A unitary operation has the property that its inverse is its conjugate transpose, $U^{-1} = U^\dagger = \overline{U}^\top$. A unitary operation, U, applied to a quantum state, $|\phi\rangle$, is written as $U|\phi\rangle$ and acts linearly, i.e., $U(\sum_k |\phi_k\rangle) = \sum_k U|\phi_k\rangle$. The application of unitaries onto a quantum state is done by dot product and written as $U_t \ldots U_2 U_1 |\phi\rangle$. Unitary operations may be performed on subsets of the system by use of the tensor product, i.e., $(U_n \otimes U_m)(|\phi\rangle \otimes |\psi\rangle) = U_n|\phi\rangle \otimes U_m|\psi\rangle$, where $|\phi\rangle \in \mathbb{C}^{2^n}, |\psi\rangle \in \mathbb{C}^{2^m}$ and U_n, U_m are respective unitary operations.

Examples of unitary operations for quantum computing are given. The controlled Z operation, CZ, applies a phase of -1 to the state $|11\rangle$ and does nothing otherwise. Another is the controlled-not operation, $CNOT$, which does nothing to $|00\rangle$ and $|01\rangle$ but changes $|10\rangle$ to $|11\rangle$ and vice versa. The first qubit is referred to as the control qubit and the second qubit is the target qubit. Their representation as unitary operations are

$$CZ = \begin{pmatrix} 1 & 0 & 0 & 0 \\ 0 & 1 & 0 & 0 \\ 0 & 0 & 1 & 0 \\ 0 & 0 & 0 & -1 \end{pmatrix} \quad \text{and} \quad CNOT = \begin{pmatrix} 1 & 0 & 0 & 0 \\ 0 & 1 & 0 & 0 \\ 0 & 0 & 0 & 1 \\ 0 & 0 & 1 & 0 \end{pmatrix}.$$

A full introduction to quantum computing can be found in [28].

2.2 Linear Optics

Linear optics is a means of implementing quantum computing on a physical device. An optical circuit has a number of *wires* (or *modes*) that photons can be sent down. Unlike quantum computing circuits, where a single qubit is sent down a single wire, a wire in an optical circuit can have several photons sent down a wire. The base components of a linear optical circuit are beam splitters (BS), wave-plates, and phase shifters. The base components introduce phases onto photons sent through them and can transfer photons from one wire to another (with some probability). These components are used to construct circuits that can be used to emulate operations used by quantum computers. The components are represented diagrammatically in Fig. 1.

Components are represented by a transfer matrix. For instance, a *phase shift* applies a shift in phase by $e^{i\phi}$ to a single wire, and a *beam splitter* acting on two wires has a transfer matrix of the form $\begin{pmatrix} \cos(\theta) & ie^{-i\phi}\sin(\theta) \\ ie^{i\phi}\sin(\theta) & \cos(\theta) \end{pmatrix}$, where θ is an angle and ϕ is the relative phase (usually $\phi = 0$ or π). For a beam splitter, $\cos^2\theta$ represents the chance of being reflected back into the same wire and $\sin^2\theta$ represents the chance of a photon being transferred to the other wire and picking up some phase. There are other components, such as wave plates, but these are not discussed.

These component operations can be combined using tensor and dot product in a similar way to quantum operations to provide a transfer matrix, $\hat{U}$, that represents the circuit (of size $\mathbb{C}^m \times \mathbb{C}^m$ for m wires). It should be noted that the transfer matrix (i) is unitary; and (ii) can be decomposed back into components via different different universal circuit designs [10,30]. For example, a pyramid of beam splitters and phase shifters can be used to decompose a transfer matrix into a circuit. This means finding a circuit is equivalent to finding the transfer matrix.

A Fock state is a state of the wire; consisting of its phase, $\alpha \in \mathbb{C}, |\alpha| \leq 1$; and the number of photons down the wire, $n \in \mathbb{Z}_+ \cup \{0\}$. We denote the Fock state of n photons and phase α as $\alpha |n\rangle_{\mathcal{F}}$. The set $\{|n\rangle_{\mathcal{F}} : n \in \mathbb{N} \cup \{0\}\}$ acts as a basis set for a wire, *i.e.*, a wire represented by the state $|\Psi\rangle$ can be written as $\sum_i \alpha_i |i\rangle_{\mathcal{F}}$, where $\alpha_i \in \mathbb{C}$ and $\sum_i |\alpha_i|^2 = 1$. Fock states can then be combined together using tensor product; for m wires a Fock basis state is written as $|n_1, n_2, \ldots, n_m\rangle_{\mathcal{F}} = |n_1\rangle_{\mathcal{F}} \otimes |n_2\rangle_{\mathcal{F}} \ldots |n_m\rangle_{\mathcal{F}}$.

To transition between basis states on a wire, the creation and annihilation operation are used.[1] The creation operator on wire i, $\hat{a}_i^\dagger$, acts on the Fock state $|n_1, \ldots, n_m\rangle_{\mathcal{F}}$ by

$$\hat{a}_i^\dagger |n_1, \ldots, n_i, \ldots, n_m\rangle_{\mathcal{F}} = \sqrt{n_i + 1} |n_1, \ldots, n_i + 1, \ldots, n_m\rangle_{\mathcal{F}}.$$

A transfer matrix, $\hat{U}$, transforms the creation operations such that

$$\hat{a}_j^\dagger \to \sum_{i=1}^{m} \hat{U}_{ij} \hat{a}_i^\dagger.$$

Further, each transfer matrix can act on Fock states through a unitary operation, $U_{\mathcal{F}}$, where

$$U_{\mathcal{F}} |n_1, \ldots, n_m\rangle_{\mathcal{F}} = \prod_{j=1}^{m} \frac{1}{\sqrt{n_j!}} \left(\sum_{i=1}^{m} \hat{U}_{ij} \hat{a}_i^\dagger \right)^{n_j} |\emptyset\rangle_{\mathcal{F}}, \tag{1}$$

with $|\emptyset\rangle_{\mathcal{F}} = |0, \ldots, 0\rangle_{\mathcal{F}}$ representing the vacuum Fock state (no photons through any wires).

---

[1] We do not discuss the annihilation operation, but to state simply $\hat{a}_i$ sends $|n_1, \ldots, n_i, \ldots, n_m\rangle_{\mathcal{F}} \to |n_1, \ldots, n_i - 1, \ldots, n_m\rangle_{\mathcal{F}}$.

Linear Optical Quantum Computing. Quantum computing is usually implemented on a linear optics device by using the dual rail system. In the dual rail system, a single qubit is implemented on two wires, where the basis states, $|0\rangle$ and $|1\rangle$, are represented by the Fock states $|1,0\rangle_{\mathcal{F}}, |0,1\rangle_{\mathcal{F}}$ respectively (*i.e.*, one photon down the first wire and none down the second wire, and vice versa). Therefore, to implement an q-qubit operation on the dual rail system, at least $2q$ wires are required to represent the operation. Normally though, a quantum operation will use extra wires, known as *auxiliary* wires, to implement an operation. Thus, a quantum operation will be implemented using $2q + m_a$ wires, where m_a is the number of auxiliary wires.

For a dual rail system with auxiliary wires, quantum computing basis states are represented by their Fock state encoding and the initial Fock states of the auxiliary wires. For instance, a two qubit system with two *vacuum* wires (no photons sent down) is represented as

$$
\begin{aligned}
|00\rangle &= |1,0,1,0\rangle_{\mathcal{F}}\,|0,0\rangle_{\mathcal{F}}\,, \quad |01\rangle = |1,0,0,1\rangle_{\mathcal{F}}\,|0,0\rangle_{\mathcal{F}}\,, \\
|10\rangle &= |0,1,1,0\rangle_{\mathcal{F}}\,|0,0\rangle_{\mathcal{F}}\,, \quad |11\rangle = |0,1,0,1\rangle_{\mathcal{F}}\,|0,0\rangle_{\mathcal{F}}\,.
\end{aligned}
\tag{2}
$$

The collection of such Fock states is called the coincidence basis, $\mathcal{C}$, and represents the desired states to be measured within a linear optics circuit. Undesirable states can be generated by the optical circuit and these are often ignored. In the setup given, examples of undesirable states include no photons detected in the wires for a quantum state $(|1,1,0,0\rangle_{\mathcal{F}}\,|0,0\rangle_{\mathcal{F}})$, multiple photons in a single wire/rail $(|0,0,2,0\rangle_{\mathcal{F}}\,|0,0\rangle_{\mathcal{F}})$, or additional photons detected in the auxiliary wires $(|0,1,0,0\rangle_{\mathcal{F}}\,|1,0\rangle_{\mathcal{F}})$. In general, all Fock states of n photons down m wires forms a basis, $\mathcal{B}_{n,m}$, and $\mathcal{C} \subset \mathcal{B}_{n,m}$.

There are two types of measurement that are of interest in linear optics (for quantum computing). The first is *post-selection*, where all wires are measured. The corresponding coincidence basis contains only the basis states in quantum computing, *e.g.*, as given in Eq. (2). Any other Fock states not in the coincidence basis are ignored.

The second measurement type is *heralded* selection, where photons are sent down the auxiliary wire and only the auxiliary wires are measured. The corresponding coincidence basis contains any valid state where the auxiliary wires have the same number of photons as they started with. Whilst quantum computing can be done with post-selection and with a higher probability of success in comparison to heralded selection; heralded is preferred over post-select because only the auxiliary wires are measured, the Fock states that represent the qubits are not measured, and therefore the quantum state can be reused to perform multiple unitary operations.

Finally, it is important to note that quantum computing operations occur probabilistically on photonic devices. This is because the quantum computing basis (the states we are interested in) only have a chance of being measured. Consider a transfer matrix, $\hat{U}$, with n photons and m wires with the goal of simulating a q qubit gate, U. Let $\mathcal{C}_{qc} \subset \mathcal{B}_{n,m}$ be the set of Fock states that represent the quantum computing basis states $(|b_1 \ldots b_q\rangle = |n_1, \ldots, n_m\rangle_{\mathcal{F}})$. How

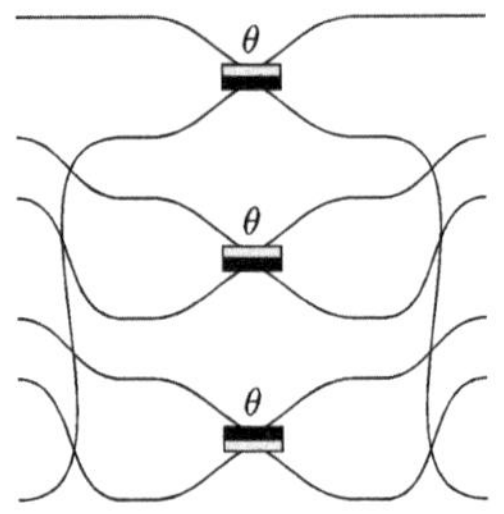

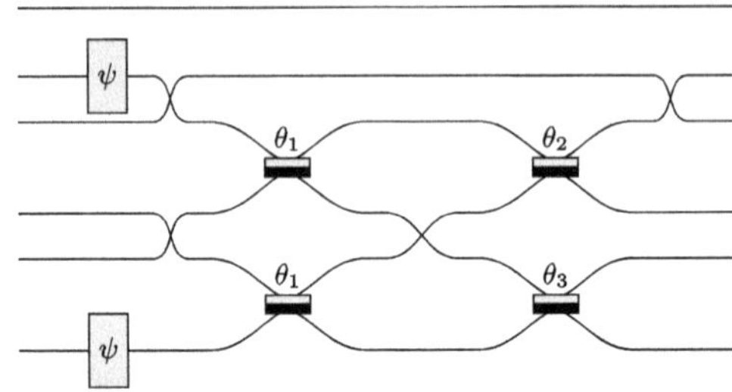

(a) Post-select scheme; θ is such that $\sin^2\theta = 1/3$ [29].

(b) Heralded scheme; $\psi = \pi$ and θ_i ($i = 1, 2, 3$) are radian values equivalent to those in [21].

Fig. 2. Optical circuits of the controlled-Z operation. The first two wires act as the control qubit, the middle two as the target qubit, and the last two are ancillary wires.

$\hat{U}$ acts on the basis Fock states in $\mathcal{B}_{n,m}$ can be seen as a matrix, $(\hat{U})_{\mathcal{F}}$, of size $|\mathcal{B}_{n,m}| \times |\mathcal{B}_{n,m}|$. Then, $\hat{U}$ implements U on its Fock state representation if

$$
(\hat{U})_{\mathcal{F}} = \begin{array}{c} |\mathcal{C}_{qc}| \\ \left\{ \end{array} \overset{\overbrace{|\mathcal{C}_{qc}|}}{\begin{pmatrix} \alpha U & M_1 \\ M_0 & M_2 \end{pmatrix}},
$$

where $\alpha \in \mathbb{C}$, $|\alpha|^2$ represents the probability of success, and M_i are block matrices (whose values we do not care about). *I.e.*, the Fock operation of $\hat{U}$ acts like U on the quantum computing basis, $\mathcal{C}_{qc}$, up to some factor, α. More details are provided in Sect. 3.2.

We now give examples of quantum computing operations implemented in the post-select and heralded schemes. In Fig. 2, the photonics circuits of the CZ operation is shown using post-selection and heralded selection. In the post-selected case, two vacuum ancillary wires (no photons sent down) are used to obtain the transfer matrix

$$
\frac{1}{3}\begin{pmatrix}
\sqrt{3} & 0 & 0 & 0 & 0 & -\sqrt{6} \\
0 & \sqrt{3} & 0 & -\sqrt{6} & 0 & 0 \\
0 & 0 & \sqrt{3} & 0 & \sqrt{6} & 0 \\
0 & \sqrt{6} & 0 & \sqrt{3} & 0 & 0 \\
0 & 0 & -\sqrt{6} & 0 & \sqrt{3} & 0 \\
-\sqrt{6} & 0 & 0 & 0 & 0 & -\sqrt{3}
\end{pmatrix}, \tag{3}
$$

which has a probability of success of $\frac{1}{9}$ [29].

In the heralded case, the circuit uses two ancillary wires with a photon sent down each ancillary wire and has a probability of success of $\frac{2}{27}$ [21]. The associated transfer matrix is

$$\begin{pmatrix} 1 & 0 & 0 & 0 & 0 & 0 \\ 0 & -\frac{1}{3} & 0 & -\frac{\sqrt{2}}{3} & \frac{\sqrt{2}}{3} & \frac{2}{3} \\ 0 & 0 & 1 & 0 & 0 & 0 \\ 0 & \frac{\sqrt{2}}{3} & 0 & -\frac{1}{3} & -\frac{2}{3} & \frac{\sqrt{2}}{3} \\ 0 & -\frac{\sqrt{3-\sqrt{6}}}{3} & 0 & \frac{\sqrt{3-\sqrt{6}}}{3} & -\frac{\sqrt{3+\sqrt{6}}}{3\sqrt{2}} & \sqrt{\frac{1}{6}-\frac{1}{3\sqrt{6}}} \\ 0 & -\frac{\sqrt{3-\sqrt{6}}}{3} & 0 & -\frac{\sqrt{3-\sqrt{6}}}{3} & -\sqrt{\frac{1}{6}-\frac{1}{3\sqrt{6}}} & -\frac{\sqrt{3+\sqrt{6}}}{3\sqrt{2}} \end{pmatrix}.$$

2.3 The Problem

To summarise, a photonics circuit of m wires can be represented by a $m \times m$ transfer matrix, $\hat{U}$. Through a procedure, this can be transformed into an operation, $\hat{U}_{\mathcal{F}}$, that works on a larger space, where we wish to ensure a particular subspace of the operation acts like a desired quantum computing operation, U, up to some factor $\alpha \in \mathbb{C}$. There is only a chance of being in this subspace, and this probability of success is $|\alpha|^2$.

The simplest problem to consider is the verification version of the problem.

Problem 1. Given a transfer matrix, $\hat{U}$, check that a quantum computing unitary operation, U, is implemented on a subspace of the Fock space, $\hat{U}_{\mathcal{F}}$, up to some factor, α.

The harder problem we are considering is the synthesis of the transfer matrix.

Problem 2. Given a quantum computing unitary operation, U, that acts on q qubits, find a photonics circuit, represented by $\hat{U}$, that implements U on its Fock space $\hat{U}_{\mathcal{F}}$.

In this general setting for the problem, one can consider various setups and different ways to encode quantum bits into Fock states. We restrict this problem into working on a specific basis and using the dual-rail encoding, where a qubit in the quantum computing space is represented by two optical wires. Additionally, it is important to maximize the likelihood of the operation is performed successfully. Thus, the problem needs to be optimized. This leaves us with the following problem.

Problem 3. Given a quantum computing unitary operation that acts on q qubits, U, $2q + m_a$ optical wires, and a coincidence basis, $\mathcal{C}$; find the transfer matrix of a dual-rail optical operation, $\hat{U}$, that maximises the likelihood of U occurring.

3 Method

3.1 Non-linear Real Arithmetic (NRA) and δ-Satisfiability

Generally, SMT solver theories are usually NP-hard but are decidable, both LIA and LRA are as such (linear integer/real arithmetic respectively). Some theories though are undecidable. One such theory, Non-linear Real Arithmetic (NRA), is the theory that will be used for solving our problem. Standard NRA mainly consists of multivariable polynomials, but can be extended to consist of non-linear functions; such as powers of variables, and trigonometric functions.

However, it was shown in [14] that by allowing a small perturbation in NRA expressions, the theory can become decidable. This perturbation comes in the form of a value $\delta \in \mathbb{Q}^+ \cup \{0\}$, which is used to weaken arithmetic expressions, $f(x) = 0$, such that they are of the form $|f(x)| \leq \delta$. This process is known as δ-weakening. Instead of returning the standard sat or unsat, an SMT solver that implements δ-weakening will instead return δ-sat, where the δ-weakening of the expressions is satisfiable, or unsat, where the expressions are unsatisfiable.

The tool dReal [15] implements δ-weakening and, when δ-sat, returns a region of values that satisfy the δ-weakened expressions. This enables it to find δ-satisfiability for (an extension of) NRA expressions. We write

$$sat, regions \leftarrow \texttt{dReal}(c, \delta),$$

to mean dReal, run with the constraints c and precision δ, returns δ-sat, unsat, or unknown (in the case of timeouts or crashes); and, if dReal returns δ-sat, return the satisfiable regions or [ ] otherwise. For example, if $\delta = 0.001$, the expressions
$$(x + 2^y = 3) \wedge (y > \sin(x)) \wedge (x > 0.5)$$

is δ-sat with $x \in [0.8685\ldots, 0.8687\ldots]$ and $y \in [1.0916\ldots, 1.0918\ldots]$. This tool forms the basis of how transfer matrices in the coincidence basis are found.

Whilst NRA is solvable using standard SMT solvers, we found that using dReal would produce a region of approximations when standard tools could not find an exact solution or dReal would find a region much faster. However, this means that instead of getting a transfer matrix that can be used, we have a region that approximately solves our constraints. In Sect. 3.4, we describe how we can find an actual transfer matrix from our region of approximations, allowing us to return an exact solution.

3.2   Encoding of Photonics Problem into NRA

Given Problem 3, we show how we can transform the problem into a problem in non-linear real arithmetic (NRA). As a reminder, we are given a quantum computing operation, U, and a coincidence basis, $\mathcal{C}$; and need to find a suitable $m \times m$ transfer matrix, $\hat{U}$, that implements U with a probability of success, $|\alpha|^2$, where m is the number of wires and α denotes the success amplitude. Thus, we consider $\hat{U}_{ij}$ and α to be real variables.

Remark 1. In general, it can be that $\alpha, \hat{U}_{ij} \in \mathbb{C}$. For ease of demonstrating the process of converting into NRA, we restrict $\hat{U}$ to be a real matrix and $\alpha \in \mathbb{R}$. However, it is possible to encode variables as two real variables consisting of the real and imaginary part, and modify the following encoding for the complex representation. We discuss the consequences of allowing variables to be complex in Sect. 4.3.

To begin with, we know that $\hat{U}$ needs to be unitary, *i.e.*, $\hat{U}\hat{U}^\dagger = I$. This can be encoded as

$$\texttt{unitary} = \bigwedge_{1 \le i \le m} \left(\sum_{1 \le k \le m} \hat{U}_{ik}\hat{U}^\dagger_{ki} = 1 \right) \wedge \bigwedge_{1 \le i \ne j \le m} \left(\sum_{1 \le k \le m} \hat{U}_{ik}\hat{U}^\dagger_{kj} = 0 \right).$$

Additionally, since $\hat{U}$ is unitary, we can restrict the values that $\hat{U}$ can take to be between -1 and 1. Further, the probability of success ($|\alpha|^2$) cannot be greater than 1. Thus, these can simply be encoded as

$$\texttt{bound} = (-1 \le \alpha) \wedge (\alpha \le 1) \wedge \left(\bigwedge_{1 \le i,j \le m} -1 \le \hat{U}_{ij} \le 1 \right).$$

The core constraint to be considered is that $\hat{U}$ implements U on its Fock matrix $(\hat{U})_{\mathcal{F}}$ up to the factor α. Our coincidence basis consists of the set of Fock states for the quantum computational basis states, $\mathcal{C}_{qc}$; and, if heralded, the set containing any other Fock state with the same number of photons in each auxiliary wires, $\mathcal{C}_h$, *i.e.*, $\mathcal{C} = \mathcal{C}_{qc}$ if using the post-select regime and $\mathcal{C} = \mathcal{C}_{qc} \cup \mathcal{C}_h$ if using the heralded regime. For $|\phi\rangle_{\mathcal{F}} \in \mathcal{C}_{qc}$, let $|b\rangle$ be the quantum computational representation of $|\phi\rangle_{\mathcal{F}}$, *e.g.*, for a vacuum auxiliary of two wires, $|00\rangle$ is the representation of $|\phi\rangle_{\mathcal{F}} = |1010\rangle_{\mathcal{F}} |00\rangle_{\mathcal{F}}$. Then, the Fock states acting on the encoded quantum computational basis states are required to be equal up to the probability amplitude, *i.e.*,

$$\texttt{fockequal} = \bigwedge_{|b\rangle \cong |\phi\rangle_{\mathcal{F}} \in \mathcal{C}_{qc}} \bigwedge_{|c\rangle \cong |\psi\rangle_{\mathcal{F}} \in \mathcal{C}_{qc}} (\alpha U_{bc} = (\hat{U}_{\mathcal{F}})_{\phi\psi}).$$

Thus, the simplest encoding of Problem 3 in NRA is

$$\texttt{core} = \texttt{bound} \wedge \texttt{unitary} \wedge \texttt{fockequal}.$$

with $\texttt{bound}$, $\texttt{unitary}$, and $\texttt{fockequal}$ having $m^2 + 2$, m^2, and $(2^q)^2$ formulae respectively, where m is the number of modes and q is the number of qubits to simulate.

Remark 2. To represent a q-qubit operation, $k = q + n_a$ photons are used, where n_a is the number of photons going down the auxiliary wires. The constraints given in $\texttt{core}$ are a combination of 2- and k-degree polynomials. The 2-degree polynomials come from the $\texttt{unitary}$ constraint and the polynomials from the $\texttt{fockequal}$ constraints are k-degree since we are using the Fock version of $\hat{U}$ (see Eq. 1). The number of photons down each auxiliary wires additionally affects the coefficients of the polynomials.

However, we can also include additional constraints based on the unitary gate required or the coincidence basis being used. For instance, for a CZ operation, we can set the first and third wire to not interact with other wires representing the qubits, *i.e.*, $\hat{U}_{1j} = \hat{U}_{j1} = 0$ for $j \in \{2,3,4\}$ and, similarly, for $\hat{U}_{3k}, \hat{U}_{k3}$ with $k \subset \{1,2,4\}$ (referred to as *non-interaction*). This can be done since the CZ

Algorithm 1: Search Algorithm using dReal

Input: unitary operation, U; coincidence basis, $\mathcal{C}$; minimum threshold, $\alpha_{min} \geq 0$; precision, $\delta > 0$; timeout > 0.

result $\leftarrow$ **unchecked**;

regions $\leftarrow$ [ ];

sat-result $\leftarrow$ δ-**sat**;

consts $\leftarrow$ $constraints(U,\mathcal{C})$;

while $sat - result = \delta\text{-}sat \wedge runtime \leq timeout$ **do**

    constraints $\leftarrow$ **consts** $\wedge\,(|\alpha|^2 \geq \alpha_{min})$;

    sat-result, regions $\leftarrow$ **dReal**(constraints, δ);

    if $sat\text{-}result = \delta\text{-}sat$ **then**

        $\alpha_{min} \leftarrow |\text{regions}[\alpha]_{ub}|^2 + \frac{1}{10}\delta$;

        result $\leftarrow$ **approximate**;

    end

end

if $result = approximate \wedge sat\text{-}result = unsat$ **then** result $\leftarrow$ δ-**optimal**;

if $result = unchecked \wedge sat\text{-}result = unsat$ **then** result $\leftarrow$ **infeasible**;

if $result = unchecked \wedge sat\text{-}result = unknown$ **then** result $\leftarrow$ **unknown**;

return result, regions

---

operation does nothing when the control qubit is in the $|0\rangle$ and, even when the state is controlled, there is no interaction with the target qubit when it is in the $|0\rangle$ state. This means that the second and fourth wires are the only wires that interact with auxiliary wires (if there are any).

Alternatively, we can also enforce constraints on the auxiliary wires, particularly vacuum auxiliary wires. We can set it such that a vacuum wire need not be used (*vacuum relaxing*, $\hat{U}_{kk} = 1 \vee \hat{U}_{kk} < 1$) or that a vacuum wire must be used (*vacuum enforcing*, $\hat{U}_{kk}\hat{U}_{kk}^{\dagger} \neq 1$). Vacuum relaxing allows the SMT solver to easily consider cases when the vacuum wire may not be used (meaning a smaller circuit can be considered) and vacuum enforcing requires the resulting circuit to use the vacuum wire in some way.

3.3   Searching Using dReal

The algorithm for how we can synthesise and find an optimal transfer matrix using dReal is described in Algorithm 1. We have shown how to make constraints for the operation we are synthesizing given a quantum computing unitary matrix, U, and coincidence basis, $\mathcal{C}$, in Sect. 3.2. Thus, we have

$$constraints(U,\mathcal{C}) = \texttt{core} \wedge \texttt{extra},$$

where $\hat{U}$ is represented by real variables and **extra** encodes any extra constraints required by the user (constraints on gates, vacuum relaxing, *etc.*.). In practice, the constraints are generated in standard SMT-LIB format [4]. The **dReal** function used in the algorithm is described in Sect. 3.1.

```
 1   (set-logic QF_NRA)
 2   (declare-fun A_0_0r () Real)
 3   ...
 4   (declare-fun A_5_5r () Real)
 5   (declare-fun alphar () Real)
 6   (assert (<= A_0_0r 1.0))
 7   (assert (>= A_0_0r (- 1.0)))
 8   ...
 9   (assert (>= alphar (- 1.0)))
10   (assert (<= alphar 1.0))
11   (assert (= (+ (* A_0_0r A_0_0r) (* A_0_4r A_0_4r) (* A_0_5r A_0_5r)) 1.0))
12   (assert (= (+ (* A_0_4r A_1_4r) (* A_0_5r A_1_5r)) 0.0))
13   (assert (= (+ (* A_0_4r A_2_4r) (* A_0_5r A_2_5r)) 0.0))
14   (assert (= (+ (* A_0_4r A_3_4r) (* A_0_5r A_3_5r)) 0.0))
15   (assert (= (+ (* A_0_0r A_5_0r) (* A_0_4r A_5_4r) (* A_0_5r A_5_5r)) 0.0))
16   ...
17   (assert (= alphar (* A_0_0r A_2_2r)))
18   (assert (= alphar (* A_0_0r A_3_3r)))
19   (assert (= alphar (* A_1_1r A_2_2r)))
20   (assert (= (* (- 1.0) alphar) (+ (* A_1_1r A_3_3r) (* A_1_3r A_3_1r))))
21   (assert (= A_1_0r 0.0))
22   (assert (= A_0_1r 0.0))
23   ...
24   (assert (not (<= (* alphar alphar) <alpha_min> )))
25   (check-sat)
26   (exit)
```

Fig. 3. Example SMT file of the photonics problem aiming to simulate a CZ gate with two vacuum wires.

Example 1. Figure 3 shows portions of a generated and simplified SMT file using the encoding described. The variables `A_i_jr` represent the values of the transfer matrix $(\hat{U}_{ij})$ that need to be found and `alphar` is the success amplitude within the Fock basis. In this file, only a real transfer matrix is considered but it can be extended to the complex domain by including twice as many variables.[2] The following lines are associated to the encodings, respectively:

- Lines 7-10, **bound** (repeated for all `A_i_jr`);
- Lines 11-16, **unitary** (equations for the first row of the identity matrix shown);
- Lines 17-20, **fockequal** (showing the 4 non-zero terms in the CZ matrix and that imaginary part of amplitude is 0);
- Lines 21-23, **extra** (in particular, the non-interaction technique is used to set some matrix values to 0);
- Line 24, represents the additional lower bound for the success probability, where `<alpha_min>` is a given constant.

---

[2] At the end of the variables, `r` represents the real part of the variable. Imaginary parts would use `i` at the end of the variable instead.

The algorithm starts with an initial probability of success to beat and tries to find a transfer matrix that has a higher probability of success. If it is able to find one, then it updates the value the probability of success needs to beat by going slightly above the upper bound of the probability of success that was found ($\alpha_{min} \leftarrow |\text{regions}[\alpha]_{ub}|^2 + \frac{1}{10}\delta$), and tries to search again. It continues polling for a set time limit or if a call returns **unsat** before stopping.

The algorithm returns one of four possible outputs:

- δ-**optimal**: a valid region was found and it is optimal up to δ (*i.e.*, the probability of success cannot get higher);
- **approximate**: a valid region was found (but it might not be optimal);
- **unknown**: no region was found within the time limit (but one may exist);
- **infeasible**: there is no valid transfer matrix with probability amplitude greater than the initial α_{min}.

If the output is δ-**optimal** or **approximate**, then the algorithm also returns the valid regions for the variables. From this, we can get out a region for $\hat{U}$ and a region for the probability amplitude α.

Remark 3. It is possible to use standard SMT solvers (*e.g.*, Z3 [11], Yices [13], cvc5 [3]) instead of dReal, which would lead the algorithm to find an exact solution. Algorithm 1 can be modified to call the appropriate solver and replace δ-**optimal** with **optimal**. However in practice, we found that beyond simple examples, dReal would be capable of finding an approximation whereas other solvers would take much longer or could not find one.

3.4   Finding a Unitary from Approximations

If the search from Sect. 3.3 was successful, two matrices will be returned, $\hat{U}_{lb}$ and $\hat{U}_{ub}$, where for any matrix A that has elements $(\hat{U}_{lb})_{ij} \leq (A)_{ij} \leq (\hat{U}_{ub})_{ij}$, A satisfies the δ-weakened expressions. However, these matrices are not unitary. This is because A satisfies the δ-weakened expressions and so $AA^{\dagger} = I + \omega$, where $|\omega_{ij}| \leq \delta$. Therefore, a suitable unitary matrix needs to be found around the region contained by $\hat{U}_{lb}$ and $\hat{U}_{ub}$ to be the final returned transfer matrix.

A unitary matrix that is close this region can be found by taking choosing a matrix within the region, A, and considering its singular value decomposition, $A = V\Sigma W$ [12,16]. Here, V and W are unitary matrices and Σ is a diagonal matrix. Since A is close to being unitary, the elements of the diagonal matrix are close to 1, *i.e.*, $(\Sigma)_{ii} \approx 1$. By replacing Σ with the identity matrix and since V and W are unitary matrices, then $\hat{A} = VIW = VW$ is a unitary matrix that is close to our original matrix A.

We apply the singular value decomposition to $\hat{U}_{lb}$ and $\hat{U}_{ub}$ to get two transfer matrices, $\hat{B}$ and $\hat{C}$ respectively, and then check to see which one more accurately describes our desired quantum computing operation U. To decide between the matrices, we take the part of the matrix that acts on the coincidence basis of $\hat{B}$ and $\hat{C}$ (B and C respectively); find their probability amplitude and scale the matrices by the inverse ($\alpha_{\hat{B}}^{-1}B, \alpha_{\hat{C}}^{-1}C$); and then compare the Frobenius norm

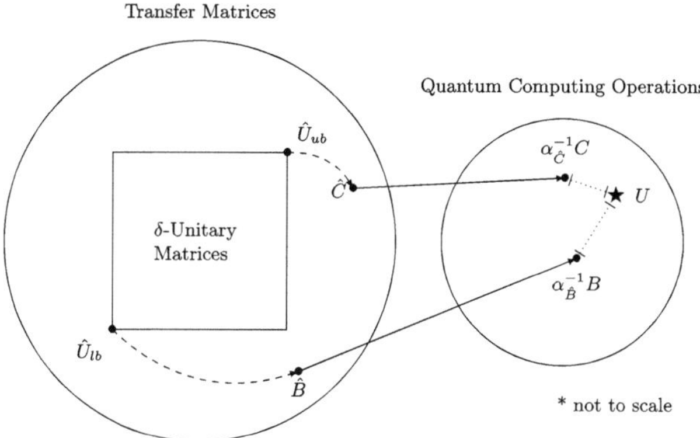

Fig. 4. Visual explanation for obtaining unitary from approximation.

against the desired unitary matrix ($\left\|U - \alpha_{\hat{B}}^{-1}B\right\|_F$, $\left\|U - \alpha_{\hat{C}}^{-1}C\right\|_F$). Whichever norm is smaller, the appropriate matrix is chosen. Figure 4 gives a visualisation.

Thus the final algorithm is relatively simple:

1. Perform the search as described in Algorithm 1 with the appropriate inputs;
2. If an approximated region was found, perform the SVD technique described on the region and return a valid transfer matrix. If no region was found, return `unsat`/`unknown`.

4   Results

The techniques described are implemented in a tool based on Python. We validate our tool using both positive and negative known results in the literature, and then explore previously unknown results. We divide this section based on whether the coincidence basis is post-selected (Sect. 4.1) or heralded (Sect. 4.2). The commands used to provide different results are provided in the artifact [24].

All experiments are performed on a laptop with an Intel(R) Core(TM) Ultra 7 165H 4.30 GHz × 16 cores processor and 32 GB of RAM using Ubuntu 24.04.3 LTS.

4.1   Post-selection

Standard Known Results. One of the main gates of importance for linear optics to implement is the CZ gate. With this gate, one can easily construct any other controlled-U operation, where U is a single qubit operation, by wrapping the CZ gate with appropriate single unitary operations on the target qubit. For instance, a controlled-not operation, $CNOT$, can be decomposed into the operations $(I \otimes H)CZ(I \otimes H)$, where H is the Hadamard operation ($H\left|0\right\rangle = \frac{\left|0\right\rangle+\left|1\right\rangle}{\sqrt{2}}$, $H\left|1\right\rangle = \frac{\left|0\right\rangle-\left|1\right\rangle}{\sqrt{2}}$).

With our tool we can verify a few known facts about the CZ operation:

- a CZ operation is `infeasible` with no auxiliary wires;
- a CZ operation is `infeasible` with a single auxiliary vacuum wire;
- a CZ operation is found with two auxiliary vacuum wires and has the same probability of success as the transfer matrix given in Eq. (3) as δ approaches 0;[3]
- and the above results hold for the $CNOT$ operation as well.

These checks can be performed in a matter of seconds. Thus, for postselection, our tool is capable of producing known results.

Different Auxiliary Wire Setups. It is important to explore what other auxiliary wire setups can be used to synthesise circuits with different probabilities of success to see if higher success rates are possible. Additionally, being able to synthesize different circuits with different auxiliary setups may demonstrate certain behaviours as limits are reached. In this section, we investigate what happens when a single auxiliary wire is used but a different number of photons are sent down the wire. This has been investigated in [2].

The results for synthesizing a CZ operation by sending different number of photons down a single auxiliary wire are visualized in Fig. 5. Our results are similar to those made in [2] (see Sect. 3 and Appendix D therein) and seem to suggest that these are close to the optimal success probabilities (for optical circuits with real phases). For example, our tool can show that it is `infeasible` for a single auxiliary wire with one photon sent down to have a success probability higher than 0.16 using only real values.

As can be seen in the results, initially a single photon sent down a single wire has a higher probability than using two vacuum wires. However, as the number of photons increase, the probability begins to decrease. There is a drop in probability to begin with, but the the success rate begins to decrease at a lower rate as the number of photons increases. It is an open question on whether this converges towards 0 or some other value.

Remark 4. The polynomials generated within the constraints will have degree $2 + n_a$, where n_a is the number of photons in the auxiliary wire ($q = 2$ for q in Remark 2). However, most of the terms have a single variable of high degree since that is where most of the photons originate from and must go, *i.e.*, the terms of the polynomials have one factor of the form $U_{ij}{}^d$ where $d \geq n_a - 2$. This and the fact that the polynomials have very few terms, due to the coincidence basis requiring a large number of photons in the auxiliary wire, is why these high degree polynomials are solvable in a short time using dReal.

---

[3] Modulo phase changes.

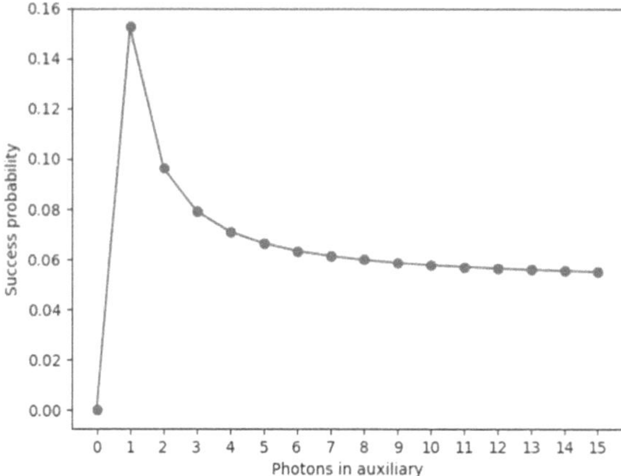

Fig. 5. Best found success probability for a CZ operation by sending numerous photons down one auxiliary wire.

Givens Rotation Gates. A Givens rotation operation is an unitary operation that performs an entanglement on the logical basis states $|01\rangle$ and $|10\rangle$ and does nothing to the $|00\rangle$ and $|11\rangle$ basis states. The Givens rotation operations form a universal set of operations for problems within quantum chemistry [1]. The family of Givens rotation operations that we investigate are of the form

$$G(\theta) = \begin{pmatrix} 1 & 0 & 0 & 0 \\ 0 & \cos(\theta) & -\sin(\theta) & 0 \\ 0 & \sin(\theta) & \cos(\theta) & 0 \\ 0 & 0 & 0 & 1 \end{pmatrix}, \tag{4}$$

with a few examples provided:

$$G(\tfrac{\pi}{4}) = \begin{pmatrix} 1 & 0 & 0 & 0 \\ 0 & \frac{1}{\sqrt{2}} & \frac{1}{\sqrt{2}} & 0 \\ 0 & \frac{1}{\sqrt{2}} & \frac{1}{\sqrt{2}} & 0 \\ 0 & 0 & 0 & 1 \end{pmatrix}, \qquad G(\tfrac{\pi}{2}) = \begin{pmatrix} 1 & 0 & 0 & 0 \\ 0 & 0 & -1 & 0 \\ 0 & 1 & 0 & 0 \\ 0 & 0 & 0 & 1 \end{pmatrix}.$$

Variations with complex factors in the entangling and non-entangling parts do exist, *e.g.*, one variation sends $|11\rangle$ to $e^{i\phi}|11\rangle$, where $\phi \in [0, 2\pi)$ is a parameter.

Our tool is capable of modelling and finding transfer matrices for Givens rotations. For example, with $\theta = \frac{\pi}{2}$, we have the following transfer matrix that uses two vacuum auxiliary wires,

$$\frac{1}{3}\begin{pmatrix} 0 & 0 & -\sqrt{3} & 0 & 0 & -\sqrt{6} \\ \sqrt{6} & 0 & 0 & -\sqrt{3} & 0 & 0 \\ \sqrt{3} & 0 & 0 & \sqrt{6} & 0 & 0 \\ 0 & \sqrt{3} & 0 & 0 & \sqrt{6} & 0 \\ 0 & -\sqrt{6} & 0 & 0 & \sqrt{3} & 0 \\ 0 & 0 & \sqrt{6} & 0 & 0 & -\sqrt{3} \end{pmatrix},$$

and it succeeds with a probability of $\frac{1}{9}$. It looks similar to a CZ gate because the operation itself is simply a permuted CZ gate.

The trend for Givens rotation gates against their found probability of success is mapped out in Fig. 6 with angles between 0 and π.[4] About one quarter of the results couldn't find a solution in the time (those with zero success probability), however running more instance or for a longer time would reveal a success probability in line with the rest of the data.

The circuits for angles 0 and 1 are exact since the related matrices are the identity and the Z gate applied to each qubit respectively (both exactly implementable). The minima of $\frac{1}{9}$ can be observed when the angle is close to 0.5. The success probability initially drops substantially but then drops at a lower rate.

It can be seen that there is some relation between the probability of success of a Givens rotation gate to its angle. There exists a similar relation for controlled phase gates [19], where there is a formula for the optimum probability of success given the angle of the phase $(e^{i\phi})$, $i.e.$, the optimal success probability at angle ϕ is determined by a function $f(\phi)$. However, the controlled phase gate observes non-monotone structure and has a minima $(\min_\phi f(\phi))$ when ϕ is between $\pi/3$ and π. The Givens gates may also have a non-monotonous structure for $0 \leq \theta \leq \frac{\pi}{2}$ depending on the formula for determining the optimal success probability parametrized by θ $(g(\theta))$, but the data gathered seems to indicate $\pi/2$ exactly being the minima $(\min_\theta g(\theta))$.

4.2   Heralded: Replication of Known Results

Now we look at some results for the heralded scheme. With the tool, we are able to prove infeasibility of some trivial and known results:

- a CZ operation is `infeasible` with no or one vacuum wire, and two vacuum wires is `infeasible` with probability at least $\frac{1}{100}$;
- a CZ operation is `infeasible` with a single photon down one auxiliary wire.[5]

Besides these results of impossibility of certain setups, we have found that our technique returns `unknown` (due to timing out) when trying to find a known instance of a CZ operation, $e.g.$, two ancillary wires with one photon down each, which has a known implementation. This is likely due to the increased degree of the polynomials and the increase in the number of equations required to solve.

---

[4] We only investigate values of θ between 0 and π since $G(\theta)^{-1} = G(-\theta)$ (as $\cos(\theta) = \cos(-\theta)$ and $\sin(\theta) = -\sin(-\theta)$). For $-\pi \leq -\theta \leq 0$, the circuit of $G(-\theta)$ is found by inverting the circuit of $G(\theta)$.

[5] Further tests suggest a single wire is `infeasible` for any number of photons.

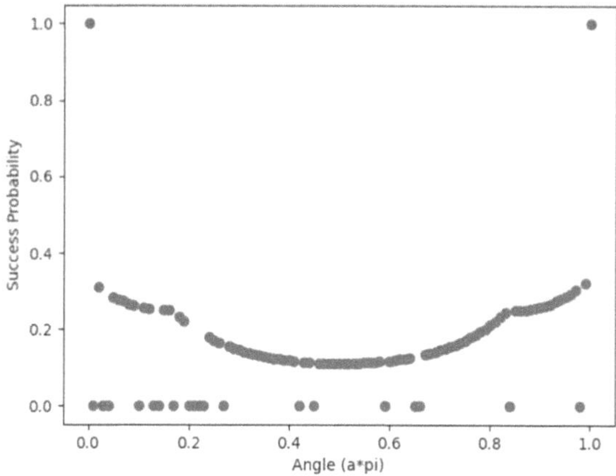

Fig. 6. Plot of found success probability for Givens operations with different angles using two vacuum wires, and a single run with a 180 s timeout.

4.3   Discussion

As can be seen with the Givens examples in Sect. 4.1, photonic quantum computing may find a better use for quantum chemistry than general quantum computing. The Givens rotation gate are more likely to work correctly than the gates used for quantum computing (at least in the post-select case), and due to the Givens rotations being universal for quantum chemistry [1], the circuits could succeed more often than circuits from quantum computing gates. Further investigations are needed for Givens gates that include complex arguments and for controlled Givens gates as well to see if the improvement in success rate is maintained.

We have focussed on results for 2-qubit quantum computing gates, however, finding circuits for gates of other sizes are of interest as well. Gates that act on a single qubit can be easily found by the tool as these gates can be implemented without any auxiliary wires. An instance of a larger gate is the Toffoli gate (CCX), which acts on three qubits, sending $|110\rangle \rightarrow |111\rangle$ (and vice versa) and performs the identity on other computational basis states. There are a few known implementations using linear optics of this and similarly sized gates [2,25,27], but it is unknown if these representations are efficient (in terms of chance of occurring or number of auxiliary/photons used). Our tool is capable of creating the assertions required to be met by circuits of different sizes, but it would take a long time to return a result/circuit for quantum gates that act on 3 or more qubits except for trivial ones (*e.g.*, it is `infeasible` to find a circuit for CCX using no auxiliary wires).

A remark regarding the number of qubits our approach can tackle is in order. The polynomials derived from the formulas shown in Sect. 3.2 are very peculiar. First, they have many variables, such as the square of the number of modes. Secondly, their monomials have a low degree. The constraint `unitary` gives monomials of degree 2, while the constraint `fockequal` gives monomials of a degree corresponding to the number of photons in the circuit. If for the post-selected

CZ of Sect. 4.1 the degree is 2, for heralded gates requiring four photons, the degree increases accordingly. From the perspective of SMT solvers, degree 2 is doable (as shown by our results), but heralded problems quickly reach the limit of what can be done following a naïve approach with existing tools.

Whilst in this paper we have focused on encoding the quantum photonics circuit $\hat{U}$ using real variables, it is possible to modify the variables to be complex using the tool and appropriate constraints can be generated. This involves representing variables as two real variables that represent the real and imaginary part, which essentially doubles the number of variables and constraints needed to specify the properties of $\hat{U}$. This can dramatically increase the time it takes to find a satisfying solution or to prove unsatisfiability. Whilst it is possible to find unitaries for single qubit operations (*e.g.*, the S gate, which does nothing to $|0\rangle$ and sends $|1\rangle$ to $i|1\rangle$), quantum computing operations with two or more qubits currently would take a long time to find a solution or remain out of reach due to the increase in the number of variables. Work would need to be done by the SMT community to develop a theory and/or an implementation capable of more efficiently solving constraints with complex variables. Alternatively, developing techniques for handling multi-variable polynomials of limited degree would be helpful for resolving the generated constraints.

Another improvement to the technique is to consider different coincidence basis states can be used, particularly in the auxiliary wires. Whilst in this paper we have focused on using non-entangled Fock states (*e.g.*, $|n_1, \ldots, n_m\rangle_{\mathcal{F}}$), some results have used entangled auxiliary wires as part of the coincidence basis. For instance, the authors in [25] use the NOON state in two auxiliary wires, which is of the form $\frac{1}{\sqrt{2}}(|N, 0\rangle_{\mathcal{F}} + |0, N\rangle_{\mathcal{F}})$ where N is a positive integer. Implementing such a feature would increase the variety of photonics circuits to consider.

One final technique to consider is to take advantage of symmetries in the quantum computing gate to reduce the number of variables needed. The idea is that multiple variables in the photonics gate can be represented by a single variable if certain symmetric properties hold. For instance, with a CZ gate if an X gate is applied to the qubits, then the $|01\rangle$ and $|10\rangle$ still fundamentally act the same. This would mean fewer variables and potentially a speed-up in searching for a satisfiable instance. More generally, this work has used a direct SMT encoding of the photonics problem. Finding an efficient SMT encoding of the photonics problem (or subsets of the problem) could alleviate the issues discussed.

5 Conclusion

In this paper, we have introduced a search technique based on δ-weakening SMT solvers for finding a linear optics circuit that implements a chosen quantum computing gate on a given setup for the circuit. We showed how any generated approximation from a (δ-weakening) SMT solver that is δ-satisfiable can be turned into an exact solution, which can be used to generate a circuit. We demonstrated the utility of our technique by demonstrating how our tool can replicate and expand upon results known in the literature of linear optics, and further how it can be used to generate new results. However, the technique faces a wall in overcoming the heralded setting of linear optics.

This paper highlights important connections between the SMT-based synthesis approach and photonic circuit design for quantum gates. To overcome the challenge presented by the heralded setting, developments in the area of SMT solving are needed. In particular, a development of a technique for handling constraints consisting of polynomials of bound degree in the NRA setting would be useful. Alternatively, development of techniques to solve Complex Arithmetic (CA) theories and implementing them in a tool would be beneficial.

Acknowledgments. This work has been partially funded by the European Union through the MSCA SE project QCOMICAL, by the French National Research Agency (ANR): projects RECIPROG ANR-21-CE48-0019, PPS ANR-19-CE48-0014, TaQC ANR-22-CE47-0012 and within the framework of "Plan France 2030", under the research projects EPIQ ANR-22-PETQ-0007, OQULUS ANR-23-PETQ-0013, HQI-Acquisition ANR-22-PNCQ-0001 and HQI-R&D ANR-22-PNCQ-0002.

Data Availability Statement. The techniques, tool, and experiments presented can be run or reproduced in a Docker artifact [24]. Sample data for some experiments are provided in Tables 1 and 2.

Table 1. Table for CZ experiment with a single auxiliary wire (Sect. 4.1): Number of Photons sent down a single auxiliary wire and the found probability of success (rounded to 4 decimal places) based on an average of successful runs of 5 runs using $\delta = 0.001$ with a 300 s timeout. The average is calculated by taking the average of the midpoint of the lower and upper bounds of the success probability for each successful run. The average time is calculated based only on the times of the successful runs.

Number of Photons	Average Success Probability	Successes	Average Time
0	0.0000	5	0.3713
1	0.1524	5	15.3623
2	0.0962	5	111.6835
3	0.0791	5	157.0195
4	0.0710	5	202.4442
5	0.0664	5	241.1577
6	0.0634	5	258.3529
7	0.0614	5	265.8271
8	0.0599	5	281.5034
9	0.0587	5	292.2920
10	0.0577	5	325.1113
11	0.0570	5	331.2662
12	0.0563	5	337.8322
13	0.0559	5	350.9504
14	0.0554	5	409.0192
15	0.0550	5	418.1825

Table 2. Sample of angles for Givens matrix using two vacuum wires (Sect. 4.1) and their average success probability (rounded down to 4 decimal places) after 5 runs with a 60 second timeout and $\delta = 0.001$. The average time is calculated based only on the times of the successful runs.

Angle ($a\pi$)	Average Success Probability	Successes	Average Time
0/12	1.0000	5	1.7663
1/12	0.2667	4	60.5386
2/12	0.2500	3	60.5095
3/12	0.1717	4	60.5354
4/12	0.1341	4	60.5279
5/12	0.1164	4	60.5187
6/12	0.1113	5	60.5557
7/12	0.1164	3	60.5107
8/12	0.1340	2	60.4988
9/12	0.1717	4	60.5040
10/12	0.2500	4	60.5152
11/12	0.2626	3	60.5392
12/12	1.0000	5	1.2744

A   Data Tables

References

1. Arrazola, J.M., Di Matteo, O., Quesada, N., Jahangiri, S., Delgado, A., Killoran, N.: Universal quantum circuits for quantum chemistry. Quantum **6**, 742 (Jun 2022). https://doi.org/10.22331/q-2022-06-20-742
2. Baldazzi, A., Pavesi, L.: Universal multiport interferometers for post-selected multi-photon gates. Advanced Quantum Technologies p. 2400418 (2024). https://doi.org/10.1002/qute.202400418
3. Barbosa, H., et al.: cvc5: a versatile and industrial-strength SMT solver. In: Fisman, D., Rosu, G. (eds.) Tools and Algorithms for the Construction and Analysis of Systems, pp. 415–442. Springer International Publishing, Cham (2022). https://doi.org/10.1007/978-3-030-99524-9_24
4. Barrett, C., Fontaine, P., Tinelli, C.: The SMT-LIB Standard: Version 2.6. Tech. rep., Department of Computer Science, The University of Iowa (2017). available at www.SMT-LIB.org
5. Barrett, C., Tinelli, C.: Satisfiability modulo theories. In: Clarke, E.M., Henzinger, T.A., Veith, H., Bloem, R. (eds.) Handbook of Model Checking, pp. 305–343. Springer International Publishing, Cham (2018). https://doi.org/10.1007/978-3-319-10575-8_11
6. Bauer-Marquart, F., Leue, S., Schilling, C.: symQV: automated symbolic verification of quantum programs. In: Chechik, M., Katoen, J.P., Leucker, M. (eds.) Formal Methods, pp. 181–198. Springer International Publishing, Cham (2023). https://doi.org/10.1007/978-3-031-27481-7_12

7. Berent, L., Burgholzer, L., Wille, R.: Towards a SAT encoding for quantum circuits: a journey from classical circuits to clifford circuits and beyond. In: Meel, K.S., Strichman, O. (eds.) 25th International Conference on Theory and Applications of Satisfiability Testing (SAT 2022). Leibniz International Proceedings in Informatics (LIPIcs), vol. 236, pp. 18:1–18:17. Schloss Dagstuhl – Leibniz-Zentrum für Informatik, Dagstuhl, Germany (2022). https://doi.org/10.4230/LIPIcs.SAT.2022.18

8. Bernardini, F., Chakraborty, A., Ordóñez, C.R.: Quantum computing with trapped ions: a beginner's guide. Eur. J. Phys. **45**(1), 013001 (2023). https://doi.org/10.1088/1361-6404/ad06be

9. Clarino, D., Seino, K., Matsuo, A., Yamashita, S.: Utilizing don't-cares to minimize CNOTs in synthesizing NNA compliant quantum circuits. In: 2024 IEEE International Conference on Quantum Computing and Engineering (QCE), vol. 02, pp. 500–501 (2024). https://doi.org/10.1109/QCE60285.2024.10375

10. Clements, W.R., Humphreys, P.C., Metcalf, B.J., Kolthammer, W.S., Walmsley, I.A.: Optimal design for universal multiport interferometers. Optica **3**(12), 1460–1465 (2016). https://doi.org/10.1364/OPTICA.3.001460

11. De Moura, L., Bjørner, N.: Z3: An efficient SMT solver. In: Proceedings of the theory and practice of software, 14th international conference on tools and algorithms for the construction and analysis of systems, pp. 337–340. TACAS'08/ETAPS'08, Springer-Verlag, Berlin, Heidelberg (2008). https://doi.org/10.1007/978-3-540-78800-3_24

12. D.M.Reich.: Characterisation and identification of unitary dynamics maps in terms of their action on density matrices, unpublished

13. Dutertre, B.: Yices 2.2. In: Biere, A., Bloem, R. (eds.) Computer Aided Verification, pp. 737–744. Springer International Publishing, Cham (2014). https://doi.org/10.1007/978-3-319-08867-9_49

14. Gao, S., Avigad, J., Clarke, E.M.: δ-complete decision procedures for satisfiability over the reals. In: Gramlich, B., Miller, D., Sattler, U. (eds.) Automated Reasoning, pp. 286–300. Springer Berlin Heidelberg, Berlin, Heidelberg (2012). https://doi.org/10.1007/978-3-642-31365-3_23

15. Gao, S., Kong, S., Clarke, E.M.: dReal: an SMT solver for nonlinear theories over the reals. In: Bonacina, M.P. (ed.) Automated Deduction – CADE-24, pp. 208–214. Springer Berlin Heidelberg, Berlin, Heidelberg (2013). https://doi.org/10.1007/978-3-642-38574-2_14

16. D Goerz, M.: Finding the closest unitary for a given matrix. https://michaelgoerz.net/notes/finding-the-closest-unitary-for-a-given-matrix/ (2014). Accessed14 Sept D2025

17. Hormozi, L., Zikos, G., Bonesteel, N.E., Simon, S.H.: Topological quantum compiling. Phys. Rev. B **75**, 165310 (2007). https://doi.org/10.1103/PhysRevB.75.165310

18. Jakobsen, A.B., Clausen, A.B., van de Pol, J., Shaik, I.: Depth-optimal quantum layout synthesis as SAT. In: Berg, J., Nordström, J. (eds.) 28th International Conference on Theory and Applications of Satisfiability Testing (SAT 2025). Leibniz International Proceedings in Informatics (LIPIcs), vol. 341, pp. 16:1–16:17. Schloss Dagstuhl – Leibniz-Zentrum für Informatik, Dagstuhl, Germany (2025). https://doi.org/10.4230/LIPIcs.SAT.2025.16

19. Kieling, K., O'Brien, J.L., Eisert, J.: On photonic controlled phase gates. New J. Phys. **12**(1), 013003 (2010). https://doi.org/10.1088/1367-2630/12/1/013003

20. Kitaev, A.: Fault-tolerant quantum computation by anyons. Ann. Phys. **303**(1), 2–30 (2003). https://doi.org/10.1016/S0003-4916(02)00018-0

21. Knill, E.: Quantum gates using linear optics and postselection. Phys. Rev. A **66**, 052306 (2002). https://doi.org/10.1103/PhysRevA.66.052306
22. Knill, E., Laflamme, R., Milburn, G.J.: A scheme for efficient quantum computation with linear optics. Nature **409**(6816), 46–52 (2001). https://doi.org/10.1038/35051009
23. Kok, P., Munro, W.J., Nemoto, K., Ralph, T.C., Dowling, J.P., Milburn, G.J.: Linear optical quantum computing with photonic qubits. Rev. Mod. Phys. **79**, 135–174 (2007). https://doi.org/10.1103/RevModPhys.79.135
24. Lewis, M., Valiron, B.: Finding photonics circuits via δ-weakening SMT. Zenodo, v1.0.1 (Sep 2025). https://doi.org/10.5281/zenodo.17233057
25. Li, Y., et al.: Quantum Fredkin and Toffoli gates on a versatile programmable silicon photonic chip. npj Quant. Inform. **8**(1), 112 (Sep 2022). https://doi.org/10.1038/s41534-022-00627-y
26. Lin, W.H., Kimko, J., Tan, B., Bjørner, N., Cong, J.: Scalable optimal layout synthesis for nisq quantum processors. In: 2023 60th ACM/IEEE Design Automation Conference (DAC) pp. 1–6 (2023). https://doi.org/10.1109/DAC56929.2023.10247760
27. Liu, W.Q., Wei, H.R.: Linear optical universal quantum gates with higher success probabilities. Adv. Quant. Technol. **6**(5), 2300009 (2023). https://doi.org/10.1002/qute.202300009
28. Nielsen, M.A., Chuang, I.L.: Quantum Computation and Quantum Information: 10th Anniversary Edition, 10th edn. Cambridge University Press, USA (2011)
29. Ralph, T.C., Langford, N.K., Bell, T.B., White, A.G.: Linear optical controlled-not gate in the coincidence basis. Phys. Rev. A **65**, 062324 (2002). https://doi.org/10.1103/PhysRevA.65.062324
30. Reck, M., Zeilinger, A., Bernstein, H.J., Bertani, P.: Experimental realization of any discrete unitary operator. Phys. Rev. Lett. **73**, 58–61 (1994). https://doi.org/10.1103/PhysRevLett.73.58
31. Seino, K., Yamashita, S.: An SMT-solver-based synthesis of NNA-compliant quantum circuits consisting of CNOT, H and T gates. In: 2023 28th Asia and South Pacific Design Automation Conference (ASP-DAC). pp. 196–201 (2023). https://doi.org/10.1145/3566097.3567931
32. Shaik, I., van de Pol, J.: Optimal layout synthesis for deep quantum circuits on NISQ Processors with 100+ Qubits. In: Chakraborty, S., Jiang, J.H.R. (eds.) 27th International Conference on Theory and Applications of Satisfiability Testing (SAT 2024). Leibniz International Proceedings in Informatics (LIPIcs), vol. 305, pp. 26:1–26:18. Schloss Dagstuhl – Leibniz-Zentrum für Informatik, Dagstuhl, Germany (2024). https://doi.org/10.4230/LIPIcs.SAT.2024.26
33. Yang, J., Kharkov, Y.A., Shi, Y., Heule, M.J.H., Dutertre, B.: Quantum circuit mapping based on incremental and parallel SAT solving. In: Chakraborty, S., Jiang, J.H.R. (eds.) 27th International Conference on Theory and Applications of Satisfiability Testing (SAT 2024). Leibniz International Proceedings in Informatics (LIPIcs), vol. 305, pp. 29:1–29:18. Schloss Dagstuhl – Leibniz-Zentrum für Informatik, Dagstuhl, Germany (2024). https://doi.org/10.4230/LIPIcs.SAT.2024.29
34. Zak, D., Mei, J., Lagniez, J.M., Laarman, A.: Reducing quantum circuit synthesis to #SAT. In: de la Banda, M.G. (ed.) 31st International Conference on Principles and Practice of Constraint Programming (CP 2025). Leibniz International Proceedings in Informatics (LIPIcs), vol. 340, pp. 38:1–38:21. Schloss Dagstuhl – Leibniz-Zentrum für Informatik, Dagstuhl, Germany (2025). https://doi.org/10.4230/LIPIcs.CP.2025.38

Forward Symbolic Execution
for Trustworthy Automation of Binary
Code Verification

Andreas Lindner[1], Karl Palmskog[2]([✉]), Scott Constable[3], Mads Dam[2],
Roberto Guanciale[2], and Hamed Nemati[2]

[1] Uppsala University, Uppsala, Sweden
`andreas.lindner@angstrom.uu.se`
[2] KTH Royal Institute of Technology, Stockholm, Sweden
`{palmskog,mfd,robertog,hnemati}@kth.se`
[3] Intel Labs, Hillsboro, OR, USA
`scott.d.constable@intel.com`

Abstract. Control flow in unstructured programs can be complex and dynamic, which makes static analysis difficult. Yet, automated reasoning about unstructured control flow is important when certifying properties of binary (machine) code in trustworthy systems, e.g., cryptographic routines. We present a theory of forward symbolic execution for unstructured programs suitable for use in theorem provers that enables automated verification of both functional and non-functional program properties. The theory's foundation is a set of inference rules where each member corresponds to an operation in a symbolic execution engine. The rules are designed to give control over the tradeoff between the preservation of precision and introduction of overapproximation. We instantiate our theory for BIR, a previously proposed intermediate language for binary analysis. We demonstrate how symbolic executors can be constructed for BIR with common optimizations such as pruning of infeasible symbolic states. We implemented our theory in the HOL4 theorem prover using the HolBA binary analysis library, obtaining machine-checked proofs of soundness of symbolic execution for BIR. We practically evaluated two applications of our theory: verification of functional properties of RISC-V binaries and verification of execution time bounds of programs running on the ARM Cortex-M0 processor. The evaluation shows that such verification can be automated with moderate overhead on medium-sized programs.

Keywords: symbolic execution · theorem proving · binary analysis

1 Introduction

Unstructured programming permits "jumps of the control to strange places" via labels and goto instructions, resulting in control flow with the appearance of a "can of spaghetti" [28]. Structured programming [17], where instructions or statements have a single exit point, is now the norm for code written by software

Y.-F. Chen et al. (Eds.): VMCAI 2026, LNCS 16417, pp. 147–172, 2026.
https://doi.org/10.1007/978-3-032-15700-3_8

developers—but code with unstructured control flow lives on in high assurance systems, e.g., as output of compilers and as hand-optimized assembly.

Static analysis of unstructured code, e.g., of RISC-V binaries [59], is difficult due to its complex, dynamic control flow. For instance, jump targets may be computed at runtime from register and memory values (*indirect jumps*). Nevertheless, trustworthy *automated* reasoning about binary code is important for certifying the correctness of high-assurance systems. In particular, binary verification may be used for *translation validation* when using untrusted compilers [53] and for directly ensuring the correctness of low-level cryptographic code [46] and security mechanisms such as software fault isolation [60].

We believe *forward symbolic execution* is well-suited for automating binary verification. Such an approach explores program paths with symbolic input, naturally captures cause-effect in control and data flow, tracks how data drives behavior without source-level types, and integrates well with control-flow reconstruction and memory-alias analysis. Based on these insights, we present a general theory of forward symbolic execution for unstructured programs suitable for use in interactive theorem provers. The basis of the theory is a set of core inference rules, each corresponding to operations in an abstract symbolic execution engine, designed to control the tradeoff between precision and overapproximation, which is important when analyzing real-world binary code (Sect. 3 and 4). The theory can be instantiated for unstructured languages by relating their concrete and symbolic semantics. We instantiate the theory for BIR, a previously proposed intermediate language for binary analysis [42], and construct sound symbolic executors with optimizations such as pruning of infeasible states. We implemented the theory in the HOL4 theorem prover [32] and instantiated it using the HolBA binary analysis library [42], providing machine-checked proofs of soundness of symbolic execution for BIR (Sect. 5).

Using our implementation, we extended HolBA with support for two novel applications: (1) automated verification of functional properties of RISC-V binaries using contracts and (2) automated verification of execution time bounds of binaries executing on the ARM Cortex-M0 processor. We call this extension HolBA-SE. Our application of HolBA-SE for functional verification of RISC-V code is based on previous work which *lifts* RISC-V instructions to BIR [42] while preserving semantics as defined by L3 [25], and provides Hoare-style contracts for binary programs [44]. To this we add trustworthy automation of BIR contract proofs and for propagating such proofs to RISC-V level (Sect. 6). We applied HolBA-SE to RISC-V case studies from two high-assurance domains, namely, cryptography and operating system kernels, indicating that we can automate translation validation and verification of hand-written assembly outside the scope of verified compilers. In our application of HolBA-SE for verification of execution time bounds (Sect. 7), we rely on the fact that in relatively simple processors such as the ARM Cortex-M0, instructions take constant time to execute. This means that we can explicitly represent execution time as a symbolic state variable, which we use to perform a collection of case studies on ARM Cortex-M0 binaries where we verify their execution time bounds. Our evaluation compares our results against an established industrial tool, aiT, showcasing the potential of our approach for trustworthy worst-case execution time (WCET) analysis.

In summary, we make the following contributions:

1. We propose a general theory of symbolic execution for unstructured programs designed and adapted for use in interactive theorem provers.
2. We provide an implementation of our theory in the HOL4, along with an instantiation for the BIR language and machine-checked soundness proofs.
3. We design symbolic executors that automatically apply our theory's inference rules for symbolic execution of BIR programs generated from RISC-V and Cortex-M0 machine code, building HOL4 progress structure theorems.
4. Using progress structures, we implement automation for functional verification of RISC-V programs and WCET verification for Cortex-M0 programs on top of the HolBA library, experimentally evaluated in several case studies.

Source code for HolBA-SE and our verification case studies is available under an open source license [43], and has been integrated into the HolBA library [33].

2   Background

We use BIR [42] as a vehicle for explaining our theory and implementation. BIR (Binary Intermediate Representation) is an intermediate language designed for binary analysis. It abstracts away ISA (Instruction Set Architecture) details while preserving the semantics of binaries, enabling architecture-independent reasoning about machine code. Beyond serving as a convenient notation in this paper, BIR provides a general framework for implementing trustworthy binary analyses, thanks to its formal semantics and implementation in the HOL4 theorem prover. The application of symbolic execution to binary code presents unique challenges. To illustrate these, we use a modular exponentiation (ModExp) routine expressed in BIR, whose control flow graph is presented in Fig. 2. The pseudocode of the corresponding modular exponentiation algorithm is shown in Fig. 1. For simplicity, the program operates only on 32-bit words instead of relying on a bignum library. The program highlights key challenges—unstructured control flow, path explosion, expression growth, and memory aliasing—that motivate our symbolic execution rules and optimizations.

2.1   The BIR Language

A BIR program consists of uniquely labeled blocks, with each block containing a sequence of statements. In the running example, the blocks consist of a single statement and are labeled by $1, \ldots, 14$. Labels $l \in L_p$ correspond to specific locations in the program p and are commonly used as the target for jump instructions. BIR statements include assignment of BIR expressions to variables (e.g., `SP := SP - 8`), unconditional and conditional jumps (e.g., `goto R3` and `if R3-32 goto 6`), a halt statement to indicate execution termination, and

```
f(e, b, m):
  [prologue]
  r := 1
  for i in 0 .. 31 {
    if e[i]
      r := (r*b)%m
    b := (b*b)%m
  }
  return r
  [epilogue]
```

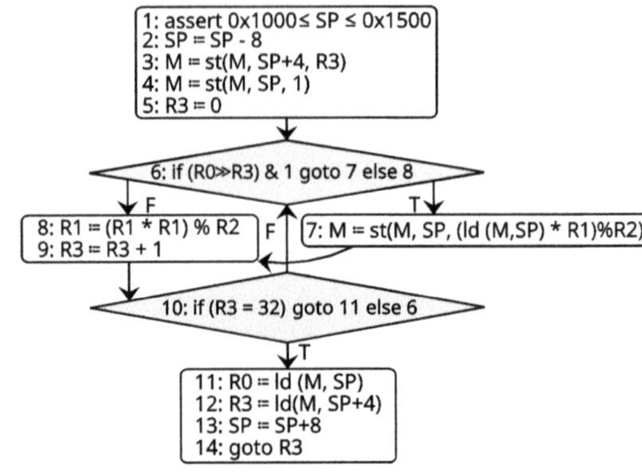

Fig. 1. ModExp pseudocode.   **Fig. 2.** BIR CFG of ModExp.

`assert` to evaluate a boolean expression and terminate execution if the assertion fails. Expressions in BIR include constants $v \in BV$, variables $x \in BVars$, conditionals, arithmetic operations, and memory operations such as *load* (`ld`) and *store* (`st`).

Fundamentally, the content of the following sections is not specific to BIR, but requires the underlying concrete model and its semantics to satisfy some properties, which indeed are satisfied by BIR. The concrete execution is modeled by a deterministic and total transition relation $\rightarrow$. In BIR, states $\perp$ and $\top$ (reached by failed assertions and halt, respectively) represent failing and successful terminating states that can only transition to themselves; every concrete state $s \in S$ consists of a program counter and a store, extracted by $pc(s)$ and $store(s)$, respectively; stores δ partially map variables $x \in BVars$ to concrete values $v \in BV$, and instructions are addressed using labels ranged over by l.

Because code is unstructured, we use sets of labels $L \subseteq L_p$ (not necessarily consecutive) to identify fragments of binary code. We define $s \leadsto_L s' = \{(s, s') \mid s \rightarrow s' \wedge pc(s) \in L\}$ as the restriction of the transition relation, where source states must belong to a fragment L, and $\rightarrow_L^n$ for the application of n transitions. Note that these relations are partial and deterministic, and if $s_0 \rightarrow_L^n s_n$ then each intermediate state has program counter in L: if $i < n$ and $s_0 \rightarrow_L^i s_i$ then $pc(s_i) \in L$.

2.2 Running Example

The BIR program in Fig. 2 implements modular exponentiation using the standard right-to-left method, computing $b^e \bmod m$. Parameters e, b, and m are passed in BIR variables R0-R2, and R3 stores the return address. The first instruction (line 1) asserts that the stack pointer (SP) is restricted to the designated stack region of memory, spanning addresses 0x1000 to 0x1500. This assumption becomes the initial path condition during symbolic execution. Lines 2–3 form the function prologue: the function allocates two words to the stack and then pushes the return address (passed in through the R3 register) onto the stack, whose top is pointed to by the stack pointer SP. In BIR, memory is modeled as a single map variable M, which is accessed and modified by the operations ld (load) and st (store). The local variable r is stored at address SP - 8, and line 4 sets its initial value to 1. The loop counter i is tracked in register R3, which is initialized to 0 at line 5. Line 6 is a conditional jump that evaluates the if condition and determines the subsequent control flow. Lines 7 and 8 implement the multiplication-and-modulo operations of the pseudocode: line 7 updates the memory location, where r is allocated on the stack, while line 8 updates the base value b directly in a register. Lines 9 and 10 implement the loop's iteration logic, incrementing the counter and branching back to line 6 until 32 iterations have been completed, after which execution proceeds to line 11. At this point, the value of r is reloaded from the stack into the return register R0. Finally, lines 12–14 constitute the epilogue: the function restores the original return address, resets the stack pointer, and executes an indirect jump via R3 to return control to the caller.

2.3   Challenges and Approach

This example highlights several core difficulties of symbolic execution for binary code: **C1 Unstructured control flow**: Indirect jumps complicate reconstruction of feasible control-flow targets, since jump addresses may depend on runtime values in registers or memory. For instance, instead of branching directly at line 6, the program could implement the conditional jump via a lookup table stored in memory, where the target address is computed from the branch condition, e.g., goto ld(M, 0x100+(((R0>>R3)&1)*4)). In this case, the jump destinations (e.g., lines 7 and 8) are retrieved from memory addresses 0x104 and 0x100, respectively. Such table-based indirection introduces an additional layer of complexity, requiring the analysis to identify all feasible addresses and corresponding memory contents to recover the possible control flow. **C2 Path explosion**: The conditional branch inside the loop leads to an exponential number of feasible execution paths, in this case 2^{32}). **C3 Expression growth**: Operations that repeatedly manipulate memory and registers, such as the modular multiplication in line 7, duplicate symbolic expressions, and cause exponential expression growth. **C4 Memory aliasing**: Since binary programs operate on flat memory, reasoning about disjointness of memory locations becomes necessary, for example, to show that internal loop operations do not overwrite the return address. **C5 Repeated fragments**: Loops and repeated function calls execute multiple times the same code fragments, leading to redundant symbolic computations if treated naively. **C6 Interval reasoning**: Many analyses require reasoning about bounded counters or index values (e.g., the loop variable), which is difficult to express without explicit support for symbolic intervals. **C7 Architectural dependencies**: Symbolic execution at the binary level is tightly coupled to the ISA and its semantics. Without an intermediate language like BIR, each symbolic executor must be tailored to a specific ISA, limiting reuse across architectures.

Our approach to addressing these challenges simultaneously is built on three key elements. **Forward exploration**: for unstructured code (C1), feasible targets of indirect jumps or table lookups can only be determined by symbolically executing the preceding instructions. **Compositional inference rules**: These rules support controlled overapproximation through the introduction of free symbols. The rules address C1 (path explosion) and C2 (expression blow-up) by merging equivalent states and bounding symbolic expression growth, while C4 (memory aliasing) is mitigated through simplification rules that identify equivalent or disjoint memory accesses. The rules also address C5 (repeated fragments) by allowing the instantiation of previously analyzed code fragments, and address C6 (interval reasoning) by manipulating path conditions and free symbols in a sound manner. **Integration with HolBA**: Binaries are lifted into BIR, an intermediate language with a formally verified transpiler in HOL4. HolBA provides an ISA-independent substrate, directly addressing C7 by allowing the same symbolic executor to reason about RISC-V, ARM, or other architectures once their lifting to BIR is defined.

Our symbolic execution data structures and soundness definitions are designed to allow incremental proofs, facilitating implementation of proof automations in interactive theorem provers. As an instance, we describe our implementation of automatic symbolic executors and semi-automatic optimization procedures within HOL4 and showcase the verification of binary code properties.

3   Symbolic Semantics

The inference rules for symbolic execution in Sect. 4 are general, but require a symbolic semantics for individual instructions of the specific model. In this section, we lay down notation for the symbolic semantics and we exemplify a few rules for the symbolic semantics of BIR, which captures the main traits of the semantics supported by our symbolic executor.

We assume symbols α that range over the set $BSyms$ and a symbolic expression language that supports at least first-order logic. Symbolic expressions are denoted by $\overline{v} \in \overline{BV}$. A symbolic store $\overline{\delta}$ is a mapping of variables to symbolic expressions. We use $\overline{\delta}_{id}$ to refer to the symbolic store that maps each variable to a different symbol, i.e., $\overline{\delta}_{id}(\mathtt{V}) = \alpha_{\mathtt{V}}$ for each $\mathtt{V}$. We use ϕ for boolean symbolic expressions, which are used, for instance, as path conditions. A symbolic state $\overline{s}$ is a triple of a concrete program counter l, a symbolic store $\overline{\delta}$, and a path condition ϕ. We define the accessor functions $pc(\overline{s}) = l$, $store(\overline{s}) = \overline{\delta}$, $path(\overline{s}) = \phi$.

The symbolic and concrete semantics are connected using a partial interpretation $H : BSyms \rightarrow BV$, mapping symbols to concrete values. Interpretation is lifted to symbolic expressions and stores, i.e., the interpretation $H(\overline{v})$ produces a concrete value, and $H(\overline{\delta})$ produces a concrete store. The resolution function $t_\phi(\overline{v}) = \{v \mid \exists H.\ H(\phi) \wedge H(\overline{v}) = v\}$ computes the set of all concretizations of the symbolic expression $\overline{v}$ under the path condition ϕ.

Definition 1. *The symbolic state $\bar{s}$ matches a concrete state s via the interpretation H, written $\bar{s} \overset{H}{\triangleright} s$, if $pc(\bar{s}) = pc(s)$, $H(store(\bar{s})) = store(s)$, and $H(path(\bar{s}))$.*

For sets of symbolic states Π, we write $\Pi \overset{H}{\triangleright} s$ if there is an $\bar{s} \in \Pi$ such that $\bar{s} \overset{H}{\triangleright} s$.

We exemplify the symbolic transition rule for conditional jumps in BIR:

$$\frac{p(l) = \text{if } e \text{ goto } e_t \text{ else } e_f \quad \phi_t = (\!|e|\!)_{\bar{\delta}} \quad \phi_f = \neg\phi_t \quad x \in \{t, f\} \quad l_b \in t_{\phi \wedge \phi_x}((\!|e_x|\!)_{\bar{\delta}})}{p \vdash (l, \bar{\delta}, \phi) \hookrightarrow (l_b, \bar{\delta}, \phi \wedge \phi_x)}$$

Note that the rule hypothesis requires knowing the current instruction. Therefore, without loss of generality, we treat the program counter as a concrete value rather than a symbolic expression. In fact, the rules are defined to account for all possible target values of the program counter via the resolution function. The rules use symbolic evaluation of the BIR expressions, which is written as $(\!|e|\!)_{\bar{\delta}}$ and substitutes in e all variables with the corresponding symbolic expression from the store $\bar{\delta}$ and results in a symbolic expression.

In the following, we use the total function $symsem(\bar{s}) = \{\bar{s}' \mid \bar{s} \hookrightarrow \bar{s}'\}$ to represent all possible successors of an input state in the symbolic semantics. Soundness means that the symbolic semantics overapproximates the concrete one, which we formulate as a simulation.

Theorem 1. *If $\bar{s} \overset{H}{\triangleright} s$ and $s \rightarrow s'$, then exists $\bar{s}' \in symsem(\bar{s})$ such that $\bar{s}' \overset{H}{\triangleright} s'$.*

Note that the same interpretation function H is used for matching both source and target states.

4 Symbolic Execution

One could build the symbolic executor by iteratively applying $symsem(\bar{s})$. However, this usually results in a large execution tree, which is infeasible to directly compute. Similarly, this naive approach neither allows for overapproximations, which is useful to speed up computation, nor to reuse previous results. The main goal of the symbolic execution rules below is to formalize sound compositions, specializations, computation reuses, simplifications, and approximations that preserve soundness throughout the analysis. While we exemplify the rules only for the symbolic semantics of BIR, the inference rule system has been proven correct for any symbolic and concrete semantics that satisfies Theorem 1.

4.1 Progress Structures

The first building block is the formalization of the intermediate results of symbolic execution. Symbolic execution iteratively builds up and composes symbolic execution trees, but only keeps track of the roots and the leaves of the trees and disregards the internal nodes that are constructed with the symbolic semantics.

We call this object a progress structure and denote it as $\bar{s} \rightsquigarrow_L \Pi$. A progress structure captures that all concrete states matched by the source state $\bar{s}$ execute through the label set L and eventually reach a concrete state that is matched by one of the target states in Π.

In order to formally define the meaning of a progress structure, we first need to introduce some auxiliary notation. The function $\Lambda(\bar{s})$ returns the set of symbols that occur in $\bar{s}$, and we overload it for sets of states as the union of symbols in the states. For a progress structure $\bar{s} \rightsquigarrow_L \Pi$, we call $bs(\bar{s} \rightsquigarrow_L \Pi) = \Lambda(\bar{s})$ its bound symbols and $fs(\bar{s} \rightsquigarrow_L \Pi) = \Lambda(\Pi) \setminus \Lambda(\bar{s})$ its free symbols. We use $H' \supseteq H$ to represent that the partial function H' extends H, i.e. $Dom(H') \supseteq Dom(H)$ and $H'(\alpha) = H(\alpha)$ for all $\alpha \in Dom(H)$.

The following two definitions support the formulation needed to overapproximate with fresh symbols in target states of progress structures.

Definition 2. *An interpretation H is minimal for $\bar{s}$, written as $\downarrow_{\bar{s}} (H)$, if $Dom(H) = \Lambda(\bar{s})$.*

This predicate is applied to source states to ensure that interpretations can be extended with arbitrary values for *free* symbols not occurring in the initial state.

Definition 3. *The symbolic state $\bar{s}'$ loosely matches the concrete state s' via H, written as $\bar{s}' \stackrel{H}{\blacktriangleright} s'$, if there is an interpretation $H' \supseteq H$ such that $\bar{s}' \stackrel{H'}{\triangleright} s'$.*

This predicate relaxes matching using interpretation extension to allow free symbols in target states $\bar{s}'$ to take on arbitrary values in the matching of concrete states s'. We then define the meaning of a progress structure via its soundness.

Definition 4. *A progress structure is sound, written $\vdash \bar{s} \rightsquigarrow_L \Pi$, if for every interpretation $\downarrow_{\bar{s}} (H)$ and state $\bar{s} \stackrel{H}{\triangleright} s$, exists $s \rightarrow^n_L s'$ where $n > 0$, and exists $\bar{s}' \in \Pi$ such that $\bar{s}' \stackrel{H}{\blacktriangleright} s'$.*

Compared with the soundness of the symbolic semantics (Theorem 1), the formulation of the soundness of the symbolic execution is different in two ways.

First, we assume minimality of the initial interpretation ($\downarrow_{\bar{s}} (H)$) and allow loose matches for final states ($\bar{s}' \stackrel{H}{\blacktriangleright} s'$), which together enable the introduction of fresh symbols. For example, suppose that the architecture has only registers R1 and R2 and we symbolically execute line 8 of Fig. 2. The symbolic semantics is $(8, \bar{\delta}_{id}, T) \hookrightarrow (9, \bar{\delta}_{id}[\text{R1} \mapsto \alpha_{\text{R1}} * \alpha_{\text{R1}} \% \alpha_{\text{R2}}], T)$. Progress structures allow *forgetting* the exact value assigned to R1, producing the sound progress structure $(8, \bar{\delta}_{id}, T) \rightsquigarrow_{\{8\}} \{(9, \bar{\delta}_{id}[\text{R1} \mapsto \alpha'], T)\}$ (as long as $\alpha' \notin \Lambda(\bar{\delta}_{id})$). Because we require any initial interpretation to be minimal (e.g., for a state $\{\text{R1} \mapsto 2, \text{R2} \mapsto 3\}$ the interpretation has to be $\{\alpha_{\text{R1}} \mapsto 2, \alpha_{\text{R2}} \mapsto 3\}$), this does not map the fresh symbol α'. Instead, the loose interpretation matching allows to choose any extended interpretation (e.g. $\{\alpha_{\text{R1}} \mapsto 2, \alpha_{\text{R2}} \mapsto 3, \alpha' \mapsto 1\}$) that maps α' to any value, including the value that the concrete execution produces. If we had not assumed the initial interpretation to be minimal, the definition would be too strict and would not allow overapproximation with fresh symbols.

$$\frac{symsem(\bar{s}) = \Pi \quad pc(\bar{s}) \in L}{\vdash \bar{s} \rightsquigarrow_L \Pi} \text{ SYMBSTEP} \qquad \frac{\vdash \bar{s} \rightsquigarrow_L \Pi \cup \{(l, \bar{\delta}, \phi)\}}{\vdash \bar{s} \rightsquigarrow_L \Pi \cup \{(l, \bar{\delta}, \phi \wedge \phi'), (l, \bar{\delta}, \phi \wedge \neg\phi')\}} \text{ CASE}$$

$$\frac{\vdash \bar{s} \rightsquigarrow_L \Pi \cup \{\bar{s}'\} \quad \forall H.\ H(\neg path(\bar{s}'))}{\vdash \bar{s} \rightsquigarrow_L \Pi} \text{ INFEASIBLE} \qquad \frac{\vdash \bar{s} \rightsquigarrow_L \Pi \quad \alpha' \notin \Lambda(\bar{s}) \cup \Lambda(\Pi)}{\vdash \bar{s}[\alpha'/\alpha] \rightsquigarrow_L \Pi[\alpha'/\alpha]} \text{ RENAME}$$

$$\frac{\vdash \bar{s} \rightsquigarrow_L \Pi \cup \{\bar{s}'\} \quad \alpha \notin \Lambda(\bar{s}) \quad \alpha' \notin \Lambda(\bar{s}) \cup \Lambda(\bar{s}')}{\vdash \bar{s} \rightsquigarrow_L \Pi \cup \{\bar{s}'[\alpha'/\alpha]\}} \text{ FREESYMB\_RENAME}$$

$$\frac{\vdash \bar{s} \rightsquigarrow_L \Pi \quad \alpha \in \Lambda(\bar{s}) \quad \Lambda(\bar{v}) \cap fs(\bar{s} \rightsquigarrow_L \Pi) = \emptyset}{\vdash \bar{s}[\bar{v}/\alpha] \rightsquigarrow_L \Pi[\bar{v}/\alpha]} \text{ SUBST}$$

Fig. 3. Overview of the symbolic-execution core inference-rules, part 1.

For example, it would require satisfying the property also for all interpretations that bind α' to some other values, including the ones that do not match the final value of R1.

Second, the concrete execution does not occur as an assumption but is a conjunct of the conclusion instead. Therefore, a sound progress structure also guarantees the termination of the concrete execution: the concrete source state must, in a finite number of transitions within L, reach a concrete target state that is loosely matched by one of the target states. We require $n > 0$ to be able to prove properties for looping code by ensuring progress when the source and target states point to the same label.

Unstructured code can be handled well with this soundness definition as arbitrary control flows are contained and abstracted as label sets, which represent code fragments. Note that indirect jumps are already handled by the symbolic semantics, which assumes knowledge about all feasible jump targets and captures those in the symbolic state set of Theorem 1.

A sound progress structure ensures that all matched concrete executions only encounter program labels in L before matching one of the targets $\bar{s}' \in \Pi$. Therefore, if $pc(\bar{s}') \notin L$ then $pc(\bar{s}')$ is only encountered once in the end. For example, the progress structure from the previous paragraph guarantees that the program counter of the target state 9 is only encountered in the final execution state, because the relation $s \rightarrow_L^n s'$ establishes that s' is reached and no transition is taken from a state pointing to $9 \notin \{8\}$. This allows establishing properties when reaching a certain label of the program for the first time. Because in Definition 4 the label set only appears in the relation $s \rightarrow_L^n s'$, which is under an existential quantifier, extending label sets is permissible. We therefore may add 9 to the label set of the previous progress structure and obtain a sound but weaker progress structure: $(8, \bar{\delta}_{id}, T) \rightsquigarrow_{\{8,9\}} \{(9, \bar{\delta}_{id}[\text{R1} \mapsto \alpha'], T)\}$. In fact, since the program counter of the target state is now in the label set, according to Definition 4, the concrete execution may go through loosely matching states several times, and thus potentially execute instructions 8 and 9 multiple times.

4.2 Inference Rules

The core set of inference rules in Fig. 3 and Fig. 4 form the foundation of the symbolic execution. Each rule is a core operation that an execution engine can carry out while guaranteeing sound results. ·

Rule SYMBSTEP. This is the base rule to obtain a progress structure from a single step of the symbolic semantics. Because the concrete transition relation and the function $symsem(\bar{s})$ are total, the simulation of Theorem 1 implies that the corresponding progress structure is sound.

Rules CASE, INFEASIBLE, RENAME, and FREESYMB_RENAME. While the symbolic semantics can already generate multiple states for individual instructions (e.g., for conditional branches), the rule CASE allows further splitting a target state by a path condition conjunct. This can be useful to analyze different cases of the same execution path individually (e.g., one representing the case that two registers point to different memory locations and the one representing memory aliasing). The rule INFEASIBLE allows dropping of infeasible paths, and RENAME allows straightforward renaming of symbols. Rule FREESYMB_RENAME allows renaming of free symbols specifically, under more relaxed assumptions.

Rule SUBST. This rule allows to derive specialized progress structures from general ones, by substituting single symbols with symbolic expressions. This is needed for reusing previous computations of binary fragments, e.g., the same code executing multiple times on several paths. By requiring that the substituted symbol is bound, the rule prevents the substitution of free symbols, to avoid missing concrete executions. Additionally, the symbols of the symbolic expression must not appear as free symbols, because they would become bound symbols otherwise, which may also lead to missing concrete executions.

Rule SIMPLIFY. This rule combines two purposes: simplifying a symbolic expression of a target state by replacing it with an equivalent one and also introducing a fresh symbol in the new symbolic expression.

Simplification requires the expressions to be equivalent under the path condition of the state that is being changed. For example, if the path condition entails $\alpha_b > 0$, then we may use this rule to replace $\alpha_a * \alpha_b > 0$ with $\alpha_a > 0$ in the store. This rule is particularly important for memory operations of binary code, by allowing for the simplification of store and load expressions under the path condition, as identified and applied in previous work [13,18,24]. In this way, load expressions can be matched with store expressions if their addresses agree in all interpretations, or bypass them if the accessed memory range is always completely disjoint. Similarly, store expressions can be removed if they are superseded by stores to the same addresses.

The introduction of a fresh symbol α requires that it does not occur in the source state, and also not in the target state where it is introduced. Otherwise, this could interfere with other usages of this symbol as per the soundness property of the progress structure. To avoid unnecessary overapproximation, the path condition of the modified state is extended with a constraint on α. Intuitively,

$$\dfrac{\vdash \overline{s} \rightsquigarrow_L \Pi \cup \{(l,\overline{\delta},\phi)\} \qquad \begin{array}{c} \forall H.\ H(\phi \wedge \alpha = \overline{v}'' \Rightarrow \overline{v} = \overline{v}') \\ \alpha \notin \Lambda(\overline{s}) \cup \Lambda((l,\overline{\delta},\phi)) \\ \Lambda(\overline{v}'') \subseteq \Lambda((l,\overline{\delta},\phi)) \end{array}}{\vdash \overline{s} \rightsquigarrow_L \Pi \cup \{(l,\overline{\delta}[x \mapsto \overline{v}'],\phi \wedge \alpha = \overline{v}'')\}} \quad \text{SIMPLIFY}$$

where $\overline{\delta}(x) = \overline{v}$ appears as a side condition under the left premise.

$$\dfrac{\begin{array}{cc} \overline{s}_1 \Rightarrow \overline{s}_1' & \vdash \overline{s}_1' \rightsquigarrow_L \Pi \cup \{\overline{s}_2\} \\ \overline{s}_2 \Rightarrow \overline{s}_2' & \Lambda(\overline{s}_1) \cap fs(\overline{s}_1' \rightsquigarrow_L \Pi \cup \{\overline{s}_2\}) = \emptyset \end{array}}{\vdash \overline{s}_1 \rightsquigarrow_L \Pi \cup \{\overline{s}_2'\}} \quad \text{CONSEQUENCE}$$

$$\dfrac{\vdash \overline{s} \rightsquigarrow_L \Pi \cup \{(l,\overline{\delta},\phi)\} \qquad \begin{array}{c} \forall H.\ H(path(\overline{s}) \Rightarrow \phi') \\ \Lambda(\phi') \subseteq \Lambda(path(\overline{s})) \end{array}}{\vdash \overline{s} \rightsquigarrow_L \Pi \cup \{(l,\overline{\delta},\phi \wedge \phi')\}} \quad \text{TRANSFER}$$

$$\dfrac{\vdash \overline{s}_A \rightsquigarrow_{L_A} \Pi_A \qquad \vdash \overline{s}_B \rightsquigarrow_{L_B} \Pi_B \qquad \Lambda(\overline{s}_A) \cap fs(\overline{s}_B \rightsquigarrow_{L_B} \Pi_B) = \emptyset}{\vdash \overline{s}_A \rightsquigarrow_{L_A \cup L_B} (\Pi_A \setminus \{\overline{s}_B\}) \cup \Pi_B} \quad \text{SEQUENCE}$$

Fig. 4. Overview of the symbolic-execution core inference-rules, part 2.

it is fixed to the value of the original symbolic expression that it replaces in the modified target state, to retain exactly the feasible set of state concretizations and not add any overapproximation in the process. For example, executing line 8 of Fig. 2 with SYMBSTEP produces the sound progress structure $(8,\overline{\delta}_{id},T) \rightsquigarrow_{\{8\}} \{(9,\overline{\delta}_{id}[\text{R1} \mapsto \overline{v}],T)\}$, where $\overline{v} = (\text{R1}*\text{R1})\%\text{R2}$. Using SIMPLIFY, we can introduce α in place of $\overline{v}$ and obtain $(8,\overline{\delta}_{id},T) \rightsquigarrow_{\{8\}} \{(9,\overline{\delta}_{id}[\text{R1} \mapsto \alpha],\alpha = \overline{v})\}$.

Rule CONSEQUENCE This rule allows to strengthen the path condition of the source state and to weaken the path condition of a target state. Free symbols of the structure $\overline{s}_1' \rightsquigarrow_L \Pi \cup \{\overline{s}_2\}$ must not occur in $\overline{s}_1$ to avoid capturing them.

We say that $\overline{s}$ is weaker than $\overline{s}'$, written as $\overline{s} \Rightarrow \overline{s}'$, if the two states have the same store and program counter and the path condition of $\overline{s}$ implies the path condition of $\overline{s}'$, i.e., $\overline{s} = (l,\overline{\delta},\phi) \wedge \overline{s}' = (l,\overline{\delta},\phi') \wedge \forall H.\ H(\phi \Rightarrow \phi')$. It is obvious that $\overline{s} \Rightarrow \overline{s}'$ implies that every state s matched by $\overline{s}$ is also matched by $\overline{s}'$.

Applying CONSEQUENCE to the example before allows removing the constraint on α from the path condition of the final state, weakening it, and obtaining $(8,\overline{\delta}_{id},T) \rightsquigarrow_{\{8\}} \{(9,\overline{\delta}_{id}[\text{R1} \mapsto \alpha],T)\}$. This results in overapproximation, since the value of R1 is now unconstrained.

Rule TRANSFER This rule is complementary to CONSEQUENCE and it allows for strengthening the target path conditions based on the source path condition. This is permitted because the interpretation H in Definition 4 is extended when loosely matching target states, and therefore the initial path condition is also satisfied in all target states.

Rule SEQUENCE Finally, this rule allows the sequential composition of two symbolic executions. This operation implicitly widens the code fragment scope by merging the two label sets. Applying this operation only makes sense in case

158   A. Lindner et al.

$\bar{s}_B$ occurs in Π_A, which also does not introduce overapproximation. We require that there is no free symbol in the second execution (B) that is bound in the first execution (A). Otherwise, the result may miss concrete executions.

To continue with the example from before, we need to first execute line 9 with SYMBSTEP to obtain $(9, \bar{\delta}_{id}[\mathsf{R1} \mapsto \alpha], T) \rightsquigarrow_{\{9\}} \{(10, \bar{\delta}_{id}[\mathsf{R1} \mapsto \alpha][\mathsf{R3} \mapsto \alpha_{\mathsf{R3}} + 1], T)\}$. The sequential composition of the two progress structures renders $(8, \bar{\delta}_{id}, T) \rightsquigarrow_{\{8,9\}} \{(10, \bar{\delta}_{id}[\mathsf{R1} \mapsto \alpha][\mathsf{R3} \mapsto \alpha_{\mathsf{R3}} + 1], T)\}$, which is allowed because the second progress structure has no free symbols.

4.3 Symbolic Execution Example

We illustrate our rules on the modular exponentiation of Fig. 2. We focus on using symbolic execution to establish control-flow graph integrity, i.e., verifying that the link register $\mathsf{R3}$ is correctly restored. We use this example to show how overapproximation helps scalability: it allows us to abstract from values irrelevant to control flow while still proving its integrity. A naïve executor that only applies SYMBSTEP and SEQUENCE would quickly become inefficient due to branching and expression growth.

We start analysing the loop body. Starting from path condition ϕ_0, executing line 6 with SYMBSTEP yields this progress structure for the conditional branch: $(6, \bar{\delta}_{id}, \phi_0) \rightsquigarrow_6 \{\bar{s}_T, \bar{s}_F\}$, where $\bar{s}_T = (7, \bar{\delta}_{id}, \phi_0 \wedge (\alpha_{\mathsf{R0}} >> \alpha_1)\&1)$ and $\bar{s}_F(8, \bar{\delta}_{id}, \phi_0 \wedge \neg((\alpha_{\mathsf{R0}} >> \alpha_1)\& 1))$. The two final states are produced directly by applying the symbolic semantics for individual instructions, rather than manually identifying the branch conditions and applying CASE.

The two paths can be analysed individually, starting from the corresponding state produced above, simply using SYMBSTEP and SEQUENCE. This results in $\bar{s}_T \rightsquigarrow_{[7...9]} \bar{s}_1$, where $\bar{s}_1 = (10, \bar{\delta}_1, \phi_0 \wedge (\alpha_{\mathsf{R0}} >> \alpha_1)\&1)$ and $\bar{\delta}_1 = \bar{\delta}_{id}[\mathsf{R3} \mapsto \alpha_1 + 1][\mathsf{R1} \mapsto \alpha_{\mathsf{R1}} * \alpha_{\mathsf{R1}}\%\alpha_{\mathsf{R2}}][\mathsf{M} \mapsto \mathsf{st}(\alpha_{\mathsf{M}}, \alpha_{\mathsf{SP}}, \mathsf{ld}(\alpha_{\mathsf{M}}, \alpha_{\mathsf{SP}}) * \alpha_{\mathsf{R1}}\%\alpha_{\mathsf{R2}})]$, and $\bar{s}_F \rightsquigarrow_{[7...9]} \bar{s}_2$, where $\bar{s}_2 = (10, \bar{\delta}_2, \phi_0 \wedge \neg(\alpha_{\mathsf{R0}} >> \alpha_1)\&1)$ and $\bar{\delta}_2 = \bar{\delta}_{id}[\mathsf{R3} \mapsto \alpha_1 + 1][\mathsf{R1} \mapsto \alpha_{\mathsf{R1}} * \alpha_{\mathsf{R1}}\%\alpha_{\mathsf{R2}}]$. These two progress structures one by one with the one obtained from line 6 via SEQUENCE renders: $(6, \bar{\delta}_{id}, \phi_0) \rightsquigarrow_{[6...9]} \{\bar{s}_1, \bar{s}_2\}$.

Overapproximating the behavior of the loop reduces the complexity of the analysis. For instance, when establishing control flow integrity and termination, the values of $\mathsf{R1}$ and the stack variable where r is stored are irrelevant. Overapproximating the former allows us to avoid duplicating the size of the expression for $\mathsf{R1}$ at each loop iteration (due to the double reference to the previous value of $\mathsf{R1}$). Overapproximating the latter allows us to merge the two symbolic states resulting from the loop execution. In both cases, the approach is similar.

For the memory operation, we first transform $\bar{s}_2$ to an identical one using simplify by explicitly storing the current memory value at α_{SP}: $(10, \bar{\delta}_2[\mathsf{M} \mapsto \mathsf{st}(\alpha_{\mathsf{M}}, \alpha_{\mathsf{SP}}, \mathsf{ld}(\alpha_{\mathsf{M}}, \alpha_{\mathsf{SP}}))], \phi_0 \wedge \neg(\alpha_{\mathsf{R0}} >> \alpha_1)\& 1)$. Then for both states, we use SIMPLIFY to introduce the fresh symbol α' for $\mathsf{ld}(\alpha_{\mathsf{M}}, \alpha_{\mathsf{SP}}) * \alpha_{\mathsf{R1}}\%\alpha_{\mathsf{R2}}$ and $\mathsf{ld}(\alpha_{\mathsf{M}}, \alpha_{\mathsf{SP}})$ respectively to make the two symbolic stores identical. Finally, we remove the equality introduced in the path condition and drop the branch condition conjuncts introduced at line 6 with CONSEQUENCE, which allows to merge the two states that are now identical and obtain the progress structure: $(6, \overline{\delta}_{id}, \phi_0) \leadsto_{[6...9]}$ $(10, \overline{\delta}_{id}[\text{R3} \mapsto \alpha_1 + 1][\text{R1} \mapsto \alpha_{\text{R1}} * \alpha_{\text{R1}} \% \alpha_{\text{R2}}][\text{M} \mapsto \text{st}(\alpha_\text{M}, \alpha_{\text{SP}}, \alpha')], \phi_0)$.

Merging two states to avoid path explosion is a common technique [40] and similar to widening for reaching fixed points in abstract interpretation [15]. A similar approach can be used to introduce a new symbol α'' that allows to abstract from the details of the expression computed for R1: $(6, \overline{\delta}_{id}, \phi_0) \leadsto_{[6...9]}$ $(10, \overline{\delta}_{id}[\text{R3} \mapsto \alpha_1 + 1][\text{R1} \mapsto \alpha''][\text{M} \mapsto \text{st}(\alpha_\text{M}, \alpha_{\text{SP}}, \alpha')], \phi_0)$.

The progress structure for the loop body can be used when creating a progress structure for the whole example function. The process for this is sequential execution from the first line, greedily instantiating the loop body progress structure whenever possible. Instantiation of progress structures requires instantiation of symbols of the state via SUBST and massaging of expressions (like reordering store operations in symbolic memory) via SIMPLIFY and RENAME. Additionally, after each instantiation and subsequent composition with SEQUENCE, it is necessary to remove the infeasible branch of the loop body progress structure with INFEASIBLE and drop the introduced conjunct with CONSEQUENCE. Using SIMPLIFY, we remove store operations that have the same symbolic address ($\alpha_{\text{SP}} - 4$), thus maintaining an overall smaller memory expression. This is a standard simplification enabled by array axiomatization [13, 18, 24]. After completing all operations, we finally obtain the overall progress structure:

$$(1, \overline{\delta}_{id}, \phi_0) \leadsto_{\{1..14\}} \left\{ \left(15, \overline{\delta}_{id} \begin{bmatrix} \text{SP} \mapsto \alpha_{\text{SP}}, \ \text{R0} \mapsto \alpha_a, \ \text{R1} \mapsto \alpha_b, \ \text{R3} \mapsto \alpha_{\text{R3}} \\ \text{M} \mapsto \text{st}(\text{st}(\alpha_\text{M}, \alpha_{\text{SP}}, \alpha_{\text{R3}}), \alpha_{\text{SP}} - 4, \alpha_a) \end{bmatrix}, \phi_0\right) \right\}$$

This progress structure has only one reachable state, which implies guaranteed termination in a matching state at line 15, at which the registers SP and R3 are restored to their initial values.

5   Implementation

This section gives an overview of our implementation in the HOL4 theorem prover of our symbolic execution theory and its instantiation for BIR.

We build on HolBA, a library for binary analysis implemented in Standard ML (SML) using the HOL4 theorem prover [32]. HolBA follows the architecture of traditional binary analysis tools such as the Binary Analysis Platform (BAP) [10] in that it *transpiles* machine code to a single intermediate language (BIR for HolBA) in which analyses are performed. In contrast to BAP, HolBA builds directly on formal ISA models as defined in their L3 specifications [25, 26], and produces HOL4 *lifting* theorems linking machine code behavior with transpiled BIR behavior via simulation relations. This significantly reduces the trusted computing base (TCB) of HolBA compared to similar tools. Fig. 5 illustrates the organization of HolBA and the relations between its components.

Our extension, HolBA-SE, adds to HolBA (1) a standalone HOL4 theory on symbolic execution and progress structures, (2) an instantiation of this theory for BIR, (3) general SML procedures automating proofs of progress structure soundness and program properties, (4) an integration of these to perform automatic verification of (i) contracts for RISC-V programs and (ii) execution time bounds for ARM Cortex-M0 programs. Thanks to the embedding of HOL4 in SML, each HolBA-SE verification task can be manifested as a custom SML program that reads files such as disassembly and then generates and checks the relevant HOL4 definitions and theorems.

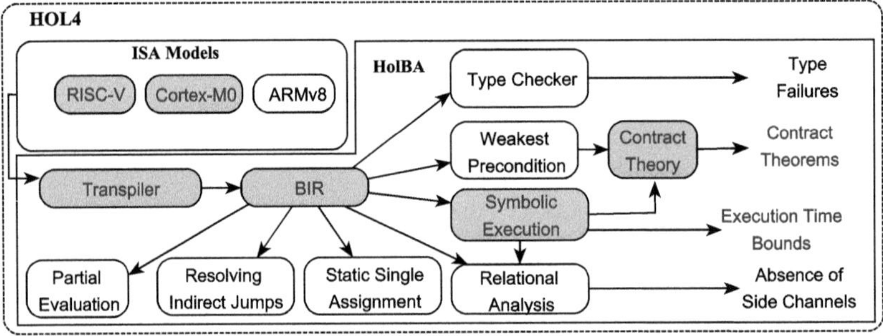

Fig. 5. Architecture of HolBA, a HOL4-based binary analysis library. Components and outputs relevant to HolBA-SE are highlighted.

5.1   Results Formalized and Proved in HOL4

Our standalone theory of symbolic execution in HOL4 encodes the definitions and proofs of all rules from Sect. 4 using predicate sets and inductive relations. To integrate this theory with BIR as represented in HolBA, we encoded the symbolic semantics as described in Sect. 3 and proved Theorem 1 in HOL4, establishing that the semantics is sound with respect to the existing concrete BIR semantics. The definitions and proofs of the symbolic execution rules consist of around 16,000 lines of HOL4 (SML) code and took 8 person months to develop.

5.2   Symbolic Executors and Progress Structures in HolBA-SE

Our approach to practical symbolic execution of BIR is to provide a framework inside HOL4 and HolBA where symbolic executors with different strategies are expressed as proof-producing SML procedures. Here, *strategy* refers to how the executor composes underlying execution blocks when run, and *proof-producing* refers to how a procedure produces a HOL4 theorem when (or if) it terminates. In executors, such theorems are witnesses that a specific progress structure ("summary") has the required properties. In a program verification workflow, progress structure theorems obtained from executors are decomposed and reused in other theorems, e.g., in Hoare-style contract proofs as described in Sect. 6.

In practice, our framework is a high-level abstraction layer (SML API) for leveraging symbolic execution rules. For a selection of rules, we provide parameterized execution blocks that perform one primitive symbolic operation and return a HOL4 theorem. When running an executor, such parameterized blocks are composed and interleaved with domain-specific post-processing steps. For example, one may apply the infeasible rule to prune branches, or simplify expressions for constant propagation and memory accesses. ISAs may require different approaches due to specific uses of expression combinations. We developed symbolic executors for RISC-V and ARM Cortex-M0 with this in mind.

5.3   Practical Evaluation of Symbolic Execution Using HolBA-SE

For application in practical automated verification tasks related to binary code, a symbolic executor needs to be fast and scalable enough to handle hundreds of instructions with complex control flow. To achieve this, we developed optimized executors for RISC-V and ARM Cortex-M0 and integrated them with the SMT solver Z3 [23], letting it discharge conditions that are BIR expressions, constant program memory lookups and symbolic value concretizations. This approach increases the TCB, but we believe the trade-off is attractive since the performance boost is considerable and the scope of discharged conditions is restricted.

Table 1. RISC-V programs and their symbolic execution running times. The number of instructions are for instructions processed during symbolic execution.

Program	Description	#Instrs	Time
aes-unopt	AES cipher, -O0 optimization	310	71.38 s
aes	AES cipher, -O1 optimization	130	6.84 s
chacha20	ChaCha20 cipher	177	46.28 s
incr	increment variable	1	0.04 s
isqrt	integer square root	6	0.12 s
kernel-trap	S3K kernel trap routines	79	9.21 s
mod2	modulo two computation	1	0.09 s
modexp	modular exponentiation	10	0.26 s
motor_set	nested calls and branching	66	3.94 s
swap	swap memory location	5	0.19 s

To investigate performance and scalability in practice, we applied our symbolic executor for RISC-V to a collection of disassembled binaries and measured the time to generate a progress structure theorem; the results are shown in Table 1. All experiments were run on an Intel Core i5-9600K with 32 GB RAM. Per the table, building HOL4 progress structure theorems for BIR programs corresponding to disassembly with hundreds of instructions generally takes minutes at most. Time increases with both the number of BIR instructions symbolically executed and the *store complexity*, which refers to how much information needs to be stored in the symbolic state. In ciphers where all output bits depend on all input bits, expressions may be replicated many times in symbolic memory.

6 Application: Automated Verification of Functional Properties of RISC-V Binaries

We automated functional verification in HolBA-SE for RISC-V binaries by integrating symbolic execution with previous work on binary lifting to BIR [42] and a theory of Hoare-style binary contracts [44]. The result is a workflow with the steps below as illustrated in Fig. 6, supporting the RV64G instruction set variant of RISC-V, a collection of uncompressed general 64-bit instructions [59]. As with most theorem prover based verification, the workflow is *incomplete* in the sense that verification steps can fail without specifications being false.

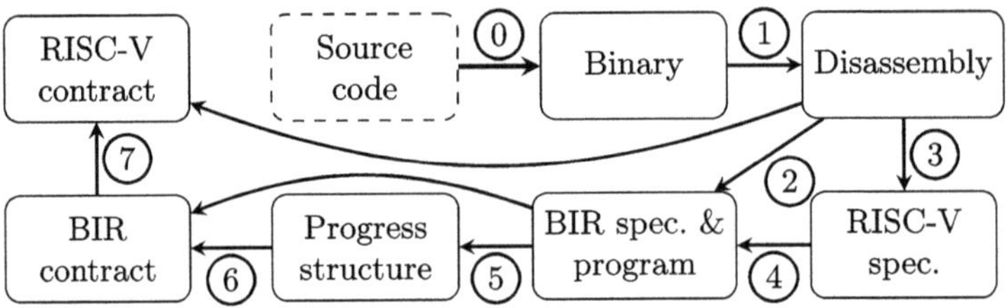

Fig. 6. Binary analysis workflow using HolBA-SE; source code is optional.

0. **Compilation:** Compile a source program to a RV64G binary. This step is *optional*, but adjustments to compilation can affect verification.
1. **Disassembly:** Disassemble the RV64G binary.
2. **Lifting:** Transpile the disassembly to BIR and generate a HOL4 lifting theorem connecting the BIR program to corresponding RV64G binary code.
3. **RISC-V specification:** Specify program boundaries inside disassembly and define pre- and post-conditions using the L3 model of RISC-V.
4. **BIR specification:** Translate RISC-V pre-conditions and post-conditions to BIR expressions and prove equivalence in HOL4.
5. **Symbolic execution:** Symbolically execute the BIR program in HOL4 according to the program boundaries and obtain a progress structure theorem.
6. **Symbolic transfer:** Prove a HOL4 contract theorem for the BIR program using the pre- and post-conditions via the progress structure theorem.
7. **Backlifting:** Prove a RISC-V contract theorem by way of the BIR contract theorem and the lifting theorem.

All steps except 3 and 4 (writing and translating specifications) are mostly or fully automated. Steps 5, 6, and 7 are novel contributions.

6.1   Contract Proof Automation Using Symbolic Execution

Steps 5 and 6 in the workflow in Fig. 6 crucially rely on sound progress structures produced by symbolic execution to prove Hoare-style contracts for BIR programs. The idea is that since our approach allows overapproximation, we can use a progress structure to establish a postcondition Q starting from a precondition P, but we cannot in general guarantee that it is the strongest one. The following theorem connects progress structures and Hoare-style contracts.

Theorem 2. *Given* $\vdash \bar{s} \leadsto_L \Pi$, *let* $\bar{L}$ *be the complement of* L. *Then, if*

- *for every* s *if* $pc(s) = pc(\bar{s})$ *and* $P(s)$ *then there exists* H *such that* $\bar{s} \overset{H}{\rhd} s$
- *for every* $\bar{s} \overset{H}{\rhd} s$, $\bar{s}' \in \Pi$, *and* $\bar{s}' \overset{H}{\blacktriangleright} s'$ *if* $P(s)$ *then* $Q(s')$
- *for every* $\bar{s}' \in \Pi$, $pc(\bar{s}') \in \bar{L}$ *holds,*

we have that for every state such that $pc(s) = l$ *and* $P(s)$ *there is a state* s' *and an* $n > 0$ *such that* $s \to_L^n s'$, $pc(s') \in \bar{L}$, *and* $Q(s')$.

```
#include <stdint.h>
uint64_t incr(uint64_t i) { return i + 1; }
```

```
incr:       file format elf64-littleriscv
Disassembly of section .text:
0000000000010488 <incr>:
   10488:        00150513                 addi    a0,a0,1
   1048c:        00008067                 ret
```

```
Definition riscv_incr_pre_def: (* RISC-V precondition *)
 riscv_incr_pre (pre_x10:word64) (m:riscv_state) : bool =
 (m.c_gpr m.procID 10w = pre_x10)
End
Definition riscv_incr_post_def: (* RISC-V postcondition *)
 riscv_incr_post (pre_x10:word64) (m:riscv_state) : bool =
 (m.c_gpr m.procID 10w = pre_x10 + 1w)
End
```

```
Theorem riscv_cont_incr: (* RISC-V contract *)
 riscv_cont incr_progbin 0x10488w {0x1048cw}
  (riscv_incr_pre pre_x10) (riscv_incr_post pre_x10)
```

Fig. 7. Example verified program that increments a 64-bit unsigned integer represented in C (top), corresponding RV64G disassembly (below top), and RISC-V specification (below disassembly) and contract theorem in HOL4 (bottom).

6.2   Workflow Example

To explain the steps in our workflow, we use the example defined in Fig. 7, which includes a simple program **incr** that increments its argument by one.

For step 0, *compilation*, our toolchain and workflow only require a RV64G binary. However, we expect many users to perform binary analysis on programs for which they have source code, enabling control over optimizations which can affect binary size. Figure 7 (top) shows the source for the example. In step 1, we *disassemble* RV64G binaries using `gnu-objdump` for further processing. The middle box in Fig. 7 shows our example's disassembly. This step is not strictly necessary, since lifting can be performed directly on the binary, but we found disassembly useful as a reference for specification. For step 2, *lifting*, HolBA-SE parses the disassembly using specifications of program bounds (address of initial instruction and address after final instruction to lift) and translates the indicated input into a BIR program inside HOL4 along with a lifting theorem connecting input and output. For step 3, *RISC-V specification*, we express binary contracts as pairs of pre- and postconditions in HOL4 using the `riscv` machine state type defined by the L3 model, as illustrated in the middle of Fig. 7, where `a0` is the standardized application binary interface (ABI) name of the register `x10`. Hence, the postcondition says that register `a0` is incremented by one.

For step 4, *BIR specification*, we manually translate RISC-V pre- and post-conditions to BIR to enable symbolic execution. Using the BIR specification, we *symbolically execute* the BIR program and perform *transfer* and *backlifting* (steps 5, 6, and 7). We use the BIR precondition as a path condition and leverage the progress structure theorem to automatically prove a BIR contract via Theorem 2, and finally obtain a RISC-V contract, as shown in the lowermost parts of Fig. 7. The contract primitive `riscv_cont` was defined in earlier work [44] and uses the HOL4 representation of the increment program disassembly (`incr_progbin`).

6.3   Case Studies

To validate that HolBA-SE can be used to verify real-world RISC-V binaries, we performed case studies for two high-assurance system software domains: cryptography (ChaCha20 encryption) and operating systems (S3K trap functions).

ChaCha20 Encryption. ChaCha20 is a stream cipher that performs twenty rounds of the ChaCha algorithm as defined by Bernstein [8,48]. We obtained a ChaCha20 binary by compiling a reference implementation in C using GCC with optimization level O1. We focused on specifying and verifying the byte encryption loop body of the ChaCha20 binary, taking inspiration from similar specifications in hacspec [9]. We abstractly define a ChaCha round in HOL4 as a transformation of 16 variables holding unsigned 32-bit words (array in the source code), built up from *lines* that update two variables at a time using word addition, XOR (`??`), and left rotation (`#<<~`):

```
chacha_line a b c d s (m : word32 -> word32) =
  let m = (a =+ (m a) + (m b)) m in
  let m = (d =+ ((m d) ?? (m a)) #<<~ s) m in m
```

Our verification of byte encryption establishes the connection between this high-level specification in HOL4 (updating the "memory function" m) and how the loop body of the RISC-V binary behaves, as in functional translation validation. The verification process took around 4 min on the machine described in Sect. 5.3.

S3K Trap Functions. S3K is an open source capability-based separation kernel targeting embedded RISC-V systems [36]. Practically, S3K manages a collection of *user processes* and ensures that they adhere to given restrictions on execution time (slices), memory access, and interprocess communication. We specified and verified two key handwritten RISC-V assembly routines from S3K: `trap_entry` and `trap_exit`. The routine `trap_entry` is entered when the user process is interrupted or an exception is encountered. It loads a pointer to the process's process control block (PCB) from the special `mscratch` RISC-V register, and then stores the user process's context (general purpose registers) to the PCB. The routine `trap_exit` is entered when the kernel resumes a user process. It loads the user process's context from the PCB, replacing register values with the values from the PCB. We obtained disassembly for the routines from the S3K author, consisting of 81 instructions. After adjusting lifting for special RISC-V instructions such as `csrrw`, we expressed routine contracts in terms of assignment of registers to memory locations (`trap_entry`), and assignments of memory locations to registers (`trap_exit`). The S3K author validated our specifications, and we then translated them to BIR manually and performed the verification process separately for each routine, which took around 7 min in total on the machine described in Sect. 5.3.

7 Application: Execution Time Bounds for Cortex-M0

The ARM Cortex-M0 processor allows for predictable execution times, due to lack of caches and a very short three-stage pipeline. According to the reference manual [2], ALU instructions take 1 cycle, while memory interactions take 1 extra cycle and control flow changes take 2 extra cycles. The Cortex-M0 formal model in HOL4 [26] maintains a clock cycle counter for this.

7.1 Representing Execution Time in BIR

To accommodate this simple timing model, BIR programs increment a cycle counter variable `c` with the execution time needed by the instructions. Throughout symbolic execution, the cycle counter is an incrementing constant value along each execution path. When paths are merged, the cycle counter is approximated as an interval, where the bounds are encoded in the path condition, linked to the symbolic store by a free symbol. For example, the cycle counter interval $[\alpha_c + 5, \alpha_c + 8]$ consists of $c \mapsto \alpha'$ in the store and the path condition conjunct $\alpha_c + 5 \leq \alpha' \leq \alpha_c + 8$. The interval for the cycle counter corresponds to the sound best-case and worst-case execution time bounds, or BCET and WCET for short.

The example program in Fig. 2 contains two conditional jumps: line 6 and line 10. The latter is for the loop iteration and is executed exactly 32 times, contributing a constant execution time. The former depends on the bits of `R0`, where taking the branch skips line 7. When executing as in Sect. 4.3, the two paths through the loop body have constant execution time and merge into an interval. During composition and instantiation, the interval bounds are added together accordingly.

Interestingly, equality of lower and upper bounds proves the absence of timing side channels. While the example program is not free of side channels, we can make its execution time independent of R0, e.g., balancing the execution times of the loop body branches by inserting nops.

7.2 Evaluation

We curated a set of evaluation programs, which we compiled with GCC unopti-mized (except for the one marked with o3 and the assembly program ldldbr8), tested on hardware, and then analyzed using both AbsInt aiT and HolBA-SE; the results are presented in Table 2. Concerning choice of programs, ldldbr8 is a manually constructed counterexample to aiT soundness, aes showcases constant-time proofs via overapproximation to avoid state size explosion due to loops, modexp is the running example from Sect. 2.2, and motor_set and pid exemplify embedded applications where execution time may be an important factor.

Testing a program consists of generating 1000 random input states and collecting observed BCET and WCET when executing on an STM32F051R8T6 chip measuring the clock cycles using the processor SysTick timer. We use AbsInt aiT [1] that is part of AbsInt a^3 for ARM Version 22.04. It is based on abstract interpretation and is the industry standard for obtaining safe WCET bounds with static analysis, but does not provide BCET values. For evaluations, aiT runs within 4s for each program. Testing and HolBA-SE count cycles as in the ARM specification of execution time for Cortex-M0 [2], starting after the first instruction has been fetched and decoded. The aiT tool instead assumes an initially empty pipeline, adding two or more clock cycles its analysis.

Program ldldbr8 consists of a sequence of 8 blocks, where each block is two loads and an unconditional branch to the next block, and a final nop. As there is trivially only one path (no splits), execution time is constant. The result illustrates that the pipeline in aiT can produce unsound results, and in contrast, the model in HOL4 is precise. Further experiments have revealed that the pipeline modeled in aiT is not equivalent to the Cortex-M0 pipeline. This has been confirmed with AbsInt along with the unsound result indicated with a † in the table, which subsequently has been fixed in aiT. Considering the different cycle counting of aiT, the programs aes and motor_set o3 also represent counterexamples to the aiT analysis. However, HolBA-SE proves constant execution time for ldldbr8 and aes, indicated using *.

Generally, the results of aiT and the symbolic execution are relatively similar. The program motor_set is as in Sect. 5.3, aes performs AES encryption with 10 rounds, modexp is the example from Sect. 2.2, and pid is a larger case study for analyzing a compiled PID controller. Since our processor does not support division and modulo, modexp uses the inefficient but easily analyzable emulation

Table 2. Per program, the number of binary instructions, where splits are conditional or indirect jumps, BCET and WCET values are in clock cycles, and aiT only provides WCET. [†] indicates unsound result and [*] indicates constant time.

Program			Testing		aiT	HolBA-SE		
name	#instrs	#splits	BCET	WCET	WCET	BCET	WCET	eval
ldldbr8	34	0	60	60	55[†]	60	60[*]	1 s
aes	543	1	3761	3761	3761[†]	3761	3761[*]	8.4 min
modexp	130	11	54358	93540	197721	43704	197560	11.4 min
- uidivmod	85	9	1362	1778	3062	1332	3060	5.9 min
motor_set	165	6	272	280	283	264	280	49 s
motor_set o3	112	12	73	91	92[†]	68	96	89 s
pid	2257	294	4589	5483	7521	1849	7529	61.8 min
- fadd	401	70	77	155	188	68	178	12.4 min
- fdiv	287	34	118	541	640	87	591	14.5 min

function `uidivmod`. Similarly, `pid` uses 11 floating-point operation emulation, provided as precompiled binaries with intricate control flows, like `fadd` and `fdiv`.

8   Related Work

Symbolic execution [5,50] was first introduced in the context of software testing by King [38] and applied to formal verification in follow-up work [30]. Subsequently, it became widely used in both testing [4,13,22,52] and verification [14,20]. Concolic execution [27,55] underapproximates for performance and targets bug-finding, whereas we target verification and maintain soundness.

Our core rules (Sect. 4.3) support standard techniques for making symbolic execution scalable. Array axiomatization [13,18,24] can flatten memory, enabling load–store bypassing and eliminating overwritten stores, curbing expression growth and solver load. Constant propagation reduces term size, while interval abstractions approximate expressions by bounded values [13,15,29]. State merging mitigates path explosion [40], akin to widening in abstract interpretation [15]. Interpolation reuses results by relaxing initial path conditions without introducing new paths [35] and value analysis can resolve indirect jumps [37].

In the context of binary verification, symbolic execution underpins translation validation [16,21,34], validation of compiler optimizations [56], analysis of side-channel absence [18,19], and test generation to validate side-channel models [11,47], and data-flow/overflow analysis [57]. Many implementations rely on intermediate languages for architectural independence [10,13,57] but do not offer the theorem prover based trustworthiness of HolBA-SE.

Mazzucato et al. [46] perform automated proofs in HOL Light of relational properties of ARMv8 and x86 machine code using Hoare logic. In their work, symbolic execution is one of many possible proof tactics; we provide a general symbolic execution theory independent of program logics and focus on proof automation for an intermediate language. Sammler et al. [51] import traces from SMT-based symbolic execution of binary code into Coq for reasoning; in contrast, our symbolic execution produces certified theorems inside HOL4. Previous works typically use weakest preconditions to verify functional properties of unstructured code [6,7,44], while our approach uses strongest postconditions.

WCET is fundamental in high assurance domains [3,12,39,49,54,58]. While we target a simple microarchitecture, prior work also abstracts complex timing features (e.g., caches, pipelines) to enable sound analysis [31,41]. Maroneze et al. extend the verified CompCert compiler with a verified loop-bound estimation [45]. However, they restrict to having a constant for WCET per instruction, which introduces overapproximation with our timing model.

9   Conclusion

We presented a theory of symbolic execution for unstructured programs suitable for theorem prover use. The theory builds on inference rules that are the basis of automated symbolic executors with common optimizations and adaptations for code using specific ISAs. We instantiated the theory for the BIR language, providing machine-checked proofs of soundness of symbolic execution of BIR in HOL4. Applications of our implementation [43] demonstrate proof automation in several case studies. Scaling further is possible via binary slicing and modular verification, but automation of such approaches is left to future work.

Acknowledgments. The authors thank Henrik Karlsson and Didrik Lundberg for their help with HolBA-SE case studies. This work was partially supported by the Wallenberg AI, Autonomous Systems and Software Program (WASP) funded by the Knut and Alice Wallenberg Foundation and a gift from Intel.

References

1. aiT Worst-Case Execution Time Analyzers. https://www.absint.com/ait/index.htm. Accessed 23 Dec 2022
2. Cortex-M0 Technical Reference Manual r0p0 - Instruction set summary. https://developer.arm.com/documentation/ddi0432/c/programmers-model/instruction-set-summary. Accessed 23 Dec 2022
3. Abella, J., et al.: WCET analysis methods: pitfalls and challenges on their trustworthiness. In: International Symposium on Industrial Embedded Systems, pp. 1–10 (2015). https://doi.org/10.1109/SIES.2015.7185039
4. Avgerinos, T., Rebert, A., Cha, S.K., Brumley, D.: Enhancing symbolic execution with Veritesting. In: International Conference on Software Engineering, pp. 1083–1094 (2014). https://doi.org/10.1145/2568225.2568293
5. Baldoni, R., Coppa, E., D'elia, D.C., Demetrescu, C., Finocchi, I.: A survey of symbolic execution techniques. ACM Comput. Surv. **51**(3) (2018). https://doi.org/10.1145/3182657

6. Barnett, M., Leino, K.R.M.: Weakest-precondition of unstructured programs. In: Workshop on Program Analysis for Software Tools and Engineering, pp. 82—-87 (2005). https://doi.org/10.1145/1108792.1108813

7. Barthe, G., Rezk, T., Saabas, A.: Proof obligations preserving compilation. In: Dimitrakos, T., Martinelli, F., Ryan, P.Y.A., Schneider, S. (eds.) FAST 2005. LNCS, vol. 3866, pp. 112–126. Springer, Heidelberg (2006). https://doi.org/10.1007/11679219_9

8. Bernstein, D.J.: Chacha, a variant of Salsa20. In: Workshop Record of SASC, vol. 8. pp. 3–5 (2008). http://cr.yp.to/chacha/chacha-20080128.pdf

9. Bhargavan, K., Kiefer, F., Strub, P.Y.: hacspec: Towards verifiable crypto standards. In: International Conference of Security Standardisation Research, pp. 1–20 (2018). https://doi.org/10.1007/978-3-030-04762-7_1

10. Brumley, D., Jager, I., Avgerinos, T., Schwartz, E.J.: BAP: a binary analysis platform. In: Gopalakrishnan, G., Qadeer, S. (eds.) CAV 2011. LNCS, vol. 6806, pp. 463–469. Springer, Heidelberg (2011). https://doi.org/10.1007/978-3-642-22110-1_37

11. Buiras, P., Nemati, H., Lindner, A., Guanciale, R.: Validation of side-channel models via observation refinement. In: International Symposium on Microarchitecture, pp. 578–591 (2021). https://doi.org/10.1145/3466752.3480130

12. Byhlin, S., Ermedahl, A., Gustafsson, J., Lisper, B.: Applying static WCET analysis to automotive communication software. In: Euromicro Conference on Real-Time Systems, pp. 249–258 (2005). https://doi.org/10.1109/ECRTS.2005.7

13. Cha, S.K., Avgerinos, T., Rebert, A., Brumley, D.: Unleashing mayhem on binary code. In: Symposium on Security and Privacy, pp. 380–394 (2012). https://doi.org/10.1109/SP.2012.31

14. Coen-Porisini, A., Denaro, G., Ghezzi, C., Pezzé, M.: Using symbolic execution for verifying safety-critical systems. In: European Software Engineering Conference, pp. 142–151 (2001). https://doi.org/10.1145/503209.503230

15. Cousot, P.: Abstract interpretation. ACM Comput. Surv. **28**(2), 324–328 (1996). https://doi.org/10.1145/234528.234740

16. Currie, D., Feng, X., Fujita, M., Hu, A.J., Kwan, M., Rajan, S.: Embedded software verification using symbolic execution and uninterpreted functions. Int. J. Parallel Prog. **34**(1), 61–91 (2006). https://doi.org/10.1007/s10766-005-0004-8

17. Dahl, O.J., Dijkstra, E.W., Hoare, C.A.R. (eds.): Structured programming. Academic Press Ltd. (1972)

18. Daniel, L.A., Bardin, S., Rezk, T.: Binsec/Rel: efficient relational symbolic execution for constant-time at binary-level. In: Symposium on Security and Privacy, pp. 1021–1038 (2020). https://doi.org/10.1109/SP40000.2020.00074

19. Daniel, L.A., Bardin, S., Rezk, T.: Hunting the Haunter–efficient relational symbolic execution for Spectre with Haunted RelSE. In: Network and Distributed Systems Security (2021). https://doi.org/10.14722/ndss.2021.24286

20. Dannenberg, R., Ernst, G.: Formal program verification using symbolic execution. IEEE Trans. Softw. Eng. **SE-8**(1), 43–52 (1982). https://doi.org/10.1109/TSE.1982.234773

21. Dasgupta, S., Dinesh, S., Venkatesh, D., Adve, V.S., Fletcher, C.W.: Scalable validation of binary lifters. In: Conference on Programming Language Design and Implementation, pp. 655–671 (2020). https://doi.org/10.1145/3385412.3385964

22. David, R., et al.: Specification of concretization and symbolization policies in symbolic execution. In: International Symposium on Software Testing and Analysis, pp. 36–46 (2016). https://doi.org/10.1145/2931037.2931048

23. De Moura, L., Bjørner, N.: Z3: an efficient SMT solver. In: International Conference on Tools and Algorithms for the Construction and Analysis of Systems, pp. 337–340 (2008). https://doi.org/10.1007/978-3-540-78800-3_24
24. Farinier, B., David, R., Bardin, S., Lemerre, M.: Arrays made simpler: an efficient, scalable and thorough preprocessing. In: International Conference on Logic for Programming, Artificial Intelligence and Reasoning, vol. 57, pp. 363–380 (2018). https://doi.org/10.29007/dc9b
25. Fox, A.: Improved tool support for machine-code decompilation in HOL4. In: Urban, C., Zhang, X. (eds.) ITP 2015. LNCS, vol. 9236, pp. 187–202. Springer, Cham (2015). https://doi.org/10.1007/978-3-319-22102-1_12
26. Fox, A.: L3: A Specification Language for Instruction Set Architectures. https://acjf3.github.io/l3. Accessed 17 Nov 2025
27. Godefroid, P., Klarlund, N., Sen, K.: DART: directed automated random testing. In: Conference on Programming Language Design and Implementation, pp. 213–223 (2005). https://doi.org/10.1145/1065010.1065036
28. Hamming, R.R.: Art of Doing Science and Engineering: Learning to Learn. CRC Press, Boca Raton (1997). https://doi.org/10.1201/9781482283198
29. Hansen, T., Schachte, P., Søndergaard, H.: State joining and splitting for the symbolic execution of binaries. In: International Workshop on Runtime Verification, pp. 76–92 (2009). https://doi.org/10.1007/978-3-642-04694-0_6
30. Hantler, S.L., King, J.C.: An introduction to proving the correctness of programs. ACM Comput. Surv. **8**(3), 331–353 (1976). https://doi.org/10.1145/356674.356677
31. Heckmann, R., Langenbach, M., Thesing, S., Wilhelm, R.: The influence of processor architecture on the design and the results of WCET tools. Proc. IEEE **91**(7), 1038–1054 (2003). https://doi.org/10.1109/JPROC.2003.814618
32. HOL development team: HOL interactive theorem prover. https://hol-theorem-prover.org. Accesed 17 Nov 2025
33. HolBA contributors: HolBA - Binary analysis in HOL. https://github.com/kth-step/HolBA. Accessed 17 Nov 2025
34. Iosif-Lazar, A.F., Al-Sibahi, A.S., Dimovski, A.S., Savolainen, J.E., Sierszecki, K., Wasowski, A.: Experiences from designing and validating a software modernization transformation. In: International Conference on Automated Software Engineering, pp. 597–607 (2015). https://doi.org/10.1109/ASE.2015.84
35. Jaffar, J., Murali, V., Navas, J.A., Santosa, A.E.: TRACER: a symbolic execution tool for verification. In: Madhusudan, P., Seshia, S.A. (eds.) CAV 2012. LNCS, vol. 7358, pp. 758–766. Springer, Heidelberg (2012). https://doi.org/10.1007/978-3-642-31424-7_61
36. Karlsson, H.A.: S3K: Capability based separation kernel for embedded RISC-V. https://github.com/kth-step/s3k. Accessed 16 Nov 2025
37. Kinder, J., Zuleger, F., Veith, H.: An abstract interpretation-based framework for control flow reconstruction from binaries. In: International Conference on Verification, Model Checking, and Abstract Interpretation, pp. 214–228 (2009). https://doi.org/10.1007/978-3-540-93900-9_19
38. King, J.C.: Symbolic execution and program testing. Commun. ACM **19**(7), 385–394 (1976). https://doi.org/10.1145/360248.360252
39. Kirner, R., Puschner, P.: Classification of WCET analysis techniques. In: International Symposium on Object-Oriented Real-Time Distributed Computing, pp. 190–199 (2005). https://doi.org/10.1109/ISORC.2005.19
40. Kuznetsov, V., Kinder, J., Bucur, S., Candea, G.: Efficient state merging in symbolic execution. In: Conference on Programming Language Design and Implementation, pp. 193–204 (2012). https://doi.org/10.1145/2254064.2254088

41. Li, X., Roychoudhury, A., Mitra, T.: Modeling out-of-order processors for WCET analysis. Real-Time Syst. **34**(3), 195–227 (2006). https://doi.org/10.1007/s11241-006-9205-5

42. Lindner, A., Guanciale, R., Metere, R.: TrABin: trustworthy analyses of binaries. Sci. Comput. Program. **174**, 72–89 (2019). https://doi.org/10.1016/j.scico.2019.01.001

43. Lindner, A., Palmskog, K., Lundberg, D.: HolBA-SE (2025). https://doi.org/10.5281/zenodo.15043917

44. Lundberg, D., Guanciale, R., Lindner, A., Dam, M.: Hoare-style logic for unstructured programs. In: de Boer, F., Cerone, A. (eds.) SEFM 2020. LNCS, vol. 12310, pp. 193–213. Springer, Cham (2020). https://doi.org/10.1007/978-3-030-58768-0_11

45. Maroneze, A., Blazy, S., Pichardie, D., Puaut, I.: A formally verified WCET estimation tool. In: International Workshop on Worst-Case Execution Time Analysis, vol. 39, pp. 11–20 (2014). https://doi.org/10.4230/OASIcs.WCET.2014.11

46. Mazzucato, D., et al.: Relational Hoare logic for realistically modelled machine code. In: Computer Aided Verification, pp. 389–413 (2025). https://doi.org/10.1007/978-3-031-98668-0_19

47. Nemati, H., Buiras, P., Lindner, A., Guanciale, R., Jacobs, S.: Validation of abstract side-channel models for computer architectures. In: Lahiri, S.K., Wang, C. (eds.) CAV 2020. LNCS, vol. 12224, pp. 225–248. Springer, Cham (2020). https://doi.org/10.1007/978-3-030-53288-8_12

48. Nir, Y., Langley, A.: ChaCha20 and Poly1305 for IETF Protocols. RFC 8439 (2018). https://doi.org/10.17487/RFC8439

49. Nowotsch, J., Paulitsch, M.: Leveraging multi-core computing architectures in avionics. In: European Dependable Computing Conference, pp. 132–143 (2012). https://doi.org/10.1109/EDCC.2012.27

50. Păsăreanu, C.S., Visser, W.: A survey of new trends in symbolic execution for software testing and analysis. Int. J. Softw. Tools Technol. Transfer **11**(4), 339–353 (2009). https://doi.org/10.1007/s10009-009-0118-1

51. Sammler, M., et al.: Islaris: verification of machine code against authoritative ISA semantics. In: International Conference on Programming Language Design and Implementation, pp. 825–840 (2022). https://doi.org/10.1145/3519939.3523434

52. Saxena, P., Poosankam, P., McCamant, S., Song, D.: Loop-extended symbolic execution on binary programs. In: International Symposium on Software Testing and Analysis, pp. 225–236 (2009). https://doi.org/10.1145/1572272.1572299

53. Sewell, T.A.L., Myreen, M.O., Klein, G.: Translation validation for a verified OS kernel. In: Conference on Programming Language Design and Implementation, pp. 471–482 (2013). https://doi.org/10.1145/2491956.2462183

54. Souyris, J., Pavec, E.L., Himbert, G., Borios, G., Jégu, V., Heckmann, R.: Computing the worst case execution time of an avionics program by abstract interpretation. In: International Workshop on Worst-Case Execution Time Analysis, vol. 1, pp. 21–24 (2007). https://doi.org/10.4230/OASIcs.WCET.2005.810

55. Stephens, N., et al.: Driller: augmenting fuzzing through selective symbolic execution. In: Network and Distributed System Security Symposium (2016). https://doi.org/10.14722/ndss.2016.23368

56. Tristan, J.B., Leroy, X.: Formal verification of translation validators: a case study on instruction scheduling optimizations. In: Symposium on Principles of Programming Languages, pp. 17–27 (2008). https://doi.org/10.1145/1328438.1328444

57. Wang, T., Wei, T., Lin, Z., Zou, W.: Intscope: automatically detecting integer overflow vulnerability in X86 binary using symbolic execution. In: Network and Distributed System Security Symposium (2009)
58. Wartel, F., et al.: Measurement-based probabilistic timing analysis: lessons from an integrated-modular avionics case study. In: International Symposium on Industrial Embedded Systems, pp. 241–248 (2013). https://doi.org/10.1109/SIES.2013.6601497
59. Waterman, A., Lee, Y., Patterson, D.A., Asanović, K.: The RISC-V instruction set manual, volume I: User-level ISA, version 2.1. Technical Report. UCB/EECS-2016-118, EECS Department, University of California, Berkeley (2016). https://www2.eecs.berkeley.edu/Pubs/TechRpts/2016/EECS-2016-118.pdf
60. Zhao, L., Li, G., De Sutter, B., Regehr, J.: ARMor: fully verified software fault isolation. In: International Conference on Embedded Software, pp. 289–298 (2011). https://doi.org/10.1145/2038642.2038687

SAT-Based Synthesis of Minimal Deterministic Real-Time Automata via 3DRTA Representation

Junjie Meng[1], Jie An[2,3]✉, Yong Li[4]✉, Andrea Turrini[4,5]✉, and Miaomiao Zhang[1]✉

[1] School of Computer Science and Technology, Tongji University, Shanghai, China
{2311452,miaomiao}@tongji.edu.cn
[2] National Key Laboratory of Space Integrated Information System,
Institute of Software Chinese Academy of Sciences, Beijing, China
anjie@iscas.ac.cn
[3] University of Chinese Academy of Sciences, Beijing, China
[4] Key Laboratory of System Software (Chinese Academy of Sciences),
Institute of Software Chinese Academy of Sciences, Beijing, China
{liyong,turrini}@ios.ac.cn
[5] Institute of Intelligent Software, Guangzhou, China

Abstract. Real-time automata (RTAs) can be viewed as a subclass of timed automata with only one clock that resets at each transition. In this paper, we propose a novel framework for learning deterministic RTAs (DRTAs) with minimal number of states from samples. Inspired by recent advances in learning deterministic finite automata, we introduce 3-valued Deterministic Real-Time Automata (3DRTAs) as an intermediate representation for the given sample set, thereby eliminating the redundancies present in existing approaches. Then, we solve the minimal DRTA learning problem from 3DRTAs by a reduction to a Boolean Satisfiability (SAT) problem. This then allows us to leverage state-of-the-art SAT solvers to find a minimal DRTA consistent with the given samples efficiently. More importantly, small DRTAs not only offer compact representation but also better interpretability of real-world systems. Experimental results demonstrate that our 3DRTA-based framework yields minimal DRTAs with significantly fewer states compared to those of existing methods. The proposed technique also opens new possibilities for scalable real-time automata learning in complex real-time domains.

Keywords: Real-time automata · Model learning · Automata learning · Passive learning · Grammatical inference · SAT solving

1 Introduction

Real-time systems are ubiquitous in modern computing, spanning from embedded controllers in automotive systems to network protocols and industrial automation. Understanding and modeling the temporal and real time behavior of such systems is crucial for ensuring correctness, safety, and performance.

© The Author(s), under exclusive license to Springer Nature Switzerland AG 2026
Y.-F. Chen et al. (Eds.): VMCAI 2026, LNCS 16417, pp. 173–196, 2026.
https://doi.org/10.1007/978-3-032-15700-3_9

Unlike traditional discrete systems that only consider the sequence of events, real-time systems must also account for the precise timing of when events occur, as timing violations can lead to system failures or safety hazards.

Timed automata (TAs) [1] have emerged as a fundamental modeling formalism for real-time systems. By extending finite automata with real-valued clocks and timing constraints, TAs provide a natural and expressive framework for specifying temporal and real-time behaviors. The success of timed automata in modeling and verification [9,10] has been demonstrated across numerous domains, from communication protocols [27] to scheduling algorithms [6] and safety-critical systems [23]. However, manually constructing timed automata models for complex real-time systems is both time-consuming and error-prone. This challenge motivates the need for automatic learning techniques that can infer timed automata models from observed system behaviors. Such learning approaches are particularly valuable when dealing with legacy systems where formal specifications are unavailable, or when validating that implementations conform to their intended temporal and real-time specifications.

The identification of timed automata from labeled samples represents a significant extension of the well-established finite automata learning problem. While automata learning has been extensively studied in the grammatical inference community [22], the addition of timing constraints introduces fundamental new challenges. Unlike discrete automata, where the primary concern is structural consistency, timed automata must satisfy both structural and timing constraints, dramatically increasing the complexity of the synthesis problem.

Recent advances in deterministic finite automata (DFAs) learning have shown the power of SAT-based synthesis approaches [21], which encode the learning problem as a Boolean satisfiability problem to find globally minimal solutions. A breakthrough in this direction is the introduction of 3-valued DFAs (3DFAs) [14], which provide a compact intermediate representation that dramatically reduces the problem size compared to the traditional Augmented Prefix Tree Acceptor (APTA) constructions. A 3DFA extends classical DFAs by allowing states to be associated with three possible outcomes: "accept", "reject", or "don't care" (undefined). This additional flexibility enables the construction of significantly smaller intermediate structures that still capture all the constraints from the labeled samples, leading to a more efficient SAT encoding with fewer variables.

The success of 3DFAs in DFA learning provides valuable insights for tackling the more complex problem of timed automata synthesis. However, as observed in [7], the introduction of clocks causes an exponential blow-up in the region graph and a sharp increase in constraint complexity, rendering synthesis and verification in the timed setting substantially harder than in the discrete case.

Among the various classes of timed automata, we focus on identifying a simple type of one-clock deterministic timed automata (1-DTAs), known as Deterministic Real-Time Automata (DRTAs) [15]. A DRTA models only the time constraints between two consecutive events, instead of arbitrary pairs of events. It can be viewed as an 1-DTA in which the clock resets at every transition. We focus on DRTAs, a restricted subclass of 1-DTAs, rather than the full class of 1-DTAs, since even the identification of DFAs is already a difficult problem:

it is NP-complete [17] and inapproximable within a polynomial [28]. Hence, it makes sense to first focus on simple extensions of DFAs. In addition, although DRTAs are a restricted subclass of 1-DTAs, their expressive power suffices for many meaningful applications like learning system models from timestamped logs [35], timing-based security or opacity analysis [37], and timed behavior verification for testing or monitoring [20].

The deterministic nature of DRTAs ensures predictable execution semantics, providing unambiguous transitions that are essential for both synthesis and verification [1]. This determinism, combined with the restriction to consecutive event timing, significantly simplifies the learning problem while maintaining sufficient expressiveness for practical real-time systems.

Several works have been conducted on learning DRTAs from samples. Existing approaches for DRTA learning [35] primarily rely on evidence-driven state-merging (EDSM) algorithms [24], which iteratively merge states in a prefix tree based on heuristics. While these methods have shown practical utility, they have their own *limitations*: (i) they often terminate at local optima without global optimality guarantees; (ii) scalability degrades as the underlying APTA grows with alphabet size, trace length, and sample count, inflating the merge search space. In addition to these EDSM-based methods, more recent work, such as the one by Tappler et al. on timed automata learning via SMT solving [32], has explored alternative paradigms. Their approach encodes the synthesis problem into an SMT instance by representing every step of each trace individually, thereby enabling the use of SMT solvers to derive candidate automata. However, this methodology exhibits two significant limitations: first, it only incorporates positive samples into the learning process, disregarding negative evidence; second, the encoding strategy introduces a large amount of redundancy, as each trajectory and each step requires an explicit representation. Consequently, with the growth in the number and length of samples, the number of encoding variables expands rapidly, which in turn exacerbates the scalability challenge and imposes substantial computational burdens on the SMT solver. We provide a detailed discussion in Sect. 5.2, after presenting our SAT-based encoding.

In this paper, to address these limitations, we propose a novel framework with an advanced intermediate representation that brings the power of modern constraint solving to the DRTA learning from positive and negative examples. Our method guarantees to return a DRTA with the minimal number of states that accepts all positive samples and rejects all negative samples.

Inspired by the success of 3DFAs in discrete automata learning, our approach introduces 3-valued Deterministic Real-Time Automata (3DRTAs) as an intermediate representation that compactly encodes both structural and timing constraints from labeled samples. The 3DRTA extends the 3DFA concept to the timed domain, allowing transitions to carry timing intervals alongside the traditional accept/reject/don't-care state semantics. This novel representation eliminates redundancies present in traditional prefix tree approaches while naturally handling the timing aspects of real-time systems.

By transforming the DRTA synthesis problem into a SAT encoding that leverages the 3DRTA structure, we enable the use of state-of-the-art SAT solvers to find globally optimal solutions. The key insight is that the compact 3DRTA representation leads to significantly smaller SAT formulations compared to direct approaches, making the synthesis problem tractable for practical applications while providing guarantees about the size of the synthesized automaton.

Contributions. Our main contributions in this paper are the following:

- In Sect. 2, after reviewing basic concepts on timed systems, we formalize the problem of learning DRTAs from labeled samples and we introduce 3DRTAs, a novel intermediate representation that extends 3DFAs to the timed domain, compactly encoding timing constraints from labeled samples while eliminating redundancies present in traditional prefix tree constructions.
- In Sect. 3, we show how to construct a 3DRTA consistent with a given set of samples, i.e., that accepts all positive samples and rejects all negative samples.
- In Sect. 4, we present a complete algorithm for synthesizing minimal DRTAs with theoretical guarantees on correctness and optimality, overcoming the limitations of heuristic-based approaches.
- In Sect. 5, we develop an efficient SAT encoding for DRTA synthesis that leverages the 3DRTA structure, transforming the learning problem into a constraint satisfaction problem solvable by modern SAT solvers with global optimality guarantees.
- In Sect. 6, we propose a region merging algorithm to reduce the excessive number of transitions that may arise in coarse DRTAs, thereby further optimizing the model size.
- In Sect. 7, we provide a comprehensive experimental evaluation, demonstrating significant efficiency improvements and superior model quality compared to existing methods.

1.1 Related Work

The most closely related work spans timed automata learning, constraint-based synthesis, and compact automata representations.

In timed automata learning, we omit the introduction to the works on Angluin's L^*-style active learning [5], in which the target is to learn automata from an oracle by queries [2–4,18,30,31,33,36]. We focus on reviewing related work on passive learning, i.e., learning automata from given samples. Verwer et al. [34] showed that one-clock deterministic timed automata are efficiently identifiable in the limit, providing theoretical learnability results for this restricted class. Building on this, Verwer et al. [35] proposed the RTI algorithm for learning DRTAs using EDSM-based approaches, demonstrating that DRTA learning can outperform sampling-based methods. Tappler et al. [32] proposed an SMT-driven methodology that encodes the problem of inferring timed automata from timestamped traces as SMT formulas and employs incremental solving along with varied search strategies. Grinchtein et al. [19] proposed passive learning algorithms for event-recording automata (ERAs), which are a specific kind of timed automata where each action is assigned a clock to record the time length from its last occurrence to the present. However, Dima [15] pointed out that RTAs are incomparable to ERAs since RTAs may accept languages consisting of two actions separated by an interval with integer length while ERAs may not.

In constraint-based synthesis for DFAs, recent work [21,38] has demonstrated the power of SAT-based approaches for finding globally minimal solutions. This inspired us to explore a SAT-based encoding for the synthesis of DRTAs consistent with the given samples.

Another important line of work focuses on compact representations of automata. The goal is to reduce the size of the intermediate structures constructed from samples while still preserving all necessary information for synthesis. Representative examples include symbolic and interval-based extensions of DFAs, which avoid explicit state explosion by encoding sets of transitions more concisely [13,16]. A recent work by Dell'Erba et al. [14] based on the 3DFA construction falls into this category, showing how 3-valued states provide a much smaller intermediate representation compared to traditional APTA. However, adapting these techniques to timed automata presents unique challenges due to the hybrid discrete-continuous nature of timing constraints.

Our work represents the first systematic attempt to bring both the theoretical insights of 3-valued automata and the practical power of constraint-based synthesis to DRTA learning. Unlike previous heuristic-based approaches, we provide theoretical guarantees on the minimality of synthesized models while achieving practical efficiency through our novel 3DRTA representation. In addition, we introduce a time-region merging algorithm to address the issue of overly coarse DRTAs containing an excessive number of transitions, thereby further reducing the model size. This combination of optimality and efficiency represents a significant advancement in timed automata learning.

2   Timed Systems and Languages

Throughout this paper, we fix a finite alphabet Σ and the set of non-negative real numbers $\mathbb{R}_{\geq 0}$ as the time domain; we denote by $[\mathbb{R}_{\geq 0}]$ the set of intervals in $\mathbb{R}_{\geq 0}$ with endpoints in $\mathbb{N} \cup \{\infty\}$.

Timed Systems and Languages. A (finite) *timed word* over $\Sigma \times \mathbb{R}_{\geq 0}$ is a finite sequence $\omega = (\sigma_1, \tau_1)(\sigma_2, \tau_2) \cdots (\sigma_n, \tau_n)$, where $\sigma_i \in \Sigma$ and $\tau_i \in \mathbb{R}_{\geq 0}$ for $1 \leq i \leq n$; $|\omega| = n$ denotes its length. The empty timed word ε satisfies $|\varepsilon| = 0$. For two timed words ω_1 and ω_2, their concatenation $\omega_1 \cdot \omega_2$ (abbreviated as $\omega_1 \omega_2$) is defined by appending ω_2 to ω_1. A *timed language* $\mathcal{L}$ is a set of timed words, i.e., $\mathcal{L} \subseteq (\Sigma \times \mathbb{R}_{\geq 0})^*$.

A *deterministic real-time transition system* (DRTS) is defined as a triple $\mathcal{T} = (Q, q_0, \Delta)$, where Q is a finite set of states, $q_0 \in Q$ is the initial state, and $\Delta \colon Q \times (\Sigma \times [\mathbb{R}_{\geq 0}]) \dashrightarrow Q$ is a transition relation with $|\Delta| < \omega$. The continuous progress of real time is tracked by the unique clock c, which resets at every transition, so the value of c represents the delay time between two actions. Thus, Δ induces the transition function $\delta\colon Q \times (\Sigma \times \mathbb{R}_{\geq 0}) \to 2^Q$ such that $\delta(q, (\sigma, \tau)) = \{\, q' \in Q \mid (q, (\sigma, I), q') \in \Delta \wedge \tau \in I \,\}$, where $I \in [\mathbb{R}_{\geq 0}]$ is the region interval having $\tau \in I$ associated with the action σ. We extend δ to timed words by setting $\delta(q, \varepsilon) = \{q\}$ and $\delta(q, (\sigma, \tau) \cdot \omega) = \bigcup_{q' \in \delta(q,(\sigma,\tau))} \delta(q', \omega)$, and subsequently to sets of states $Q' \subseteq Q$ by $\delta(Q', \omega) = \bigcup_{q \in Q'} \delta(q, \omega)$.

The run of $\mathcal{T}$ on a finite word $\omega = (\sigma_1, \tau_1) \cdots (\sigma_n, \tau_n)$ is the sequence of states $\rho = q_0 q_1 \cdots q_n \in Q^+$ such that, for every $0 \leq i < n$, the subsequent state q_{i+1} is determined by $q_{i+1} = \delta(q_i, (\sigma_{i+1}, \tau_{i+1}))$.

Definition 1. *A Deterministic Real-Time Automaton (DRTA) is a pair $\mathcal{A} = (\mathcal{T}, A)$ with $\mathcal{T} = (Q, q_0, \Delta)$ being a DRTS and $A \subseteq Q$ the set of accepting states.*

For every state $q \in Q$ and input symbol $a \in \Sigma$, the set of time intervals of outgoing transitions labeled by a must be *pairwise disjoint*, that is, for any two transitions $(q, (a, I_1), q_1), (q, (a, I_2), q_2) \in \Delta$, we have $I_1 \cap I_2 = \emptyset$ whenever $q_1 \neq q_2$ and $I_1 = I_2$ whenever $q_1 = q_2$. This ensures that for each time value τ, at most one transition labeled a is enabled from state q.

A timed word $\omega \in (\Sigma \times \mathbb{R}_{\geq 0})^*$ is accepted by $\mathcal{A}$ if there exists a run on ω whose last state is in A; the language $\mathcal{L}(\mathcal{A})$ of $\mathcal{A}$ is the set of all words accepted by $\mathcal{A}$. As real-time automata are determinable [15], we say that a timed language is a *real-time language* if it can be recognized by a DRTA.

Now, we can formalize the problem of learning minimal DRTAs from labeled samples as follows.

Definition 2 (Minimal DRTA synthesis problem). *Given a finite set of labeled timed words $S = (S^+, S^-) \subseteq (\Sigma \times \mathbb{R}_{\geq 0})^*$, construct a DRTA $\mathcal{A} = (\mathcal{T}, A)$ that satisfies the following conditions:*

(i) ***Consistency:*** *for each $\omega^+ \in S^+$, $\omega^+ \in \mathcal{L}(\mathcal{A})$ and for each $\omega^- \in S^-$, $\omega^- \notin \mathcal{L}(\mathcal{A})$;*
(ii) ***Minimality:*** *among all DRTAs consistent with S, $\mathcal{A}$ has the smallest number of states.*

To solve the above problem, we introduce 3-valued deterministic real-time automata, which extend DRTAs to process languages with "don't-care" words.

Definition 3. *A 3-valued Deterministic Real-Time Automaton (3DRTA) is defined as a triple $\mathcal{A} = (\mathcal{T}, A, R)$ where $\mathcal{T} = (Q, q_0, \Delta)$ is a DRTS and A and R, together with $D = Q \setminus (A \cup R)$, form a partition of the set of states Q. $A \subseteq Q$ is the set of* accepting *states, $R \subseteq Q$ is the set of* rejecting *states, and the remaining states D are called* don't-care *states.*

A run of $\mathcal{A} = (\mathcal{T}, A, R)$ is *accepting* (respectively, *rejecting*) if it ends in an accepting (resp. rejecting) state. A timed word ω is *accepted* (resp. *rejected*) by $\mathcal{A}$ if it has an accepting (resp. rejecting) run on ω. 3DRTAs map all words in $(\Sigma \times \mathbb{R}_{\geq 0})^*$ to *three* values: accepting $(+)$, rejecting $(-)$, and don't-care $(?)$, where they are accepting if they have an accepting run, rejecting if they have a rejecting run, and don't-care otherwise. Note that DRTAs are a special type of 3DRTAs with only accepting and rejecting states, that is, $D = \emptyset$ and $A \cup R = Q$. As for DRTAs, we denote the language of a 3DRTA $\mathcal{A}$ by $\mathcal{L}(\mathcal{A})$, i.e., the set of all words accepted by $\mathcal{A}$.

Following An et al.'s development of a Myhill–Nerode style theorem for real-time languages [4], we know that there exists a unique minimal DRTA recognizing a given real-time language. Similarly, as suggested by Chen et al. [11] for DFAs, we can identify equivalent words that reach the same state in the 3DRTA recognizing a function $L\colon (\Sigma \times \mathbb{R}_{\geq 0})^* \to \{+, -, ?\}$. We define the equivalence relation $\sim_L \subseteq (\Sigma \times \mathbb{R}_{\geq 0})^* \times (\Sigma \times \mathbb{R}_{\geq 0})^*$ over timed words as follows:

$$\omega_1 \sim_L \omega_2 \quad \Longleftrightarrow \quad \forall \omega' \in (\Sigma \times \mathbb{R}_{\geq 0})^* : L(\omega_1 \cdot \omega') = L(\omega_2 \cdot \omega').$$

Here, ω' denotes a possible suffix extension that can follow either ω_1 or ω_2. We denote by $|\sim_L|$ the *index* of the equivalence relation $\sim_L$, that is, the number of its equivalence classes.

Given a finite set of labeled samples $S = (S^+, S^-)$ in $(\Sigma \times \mathbb{R}_{\geq 0})^*$, we interpret S as a classification function that induces an equivalence relation $\sim_S$. For any timed word $\omega \in (\Sigma \times \mathbb{R}_{\geq 0})^*$, the classification of ω is given by:

$$S(\omega) = \begin{cases} + & \text{if } \omega \in S^+, \\ - & \text{if } \omega \in S^-, \\ ? & \text{otherwise.} \end{cases}$$

Finally, as S is a finite set, suppose it can be recognized by a DRTA $\mathcal{A}$ and so by a 3DRTA, we conclude with a straightforward proposition: since the number of equivalence classes $|\sim_S|$ induced by the sample set S is bounded by the number of equivalence classes $|\sim_{\mathcal{L}(\mathcal{A})}|$ of the potential target real-time language $\mathcal{L}(\mathcal{A})$, which is finite [4], it follows that $\sim_S$ has only finitely many equivalence classes.

Region Words and Regionization. Referring to the Myhill-Nerode theorem for real-time languages [4], we now introduce the region abstraction mapping time points to time intervals. As a DRTA can be viewed as a one-clock timed automaton where the clock resets after each action, the time domain can be partitioned into finitely many equivalence classes called *regions*. Given a constant $\kappa \in \mathbb{N}$, we define the region $[\![v]\!]_\kappa$ containing $v \in \mathbb{R}_{\geq 0}$ as follows: if $v \leq \kappa$, then $[\![v]\!]_\kappa = [v, v]$ if $v \in \mathbb{N}$, and $[\![v]\!]_\kappa = (\lfloor v \rfloor, \lfloor v \rfloor + 1)$ otherwise, where $\lfloor v \rfloor$ is the integer part of v; if $v > \kappa$, then $[\![v]\!]_\kappa = (\kappa, \infty)$. The constant κ induces a finite partition of the non-negative real line into a set of disjoint regions, each corresponding to either an integer point or an open interval between two consecutive integers. Formally, $\mathbb{R}_{\geq 0}$ is partitioned into $2\kappa + 2$ regions: $[n, n]$ for $0 \leq n \leq \kappa$, $(n, n+1)$ for $0 \leq n < \kappa$, and (κ, ∞), with $n \in \mathbb{N}$. We let $[\![\mathbb{R}_{\geq 0}]\!]_\kappa$ denote the set of all regions.

Definition 4 (Region word). *Given a timed word* $\omega = (\sigma_1, \tau_1) \cdots (\sigma_n, \tau_n)$, *the* region word *of* ω *is defined as* $[\![\omega]\!]_\kappa = (\sigma_1, [\![\tau_1]\!]_\kappa) \cdots (\sigma_n, [\![\tau_n]\!]_\kappa)$, *where* $\kappa = \lceil \max\{\tau_i \mid 1 \leq i \leq n\} \rceil$.

To make the construction below consistent with timed automata semantics, we assume that the labeled samples are *region-consistent*: for any two timed words ω_1, ω_2, if $[\![\omega_1]\!]_\kappa = [\![\omega_2]\!]_\kappa$ then it is not the case that $\omega_1 \in S^+$ and $\omega_2 \in S^-$. Equivalently, no two samples that differ only by choosing timestamps within the *same* region may receive conflicting labels. For example, if $\kappa \geq 7$, then $(a, 6.4)$ and $(a, 6.9)$ both lie in the region $(6, 7)$, so having $(a, 6.4) \in S^+$ and $(a, 6.9) \in S^-$ is disallowed. Timed automata with integer-bounded constraints on their transitions generate region-consistent timed words.

Given a timed word ω, we denote by $\omega[i]$ the i-th timed action (σ_i, τ_i), by $\omega[i, k]$ the subword starting at the i-th action and ending at the $(k-1)$-th action if $1 \leq i < k$, and by ε otherwise. Similarly, $\omega[i \cdots]$ represents the suffix starting at the i-th action if $i < |\omega|$, and ε otherwise. A timed word ω' is a *prefix* of ω if $\omega = \omega' \cdot \omega''$ for some $\omega'' \in (\Sigma \times \mathbb{R}_{\geq 0})^*$. The set of all prefixes of ω is denoted as $\mathbf{prefixes}(\omega)$, and for a set S of timed words, $\mathbf{prefixes}(S) = \bigcup_{\omega \in S} \mathbf{prefixes}(\omega)$.

We next establish the relation between a timed word ω and its region word $[\![\omega]\!]_\kappa$, in particular how regionization preserves the prefix structure.

Lemma 1 (Prefix preservation under regionization). *For every timed word $\omega = (\sigma_1, \tau_1) \cdots (\sigma_n, \tau_n)$ and every $1 \leq j \leq |\omega|$, the regionization operator $[\![\cdot]\!]_\kappa$ commutes with the prefix operator: $[\![\omega[1, j]]\!]_\kappa = [\![\omega]\!]_\kappa[1, j]$.*

Proof. By Definition 4, the operator $[\![\cdot]\!]_\kappa$ acts letter-wise and it is length-preserving; also, each timed action (σ, τ) is mapped to $(\sigma, [\![\tau]\!]_\kappa)$ without altering the order of actions. Hence, the image of the first j actions of ω under $[\![\cdot]\!]_\kappa$ is exactly the first j actions of $[\![\omega]\!]_\kappa$. Formally, if $\omega = (\sigma_1, \tau_1) \cdots (\sigma_n, \tau_n)$, then $[\![\omega[1, j]]\!]_\kappa = (\sigma_1, [\![\tau_1]\!]_\kappa) \cdots (\sigma_j, [\![\tau_j]\!]_\kappa) = [\![\omega]\!]_\kappa[1, j]$. Thus, the claim follows. $\square$

Corollary 1 (Prefixes of sets under regionization). *For any set of timed words $S \subseteq (\Sigma \times \mathbb{R}_{\geq 0})^*$, we have*

$$[\![\mathbf{prefixes}(S)]\!]_\kappa = \bigcup_{\omega \in S} \mathbf{prefixes}([\![\omega]\!]_\kappa).$$

Proof. By definition, $\mathbf{prefixes}(S) = \bigcup_{\omega \in S} \mathbf{prefixes}(\omega)$. Applying Lemma 1, for each $\omega \in S$ we obtain $[\![\mathbf{prefixes}(\omega)]\!]_\kappa = \mathbf{prefixes}([\![\omega]\!]_\kappa)$. Taking the union over all $\omega \in S$ yields the result. $\square$

This equivalence formally clarifies the relation between ω and $[\![\omega]\!]_\kappa$: reasoning about prefixes can be carried out either at the timed-word level or at the region-word level, without loss of generality.

3   Construction of 3DRTAs

Given a set of samples $S = (S^+, S^-)$, that we also write as $S = S^+ \uplus S^-$, our aim is to construct a DRTA $\mathcal{A}$ consistent with S. In the RTI algorithm proposed by Verwer et al. [35], the learning process requires the construction of a timed Augmented Prefix Tree Acceptor (APTA) prior to performing state merging.

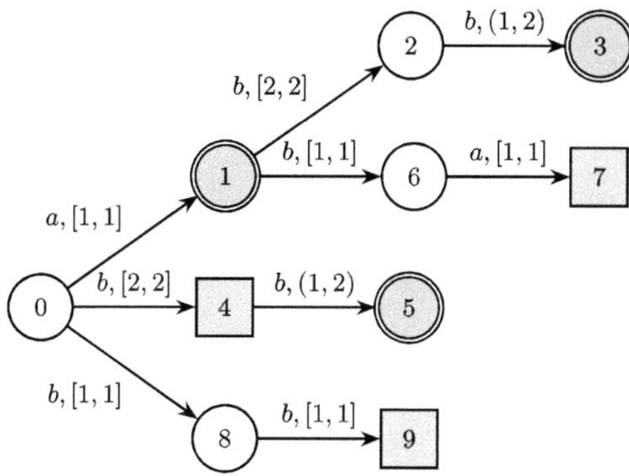

Fig. 1. The TAPTA for the sample set of timed words $S = (S^+, S^-)$, where $S^+ = (\{(a, 1), (a, 1)(b, 2)(b, 1.1), (b, 2)(b, 1.6)\}$ and $S^- = \{(a, 1)(b, 1)(a, 1), (b, 2), (b, 1)(b, 1)\})$. Blue double-circled round nodes are accepting states while red square nodes are rejecting states.

Similarly, in the approach of Tappler et al. [32], each step of every trace must be explicitly encoded into SMT formulas, which essentially correspond to a larger, unmerged timed APTA structure. Inspired by Verwer's timed APTA construction, we propose a related algorithm, termed TAPTA, outlined in Algorithm 1. The key distinction lies in how prefixes are treated: Verwer's timed APTA disregards timing information when identifying prefixes, which necessitates the subsequent use of a split operation to separate accepting and rejecting states. The split operation divides a transition $(q, (\sigma, I), q')$ into two or more sub-transitions whenever the time interval I contains both accepting and rejecting samples. A new boundary point is chosen to partition I into disjoint sub-intervals $I_1, I_2, \ldots$, and new target states are introduced so that accepting and rejecting behaviors are separated accordingly. In contrast, our TAPTA incorporates timing into the prefix representation, thereby eliminating the need for an additional split step and allowing the structure to be encoded directly.

A TAPTA for the set of samples $S = (S^+, S^-)$ requires consideration of both identical *action prefixes* σ and *timed region prefixes* $[\![\tau]\!]_\kappa$ in the DRTA, where $\kappa = \lceil \max\{\tau \in \mathbb{R}_{\geq 0} \mid (\sigma, \tau) \text{ appears in } S\}\rceil$. We first map each word ω to its corresponding region word $[\![\omega]\!]_\kappa$. In a TAPTA, each state corresponds to a unique prefix of a region word in S. First, define a function $f\colon \mathbf{prefixes}(S) \to \mathbb{N}_S$, where $\mathbb{N}_S = \{0, 1, \ldots, |\mathbf{prefixes}(S)| - 1\}$. Intuitively, f maps each prefix $u \in \mathbf{prefixes}(S)$ to a state in the TAPTA represented by a unique number in $\mathbb{N}_S$. Formally, a TAPTA $\mathcal{P}$ for S is a 3DRTA $(\mathcal{T}, A, R)$ where the DRTS $\mathcal{T}$ consists of the state set $\mathbb{N}_S$, initial state $f(\varepsilon)$, and transition function Δ defined as $\Delta(i, (a, I)) = j$ if $f([\![\omega]\!]_\kappa) = i$ and $f([\![\omega]\!]_\kappa(a, I)) = j$, where $[\![\omega]\!]_\kappa, [\![\omega]\!]_\kappa(a, I) \in \mathbf{prefixes}(S)$, $a \in \Sigma$, and $I \in [\![\mathbb{R}_{\geq 0}]\!]_\kappa$; we define $A = \{i \in \mathbb{N}_S \mid f(u) = i \text{ for some } u \in S^+\}$ and $R = \{i \in \mathbb{N}_S \mid f(u) = i \text{ for some } u \in S^-\}$.

For example, the TAPTA shown in Fig. 1 is generated from the sample set $S = (S^+, S^-)$, where S^+ contains the three accepted timed words $(a, 1)$,

---

Algorithm 1: BUILDTIMEDAPTA

Input: $S = \{S^+, S^-\}$: labeled timed traces
Output : $\mathcal{P}$: a deterministic real-time automaton (TAPTA)

1   $\kappa \leftarrow \lceil\max\{\tau \in \mathbb{R}_{\geq 0} \mid (\sigma, \tau) \text{ appears in } S\}\rceil$;   ▷ Determine maximum constant
2   $[\![\mathbb{R}_{\geq 0}]\!]_\kappa \leftarrow \{[0,0], (0,1), [1,1], \ldots, [\kappa,\kappa], (\kappa,\infty)\}$;   ▷ Time regions
3   $\mathcal{P} \leftarrow (\mathcal{T} = (Q = \{0\}, q_0 = 0, \Delta = \emptyset), A = \emptyset, R = \emptyset)$;   ▷ Initialize TAPTA
4   **foreach** *timed trace* $\omega = (\sigma_1, \tau_1) \cdots (\sigma_n, \tau_n) \in S$ **do**
5      $q \leftarrow 0$;   ▷ Start from initial state
6   **for** $i \leftarrow 1$ **to** n **do**
7          $\tau \leftarrow \tau_i, \sigma \leftarrow \sigma_i$;
8   **if** *there is no transition* $(q, \sigma, [\![\tau]\!]_\kappa, q')$ *in* Δ **then**
9              $q_{\text{new}} = |Q|$;   ▷ Create a new, fresh state
10             $Q \leftarrow Q \cup \{q_{\text{new}}\}$;   ▷ Add it to the TAPTA
11             $\Delta \leftarrow \Delta \cup \{q \xrightarrow{\sigma, [\![\tau]\!]_\kappa} q_{\text{new}}\}$;   ▷ Add the corresponding transition
12          $q \leftarrow \Delta(q, (\sigma, [\![\tau]\!]_\kappa))$;   ▷ Move to next state
13   **if** $\omega \in S^+$ **then**
14          $A \leftarrow A \cup \{q\}$;   ▷ Final state is accepting
15   **else**
16          $R \leftarrow R \cup \{q\}$;   ▷ Final state is rejecting
17 **return** $\mathcal{P}$;   ▷ Return the TAPTA

$(a,1)(b,2)(b,1.1)$, and $(b,2)(b,1.6)$; the three rejected timed words in S^- are $(a,1)(b,1)(a,1)$, $(b,2)$, and $(b,1)(b,1)$; and we have $f = \{\varepsilon \mapsto 0, (a,[1,1]) \mapsto 1, (a,[1,1])(b,[2,2]) \mapsto 2, (a,[1,1])(b,[2,2])(b,(1,2)) \mapsto 3, \ldots\}$. The set of accepting states is $A = \{1,3,5\}$ and the set of rejecting states is $R = \{4,7,9\}$.

3DRTA Construction. For simplicity of presentation, we assume that the full TAPTA $\mathcal{P}$ from S is given. Note that our 3DRTA can also be constructed on-the-fly from S using the same techniques presented in [12,14]. Our reduction process works in a backward manner as follows: as data structure, it uses a hash map called `Register` mapping equivalent states to the same representative state.

Initially, we map all accepting states without outgoing transitions into one single representative state and store them in the `Register`. Since each rejecting state is not equivalent to other states, each rejecting state is stored as its own representative in the `Register`. Then our reduction procedure iteratively traverses the TAPTA states in a backward manner from the leaves towards the root and processes the 3DRTA as follows. In each iteration, we first collect the states whose all successors are representative states in `Register`, and then identify equivalent states. Based on the definition of $\sim_S$, we define that two states $p, q \in Q$ are equivalent, denoted $p \equiv q$ if, and only if:

– both states share identical acceptance properties (i.e., both are accepting, rejecting, or don't-care states);
– for every input symbol $\sigma \in \Sigma$ and time region $I \in [\![\mathbb{R}_{\geq 0}]\!]_\kappa$, they either both lack transitions or share the same successor state in the `Register`.

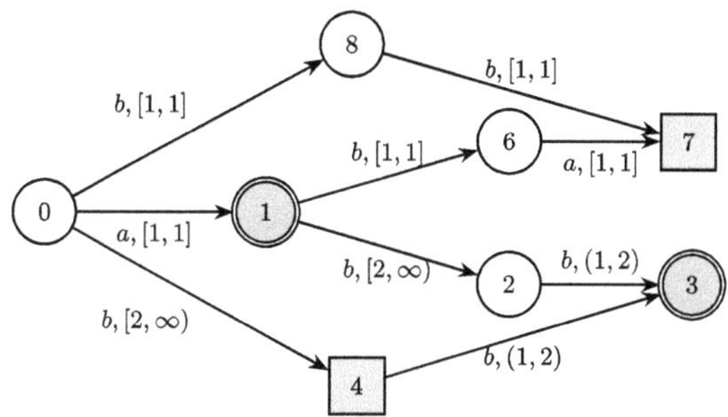

Fig. 2. The 3DRTA $\mathcal{A} = (\mathcal{T}, A, R)$ constructed from the TAPTA in Fig. 1, where $A = \{1, 3\}$ and $R = \{4, 7\}$.

A representative state is then created for each equivalence class and stored in the `Register`. All equivalent states in the DRTA are replaced by their representative. This reduction iterates until all states, including the initial state q_0, are processed, yielding a minimal DRTA consistent with the labeled timed sample set S. The temporal consistency enforced by time regions ensures that merged states preserve both symbolic and real-time behaviors of the original automaton.

Theorem 1. *Given a set of samples $S = (S^+, S^-)$, the 3DRTA construction produces a 3DRTA consistent with S.*

Proof. From the samples $S = (S^+, S^-)$, first the TAPTA for it is constructed. By a simple inspection of the construction and by Lemma 1, it is easy to observe that the TAPTA is consistent with S, as any (regionized) sample word s in S leads to a unique state that is accepting (rejecting, resp.) if and only if $s \in S^+$ ($s \in S^-$, resp.); inner states reached by prefixes not in S are "don't-care" states.

Similarly, in the 3DRTA construction, only leaf states of the same type are merged, and inner states are merged only if there is no outgoing transition that can distinguish them (by e.g. having a different input symbol σ, or a different time region I, or a different target state). This, together with Lemma 1 and Corollary 1, ensures that consistency with S is preserved. □

Note that our 3DRTA can also be constructed *on the fly* from S in the same manner as in [14] provided the sample words are taken out from S in the standard lexicographical order.

As an example, this construction process can merge the states 3 and 5 of the TAPTA in Fig. 1 reached by positive samples into the representative state 3, and merge the states 7 and 9 reached by negative samples into the representative state 7, as shown in Fig. 2. As clarified in Definition 4, every timestamp in a timed word is first mapped to its corresponding region form. Consequently, precise integer points such as 1 remain as single-point regions $[1, 1]$. In contrast, if the maximum integer delay κ actually occurs in the sample set, the corresponding timestamps $\tau = \kappa$ still belongs to the singleton region $[\kappa, \kappa]$; the tail region (κ, ∞) is reserved exclusively for delays strictly greater than κ. Accordingly, the transition in Fig. 2 from state 1 to state 2 labeled with $(b, [2, 2])$ should be

---

Algorithm 2: SAT-based DRTA synthesis

Input: $S = (S^+, S^-)$: positive and negative timed trace samples
Output : $\mathcal{A}$: a deterministic real-time automaton (DRTA)

1   $\mathcal{A} \leftarrow \text{BuildTimed3DRTA}(S^+, S^-)$;   ▷ Construct 3DRTA structure
2   $n \leftarrow 2$;   ▷ Initialize minimum state count
3   **while true do**
4      $\Phi_n \leftarrow \text{Encode}(\mathcal{A}, n)$;   ▷ Generate SAT constraints
5      $SAT, \mathcal{D} \leftarrow \text{Solve}(\Phi_n)$;   ▷ Check Φ_n satisfiability
6   **if** SAT **then**
7          $\mathcal{D}' \leftarrow \text{OptimizeTransitions}(\mathcal{D})$;   ▷ Apply region merging optimization
8   **return** $\mathcal{D}'$;   ▷ Return synthesized DRTA
9   **else**
10          $n \leftarrow n + 1$;   ▷ Increment state count and retry

---

interpreted as covering all delays $\tau \geq 2$ by merging $[2, 2]$ and $(2, \infty)$ in the final simplification step presented in Sect. 6.

As can be seen from the result, the obtained 3DRTA can be significantly smaller than the original TAPTA. Therefore, by using 3DRTAs instead of the TAPTA in our encoding, we can substantially reduce the number of required variables and constraints, resulting in a more manageable SAT problem and overall faster solving performance, as also confirmed by the experiments in Sect. 7.

4   SAT-Based DRTA Synthesis Framework

In this section we develop a comprehensive SAT-based synthesis framework for DRTAs that integrates minimal automaton construction, constraint encoding, iterative state space exploration, and verification processes. This framework addresses the fundamental challenge of learning temporal automata from timed traces while ensuring both minimality and correctness properties. The overall algorithm is given as Algorithm 2.

The synthesis process begins with the construction of a minimal 3DRTA structure from positive and negative timed trace samples $S = (S^+, S^-)$, where each sample s_i represents a sequence of timed events $(\sigma_1, \tau_1) \cdots (\sigma_l, \tau_l) \in (\Sigma \times \mathbb{R}_{\geq 0})^*$. We assume that S is not trivial, i.e., $S^+ \neq \emptyset$ and $S^- \neq \emptyset$; for trivial cases, DRTAs with universal and empty language can be directly produced.

The synthesis framework employs the BuildTimed3DRTA procedure to generate an initial 3DRTA structure $\mathcal{A} = (\mathcal{T}, A, R)$. We formalize the state space minimization search strategy as an optimization problem $\min_n n$, where n represents the minimum number of states needed by a DRTA consistent with S. The algorithm begins with $n = 2$ (minimum number of states for non-trivial DRTAs) and systematically increments this bound until a satisfiable solution is discovered, ensuring the synthesis of a minimal deterministic automaton.

For each candidate state set cardinality n, the algorithm calls the Encode procedure (detailed in Sect. 5) to transform the temporal learning problem into a Boolean satisfiability instance. The encoding process generates a comprehensive constraint system Φ_n that is then submitted to a SAT solver to determine whether Φ_n encodes a valid DRTA with n states consistent with S. If the formula Φ_n is satisfiable, then the algorithm calls the OPTIMIZETRANSITIONS procedure (given in Sect. 6) providing the region merging optimization: it applies heuristic methods to minimize the number of temporal regions while preserving determinism and completeness. Otherwise, the number of states is increased to $n+1$ since no DRTA with n states is consistent with S.

5   3DRTA to SAT Encoding for DRTA Synthesis

We now present the SAT encoding Φ_n returned by the ENCODE$(\mathcal{A}, n)$ procedure for solving the DRTA synthesis problem. Given the initial 3DRTA structure $\mathcal{A} = (\mathcal{T} = (Q, q_0, \Delta), A, R)$ consistent with the sample set $S = (S^+, S^-)$, we are looking for a DRTA $\mathcal{A}_s$ with n states that accepts all positive samples in S^+ and rejects all negative samples in S^-. States, transitions, and accepting states of $\mathcal{A}_s$ are obtained by the model of Φ_n, provided it is satisfiable.

5.1   SAT Encoding

To encode the DRTA synthesis problem, we use the following four types of variables, where $[n]$ denotes the set $[n] = \{0, 1, \ldots, n-1\}$; to simplify the presentation, we refer to states and transitions of $\mathcal{A}$ as nodes and edges, respectively, and we use the terms states and transitions for $\mathcal{A}_s$:

- **Node state variables** $x_{v,i}$, where $v \in Q$ and $i \in [n]$. $x_{v,i} \equiv 1$ if and only if the node v of $\mathcal{A}$ is mapped to the state i of $\mathcal{A}_s$.
- **Transition relation variables** $e_{i,\sigma,I,j}$, where $\sigma \in \Sigma$, $I \in [\![\mathbb{R}_{\geq 0}]\!]_\kappa$, and $i, j \in [n]$. $e_{i,\sigma,I,j} \equiv 1$ if and only if $\mathcal{A}_s$ has a transition from state i to state j over symbol σ within time region I.
- **Acceptance variables** z_i, where $i \in [n]$. $z_i \equiv 1$ if and only if state i of $\mathcal{A}_s$ is an accepting state.

We now give the list of constraints for the SAT encoding. First of all, we require that the synthesized $\mathcal{A}_s$ must be a valid DRTA. That is, it must be deterministic, complete, and satisfy the temporal constraints for all sample traces. For this, the following constraints must be enforced:

1. **Positive sample acceptance**: every positive timed trace in S^+ must be accepted by $\mathcal{A}_s$:
$$\bigwedge_{v \in A} \bigwedge_{i \in [n]} x_{v,i} \implies z_i.$$

2. **Negative sample rejection**: every negative timed trace in S^- must be rejected by $\mathcal{A}_s$:
$$\bigwedge_{v \in R} \bigwedge_{i \in [n]} x_{v,i} \longrightarrow \neg z_i.$$

3. **State-region mapping**: each node v of $\mathcal{A}$ must be represented in $\mathcal{A}_s$:

$$\bigwedge_{v \in Q} \bigvee_{i \in [n]} x_{v,i}$$

4. **Unique state–region mapping**: let $M = \{\, v \in Q \mid \exists u, u' \in \mathbf{prefixes}(S) : u \neq u' \wedge v = \Delta(r, u) = \Delta(r, u') \,\}$ be the set of nodes reached by multiple different prefixes; only states with unique prefixes must have a single representative:

$$\bigwedge_{v \in Q \setminus M} \bigwedge_{i,j \in [n],\, i<j} \neg(x_{v,i} \wedge x_{v,j}).$$

5. **Transition relation**: for each $v \in Q$, $\sigma \in \Sigma(v)$, $I \in [\![\mathbb{R}_{\geq 0}]\!]_\kappa$, and $i, j \in [n]$, if state i represents node v and there exists a transition from i to j over (a, I), then j represents node $\Delta(v, (a, I))$:

$$\bigwedge_{v \in Q} \bigwedge_{a \in \Sigma(v)} \bigwedge_{I \in [\![\mathbb{R}_{\geq 0}]\!]_\kappa} \bigwedge_{i,j \in [n]} (x_{v,i} \wedge e_{i,a,I,j}) \implies x_{\delta(v,(a,I)),j}$$

6. **Deterministic transitions**: for the same source state i, when receiving the same symbol σ and two time regions intersect, at most only one marked state can be the successor state:

$$\bigwedge_{\substack{\sigma \in \Sigma \\ }} \bigwedge_{\substack{I_1, I_2 \in [\![\mathbb{R}_{\geq 0}]\!]_\kappa \\ I_1 \cap I_2 \neq \emptyset}} \bigwedge_{\substack{i,j,t \in [n] \\ j \neq t}} \neg(e_{i,\sigma,I_1,j} \wedge e_{i,\sigma,I_2,t})$$

7. **Initial node constraint**: the initial node q_0 of $\mathcal{A}$ is mapped to the initial state 0 of $\mathcal{A}_s$:

$$x_{q_0,0}$$

8. **Original edge covering constraint**: each edge of $\mathcal{A}$ must have a corresponding transition in $\mathcal{A}_s$:

$$\bigwedge_{(v,\sigma,I,u) \in \Delta} \bigvee_{i,j \in [n]} x_{v,i} \wedge e_{i,\sigma,I,j} \wedge x_{u,j}$$

Encoding Size. Regarding the number of clauses in the SAT encoding, the dominant factors are: (i) $\mathcal{O}((|A| + |R|) \cdot n)$ from positive/negative sample consistency (constraints 1 and 2), (ii) $\mathcal{O}(|Q| \cdot n)$ from mapping (constraint 3), (iii) $\mathcal{O}(|Q \setminus M| \cdot n^2)$ from uniqueness (constraint 4), (iv) $\mathcal{O}(|Q| \cdot |\Sigma| \cdot |[\![\mathbb{R}_{\geq 0}]\!]_\kappa| \cdot n^2)$ from transition consistency (constraint 5), (v) $\mathcal{O}(|\Sigma| \cdot |[\![\mathbb{R}_{\geq 0}]\!]_\kappa|^2 \cdot n^2)$ from determinism (constraint 6), and (vi) $\mathcal{O}(|\Delta| \cdot n^2)$ from edge covering (constraint 8). Since $|[\![\mathbb{R}_{\geq 0}]\!]_\kappa| = 2\kappa + 2$, the overall number of constraints is

$$\mathsf{size}(\Phi_n) = \mathcal{O}\big(n^2(|Q||\Sigma|\kappa + |\Sigma|\kappa^2 + |\Delta|) + n(|A| + |R|)\big)$$

that is thus polynomial in the sizes $|Q|$, $|\Sigma|$, κ, $|\Delta|$ of the 3DRTA $\mathcal{A}$, and the number of states n we allow the DRTA synthesis procedure to use.

Correctness of the Encoding and Minimality. We formalize the connection between satisfiability of the encoding and the existence of a consistent DRTA, and then derive minimality from the size-increasing search over n.

Theorem 2. *Let $S = (S^+, S^-)$ be a finite, internally consistent sample set. For any $n \in \mathbb{N}_{\geq 1}$, the formula $\Phi_n(S)$ produced by* Encode *is satisfiable if and only if there exists a DRTA with n states that is consistent with S.*

Proof. ($\Rightarrow$) Suppose $\Phi_n(S)$ is satisfiable with a model M. Decoding M yields a DRTA $\mathcal{A}_n$ with n states: (i) the determinism and completeness clauses of Φ_n induce a well-formed transition structure over time regions; (ii) structural clauses fix initial/accepting status; and (iii) trace clauses ensure that every $\omega \in S^+$ is accepted and every $\omega \in S^-$ is rejected. Hence $\mathcal{A}_n$ is a DRTA consistent with S.

($\Leftarrow$) Conversely, let $\mathcal{A}_n$ be a DRTA with n states consistent with S. Assign the variables of $\Phi_n(S)$ to match the transitions, time regions, and accepting states of $\mathcal{A}_n$. By construction, determinism/completeness/structure constraints are satisfied, and each labeled trace in S is evaluated in accordance with $\mathcal{A}_n$, so $\Phi_n(S)$ is satisfiable. $\qquad\square$

Corollary 2. *Consider the procedure that checks $\Phi_n(S)$ in increasing order of n and returns the first n^* for which $\Phi_{n^*}(S)$ is satisfiable. Then n^* is the minimal number of states among all DRTAs consistent with S.*

Proof. By Theorem 2, $\Phi_n(S)$ is satisfiable exactly when an n-state consistent DRTA exists. Hence the first n^* with $\Phi_{n^*}(S)$ satisfiable is the smallest such n. Starting from $n = 1$ covers trivial cases (empty or universal languages) using a single-state DRTA. $\qquad\square$

Hence, the algorithm ensures both theoretical soundness and practical efficiency when synthesizing minimal timed automata from complex timed behavioral specifications, automatically determining the optimal state space cardinality while maintaining correctness properties.

Existing passive DRTA learning, such as the RTI algorithm [35], relies on heuristic state merging and provides no global minimality guarantees. To the best of our knowledge, no prior work is able to construct, from finite samples, a provably minimal DRTA with a complete decision procedure. Our framework fills this gap: the existence and uniqueness of minimal DRTAs is guaranteed by the Myhill–Nerode theorem for real-time languages [4], and we are the first to leverage this result to actually synthesize state-minimal DRTAs from finite labeled samples.

5.2   Comparison with Tappler's SMT-Based Approach

We now compare our method, based on the 3DRTA representation, with the SMT-based approach for timed automata synthesis proposed by Tappler et al. [32]. While both approaches aim at synthesizing timed models, they differ fundamentally in modeling assumptions, data usage, and constraint formulation.

Introduction to Basic Situation. Following Tappler et al. [32], the action alphabet is partitioned into *inputs* and *outputs*, $\Sigma = \Sigma_I \cup \Sigma_O$. By convention, input labels carry a "?" (e.g., a?) and output labels a "!" (e.g., b!). Determinism is enforced on outputs: for any state ℓ, the guards of outgoing *output* edges from ℓ are pairwise disjoint, so at most one output can be enabled at any time. Moreover, an upper bound on the delay before producing an output is required (the *k-urgency* assumption). In contrast, our DRTA framework does not separate Σ into inputs/outputs; traces are classified as accepting/rejecting via positive/negative samples. Hence our raw traces (without "?/!" annotations) cannot be used directly in Tappler's implementation unless converted to pure-input sequences.

Data Usage. Tappler's method relies exclusively on positive samples, motivated by the fact that real system testing can only observe the executions produced by the system, without access to negative traces. To avoid overgeneralization in such a setting, their method introduces an additional *k-urgency* assumption, i.e., the requirement that every output edge must be associated with an upper bound on its timing interval. Our framework, by contrast, directly exploits both positive and negative samples, enabling us to eliminate spurious behaviors without additional assumptions. This fundamental difference implies that our negative-sample data cannot be directly input into their algorithm.

Constraint Formulation. Tappler's method explicitly encodes every step of each trace into SMT formulas. While this enables the use of powerful SMT solvers, it also introduces substantial redundancy, causing the number of encoding variables to increase rapidly with the number and length of the samples. In contrast, our 3DRTA-based encoding eliminates prefix redundancies through region-based reasoning and achieves the same expressive power with significantly fewer constraints, while still preserving determinism and completeness.

Practical Implications. Given these differences, our experimental samples cannot be directly applied to Tappler's method: the lack of negative-sample handling, the input/output separation, and the *k-urgency* requirement create fundamental mismatches in problem settings. Considering these issues, our experiments in Sect. 7 do not compare our approach with Tappler's method. Therefore, instead, we compare the number of constraints in the encoded formula. For a fair comparison of the constraint sizes, we reformulate our data into the input format supported by their algorithm and align the number of states. As shown in Table 1, our 3DRTA-based encoding consistently yields a substantially smaller number of constraints compared to Tappler's SMT encoding across all sample sizes.

With increasing sample scale, the gap between the two approaches becomes more pronounced, highlighting the advantage of our method in producing more compact encoding and enabling more efficient solving in practice.

Table 1. Constraint count comparison between Tappler's SMT encoding and our 3DRTA-based encoding across sample sizes.

Samples	10	15	20	25	50	75	100	150	300	500	1000
Tappler	225	394	460	539	1180	1834	2265	3920	6755	11635	22620
3DRTA	125	259	305	352	767	1164	1396	2274	3523	5673	9177

Algorithm 3: Region Merging for DRTA

Input: Δ: The set of all transitions (q, σ, I, q')
Output : Opt: optimized mapping $(q, \sigma) \mapsto \{q' \mapsto \mathcal{I}\}$
1 Opt $\leftarrow \emptyset$;
2 $G \leftarrow \text{GROUPBY}(\Delta, \text{keys} = (q, \sigma))$;
3 **foreach** $(q, \sigma) \in \text{keys}(G)$ **do**
4      $B \leftarrow \text{GROUPBY}(G[(q, \sigma)], \text{keys} = q')$;
5   **if** $|B| = 1$ **then** $\triangleright$ Only one target
6   Opt$[(q, \sigma)] \leftarrow \{q' \mapsto \{[0, \infty)\}\}$;
7   **else**
8          $\mathfrak{P} \leftarrow \text{EXTRACTPROTECTIONPOINTS}(B)$; $\triangleright$ Get protection points
9          $B \leftarrow \text{OPTIMIZATION}(B, \mathfrak{P})$;
10         $H \leftarrow \text{FINDGAPS}(B)$; $\triangleright$ Identify uncovered time intervals
11         $B \leftarrow \text{FILLGAPS}(B, H, \mathfrak{P})$; $\triangleright$ Fill missing regions
12   Opt$[(q, \sigma)] \leftarrow B$;
13 **return** Opt

---

6 Region Merging for DRTAs

We address the optimization of a DRTA $\mathcal{A}$ by merging time regions on transitions while preserving language semantics. Let $\Delta \subseteq Q \times (\Sigma \times [\mathbb{R}_{\geq 0}]) \times Q$ be the transition relation of $\mathcal{A}$, with $|\Delta| < \infty$. Given $q, q' \in Q$ and $\sigma \in \Sigma$, we define $\Delta(q, \sigma, q') = \{I \mid (q, \sigma, I, q') \in \Delta\}$ and $\Delta(q, \sigma) = \{(I, q') \mid (q, \sigma, I, q') \in \Delta\}$. Our goal is to minimize the number of regions used by Δ while ensuring (i) determinism, (ii) completeness, and (iii) protection, introduced below.

Constraints. Let $q \in Q$ and $\sigma \in \Sigma$; then we enforce the following constraints. (i) *Determinism*: the sets of time regions assigned to distinct targets are pairwise disjoint: for all $q'_1 \neq q'_2$, $I_1 \in \Delta(q, \sigma, q'_1)$, and $I_2 \in \Delta(q, \sigma, q'_2)$, we require $I_1 \cap I_2 = \emptyset$. (ii) *Completeness*: $\bigcup_{(I, q') \in \Delta(q, \sigma)} I = \mathbb{R}_{\geq 0}$. (iii) *Protection*: let $\mathfrak{P}(q, \sigma) = \{\tau \in \mathbb{N} \mid \exists q' \in Q : (q, \sigma, [\tau, \tau], q') \in \Delta\}$ be the set of *protection points*, i.e., timestamps τ such that $(q, \sigma, [\tau, \tau], q') \in \Delta$ for some state q'. These protection points are integer timestamps that must remain mapped to their original successor state during region merging, i.e., the exact transition $(q, \sigma, [\tau, \tau], q')$ cannot be reassigned to a different $q'' \neq q'$.

Workflow. The overall framework of the region merging algorithm is shown in Algorithm 3. For each pair (q, σ) we aggregate $\Delta(q, \sigma)$ and work locally. If $\Delta(q, \sigma)$ has a single target q', we can replace all its regions by the maximal region $[0, \infty)$ assigned to q' (this is safe by determinism and preserves completeness). Otherwise, we (1) compute the protection map $\pi \colon \mathfrak{P}(q, \sigma) \to Q$; each protection point is mapped to its unique target state q', and it must not be included in any transition leading to a state other than q'. (2) We run a protection-aware merging routine within each target q' to coalesce adjacent or overlapping regions whenever doing so does not capture a foreign protection point. We *split* any region I that straddles a protected point $\tau \in \mathfrak{P}(q, \sigma)$ into subregions whose interiors exclude τ, so that no subsequent merge can violate protection. Finally, (3) we *fill gaps* to restore completeness by assigning uncovered intervals to adjacent targets according to a priority heuristic.

Protection-Aware Merging. Given $\mathfrak{P}(q, \sigma)$, fix a target $s \in Q$ and consider two candidate regions $R_1, R_2 \in \Delta(q, \sigma, s)$ with $R_1 \cap R_2 \neq \emptyset$ or sharing endpoints. Define $R_{\mathrm{m}} = \mathrm{hull}(R_1 \cup R_2)$, where hull denotes the interval hull, i.e., the smallest interval covering the two regions. We allow the merge if and only if no protection point belonging to a different target would be newly engulfed:

$$\mathrm{CanMerge}(R_1, R_2) := \nexists p \in \mathfrak{P}(q, \sigma) : p \in R_{\mathrm{m}} \wedge p \notin (R_1 \cup R_2) \wedge \pi(p) \neq s.$$

If CanMerge holds, we replace R_1 and R_2 by R_{m}; otherwise, we remove this protection point from the merged region R_{m}.

Gap Filling. To satisfy completeness, let $\mathsf{Gaps}(q, \sigma)$ denote the (finite) set of *uncovered intervals*, i.e., maximal contiguous sub-intervals of $\mathbb{R}_{\geq 0}$ that are not contained in any interval I with $(I, q') \in \Delta(q, \sigma)$. For a gap g and a candidate target s we define their priority as

$$\mathrm{Priority}(g, s) = \begin{cases} 2 & \text{if } g \text{ is lower-adjacent to some } I \in \Delta(q, \sigma, s), \\ 1 & \text{if } g \text{ is upper-adjacent to some } I \in \Delta(q, \sigma, s), \\ \frac{1}{\mathrm{dist}(g,s)+\varepsilon} & \text{otherwise,} \end{cases}$$

where $\varepsilon > 0$ is a fixed small constant (e.g., 10^{-6}), used only to avoid division by zero when $\mathrm{dist}(g, s) = 0$. Its precise value does not affect the relative ordering of priorities, as long as it is sufficiently small. The distance dist is defined as

$$\mathrm{dist}(g, s) = \min\{ |e - e'| \mid e \in \mathrm{E}(g) \wedge e' \in \mathrm{E}(\Delta(q, \sigma, s)) \}$$

where $\mathrm{E}(g)$ denotes the set of the two endpoints of the gap g and $\mathrm{E}(\Delta(q, \sigma, s))$ denotes the set of all endpoints of intervals $I \in \Delta(q, \sigma, s)$. Thus, $\mathrm{dist}(g, s)$ is the minimal absolute distance between any endpoint of g and any endpoint of an interval associated with target state s. We assign each g to the target s having maximum priority. In case of ties, we apply the *minimum-color-first rule*, meaning that among candidate targets we select the one that currently has the smallest number of assigned intervals (i.e., the "least colored" state); if a tie still occurs, we resolve it by taking arbitrarily one the targets in the tie. This ensures a balanced allocation of gaps across targets and avoids unnecessary region proliferation. Assignments never include a protected point with $\pi(p) \neq s$.

For example, suppose $\Delta(q, \sigma)$ relates $[1, 3]$, $[2, 4)$, $[3, 5]$ to q_1 and the point $[4, 4]$ to q_2. The procedure protects 4, splits q_1's regions around 4, merges the remainder into $[0, 4) \cup (4, \infty)$ for q_1, and keeps $[4, 4]$ for q_2, thereby preserving determinism and completeness with a minimal number of regions.

Complexity. Let n denote the number of initial regions for a fixed (q, σ). With sorting and linear scans inside each target, the overall running time is $O(n^2)$ in the worst case (dominated by guarded merge checks) and $O(n)$ space. Empirically, this yields a substantial reduction in the number of regions while maintaining the original transition semantics.

7   Experimental Evaluation

We have implemented our approach in a Python-based prototype[1] and generated randomly timed automata of different sizes that we used to sample random traces then used as input benchmarks. The automata have between 3 and 30 states and between 2 and 8 letters; the produced benchmarks have between 5 and 1000 samples, for a total of 6972 instances. Let FLEXFRINGE and RTA denote the tools implementing the original RTI [35] and our RTA approaches, respectively. We ran the tools on a desktop machine with an i5-12400F CPU and 16 GB of memory running Ubuntu 24.04.3; we used BENCHEXEC [8] to get reliable benchmark outcomes.

In general, FLEXFRINGE is faster than RTA, but it synthesizes DRTAs that are much larger. Regarding the running time, both tools were able to generate DRTAs for the benchmarks well within the 30 minutes time limit we imposed in BENCHEXEC; when the generated 3DRTA contains a large number of states, the SAT encoding from Sect. 5.1 contains a huge number of variables and constraints, which would take Z3 [25], the SAT solver we used, a large amount of time to check its satisfiability. Still, all benchmarks have been completed in less than 15 minutes each.

Regarding the size of the synthesized DRTAs, the plots in Fig. 3 show the comparison between the number of states produced by our RTA and the original FLEXFRINGE, as well their state distributions; we use logarithmic axes in both plots. As we can see from the scatter plot on the left part of Fig. 3, RTA consistently produces fewer states than FLEXFRINGE, since there is not a single mark on or below the dashed diagonal line. Moreover, RTA is able to generate DRTAs with much fewer states than FLEXFRINGE: for instance, on one benchmark RTA needed just 2 states while FLEXFRINGE used 577 states while on

---

[1] See https://github.com/Sakura-0407/VMCAI_Synthesis-of-Minimal-DRTA.

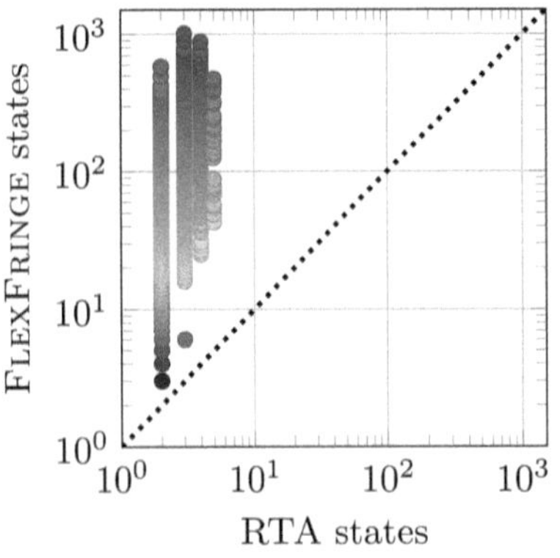 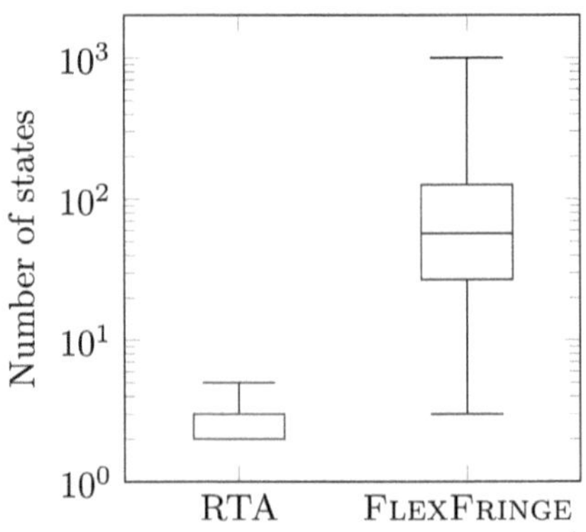

Fig. 3. State comparison between RTA and FLEXFRINGE

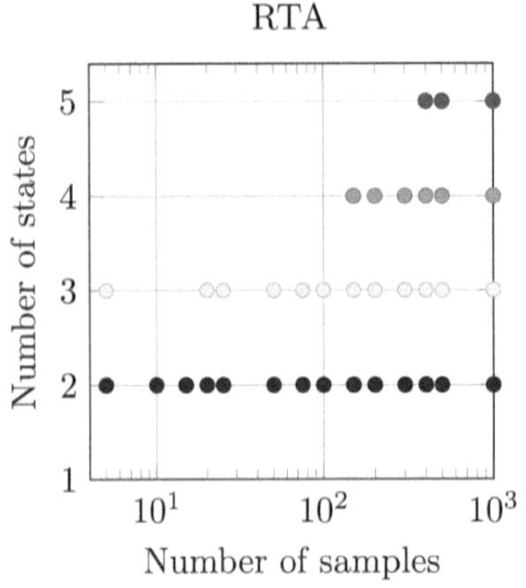 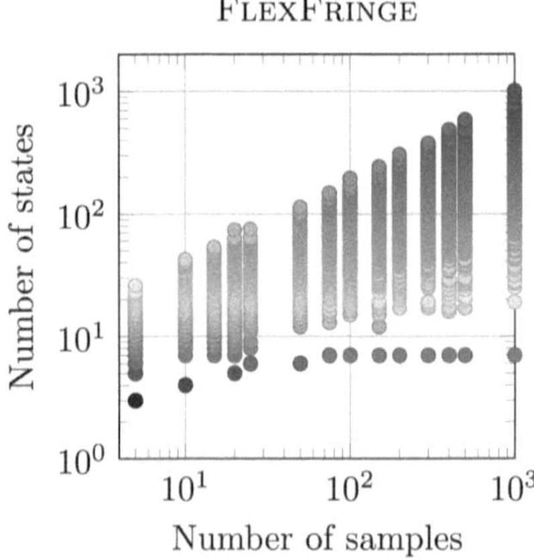

Fig. 4. State distribution vs. number of samples by RTA and FLEXFRINGE

another benchmark the states were 3 and 1002, respectively. This is confirmed by the box plots on the right part of Fig. 3: all solved cases by RTA needed at most 5 states, while on the same benchmarks FLEXFRINGE reached 1002 states, with the central 50% of the synthesized DRTAs having between 27 and 126 states.

Figure 4 shows the distribution of the number of states with respect to the number of samples. As one would expect, and as confirmed by the plots, when we have more samples, usually we need more states to make the DRTA consistent with them. However, RTA is able to achieve this with much fewer states than FLEXFRINGE, even with very few samples like 5 or 10.

These experiments confirm that the synthesis procedure we proposed in Sects. 3–6 yields DRTAs that are much smaller than the ones produced by FLEXFRINGE based on heuristics.

8 Conclusion and Future Work

In this paper, we presented a systematic framework for synthesizing minimal deterministic real-time automata (DRTAs) from positive and negative timed traces. Our approach is centered around three main contributions. First, we introduced the 3DRTA as a novel intermediate representation that extends the idea of 3-valued automata to the timed setting, allowing timing constraints to be encoded compactly while avoiding the redundancies inherent in prefix-tree constructions. Second, we developed a SAT-based synthesis procedure that incrementally explores the state space and guarantees minimality of the resulting DRTA. Third, we proposed a region merging algorithm that reduces the number of temporal regions in transitions without violating determinism, completeness, or exact-point semantics. Through extensive experiments, we demonstrated that our 3DRTA-based encoding generates substantially fewer constraints than Tappler's SMT-based method, thereby yielding a more compact encoding and significantly improving solving efficiency. In addition, compared to the RTI algorithm, which relies on heuristic evidence-driven state merging, our approach is able to synthesize DRTAs with much smaller state spaces. These results highlight the advantage of combining the 3DRTA representation with SAT-based synthesis, offering both theoretical minimality guarantees and practical scalability.

Although our work provides a solid theoretical and empirical foundation, several directions remain open for future research. One promising avenue is to improve robustness against noisy or incomplete data, where the strict separation of positive and negative samples may not hold in practice. Another direction is to extend our framework to richer classes of timed automata, such as event-recording automata or multi-clock models, and to investigate how the principles of 3DRTA encoding can be generalized. Further work may also focus on tighter integration with modern SAT/SMT solvers, for example by adopting incremental encodings or solver-guided heuristics to accelerate synthesis. Adapting the existing divide-and-conquer strategies for logic learning [26,29] to accelerate automata learning is also an interesting direction. Finally, large-scale case studies in real-world verification and system identification would provide valuable insights into the applicability and performance of our method in industrial contexts.

Acknowledgments. We thank the anonymous reviewers for their useful remarks that helped us improve the quality of the paper.

Work supported in part by the National Key R&D Program of China (Grant No. 2022YFA1005101), by the National Natural Science Foundation of China (NSFC) (Grants Nos. W2511064, 62472316, 62192732, 62032024, and 62032019), by the ISCAS Basic Research (Grant Nos. ISCAS-JCZD-202406 and ISCAS-JCZD-202302), by the CAS Project for Young Scientists in Basic Research (Grant No. YSBR-040), and by the ISCAS New Cultivation Project ISCAS-PYFX-202201. This project is part of the European Union's Horizon 2020 research and innovation programme under the Marie Skłodowska-Curie grant no. 101008233.

References

1. Alur, R., Dill, D.L.: A theory of timed automata. Theor. Comput. Sci. **126**(2), 183–235 (1994). https://doi.org/10.1016/0304-3975(94)90010-8
2. An, J., Chen, M., Zhan, B., Zhan, N., Zhang, M.: Learning one-clock timed automata. In: TACAS 2020. LNCS, vol. 12078, pp. 444–462. Springer, Cham (2020). https://doi.org/10.1007/978-3-030-45190-5_25
3. An, J., Wang, L., Zhan, B., Zhan, N., Zhang, M.: Learning real-time automata. Sci. China Inf. Sci. **64**(9), 1–17 (2021). https://doi.org/10.1007/s11432-019-2767-4
4. An, J., Zhan, B., Zhan, N., Zhang, M.: Learning nondeterministic real-time automata. ACM Trans. Embed. Comput. Syst. **20**(5s), 99:1–99:26 (2021). https://doi.org/10.1145/3477030
5. Angluin, D.: Learning regular sets from queries and counterexamples. Inf. Comput. **75**(2), 87–106 (1987). https://doi.org/10.1016/0890-5401(87)90052-6
6. Behrmann, G., Larsen, K.G., Rasmussen, J.I.: Priced timed automata: algorithms and applications. In: de Boer, F.S., Bonsangue, M.M., Graf, S., de Roever, W.-P. (eds.) FMCO 2004. LNCS, vol. 3657, pp. 162–182. Springer, Heidelberg (2005). https://doi.org/10.1007/11561163_8
7. Bengtsson, J., Yi, W.: Timed automata: semantics, algorithms and tools. In: Desel, J., Reisig, W., Rozenberg, G. (eds.) ACPN 2003. LNCS, vol. 3098, pp. 87–124. Springer, Heidelberg (2004). https://doi.org/10.1007/978-3-540-27755-2_3
8. Beyer, D., Löwe, S., Wendler, P.: Reliable benchmarking: requirements and solutions. Int. J. Softw. Tools Technol. Transfer **21**(1), 1–29 (2017). https://doi.org/10.1007/s10009-017-0469-y
9. Bouyer, P., Laroussinie, F., Reynier, P.-A.: Diagonal constraints in timed automata: forward analysis of timed systems. In: Pettersson, P., Yi, W. (eds.) FORMATS 2005. LNCS, vol. 3829, pp. 112–126. Springer, Heidelberg (2005). https://doi.org/10.1007/11603009_10
10. Cassez, F., Jensen, P.G., Larsen, K.G.: Verification and parameter synthesis for real-time programs using refinement of trace abstraction. Fund. Inf. **178**(1–2), 31–57 (2021). https://doi.org/10.3233/FI-2021-1997
11. Chen, Y.-F., Farzan, A., Clarke, E.M., Tsay, Y.-K., Wang, B.-Y.: Learning minimal separating DFA's for compositional verification. In: Kowalewski, S., Philippou, A. (eds.) TACAS 2009. LNCS, vol. 5505, pp. 31–45. Springer, Heidelberg (2009). https://doi.org/10.1007/978-3-642-00768-2_3
12. Daciuk, J., Mihov, S., Watson, B.W., Watson, R.E.: Incremental construction of minimal acyclic finite state automata. Comput. Linguist. **26**(1), 3–16 (2000). https://doi.org/10.1162/089120100561601
13. D'Antoni, L., Veanes, M.: Minimization of symbolic automata. In: Proceedings of the 41st Annual ACM SIGPLAN-SIGACT Symposium on Principles of Programming Languages, POPL 2014, pp. 541–553. ACM (2014). https://doi.org/10.1145/2535838.2535849
14. Dell'Erba, D., Li, Y., Schewe, S.: DFAMiner: mining minimal separating DFAs from labelled samples. In: 26th International Symposium on Formal Methods, FM 2024 (2). LNCS, vol. 14934, pp. 48–66. Springer, Heidelberg (2024). https://doi.org/10.1007/978-3-031-71177-0_4
15. Dima, C.: Real-time automata. J. Automata Lang. Comb. **6**(1), 3–23 (2001). https://doi.org/10.25596/jalc-2001-003

16. Fisman, D., Frenkel, H., Zilles, S.: Inferring symbolic automata. Logical Methods Comput. Sci. **19**(2), 5:1–5:37 (2023). https://doi.org/10.46298/LMCS-19(2:5)2023
17. Gold, E.M.: Complexity of automaton identification from given data. Inf. Control **37**(3), 302–320 (1978). https://doi.org/10.1016/S0019-9958(78)90562-4
18. Grinchtein, O., Jonsson, B., Leucker, M.: Learning of event-recording automata. Theor. Comput. Sci. **411**(40–42), 4029–4054 (2010). https://doi.org/10.1016/J.TCS.2010.07.008
19. Grinchtein, O., Leucker, M., Piterman, N.: Inferring network invariants automatically. In: Furbach, U., Shankar, N. (eds.) IJCAR 2006. LNCS (LNAI), vol. 4130, pp. 483–497. Springer, Heidelberg (2006). https://doi.org/10.1007/11814771_40
20. Hessel, A., Larsen, K.G., Nielsen, B., Pettersson, P., Skou, A.: Time-optimal real-time test case generation using UPPAAL. In: Petrenko, A., Ulrich, A. (eds.) FATES 2003. LNCS, vol. 2931, pp. 114–130. Springer, Heidelberg (2004). https://doi.org/10.1007/978-3-540-24617-6_9
21. Heule, M.J.H., Verwer, S.: Exact DFA identification using SAT solvers. In: Sempere, J.M., García, P. (eds.) ICGI 2010. LNCS (LNAI), vol. 6339, pp. 66–79. Springer, Heidelberg (2010). https://doi.org/10.1007/978-3-642-15488-1_7
22. de la Higuera, C.: A bibliographical study of grammatical inference. Pattern Recogn. **38**(9), 1332–1348 (2005). https://doi.org/10.1016/J.PATCOG.2005.01.003
23. Jiang, Z., Pajic, M., Moarref, S., Alur, R., Mangharam, R.: Modeling and verification of a dual chamber implantable pacemaker. In: Flanagan, C., König, B. (eds.) TACAS 2012. LNCS, vol. 7214, pp. 188–203. Springer, Heidelberg (2012). https://doi.org/10.1007/978-3-642-28756-5_14
24. Lang, K.J., Pearlmutter, B.A., Price, R.A.: Results of the abbadingo one DFA learning competition and a new evidence-driven state merging algorithm. In: 4th International Colloquium on Grammatical Inference, ICGI 1998. LNCS, vol. 1433, pp. 1–12. Springer, Heidelberg (1998). https://doi.org/10.1007/BFB0054059
25. de Moura, L., Bjørner, N.: Z3: an efficient SMT solver. In: Ramakrishnan, C.R., Rehof, J. (eds.) TACAS 2008. LNCS, vol. 4963, pp. 337–340. Springer, Heidelberg (2008). https://doi.org/10.1007/978-3-540-78800-3_24
26. Neider, D., Gavran, I.: Learning linear temporal properties. In: 2018 Formal Methods in Computer Aided Design, FMCAD 2018, pp. 1–10. IEEE (2018). https://doi.org/10.23919/FMCAD.2018.8603016
27. Ouaknine, J., Worrell, J.: On the decidability of metric temporal logic. In: 20th IEEE Symposium on Logic in Computer Science, LICS 2005, pp. 188–197. IEEE Computer Society (2005). https://doi.org/10.1109/LICS.2005.33
28. Pitt, L., Warmuth, M.K.: The minimum consistent DFA problem cannot be approximated within any polynomial. In: 21st Annual ACM Symposium on Theory of Computing, STOC 1989, pp. 421–432. ACM (1989). https://doi.org/10.1145/73007.73048
29. Riener, H.: Exact synthesis of LTL properties from traces. In: 2019 Forum for Specification and Design Languages, FDL 2019, pp. 1–6. IEEE (2019). https://doi.org/10.1109/FDL.2019.8876900
30. Shen, W., An, J., Zhan, B., Zhang, M., Xue, B., Zhan, N.: PAC learning of deterministic one-clock timed automata. In: Lin, S.-W., Hou, Z., Mahony, B. (eds.) ICFEM 2020. LNCS, vol. 12531, pp. 129–146. Springer, Cham (2020). https://doi.org/10.1007/978-3-030-63406-3_8
31. Tang, X., Shen, W., Zhang, M., An, J., Zhan, B., Zhan, N.: Learning deterministic one-clock timed automata via mutation testing. In: 20th International Symposium

on Automated Technology for Verification and Analysis, ATVA 2022. LNCS, vol. 13505, pp. 233–248. Springer, Heidelberg (2022). https://doi.org/10.1007/978-3-031-19992-9_15

32. Tappler, M., Aichernig, B.K., Lorber, F.: Timed automata learning via SMT solving. In: 14th International Symposium on NASA Formal Methods, NFM 2022). LNCS, vol. 13260, pp. 489–507. Springer, Heidelberg (2022). https://doi.org/10.1007/978-3-031-06773-0_26

33. Teng, Y., Zhang, M., An, J.: Learning deterministic multi-clock timed automata. In: Ábrahám, E., Jr., M.M. (eds.) 27th ACM International Conference on Hybrid Systems: Computation and Control, HSCC 2024, pp. 6:1–6:11. ACM (2024). https://doi.org/10.1145/3641513.3650124

34. Verwer, S., de Weerdt, M., Witteveen, C.: The efficiency of identifying timed automata and the power of clocks. Inf. Comput. **209**(3), 606–625 (2011). https://doi.org/10.1016/J.IC.2010.11.023

35. Verwer, S., de Weerdt, M., Witteveen, C.: Efficiently identifying deterministic real-time automata from labeled data. Mach. Learn. **86**(3), 295–333 (2012). https://doi.org/10.1007/S10994-011-5265-4

36. Waga, M.: Active learning of deterministic timed automata with Myhill-Nerode style characterization. In: Enea, C., Lal, A. (eds.) CAV 2023. Lecture Notes in Computer Science, vol. 13964, pp. 3–26. Springer, Heidelberg (2023). https://doi.org/10.1007/978-3-031-37706-8_1

37. Wang, L., Zhan, N., An, J.: The opacity of real-time automata. IEEE Trans. Comput. Aided Des. Integr. Circuits Syst. **37**(11), 2845–2856 (2018). https://doi.org/10.1109/TCAD.2018.2857363

38. Zakirzyanov, I., Shalyto, A., Ulyantsev, V.: Finding All minimum-size DFA consistent with given examples: SAT-based approach. In: Cerone, A., Roveri, M. (eds.) SEFM 2017. LNCS, vol. 10729, pp. 117–131. Springer, Cham (2018). https://doi.org/10.1007/978-3-319-74781-1_9

Try-Mopsa:
Relational Static Analysis in Your Pocket

Raphaël Monat[(✉)]

Univ. Lille, Inria, CNRS, Centrale Lille,
UMR 9189 CRIStAL, 59000 Lille, France
`raphael.monat@inria.fr`

Abstract. Static analyzers are complex pieces of software with large dependencies. They can be difficult to install, which hinders adoption and creates barriers for students learning static analysis. This work introduces Try-Mopsa: a scaled-down version of the Mopsa static analysis platform, compiled into JavaScript to run purely as a client-side application in web browsers. Try-Mopsa provides a responsive interface that works on both desktop and mobile devices. Try-Mopsa features all the core components of Mopsa. In particular, it supports relational numerical domains. We present the interface, changes and adaptations required to have a pure JavaScript version of Mopsa. We envision Try-Mopsa as a convenient platform for onboarding or teaching purposes.

Keywords: Static Analysis · Abstract Interpretation · Usability · Teaching

1 Introduction

Static analyzers are complex pieces of software, usually building on a large number of dependencies. For example, the Mopsa static analysis platform [24] requires among others: two parsing libraries (Menhir and libclang), the Zarith library to handle arbitrary precision arithmetic, and the Apron library to handle relational numerical domains. When facing a large number of users (or students), there is always a chance to encounter some installation issues. While good packaging or containerization can certainly limit those, installing a new tool still consumes time and resources. In any case, the installation process hinders both testing and adoption of new static analysis tools.

One remedy to this issue is to provide zero-install ways to try a software, for example by enabling tool usage through a web browser. This article presents Try-Mopsa, a scaled-down version of the Mopsa static analysis platform that runs entirely as a client-side web application. It relies on a responsive interface to

This work is partially supported by grant agreement ANR-24-CE25-7956-01 RAISIN from the French Agence Nationale de la Recherche, and by an Amazon Research Award, Fall 2024.

support a wide variety of devices (from smartphones to computers), and supports all core features of Mopsa, including relational domains and an interactive engine acting as an abstract debugger. Try-Mopsa is available online [38]. Thanks to its purely client-side implementation, Try-Mopsa is unaffected by the scalability issues of server-side implementations serving a large number of concurrent users.

We provide a brief overview of Mopsa in Sect. 2. Then, we describe in Sect. 3 how we adapted Mopsa to make it runnable in a web page. Section 4 provides an overview of the resulting web interface, Sect. 5 evaluates several features of the implementation, and Sect. 6 discusses related work.

2   A Brief Overview of Mopsa

Mopsa is a Modular Open Platform for Static Analysis, rooted within the abstract interpretation framework [10]. It aims at providing a convenient platform for static analysis learners, developers and users. Although Mopsa explores some new perspectives for the design of static analyzers, it is stable and precise enough to be on-par with state-of-the-art academic program analyzers participating to the Software-Verification Competition [4,42]. Journault et al. [24] describe the core of Mopsa's principles, and Monat [37, Chapter 3] provides an in-depth introduction to Mopsa's architecture. We briefly describe three features of interest for this article:

Multilanguage support. Mopsa supports the analysis of multiple programming languages. Currently, it supports the analysis of an in-house toy imperative language (called "Universal"), of C [49] and of Python [40,43].

User-defined analysis combination. As an analysis platform, Mopsa offers a wide variety of abstract domains to choose from. Users define in a configuration file which abstract domains they want to enable, and how they should be combined (reduced product, ...).

Tailored for relational domains. Relational domains greatly improve the expressiveness of analyses by being able to infer constraints between variables. In addition, every abstract domain can introduce ghost variables and add constraints on those, which can be handled by an underlying relational domain.

We show in Fig. 1 the main components of Mopsa, and describe them below. All components are supported by Try-Mopsa, except those filled in gray. We discuss unsupported components and potential replacements in Sect. 3.

The frontend handles the parsing of a program into an abstract syntax tree (AST). As we mentioned above, Mopsa currently supports the analysis of three programming languages.

The analysis builder takes a JSON configuration file describing the choice of abstract domains and their combinators, makes sure it is valid, and enables the corresponding abstract domains in the toplevel analysis. It also sets passed options (either for the framework or for the enabled abstract domains).

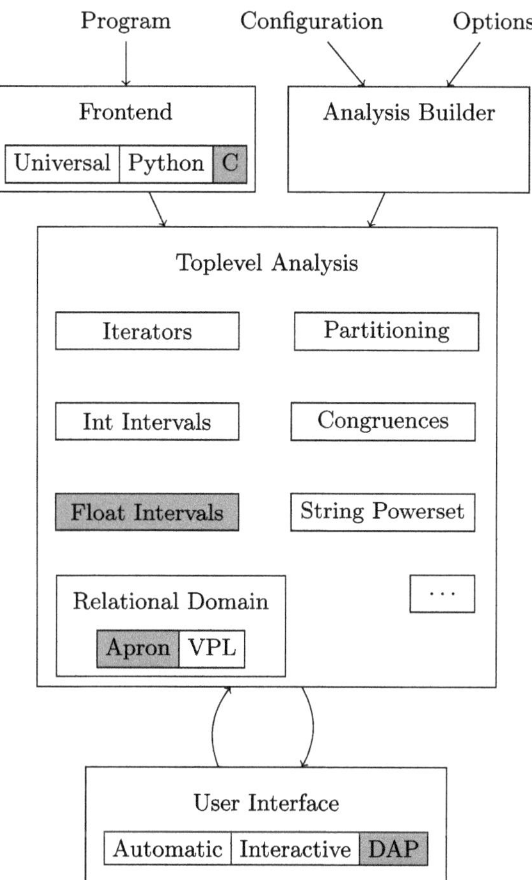

Fig. 1. Components of Mopsa. All components are supported by Try-Mopsa, except those filled in gray.

Once the parsing is done and the abstract domains are combined according to the specified configuration, the toplevel analysis starts. It runs the combination of chosen abstract domains, such as iterators handling loops and function calls, trace and state partitioning, numerical abstract domains and string abstractions.

Different user interfaces are offered by Mopsa. The automatic interface is the classic one: the analysis runs to completion and then displays the results. The interactive engine lets user navigate the abstract execution of the program, where analysis computations are performed on-the-fly accordingly. This interface acts as a **gdb**-like abstract debugger, and supports breakpoints, program navigation, printing of abstract states, ... The DAP interface provides a debug adapter protocol interface, similar to the interactive engine but allowing interactions with IDEs supporting this protocol. Monat et al. [41, Section 5] describe the former two interfaces in more details.

3   Try-Mopsa: Under the Hood

This section describes the implementation of Try-Mopsa. Section 3.1 focuses on the backend part, explaining how we adapted Mopsa to be able to compile it to JavaScript. Section 3.2 shows how this compiled JavaScript is integrated within a web page to provide a user-friendly interface.

3.1   Compiling Mopsa to JavaScript

According to `cloc` [12], the current version of Mopsa consists of 87,856 lines of code. The overwhelming majority of the codebase (89%) is written in OCaml. We rely on the Js_of_ocaml (JSOO) compiler [56] to compile these components of Mopsa to JavaScript. This compiler option is selected in the build system as a drop-in replacement to the standard OCaml compiler creating executables. This compilation process already supports most Mopsa features, including the parsing of user-defined analysis combination, fixpoint and interprocedural iterators, heap and string abstractions, trace and state partitioning.

However, some crucial features of Mopsa rely on external dependencies: we use Menhir or libclang to parse programs, the Zarith library to handle arbitrary precision arithmetic, and the Apron library to handle relational numerical domains. We now discuss how these components have been adapted to compile to JavaScript.

Parsing Libraries. Mopsa relies on the Menhir library [50] to implement parsers for the Universal and the Python programming languages. As Menhir is written in pure OCaml, JSOO natively supports compiling it to JavaScript.

Try-Mopsa currently does not support C parsing. Indeed, Mopsa depends on LLVM and its libclang library to parse C programs, and our C parser contains 4,182 lines of C++ glue code. While JSOO can interface with manually written JavaScript stubs, and some prototype versions of LLVM have been compiled to WebAssembly, we believe the current level of support for these processes would require too much manual work.

Arbitrary-Precision Integer Arithmetic. Our integer abstractions (intervals, congruences) rely on arbitrary-precision arithmetic to avoid overflows and unsoundness. Mopsa relies on the Zarith [33] library to implement these mathematical integers. Try-Mopsa makes use of a third-party alternative implementation in JavaScript of the Zarith interface, called Zarith_stubs_js [20] and natively supported by the compilation process of JSOO.

Relational Domains. Mopsa depends on the Apron library [23] to support relational numeric domains such as octagons [31] and polyhedra [11]. However, Apron abstract domains are written in C/C++: compiling them to JavaScript, just like our C frontend, would be quite cumbersome. For Try-Mopsa, we chose to replace Apron with the alternative Verified Polyhedra Library (VPL) [5], which provides different implementations of polyhedra. Given that some domains of the

VPL are pure OCaml implementations, JSOO is able to compile it to JavaScript, enabling relational abstract domains to run within web browsers. We wrote a new VPL wrapper, making it compatible with Mopsa's interface, and adding support for the `fold/expand` operators [18] required by Mopsa but not built in the VPL.

Floating-Point Abstractions. Our current implementation of floating-point intervals sets various rounding modes in the floating-point unit to be sound [30]. Since this is not supported by JavaScript, these intervals are currently not supported by Try-Mopsa. Alternative implementations of floating-point intervals, not relying on specific rounding modes could be developed in future work to alleviate this limitation.

3.2   Handling Web Interactions

Try-Mopsa is split between a toplevel, handling the page interaction with the user and a web worker, handling the actual computations of the analysis. Thanks to this decoupling, the web worker and the page rendering lie within different threads of the web browser, so longer computations do not freeze the browser's rendering process.

Web Worker. The web worker relies primarily on the JavaScript executable obtained through the process described in the previous section.

The compiled version of Mopsa relies on standard output to interact with the user. Additionally, the interactive engine reads user commands from the standard input. We rely on JSOO's utilities to set up callbacks intercepting any standard input/output and redirecting them to the toplevel. Thanks to this approach, Mopsa did not require any modification to work in the browser.

We rely on the virtual filesystem mechanism of JSOO to embed internal files Mopsa may expect to find during its analysis. This currently concerns Python stubs for various library modules.

Toplevel. The toplevel renders the initial dynamic elements of the page and updates them in response to user action.

It uses the Ace editor [52], and the Ezjs_ace bindings [48] to display the input program and the analysis results. We have extended the Ace editor to provide syntax highlighting for the Universal language, to allow collapsing of printed elements within the analysis result, and to render ANSI escape codes in HTML, as those are used by Mopsa to print in color.

The toplevel also draws a form so that user can select options. The form is automatically generated from the option metadata encoded within Mopsa (which are fetched through a query to the web worker). The options depend on which abstract domains are activated, and thus on the configuration chosen by the user. The toplevel ensures that the option form is redrawn whenever the configuration is changed.

When a user starts an analysis, the toplevel creates a web worker to perform it. The user may choose to interrupt the analysis, in which case the toplevel then signals the web worker to stop.

Interactive Engine. The interactive engine of Mopsa follows an interaction loop by awaiting user input (in a synchronous, blocking fashion) and providing corresponding results. However, web interfaces typically rely on asynchronous processes. We explored options such as implementing an asynchronous interactive engine. This would have required the decoupling of the interactive interface from the analysis computations, which would have been a large implementation effort introducing breaking changes. In the end, we chose to rely on the sync-message [35] JavaScript library enabling synchronous communication between the web worker to the toplevel, should some user input be required. Thanks to this embedding, all user-interface features of Mopsa are supported in Try-Mopsa.

4   Try-Mopsa: Interface Overview

The interface of Try-Mopsa is displayed in Figs. 2, 4 and 5 and can be tried online [38]. The interface is responsive, to ensure it can be used on a wide variety of screens, from smartphones to desktops.

The landing page shown in Fig. 2 provides a code editor and displays analysis results. By default, input programs are written in our toy imperative language. Users can also switch to the Python analysis if they wish to, and they can load some example programs. The analysis output consists of abstract states displayed when the `print()` instruction is analyzed, as well as a report on the runtime errors potentially detected by the analyzer. If the interactive engine has been enabled, the interaction happens in this same box. Note that on wider screens, the responsive design displays the code editor and analysis results side-by-side.

In Fig. 2, we loaded the example program `str_alphabet2.u`, reproduced textually in Fig. 3. Note that in the web editor, the definition of `to_string`, acting as a cast, is folded to highlight the editor capabilities.

The program starts with the string `s`, initialized to the letter "a". It then runs a loop until the number defined by `'a' + i` reaches the end of the lowercase alphabet. Similarly to C, in Universal, characters (between single quotes) are interpreted as integers through their ASCII code. This loop may stop non-deterministically at any iteration (line 10). At each loop iteration, the string `s` is concatenated (with the @ operator) with the singleton string whose character corresponds to the integer `'a' + i`. Thus the program non-deterministically computes a string `s` among $S = \{$ "a", "ab", "abc", ..., "abc$\cdots$xyz" $\}$. The assertion at line 18 checks a simpler property: that `s` belongs to S but up to any permutation of the characters.

Figure 4 shows the configuration editor, letting users specify their choice of abstract domains and combinators. Similarly to the code editor, users can also load a configuration from a pre-defined list. We comment on the loaded configuration, `string_product_relational.json`. This configuration features three

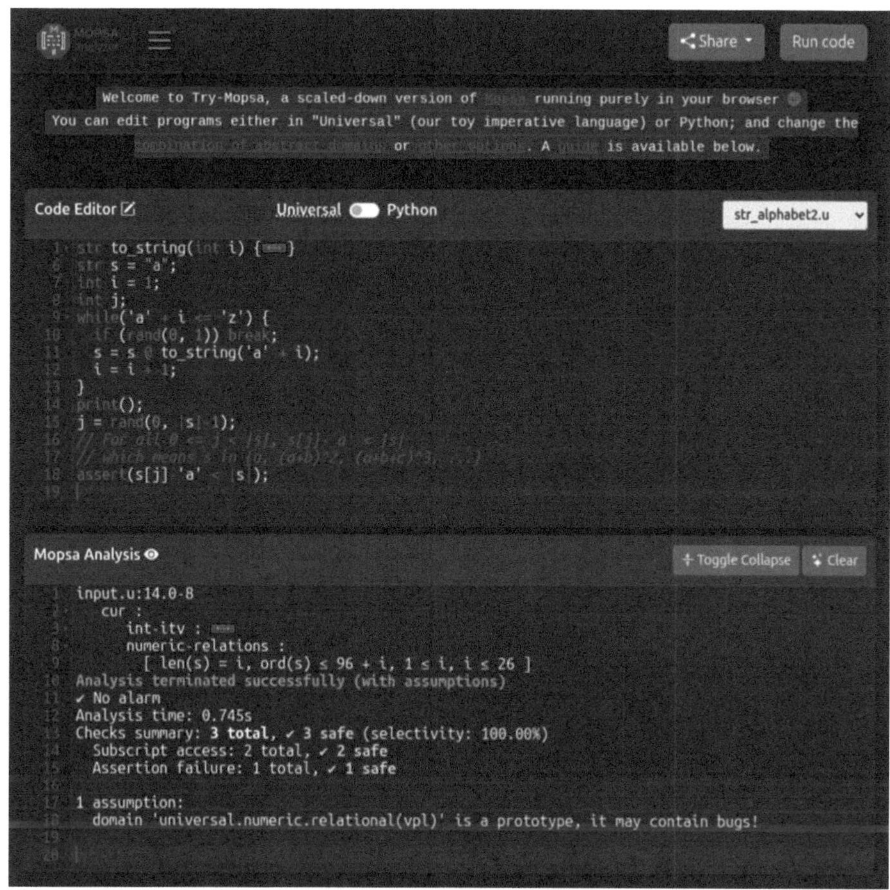

Fig. 2. Interface of Try-Mopsa: landing page, with program editor and analysis output.

```
 1   str to_string(int i) { /* omitted */ }
 6   str s = "a";
 7   int i = 1;
 8   int j;
 9   while('a' + i <= 'z') {
10     if (rand(0, 1)) break;
11     s = s @ to_string('a' + i);
12     i = i + 1;
13   }
14   print();
15   j = rand(0, |s|-1);
16   // For all 0 <= j < |s|, s[j]-'a' < |s|
17   // which means s in {a, (a+b)^2, (a+b+c)^3, ...}
18   assert(s[j]-'a' < |s|);
```

Fig. 3. Program `str_alphabet2.u` used as example.

```json
{
    "language": "universal",
    "domain": {
        "switch": [
            // Iterators
            "universal.iterators.program",
            "universal.iterators.intraproc",
            "universal.iterators.loops",
            "universal.iterators.interproc.inlining",
            "universal.iterators.unittest",
            {
                "product": [
                    "universal.toy.string_length",
                    "universal.toy.string_summarization"
                ],
                "reductions": [
                    "universal.toy.string_reduction"
                ]
            },
            // Numeric abstraction
            {
                "product": [
                    {
                        "nonrel": "universal.numeric.values.intervals.integer"
                    },
                    "universal.numeric.relational"
                ],
                "reductions": [
                    "universal.numeric.reductions.intervals_rel"
                ]
            }
        ]
    }
}
```

Fig. 4. Interface of Try-Mopsa: configuration editor, allowing to specify the choice of abstract domains and their combinators.

notable abstract domains, working together. The string length domain handles ghost variables representing the length of each string. The string summarization handles ghost variables representing a summary of the contents of each string, through the ASCII codes of the contained characters. These two domains are defined as a reduced product as they infer complementary pieces of information [37]. Finally, a relational numerical domain is enabled. It handles the constraints passed by the aforementioned string domains, which allows it to infer relations between those different ghost variables, and program variables.

The result of analyzing `str_alphabet2.u` using `string_product_relational.json` is displayed at the bottom of Fig. 2. In this case, Try-Mopsa is able to prove the complex assertion at line 18, which mixes information about the contents and the length of string `s`. Let us briefly comment on the abstract state inferred after the loop and printed at lines 1–9 of the analysis result in Fig. 2. We can see that the relational domain inferred equality between the length of the string `s` and the integer variable `i`. We can further notice an interesting relation

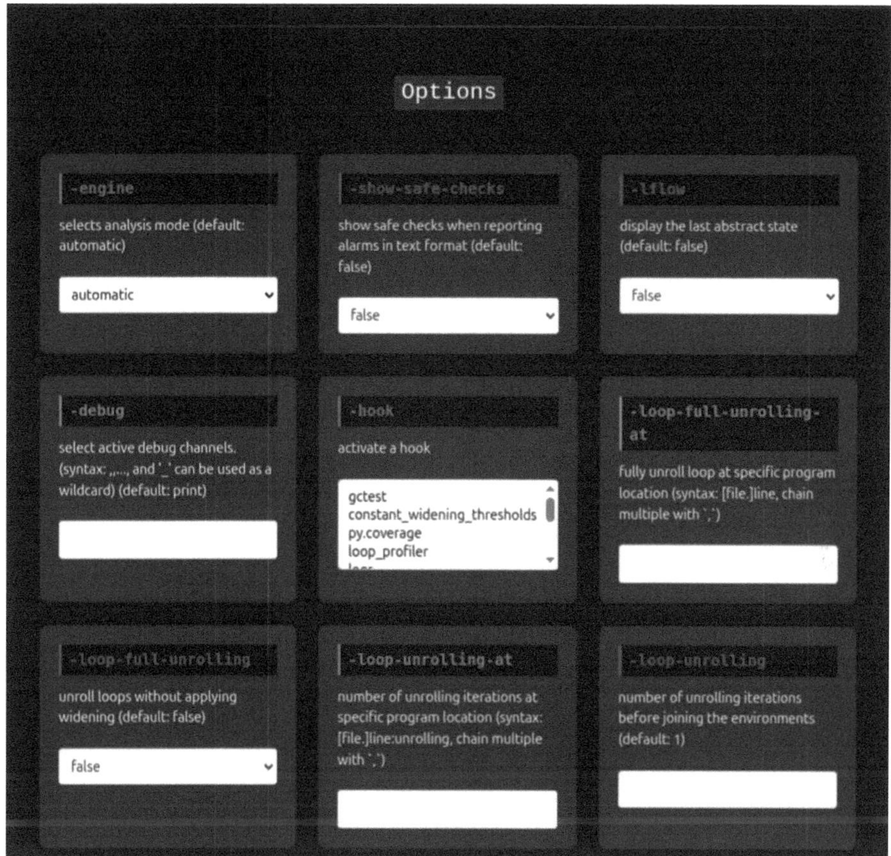

Fig. 5. Interface of Try-Mopsa: option selection.

between the contents of s (ord(s)) and its length: ord(s) $\leq$ 'a' + len(s) $-$ 1. This invariant has been obtained thanks to the cooperation between the string abstract domains and the relational abstract domain. Note that the second inequality is exactly the invariant proved by Mopsa at line 18.

Figure 5 displays various options to tweak the analysis. As we mentioned in the previous section, these options are dynamically generated by the toplevel, depending on which abstract domains are activated. In this example, the first five options are defined at the framework level, while the last four options customize the behavior of the loop iterator. Note that the option's argument can be selected through different interfaces depending on the expected output (e.g., choice among fixed list, integer, string).

Table 1. Preliminary evaluation of analysis times with relational abstract domains enabled. The mean and standard deviation have been computed over 10 runs.

Program	Binary execution	Firefox execution
attributes.py	0.09 s ± 0.00	0.52 s ± 0.03
fspath.py	0.09 s ± 0.00	0.49 s ± 0.02
list.py	0.15 s ± 0.01	0.88 s ± 0.03
loop.py	0.12 s ± 0.01	0.72 s ± 0.03
recency.py	0.08 s ± 0.01	0.63 s ± 0.03
str_alphabet.u	0.04 s ± 0.01	0.20 s ± 0.01
str_alphabet2.u	0.23 s ± 0.01	0.81 s ± 0.01
str_conc_loop.u	0.08 s ± 0.01	0.29 s ± 0.01
str_conc_loop2.u	0.04 s ± 0.01	0.15 s ± 0.01

To further improve its usability, Try-Mopsa can also be used as a progressive web app, meaning that it can be installed locally on a device and run without relying on the network. In addition, a share button (in the top-right corner of Fig. 2) lets users share their current program, configuration and option choices by encoding it into a URL. This can be useful for example to simplify practical sessions where no manual loading is required.

5 Implementation Discussion

This section briefly discusses the implementation size, maintainability, performance and browser compatibility of Try-Mopsa.

Implementation. The webpage, including extensions of the Ace editor for our purposes, is written in around 1,500 lines of HTML, CSS and JavaScript. The toplevel and worker of Try-Mopsa consist of 670 lines of OCaml code. The implementation within Mopsa of the VPL binder adds around 500 lines of OCaml code. Once compiled with full optimizations, the worker and all the Mopsa code relying on it are compiled into a relatively compact JavaScript file of 3.1 megabytes.

Maintainability. We argue that the implementation of Try-Mopsa is maintainable with respect to the standard implementation of Mopsa. Indeed, Try-Mopsa does not introduce breaking changes, meaning future development can be shared in the same repository.

In addition, most components of Mopsa were reused as-is, thanks in particular to the reuse of the same input-output loop as a terminal user of Mopsa would. The only major component that has been added is the support for relational abstract domains through the VPL library. This implementation provides adequate results on the toy examples we provide along with Try-Mopsa.

Performance. The goal of Try-Mopsa is to ease tool demonstrations and testing without installation, where raw analysis performance is not a priority. Nevertheless, we ran a preliminary evaluation comparing the analysis times between the natively generated executable, and the same code running in Firefox. We ran this comparison on the toy, default examples of Try-Mopsa, with configurations using relational abstract domains. We show in Table 1 the nine most significant programs to analyze. On average, Try-Mopsa is five times slower than native binary execution. While this is a noticeable slowdown, we believe the analysis times – below one second – make Try-Mopsa still usable for demonstration or teaching purposes. Future work could focus on using JSOO's recent WebAssembly code generation features to obtain more efficient code.

Browser Compatibility. We use the Playwright framework [53] to test the browser compatibility of Try-Mopsa. The tests run multiple usage scenarios and check the results are visible and as expected. These tests run during our continuous integration process and have helped identify several usability issues, particularly on mobile browsers. We currently test three desktop browsers (Chrome, Firefox, Safari), and two mobile browsers (Chrome, Safari), using different viewports corresponding to popular iPhone and Android devices, either used in portrait or in landscape orientation. There are currently 5 test scenarios, resulting in 35 tests depending on the chosen browser and viewport. Additional tests ensure that all examples programs can be analyzed in the relevant example configurations proposed in our interface. We currently provide 13 examples programs for Universal (resp. 5 for Python), and 11 configurations for Universal (resp. 6 for Python).

6   Related Work

In previous works, developers relied on server-side computations to provide web interfaces for static analyzers relying on relational numerical domains. This includes for example Interproc [21,22], Banal [29,32] and FuncTion [54,55]. While client-server implementations require less work to adapt a static analyzer to the web, these implementations raise more security concerns and can be less scalable in the number of users. In addition, these implementations need to maintain a compatible server service over the years. This proves to be difficult: at the time of writing, the majority of these approaches are no longer operational.

More recently, pure client-side web interfaces have been developed for static analyzers. These scale much better with the number of concurrent users, and their "only" dependency is a web browser – one of the most widespread pieces of software. Lermusiaux and Montagu [26,27] provide a demo web interface for their Salto analyzer of OCaml code, following previous demonstrators accompanying works around the static analysis of functional languages [44,45]. Ferreiro et al. [16] and Morales et al. [46] provide a WebAssembly version of Ciao Prolog, embedding the CiaoPP analyzer, which also supports interactive user inputs. In their BINSEC tutorial at PLDI 2025, Recoules and Bardin [51] provide a web tutorial where users can choose between running BINSEC [13] in CLI or running analysis scenarios directly in their browser. To the best of our knowledge, Try-Mopsa is the only web interface supporting the analysis of multiple languages (our toy imperative language and Python), numerical relational domains, and an interactive exploration of the analysis.

Note that another lightweight approach for trying out some new software is the use of Docker containers. We actually provide Docker containers with each release of Mopsa [34]; these containers supports all features of Mopsa (e.g., including the C analysis).

Negrini et al. [47] describe how they use their static analysis platform LiSA within their static analysis courses. Students are tasked with installing LiSA, and implementing several simple abstract domains. Try-Mopsa aims at providing a zero-install static analyzer relying on complex abstract domains – such as polyhedra – to illustrate how such a tool works. Evaluating the impact of Try-Mopsa on real teaching cohorts is left as future work.

Following studies about developer use of static analyzers [7,15], there are quite a few works discussing interfaces for static analyzers. MagpieBridge [28] aims at simplifying the display of analysis results within IDEs supporting the Language-Server Protocol. Several interfaces have also been developed to debug static analyzers [14], provide an abstract debugger [19], or mix concrete and abstract debuggers [36]. Correnson [9] recently introduced a new graphical interface for Frama-C users.

Outside of the abstract interpretation community, Kosmatov et al. [25] provide a unit testing as a service platform. Both the Alt-Ergo SMT Solver [8] and the Why3 deductive verification platform [17] leverage JSOO to provide zero-install web versions [1,3]. Canou et al. [6] describe the technical components required to run an OCaml MOOC (Massive Open Online Course) who attracted thousands of learners. It relies in particular on a pure client-side development environment avoiding installation woes students could otherwise face. Arias et al. [2] also provide a pure JavaScript version for the Rocq proof assistant, which aims at improving literate programming formatting and reducing installation burdens, in an educational context.

7   Conclusion

This article introduced Try-Mopsa, a pure client-side implementation of Mopsa running in the browser or as a progressive web app. Try-Mopsa supports all the main core components of Mopsa, including support for relational numerical domains and its abstract debugger. Try-Mopsa is designed to work on a wide range of devices, from smartphones to desktops, thanks to a responsive design and continuous testing of the support of popular browsers and various resolutions. We envision Try-Mopsa to be a convenient tool to demonstrate static analysis capabilities, for teaching purposes or more generally to attract new users.

Acknowledgments. We thank the anonymous VMCAI reviewers and the participants of Dagstuhl Seminar #25421 for their valuable feedback. We are grateful to Aymane Ismail for prototyping an early version of Try-Mopsa; and to Alexandre Maréchal for his help in using the VPL. We thank Aymeric Fromherz, Louis Gesbert, Vincent Botbol, Louis Rustenholz, Antoine Miné, Abdelraouf Ouadjaout and the whole Mopsa team for their helpful discussion and comments on this work.

Data Availability Statement. This paper is accompanied by an artefact [39], which has been peer-reviewed as "available, functional and reusable".

References

1. Alt-Ergo Developers. Try Alt-Ergo. https://alt-ergo.ocamlpro.com/try.html
2. Arias, E.J.G., Pin, B., Jouvelot, P.: jsCoq: towards hybrid theorem proving interfaces. In: UITP (2016). https://doi.org/10.4204/EPTCS.239.2
3. Becker, B., et al.: Try Why3 (2025). https://www.why3.org/try/
4. Beyer, D.: Competition on software verification and witness validation: SV-COMP 2023. In: TACAS (2023). https://doi.org/10.1007/978-3-031-30820-8_29
5. Boulmé, S., Maréchal, A., Monniaux, D., Périn, M., Yu, H.: The verified polyhedron library: an overview. In: SYNASC (2018). https://doi.org/10.1109/SYNASC.2018.00014
6. Canou, B., Di Cosmo, R., Henry, G.: Scaling up functional programming education: under the hood of the OCaml MOOC. In: Proc. ACM Program. Lang. ICFP (2017). https://doi.org/10.1145/3110248
7. Christakis, M., Bird, C.: What developers want and need from program analysis: an empirical study. In: ASE (2016). https://doi.org/10.1145/2970276.2970347
8. Conchon, S., Coquereau, A., Iguernlala, M., Mebsout, A.: Alt-Ergo 2.2. In: SMT Workshop, Oxford, United Kingdom (2018). https://inria.hal.science/hal-01960203
9. Correnson, L.: Ivette: a modern GUI for frama-C. In: SEFM (2022). https://doi.org/10.1007/978-3-031-26236-4_10
10. Cousot, P., Cousot, R.: Abstract interpretation: a unified lattice model for static analysis of programs by construction or approximation of fixpoints. In: POPL (1977). https://doi.org/10.1145/512950.512973
11. Cousot, P., Halbwachs, N.: Automatic discovery of linear restraints among variables of a program. In: POPL (1978). https://doi.org/10.1145/512760.512770
12. Danial, A.: cloc: v2.06 (2025). https://github.com/AlDanial/cloc/
13. Djoudi, A., Bardin, S.: BINSEC: binary code analysis with low-level regions. In: TACAS (2015). https://doi.org/10.1007/978-3-662-46681-0_17
14. Do, L.N.Q., Krüger, S., Hill, P., Ali, K., Bodden, E.: Debugging static analysis. IEEE Trans. Softw. Eng. **7**, 697–709 (2020). https://doi.org/10.1109/TSE.2018.2868349
15. Do, L.N.Q., Wright, J.R., Ali, K.: Why do software developers use static analysis tools? A user-centered study of developer needs and motivations. IEEE Trans. Softw. Eng. **3**, 835–847 (2022). https://doi.org/10.1109/TSE.2020.3004525
16. Ferreiro, D., Morales, J.F., Abreu, S., Hermenegildo, M.V.: Demonstrating (hybrid) active logic documents and the ciao prolog playground, and an application to verification tutorials. In: ICLP (2023). https://doi.org/10.4204/EPTCS.385.33

17. Filliâtre, J.C., Paskevich, A.: Why3 - where programs meet provers. In: ESOP (2013). https://doi.org/10.1007/978-3-642-37036-6_8
18. Gopan, D., DiMaio, F., Dor, N., Reps, T.W., Sagiv, S.: Numeric Domains with Summarized Dimensions. In: TACAS (2004). https://doi.org/10.1007/978-3-540-24730-2_38
19. Holter, K., Hennoste, J.O., Lam, P., Saan, S., Vojdani, V.: Abstract debuggers: exploring program behaviors using static analysis results. In: Onward! (2024). https://doi.org/10.1145/3689492.3690053
20. Jane Street Group, LLC. JavaScript stubs for the Zarith Library. Version 0.17.0 (2024). https://github.com/janestreet/zarith_stubs_js
21. Jeannet, B.: Relational interprocedural verification of concurrent programs. Softw. Syst. Model **2** (2013). https://doi.org/10.1007/S10270-012-0230-7
22. Jeannet, B.: Web Interface for the Interproc Analyzer (2009). https://pop-art.inrialpes.fr/interproc/interprocweb.cgi.html
23. Jeannet, B., Miné, A.: Apron: a library of numerical abstract domains for static analysis. In: CAV (2009). https://doi.org/10.1007/978-3-642-02658-4_52
24. Journault, M., Miné, A., Monat, R., Ouadjaout, A.: Combinations of reusable abstract domains for a multilingual static analyzer. In: VSTTE (2019). https://doi.org/10.1007/978-3-030-41600-3_1
25. Kosmatov, N., Williams, N., Botella, B., Roger, M.: Structural unit testing as a service with pathcrawler-online.com . In: SOSE (2013). https://doi.org/10.1109/SOSE.2013.78
26. Lermusiaux, P., Montagu, B.: Detection of uncaught exceptions in functional programs by abstract interpretation. In: ESOP (2024). https://doi.org/10.1007/978-3-031-57267-8_15
27. Lermusiaux, P., Montagu, B.: Web Demo for the Salto Analyzer (2025). https://salto.gitlabpages.inria.fr/demo/salto.html
28. Luo, L., Dolby, J., Bodden, E.: MagpieBridge: a general approach to integrating static analyses into IDEs and editors (tool insights paper). In: ECOOP (2019). https://doi.org/10.4230/LIPICS.ECOOP.2019.21
29. Miné, A.: Inferring sufficient conditions with backward polyhedral under-approximations. In: NSAD@SAS (2012). https://doi.org/10.1016/J.ENTCS.2012.09.009
30. Miné, A.: Relational abstract domains for the detection of floating- point run-time errors. In: ESOP (2004). https://doi.org/10.1007/978-3-540-24725-8_2
31. Miné, A.: The octagon abstract domain. High. Order Symb.Comput. **1** (2006). https://doi.org/10.1007/S10990-006-8609-1
32. Miné, A.: Web Interface for Sufficient Condition Polyhedral Prototype Analyzer (2012). https://mine.perso.lip6.fr/banal/
33. Miné, A., Leroy, X., Cuoq, P.: Zarith: arithmetic and logical operations over arbitrary-precision integers. Version 1.14 (2024). https://github.com/ocaml/Zarith
34. Miné, A., et al.: Docker Containers for Each Mopsa Release (2025). https://gitlab.com/mopsa/mopsa-analyzer/container_registry/6390468
35. Mojaki, A.: sync-message. Version 0.0.12 (2023). https://github.com/alexmojaki/sync-message
36. Van Molle, M., Vandenbogaerde, B., De Roover, C.: Cross-level debugging for static analysers. In: SLE (2023). https://doi.org/10.1145/3623476.3623512
37. Monat, R.: Static Type and Value Analysis by Abstract Interpretation of Python Programs with Native C Libraries. PhD thesis. Sorbonne Université, France (2021)
38. Monat, R.: Try-Mopsa (2025). https://try-mopsa.rmonat.fr

39. Monat, R.: Try-mopsa: relational static analysis in your pocket (artefact). Sept. (2025). https://doi.org/10.5281/zenodo.17157478
40. Monat, R., Ouadjaout, A., Miné, A.: A multilanguage static analysis of python programs with native C extensions . In: SAS (2021). https://doi.org/10.1007/978-3-030-88806-0_16
41. Monat, R., Ouadjaout, A., Miné, A.: Easing maintenance of academic static analyzers. Int. J. Softw. Tools Technol. Transf. **6**, 673–686 (2024). https://doi.org/10.1007/S10009-024-00770-1
42. Monat, R., Ouadjaout, A., Miné, A.: Mopsa-C: modular domains and relational abstract interpretation for C programs (competition contribution). In: TACAS (2023). https://doi.org/10.1007/978-3-031-30820-8_37
43. Monat, R., Ouadjaout, A., Miné, A.: Static type analysis by abstract interpretation of python programs. In: ECOOP (2020). https://doi.org/10.4230/LIPICS.ECOOP.2020.17.
44. Montagu, B., Jensen, T.P.: Stable relations and abstract interpretation of higher-order programs. In: Proceedings of ACM Programming Language. ICFP (2020). https://doi.org/10.1145/3409001
45. Montagu, B., Jensen, T.P.: Trace-based control-flow analysis. In: PLDI (2021). https://doi.org/10.1145/3453483.3454057
46. Morales, J.F., Pradales, G., Hermenegildo, M.: Ciao Prolog Playground (2022). https://ciao-lang.org/playground/
47. Negrini, L., Arceri, V., Olivieri, L., Cortesi, A., Ferrara, P.: Teaching through practice: advanced static analysis with LiSA. In: FMTea (2024). https://doi.org/10.1007/978-3-031-71379-8_3
48. OCamlPro. Ezjs_ace: bindings for the Ace editor. Version 0.1.1 (2020). https://github.com/ocamlpro/ezjs_ace
49. Ouadjaout, A., Miné, A.: A library modeling language for the static analysis of C programs. In: SAS (2020). https://doi.org/10.1007/978-3-030-65474-0_11
50. Pottier, F., Régis-Gianas, Y.: Menhir: an LR(1) parser generator for OCaml. Version 20250903 (2025). https://gitlab.inria.fr/fpottier/menhir
51. Recoules, F., Bardin, S.: BINSEC Tutorial at PLDI'25: Adapting Symbolic Execution for Binary-level Security (2025). https://binsec.github.io/tutorial-pldi2025/
52. The Ace Developers. Ace (Ajax.org Cloud9 Editor). Version 1.43.3 (2025). https://github.com/ajaxorg/ace
53. The Playwright Development Team. Playwright: framework for Web Testing and Automation. Version 1.55.0 (2025). https://github.com/microsoft/playwright
54. Urban, C.: FuncTion: an abstract domain functor for termination - (competition contribution). In: TACAS (2015). https://doi.org/10.1007/978-3-662-46681-0_46
55. Urban, C.: Web Interface for FuncTion (2015). https://www.di.ens.fr/~urban/FuncTion.html
56. Vouillon, J., Balat, V.: From bytecode to javascript: the Js_of_ocaml compiler. Softw. Pract. Exp. **8** (2014). https://doi.org/10.1002/SPE.2187

Termination Resilience Static Analysis

Naïm Moussaoui Remil[(✉)] [iD] and Caterina Urban [iD]

Inria & ENS | PSL, Paris, France
{naim.moussaoui-remil,caterina.urban}@inria.fr

Abstract. We present a novel abstract interpretation-based static analysis framework for proving Termination Resilience, the absence of Robust Non-Termination vulnerabilities in software systems. Robust Non-Termination characterizes programs where an untrusted (e.g., externally-controlled) input can force infinite execution, independently of other trusted (e.g., controlled) variables.Our framework is a semantic generalization of Cousot and Cousot's abstract interpretation-based ranking function derivation, and our sound static analysis extends Urban and Miné's decision tree abstract domain in a non-trivial way to manage the distinction between untrusted and trusted program variables. Our approach is implemented in an open-source tool and evaluated on benchmarks sourced from SV-COMP and modeled after real-world software, demonstrating practical effectiveness in verifying Termination Resilience and detecting potential Robust Non-Termination vulnerabilities.

1 Introduction

Termination is one of the oldest and most fundamental properties of programs. Yet, in practice, requiring termination in every possible case can be unnecessarily restrictive. Indeed, not all program variables play the same role in program termination. Some variables are *untrusted*: their values may be externally controlled, for instance by user input, an operating system scheduler, or a third-party library. If controlling such variables alone is sufficient to enforce divergence, we say that the program is vulnerable to *Robust Non-Termination*, which represents a serious concern in practice. By contrast, other variables are *trusted*: their values stem from non-deterministic choices, internal computations, or library calls under the control of the programmer. Non-termination that may be avoided through these trusted choices may be acceptable, as it cannot be reliably exploited by an adversary; it reflects internal system behavior rather than external influence. From a security perspective, this distinction is crucial: divergence caused by untrusted inputs constitutes a potential denial-of-service vulnerability, whereas divergence caused by trusted variables does not. From a software engineering standpoint, the distinction is equally important as it hints at a principled way to triage non-termination alarms, helping developers prioritize the more critical cases.

This distinction motivates the property of *Termination Resilience*; for every possible (sequence of) untrusted input(s), there exists at least one terminating execution. In semantic terms, we treat untrusted variables demonically, since

Y.-F. Chen et al. (Eds.): VMCAI 2026, LNCS 16417, pp. 212–236, 2026.
https://doi.org/10.1007/978-3-032-15700-3_11

```
¹x := input;
²z := 10;
if ³(...) {
    while ⁴(z >= 0) {
       ⁵z := z - x;
    }
} else {
    ⁶c := [-2, 1];
    while ⁷(z >= x) {
       ⁸z := z + c;
    }
}⁹
```

Fig. 1. Program P.

termination must be ensured under all value possibilities, while trusted variables are treated angelically, since it suffices that some of their values enable termination. Besides, we support both scenarios where untrusted inputs have full and partial knowledge on trusted variables. Consider program P (inspired from our benchmarks and written in the small programming language that we use for illustration in the paper) in Fig. 1: the variable x is an untrusted input (cf. program label 1), while the value of c is non-deterministically chosen between -2 and 1 (cf. label 6). Both **while** loops in the program are non-terminating, the first when $x \leq 0$, and the second when $(z \geq x$ and$)$ $c \geq 0$. However, only for the first loop (non-)termination depends on untrusted variables, while the second loop satisfies Termination Resilience. Termination Resilience thus provides a nuanced alternative to Termination. Note that in this program the untrusted input is the first to choose a value (cf. label 1), hence it has no knowledge about the trusted variables (c and z) values.

We design *an abstract interpretation-based static analysis framework to prove Termination Resilience*. Our framework is a generalization of Cousot and Cousot's idea of computing a ranking function by abstract interpretation [12]. Specifically, we extend their semantic definitions to accommodate the asymmetric treatment of untrusted and trusted program variables. We derive a sound static analysis for Termination Resilience by building upon Urban and Miné's decision tree abstract domain [34], enhancing it with novel and non-trivial transformer operators to effectively manage this mixed settings. When unable to prove that a program always satisfies Termination Resilience, *our static analysis automatically infers sufficient preconditions that ensure Termination Resilience*. For example, for program P in Fig. 1, our analysis proves Termination Resilience for the second loop and automatically infers the precondition $x > 0$ for the first loop.

Our analysis is implemented in an open-source tool, as an extension of the FUNCTION static analyzer [32], and evaluated on benchmarks drawn from the International Competition on Software Verification (SV-COMP) and modeled after real-world software. The experimental results demonstrate the practical usefulness of the analysis as a means to triage non-termination alarms based on whether they depend on untrusted or trusted variables, provings developers with a systematic way to prioritize the most critical cases.

2   Termination Resilience

In this section, we give a mathematical characterization of the behavior of a program, and we formally define our property of interest, *Termination Resilience*.

Program Semantics. We model the operational behavior of a program as a *transition system* $\langle \Sigma, \tau \rangle$ where Σ is a (potentially infinite) set of program states, and the transition relation $\tau \subseteq \Sigma \times \Sigma$ describes possible transition between states [9,10]. Specifically, for our purposes, $\tau \overset{\text{def}}{=} \tau^i \uplus \tau^r$ is the disjoint union of *input transitions* in τ^i, which are transitions that consume an untrusted input, and *regular transitions* in τ^r. For simplicity, without loss of generalization, we assume that all transitions $\langle s, s' \rangle \in \tau$ from a state $s \in \Sigma$ are of the same kind, i.e., either all input transition or all regular transitions. The set of final program states is $\Omega_\tau \overset{\text{def}}{=} \{s \in \Sigma \mid \forall s' \in \Sigma : \langle s, s' \rangle \notin \tau\}$.

Given a transition system $\langle \Sigma, \tau \rangle$, the function $\mathrm{pre}_{\tau^r} : \mathcal{P}(\Sigma) \to \mathcal{P}(\Sigma)$ maps a given set of states $X \subseteq \Sigma$ to the set of their predecessors with respect to τ^r: $\mathrm{pre}_{\tau^r}(X) \overset{\text{def}}{=} \{s \in \Sigma \mid \exists s' \in X : \langle s, s' \rangle \in \tau^r\}$, and the function $\widetilde{\mathrm{pre}}_{\tau^i} : \mathcal{P}(\Sigma) \to \mathcal{P}(\Sigma)$ maps a set of states $X \subseteq \Sigma$ to the set of states whose successors with respect to τ^i are all in X: $\widetilde{\mathrm{pre}}_{\tau^i}(X) \overset{\text{def}}{=} \{s \in \Sigma \mid \forall s' \in \Sigma : \langle s, s' \rangle \in \tau^i \Rightarrow s' \in X\}$.

In the following, given a set S, let ε be an empty sequence of elements in S, S^+ be the set of all non-empty finite sequences of elements in S, S^ω be the set of all infinite sequences, $S^{+\infty} \overset{\text{def}}{=} S^+ \cup S^\omega$ be the set of all non-empty finite or infinite sequences of elements in S, and $S^{*\infty} \overset{\text{def}}{=} S^{+\infty} \cup \{\varepsilon\}$. We write $T^+ \overset{\text{def}}{=} T \cap S^+$, $T^\omega \overset{\text{def}}{=} T \cap S^\omega$ for the selection of the non-empty finite sequences and the infinite sequences of $T \in \mathcal{P}(S^{+\infty})$, and $T; T' = \{tst' \in S^{+\infty} \mid s \in S \wedge ts \in T \wedge st' \in T'\}$ for the merging of two sets of sequences $T \in \mathcal{P}(S^+)$ and $T' \in \mathcal{P}(S^{+\infty})$, when a finite sequence in T terminates with the initial element of a sequence in T'.

A *trace* $\sigma \in \Sigma^{+\infty}$ is a non-empty sequence of program states where each pair of consecutive states $s, s' \in \Sigma$ is described by the transition relation τ, i.e., $\langle s, s' \rangle \in \tau$. We write σ_n to denote the n-th state of a trace $\sigma \in \Sigma^{+\infty}$ (with $n < |\sigma|$), σ^i (resp. σ^r) for the (possibly empty) sequence of input (resp. regular) transitions (i.e., pairs of consecutive states) in σ, and T^i for the set $\{\sigma^i \mid \sigma \in T\}$ of sequences of input transitions in traces in a set $T \in \mathcal{P}(\Sigma^{+\infty})$ (resp. T^r for the set $\{\sigma^r \mid \sigma \in T\}$ of sequences of regular transitions). The *trace semantics* $\Lambda \in \mathcal{P}(\Sigma^{+\infty})$ generated by a transition system $\langle \Sigma, \tau \rangle$ is the union of all non-empty finite traces terminating in Ω_τ, and all infinite traces. It can be expressed as a least fixpoint in the complete lattice $\langle \mathcal{P}(\Sigma^{+\infty}), \sqsubseteq, \sqcup, \sqcap, \Sigma^\omega, \Sigma^+ \rangle$ [9]:

$$\Lambda = \mathsf{lfp}^{\sqsubseteq} \Theta$$
$$\Theta(T) \overset{\text{def}}{=} \Omega_\tau \cup (\tau; T) \tag{1}$$

(a) The transition system $\langle \Sigma_1, \tau_1 \rangle$.   (b) The transition system $\langle \Sigma_2, \tau_2 \rangle$.

Fig. 2. The transition systems of Examples 1 and 2. Input transitions are represented with dashed red lines, while solid black lines represent regular transitions.

where the computational order is $T_1 \sqsubseteq T_2 \overset{\text{def}}{\Leftrightarrow} T_1^+ \subseteq T_2^+ \wedge T_1^\omega \supseteq T_2^\omega$. In the following, we write $\Lambda[\![P]\!]$ to denote the trace semantics of a given program P. Given an initial set of states $I \subseteq \Sigma$, we define $\Lambda[\![P]\!]I \overset{\text{def}}{=} \{\sigma \in \Lambda[\![P]\!] \mid \sigma_0 \in I\}$ as the restriction of the trace semantics of P to traces starting in I.

Example 1. Let $\langle \Sigma_1, \tau_1 \rangle$ be a transition system with $\Sigma_1 = \{a, b, c\}$ and $\tau_1 = \tau_1^i \cup \tau_1^r$ where $\tau_1^i = \{\langle a, b \rangle, \langle a, c \rangle\}$ and $\tau_1^r = \{\langle c, c \rangle\}$ (cf. Figure 2a). The trace semantics generated by $\langle \Sigma_1, \tau_1 \rangle$ and starting in $I = \{a\}$ is $\Lambda_1 = \{ab, ac^\omega\}$.   ◁

Example 2. Let $\langle \Sigma_1, \tau_1 \rangle$ be the transition system from Example 1 and let $\langle \Sigma_2, \tau_2 \rangle$ be another transition system with $\Sigma_2 = \Sigma_1 \cup \{d, e\}$ and $\tau_2 = \tau_1^i \cup \tau_2^r$ where $\tau_2^r = \tau_1^r \setminus \{\langle c, c \rangle\} \cup \{\langle b, d \rangle, \langle b, e \rangle, \langle e, e \rangle\}$ (cf. Figure 2b). The trace semantics generated by $\langle \Sigma_2, \tau_2 \rangle$ and starting in $I = \{a\}$ is $\Lambda_2 = \{abd, abe^\omega, ac\}$.   ◁

Program Property. As customary in abstract interpretation [10], we represent properties of entities in a universe $\mathbb{U}$ by a subset of this universe. We formally define the Robust Non-Termination program property.

Definition 1 (Robust Non-Termination). *The* Robust Non-Termination *program property is the set of sets of traces for which a (possibly empty) sequence of input transitions only yields diverging traces:*

$$\mathcal{RNT} \overset{\text{def}}{=} \{T \in \mathcal{P}(\Sigma^{+\infty}) \mid \exists j \in T^i \, \forall \sigma \in T : \sigma^i = j \Rightarrow \sigma \in T^\omega\} \tag{2}$$

Note that any set $T \subseteq \mathcal{P}(\Sigma^\omega)$ of infinite traces always belongs to $\mathcal{RNT}$, even if the traces do not consume input values (in such case $j = \varepsilon$), i.e., an always non-terminating program is always vulnerable to non-termination exploits.

Example 3. Let $\langle \Sigma_1, \tau_1 \rangle$ be the transition system from Example 1. The input transition $\langle a, c \rangle$ only yields the diverging trace $ac^\omega \in \Lambda_1$. Thus $\Lambda_1 \in \mathcal{RNT}$.   ◁

The negation of Robust Non-Termination is our property of interest. It expresses the ability of the program to (possibly) defend itself from non-termination exploits. Formally, we call this property *Termination Resilience*.

Definition 2 (Termination Resilience). *The* Termination Resilience *program property is the set of sets of program traces for which there exists a terminating trace for all sequence of input transitions:*

$$\mathcal{TR} \stackrel{def}{=} \neg\mathcal{RNT} = \{T \in \mathcal{P}(\Sigma^{+\infty}) \mid \forall j \in T^i \; \exists \sigma \in T^+ : \sigma^i = j\} \tag{3}$$

Example 4. Let $\langle \Sigma_2, \tau_2 \rangle$ be the transition system from Example 2. For any sequence of input transitions in $(\Lambda_2)^i = \{\langle a, b \rangle, \langle a, c \rangle\}$, there is at least one terminating trace in Λ_2, i.e., abd for $\langle a, b \rangle$, and ac for $\langle a, c \rangle$. Thus, $\Lambda_2 \in \mathcal{TR}$. ◁

The trace semantics $\Lambda[\![P]\!]$ of a program P, fully describing the behavior of P, exactly characterizes Termination Resilience of P (and its negation, Robust Non-Termination), for an initial set of states $I \subseteq \Sigma$:

$$P \models_I \mathcal{TR} \Leftrightarrow \Lambda[\![P]\!]I \in \mathcal{TR} \Leftrightarrow \Lambda[\![P]\!]I \notin \mathcal{RNT} \Leftrightarrow P \not\models_I \mathcal{RNT}. \tag{4}$$

In the next two sections, we abstract the trace semantics Λ to a special-purpose sound and complete (but not computable) semantics Λ_{tr} that forgets details of the program behavior that are irrelevant for reasoning about Termination Resilience. Next, we further abstract this semantics into a sound but computable semantics $\Lambda_{\mathsf{tr}}^{\natural}$ that yields a static analysis for Termination Resilience.

3   Termination Resilience Semantics

We derive our concrete semantics Λ_{tr} by abstract interpretation of the trace semantics Λ. The abstraction eliminates program traces that are potentially branching to non-termination through different untrusted inputs, and proves Termination Resilience for the remaining traces by associating a well-founded quantity [13,31] to program states belonging to terminating traces. Specifically, we abstract the trace semantics $\Lambda \in \mathcal{P}(\Sigma^{+\infty})$ into the *termination resilience semantics* $\Lambda_{\mathsf{tr}} \colon \Sigma \rightharpoonup \mathbb{O}$ (where $\mathbb{O}$ is the set of ordinals), which is the best (potential) ranking function [12,33] for the program states in the traces in Λ.

We define the termination resilience abstraction $\alpha_{\mathsf{tr}} \colon \mathcal{P}(\Sigma^{+\infty}) \rightarrow (\Sigma \rightharpoonup \mathbb{O})$ to map sets of program traces to a (potential) ranking function as:

$$\alpha_{\mathsf{tr}}(T) \stackrel{def}{=} \alpha_{\mathsf{v}} \circ \vec{\alpha}(T) \tag{5}$$

where $\vec{\alpha} \colon \mathcal{P}(\Sigma^{+\infty}) \rightarrow \mathcal{P}(\Sigma \times \Sigma)$ extracts the smallest transition relation $\tau \subseteq \Sigma \times \Sigma$ that generates a given set of traces T: $\vec{\alpha}(T) \stackrel{def}{=} \{\langle s, s' \rangle \mid \exists \sigma, \sigma' : \sigma s s' \sigma' \in T\}$, and $\alpha_{\mathsf{v}} \colon \mathcal{P}(\Sigma \times \Sigma) \rightarrow (\Sigma \rightharpoonup \mathbb{O})$ ranks the elements in the domain of r:

$$\alpha_{\mathsf{v}}(\emptyset) \stackrel{def}{=} \dot{\emptyset}$$

$$\alpha_{\mathsf{v}}(r)s \stackrel{def}{=} \begin{cases} 0 & s \in \Omega_r \\ \sup\left\{\alpha_{\mathsf{v}}(r)s' + 1 \,\middle|\, \langle s, s' \rangle \in r^i\right\} & s \in \widetilde{\mathsf{pre}}_{r^i}(\mathrm{dom}(\alpha_{\mathsf{v}}(r))) \\ \inf\left\{\alpha_{\mathsf{v}}(r)s' + 1 \,\middle|\, \mathrm{dom}(\alpha_{\mathsf{v}}(r)) \wedge \langle s, s' \rangle \in r^r\right\} & s \in \mathsf{pre}_{r^r}(\mathrm{dom}(\alpha_{\mathsf{v}}(r))) \\ \text{undefined} & \text{otherwise} \end{cases}$$

with $\dot{\emptyset}$ representing the totally undefined function. The (potential) ranking function is defined incrementally, starting from the final states in Ω_r, where the function has value 0 (and is undefined elsewhere). Then, the domain of the function grows by retracing the program (transitions) backwards and counting the number of performed program steps as value of the function. We model the non-determinism arising from input transitions *demonically*, requiring all successor states to already be in the domain of the function $(s \in \widetilde{\mathsf{pre}}_{\tau^i}(\mathsf{dom}(\alpha_{\mathsf{v}}(r))))$, and count the *maximum* number of performed program steps. Instead, non-determinism arising from regular transitions is treated *angelically*, and the potential ranking functions counts the minimum number of steps to termination.

We can now formally define the termination resilience semantics Λ_{tr}.

Definition 3 (Termination Resilience Semantics). *Let $\langle \Sigma, \tau \rangle$ be a transition system. Its termination resilience semantics $\Lambda_{\mathsf{tr}} \in (\Sigma \rightharpoonup \mathbb{O})$ is:*

$$\Lambda_{\mathsf{tr}} \stackrel{def}{=} \alpha_{\mathsf{tr}}(\Lambda) = \mathsf{lfp}^{\sqsubseteq}_{\dot{\emptyset}} \Theta_{\mathsf{tr}}$$

$$\Theta_{\mathsf{tr}}(f) \stackrel{def}{=} \lambda s. \begin{cases} 0 & s \in \Omega_\tau \\ \sup\left\{ f(s') + 1 \mid \langle s, s' \rangle \in \tau^i \right\} & s \in \widetilde{\mathsf{pre}}_{\tau^i}(\mathsf{dom}(f)) \\ \inf\left\{ f(s') + 1 \mid s' \in \mathsf{dom}(f), \langle s, s' \rangle \in \tau^r \right\} & s \in \mathsf{pre}_{\tau^r}(\mathsf{dom}(f)) \\ undefined & otherwise \end{cases}$$

where $\Lambda \in \mathcal{P}(\Sigma^{+\infty})$ is the trace semantics (cf. Equation 2).

We write $\Lambda_{\mathsf{tr}}\llbracket P \rrbracket$ for the termination resilience semantics of a program P.

Example 5. Let $\langle \Sigma_2, \tau_2 \rangle$ and Λ_{t2} be the transition system from Example 2 and its termination resilience trace semantics, respectively. The fixpoint iterates of its termination resilience semantics $\Lambda_{\mathsf{tr}2} \stackrel{def}{=} \alpha_{\mathsf{tr}}(\Lambda_{\mathsf{t2}})$ are the following:

Note that $\Lambda_{\mathsf{tr}2}$ is defined over the state b even if a transition can lead to the diverging trace be^ω because the non-deterministic choice is treated angelically: there is at least one choice that ensures termination (via the trace bd).   ◁

The termination resilience semantics $\Lambda_{\mathsf{tr}}\llbracket P \rrbracket$ of a program P is sound and complete for verifying Termination Resilience, for an initial set of states $I \subseteq \Sigma$:

Theorem 1 (Soundness&Completeness). $\Lambda\llbracket P \rrbracket I \in \mathcal{TR} \iff I \subseteq \mathsf{dom}(\Lambda_{\mathsf{tr}}\llbracket P \rrbracket)$

```
 1
 x := input;
 2
 z := 10;


while  3(z >= 0) {
    4
    z := z - x;
} 5
```

```
 1
 x = input;
 2
 z := 10;
 3
 c := [-2, 1];
while  4(z >= x) {
    5
    z := z + c;
} 6
```

(a) Program P_1 (b) Program P_2

Fig. 3. Programs akin to the transitions systems in Example 1 (a) and 2 (b).

4   Denotational Termination Resilience Semantics

The formal treatment given in the previous section is language independent. In the following, for simplicity, we introduce a small sequential programming language that we use for illustration throughout the rest of the paper:

$$\text{AExp} \ni \mathsf{a} ::= x \mid c \mid -\mathsf{a} \mid \mathsf{a} \diamond \mathsf{a}$$

$$\text{Stmt} \ni \mathsf{s} ::= {}^l\text{skip} \mid {}^l x := \text{input} \mid {}^l x := [c_1, c_2] \mid {}^l x := \mathsf{a}$$

$$\mid \text{if}^l\, \mathsf{a} \bowtie 0\, \{\mathsf{s}\} \mid \text{while}^l \mathsf{a} \bowtie 0\, \{\mathsf{s}\} \mid \mathsf{s}; \mathsf{s}$$

$$\text{Prog} \ni \mathsf{p} ::= \mathsf{s}^l$$

where $x \in \mathbb{X}$ ranges over program variables, $c \in \mathbb{Z}$, $c_1 \in \mathbb{Z} \cup \{-\infty\}$, $c_2 \in \mathbb{Z} \cup \{+\infty\}$ range over mathematical integer values possibly extended to include (negative or positive) infinity, $\diamond \in \{+, -, *, \dots\}$, and $\bowtie \in \{=, \neq, \leq, \dots\}$. A unique label $l \in \mathcal{L}$ appears within each instruction statement. Variable assignments can involve an untrusted input (${}^l x := \text{input}$), a (trusted) non-deterministic choice (${}^l x := [c_1, c_2]$) or an arithmetic expression (${}^l x := \mathsf{a}$). A program is an instruction statement $\mathsf{s} \in \text{Stmt}$ followed by a label $l \in \mathcal{L}$.

Figure 3 shows two programs written in our small language, both slices of program P from Fig. 1. Program P_1 (cf. Figure 3a) is akin to the transition system in Example 1: the input $x \leq 0$ leads to non-termination (state c in Fig. 2a), while for $x > 0$ (state b) the program always terminates. Instead, P_2 (cf. Figure 3b) is akin to Example 2: for the input $x \leq 10$ (state b in Fig. 2b), termination is possible with $c < 0$ (state d).

In this section, we instantiate the definition of our termination resilience semantics Λ_{tr} with respect to programs $\mathsf{p} \in \text{Prog}$ written in our small language. Then, in the next section, we will define its abstraction $\Lambda_{\mathsf{tr}}^{\natural}$.

Taint Semantics. Let $\mathcal{E} = \mathbb{X} \to \mathbb{Z}$ denote the set of environments, where each $\rho \in \mathcal{E}$ maps variables to their integer values. We adopt the standard semantics of expressions $[\![e]\!] : \mathcal{E} \rightharpoonup \mathcal{P}(\mathbb{Z})$ mapping an expression $e \in \text{AExp} \cup \{\text{input}, [c_1, c_2]\}$ to the set of all its possible values in a given environment.

$$\mathcal{T}[\![^l \, \mathsf{skip}]\!]\mathrm{T} \overset{\mathrm{def}}{=} \mathrm{T}$$

$$\mathcal{T}[\![^l \, x := \mathsf{input}]\!]\mathrm{T} \overset{\mathrm{def}}{=} \mathrm{T} \cup \{x\}$$

$$\mathcal{T}[\![^l \, x := [c_1, c_2]]\!]\mathrm{T} \overset{\mathrm{def}}{=} \mathrm{T} \setminus \{x\}$$

$$\mathcal{T}[\![^l \, x := \mathsf{a}]\!]\mathrm{T} \overset{\mathrm{def}}{=} \begin{cases} \mathrm{T} \cup \{x\} & \mathcal{T}[\![\mathsf{a}]\!]\mathrm{T} \\ \mathrm{T} \setminus \{x\} & \text{otherwise} \end{cases}$$

$$\mathcal{T}[\![\mathsf{if}^l \, \mathsf{a} \bowtie 0 \, \{\mathsf{s}\}]\!]\mathrm{T} \overset{\mathrm{def}}{=} \begin{cases} \mathcal{T}[\![\mathsf{s}]\!]\mathrm{T} \cup \mathrm{ASSIGNED}(\mathsf{s}) \cup \mathrm{T} & \mathcal{T}[\![\mathsf{a}]\!]\mathrm{T} \\ \mathcal{T}[\![\mathsf{s}]\!]\mathrm{T} \cup \mathrm{T} & \text{otherwise} \end{cases}$$

$$\mathcal{T}[\![\mathsf{while}^l \, \mathsf{a} \bowtie 0 \, \{\mathsf{s}\}]\!]\mathrm{T} \overset{\mathrm{def}}{=} \mathsf{lfp}^{\subseteq}_{\mathrm{T}} \, \mathcal{T}[\![\mathsf{if}^l \, \mathsf{a} \bowtie 0 \, \{\mathsf{s}\}]\!]$$

$$\mathcal{T}[\![\mathsf{s}_1; \mathsf{s}_2]\!]\mathrm{T} \overset{\mathrm{def}}{=} \mathcal{T}[\![\mathsf{s}_2]\!](\mathcal{T}[\![\mathsf{s}_1]\!]\mathrm{T})$$

Fig. 4. Taint semantics $\mathcal{T}[\![\mathsf{s}]\!] : \mathcal{P}(\mathbb{X}) \to \mathcal{P}(\mathbb{X})$, where $\mathrm{ASSIGNED}(\mathsf{s})$ is the set of variables assigned with an arithmetic expression within the statement s.

Program states are pairs $\mathcal{L} \times \mathcal{E}$ of a label $l \in \mathcal{L}$ and an environment $\rho \in \mathcal{E}$ defining the values of the program variables at the program point designated by l. To track the effect of input transitions on program states, we need a *taint semantics* $\mathcal{T}[\![\mathsf{s}]\!] : \mathcal{P}(\mathbb{X}) \to \mathcal{P}(\mathbb{X})$ that collects variables that depend on untrusted inputs, i.e., that are *tainted*, after the execution of a program statement (cf. Figure 4). The taint semantics $\mathcal{T}[\![\mathsf{a}]\!] : \mathcal{P}(\mathbb{X}) \to \{\mathsf{true}, \mathsf{false}\}$ of expressions is:

$$\mathcal{T}[\![\mathsf{a}]\!]\mathrm{T} \overset{\mathrm{def}}{=} \exists \rho \in \mathcal{E} : \exists V \in \mathbb{Z}^{|\mathbb{X}|} : V \in [\![\mathsf{a}]\!]\rho \wedge \forall \rho' \neq_{\mathrm{T}} \rho : V \notin [\![\mathsf{a}]\!]\rho' \qquad (6)$$

where $\rho' \neq_{\mathrm{T}} \rho \Leftrightarrow \exists \emptyset \subset \mathrm{T}' \subseteq \mathrm{T} : (\forall x \notin \mathrm{T}' : \rho(x) = \rho'(x)) \wedge (\forall x \in \mathrm{T}' : \rho(x) \neq \rho'(x))$ denotes program environments that differ only on (a non-empty subset of) the tainted variables in T. An expression is tainted if there exists a value of its semantics $[\![\mathsf{a}]\!]$ that is only possible in an environment ρ with a certain value V for the set T of tainted variables, i.e., the value of the expression depends on untrusted program inputs. Note that this definition differs from a classical non-interference-like definition (e.g., $\mathcal{T}[\![\mathsf{a}]\!]\mathrm{T} \overset{\mathrm{def}}{=} \forall \rho, \rho' \in \mathcal{E} : \forall x \notin \mathrm{T} : \rho(x) = \rho'(x) \wedge [\![\mathsf{a}]\!]\rho \neq [\![\mathsf{a}]\!]\rho')$ to also take into account the non-determinism arising from trusted variables (i.e., $[\![\mathsf{a}]\!]\rho \neq [\![\mathsf{a}]\!]\rho'$ could be the consequence of non-deterministic value choices rather than untrusted inputs). For simplicity, in this definition we deliberately do not take into account the reachable environments at each program label. A more precise definition would be straightforward but needlessly cumbersome for our purposes. The taint semantics $\mathcal{T}[\![\mathsf{p}]\!] \in \mathcal{P}(\mathbb{X})$ of a program $\mathsf{p} \in \mathrm{Prog}$ is $\mathcal{T}[\![\mathsf{p}]\!] = \mathcal{T}[\![\mathsf{s}^l]\!] \overset{\mathrm{def}}{=} \mathcal{T}[\![\mathsf{s}]\!]\emptyset$. By pointwise-lifting of $\mathcal{T}$ with respect to the program labels $\mathcal{L}$, we obtain the lifted taint semantics $\dot{\mathcal{T}} : \mathcal{L} \to \mathcal{P}(\mathbb{X})$ mapping each label to the set of variables that are can be tainted at that label.

Denotational Termination Resilience Semantics. We can now define in Fig. 5 the termination resilience semantics $\Lambda_{\mathsf{tr}}[\![\mathsf{s}]\!] : (\mathcal{E} \rightharpoonup \mathbb{O}) \to (\mathcal{E} \rightharpoonup \mathbb{O})$ for each

$$\Lambda_{\mathsf{tr}}[\![^{l}\,\mathsf{skip}]\!]f \stackrel{\mathrm{def}}{=} \lambda\rho \in dom(f).f(\rho) + 1$$

$$\Lambda_{\mathsf{tr}}[\![^{l}\,x := \mathsf{e}]\!]f \stackrel{\mathrm{def}}{=} \lambda\rho. \begin{cases} \sup\{f(\rho[x \leftarrow v]) + 1 \mid v \in [\![\mathsf{e}]\!]\rho\} & \mathsf{e} = \mathsf{input} \vee A \\[6pt] \inf\left\{ f(\rho[x \leftarrow v]) + 1 \;\middle|\; \begin{array}{l} v \in [\![\mathsf{e}]\!]\rho\,\wedge \\ \rho[x \leftarrow v] \in \mathsf{dom}(f) \end{array} \right\} & \mathsf{e} = [c_1, c_2] \vee B \\[12pt] \mathsf{undefined} & \mathsf{otherwise} \end{cases}$$

$$A \Leftrightarrow (\mathsf{e} = \mathsf{a} \wedge \mathfrak{T}[\![\mathsf{a}]\!](\dot{\mathfrak{T}}(l)) \wedge \exists v \in [\![\mathsf{a}]\!]\rho \wedge \forall v \in [\![\mathsf{a}]\!]\rho, \rho[x \leftarrow v] \in \mathsf{dom}(f))$$

$$B \Leftrightarrow (\mathsf{e} = \mathsf{a} \wedge \neg\mathfrak{T}[\![\mathsf{a}]\!](\dot{\mathfrak{T}}(l)) \wedge \exists v \in [\![\mathsf{a}]\!]\rho : \rho[x \leftarrow v] \in \mathsf{dom}(f))$$

$$\Lambda_{\mathsf{tr}}[\![\mathsf{if}^{l}\,\mathsf{a} \bowtie 0\,\{\mathsf{s}\}]\!]f \stackrel{\mathrm{def}}{=} \Theta_{\mathsf{tr}}(f)$$

$$\Theta_{\mathsf{tr}} \stackrel{\mathrm{def}}{=} \lambda X.\lambda\rho. \begin{cases} \Lambda_{\mathsf{tr}}[\![\mathsf{s}]\!]X(\rho) + 1 & \forall v \in [\![\mathsf{a}]\!]\rho : v \bowtie 0 \wedge \rho \in \mathsf{dom}(\Lambda_{\mathsf{tr}}[\![\mathsf{s}]\!]X(\rho)) \\ f(\rho) + 1 & \forall v \in [\![\mathsf{a}]\!]\rho : v \not\bowtie 0 \wedge \rho \in \mathsf{dom}(f(\rho)) \\ \mathsf{undefined} & \mathsf{otherwise} \end{cases}$$

$$\Lambda_{\mathsf{tr}}[\![\mathsf{while}^{l}\,\mathsf{a} \bowtie 0\,\{\mathsf{s}\}]\!]f \stackrel{\mathrm{def}}{=} \mathsf{lfp}_{\dot{\emptyset}}^{\sqsubseteq} \Theta_{\mathsf{tr}}$$

$$\Lambda_{\mathsf{tr}}[\![\mathsf{s}_1; \mathsf{s}_2]\!]f \stackrel{\mathrm{def}}{=} \Lambda_{\mathsf{tr}}[\![\mathsf{s}_1]\!](\Lambda_{\mathsf{tr}}[\![\mathsf{s}_2]\!]f)$$

Fig. 5. Termination resilience semantics $\Lambda_{\mathsf{tr}}[\![\mathsf{s}]\!] : (\mathcal{E} \rightharpoonup \mathbb{O}) \rightarrow (\mathcal{E} \rightharpoonup \mathbb{O})$. Since $[\![\mathsf{a}]\!]\rho$ is always a singleton $\{v\}$, we abuse notation and write $[\![\mathsf{a}]\!]\rho$ to directly denote v.

program statement. Each transfer function $\Lambda_{\mathsf{tr}}[\![\mathsf{s}]\!]$ takes as input a (potential) ranking function that is valid *after* the execution of the statement s and returns a (potential) ranking function that is valid *before* the execution of s.

If the right-hand side of an assignment is an untrusted input ($^{l}\,x := \mathsf{input}$) or an expression a ($^{l}\,x := \mathsf{a}$) that depends on tainted variables ($\mathfrak{T}[\![\mathsf{a}]\!](\dot{\mathfrak{T}}(l))$), the termination resilience semantics $\Lambda_{\mathsf{tr}}[\![^{l}\,x := \mathsf{a}]\!]$ returns a (potential) ranking function only defined over the environments ρ that, when subject to the assignment ($\rho[x \leftarrow v]$ with $v \in [\![\mathsf{a}]\!]\rho$), *always* belong to the domain of the given (potential) ranking function f (case A). The value of f for these environments is increased by one, to take into account another program execution step before termination, and the value of the resulting ranking function is the least upper bound of these values (to count the maximum number of steps to termination). Otherwise – if the right-hand side of the assignment is a non-deterministic choice ($^{l}\,x := [c_1, c_2]$) or an expression a that does not depend on tainted variables ($\neg\mathfrak{T}[\![\mathsf{a}]\!](\dot{\mathfrak{T}}(l))$) – the resulting function is defined over the environments ρ for which *at least one* value $v \in [\![\mathsf{a}]\!]\rho$ yields an environment $\rho[x \leftarrow v]$ in $\mathsf{dom}(f)$ (case B). Its value is the greatest lower bound of the value of f on these environments plus one (to count the minimum number of steps to termination).

The termination resilience semantics $\Lambda_{\mathsf{tr}}[\![\mathsf{if}^{l}\,\mathsf{a} \bowtie 0\,\{\mathsf{s}\}]\!]$ for if statements, via Θ_{tr}, defines a potential ranking function over all environments in $\mathsf{dom}(\Lambda_{\mathsf{tr}}[\![\mathsf{s}]\!]f)$ or $\mathsf{dom}(f)$, i.e., all environments potentially leading to a terminating execution, depending on whether they satisfy or not satisfy the guard $\mathsf{a} \bowtie 0$. The termination resilience $\Lambda_{\mathsf{tr}}[\![\mathsf{while}^{l}\,\mathsf{a} \bowtie 0\,\{\mathsf{s}\}]\!]$ for while statements is the least fixpoint of Θ_{tr} with respect to the computational order $f_1 \sqsubseteq f_2 \overset{\text{def}}{\Leftrightarrow} \mathsf{dom}(f_1) \subseteq \mathsf{dom}(f_2) \wedge \forall x \in \mathsf{dom}(f_1) : f_1(x) \leq f_2(x)$.

The termination resilience semantics $\Lambda_{\mathsf{tr}}[\![\mathbf{p}]\!] \in (\mathcal{E} \rightharpoonup \mathbb{O})$ of a program $\mathbf{p} \in \mathrm{Prog}$ is determined by $\Lambda_{\mathsf{tr}}[\![\mathbf{s}]\!]$ taking as input the constant function equal to zero.

Definition 4 (Denotational Termination Resilience Semantics). *Let* $\mathbf{p} \in$ *Prog be a program. Its denotational termination resilience semantics is:*

$$\Lambda_{\mathsf{tr}}[\![\mathbf{p}]\!] = \Lambda_{\mathsf{tr}}[\![\mathbf{s}^l]\!] \overset{\text{def}}{=} \Lambda_{\mathsf{tr}}[\![\mathbf{s}]\!](\lambda\rho.0). \tag{7}$$

5 Abstract Termination Resilience Semantics

In this section, we propose a sound abstraction $\Lambda_{\mathsf{tr}}^{\natural}[\![\mathbf{p}]\!]$ of the denotational termination resilience semantics $\Lambda_{\mathsf{tr}}[\![\mathbf{p}]\!]$ defined in Sect. 4 by leveraging (and extending) Urban and Miné's decision tree numerical abstract domain $\mathcal{T}$ [34] to abstract potential ranking functions by means of piecewise-defined partial functions. Specifically, we consider the concretization-based abstraction [11] $\langle \mathcal{E} \rightharpoonup \mathbb{O}, \preceq \rangle \xleftarrow{\gamma} \langle \mathcal{T}, \preceq_{\mathcal{T}} \rangle$ where the approximation order $\preceq$ is defined as follows:

$$f_1 \preceq f_2 \Leftrightarrow \mathsf{dom}(f_1) \supseteq \mathsf{dom}(f_2) \wedge \forall x \in \mathsf{dom}(f_2) : f_1(x) \leq f_2(x). \tag{8}$$

We define $\Lambda_{\mathsf{tr}}^{\natural}[\![\mathbf{p}]\!]$ so $\Lambda_{\mathsf{tr}}[\![\mathbf{p}]\!] \preceq \gamma(\Lambda_{\mathsf{tr}}^{\natural}[\![\mathbf{p}]\!])$ meaning that $\Lambda_{\mathsf{tr}}^{\natural}[\![\mathbf{p}]\!]$ *over-approximates* the value of $\Lambda_{\mathsf{tr}}[\![\mathbf{p}]\!]$ and *under-approximates* its domain $\mathsf{dom}(\Lambda_{\mathsf{tr}}[\![\mathbf{p}]\!])$. In this way, the abstraction $\Lambda_{\mathsf{tr}}^{\natural}[\![\mathbf{p}]\!]$ provides *sufficient preconditions* for termination resilience.

Remark 1. Note that Urban and Miné's abstract termination semantics [34]—let us denote it $\Lambda_{\mathsf{t}}^{\natural}[\![\mathbf{p}]\!]$—is already a sound but coarse over-approximation of $\Lambda_{\mathsf{tr}}^{\natural}[\![\mathbf{p}]\!]$: $\Lambda_{\mathsf{tr}}[\![P]\!] \preceq \gamma(\Lambda_{\mathsf{t}}^{\natural}[\![\mathbf{p}]\!])$. Our objective in this section is to build upon and relax $\Lambda_{\mathsf{t}}^{\natural}[\![\mathbf{p}]\!]$ for reasoning about Termination Resilience.

5.1 Decision Tree Abstract Domain

In the following, we recall the decision tree abstract domain only insofar as it is necessary for our analysis. We do not redefine aspects of the domain that remain unchanged, and refer to [33] for the full formal definitions.

An element $t \in \mathcal{T}$ is a piecewise-defined partial function represented by a full binary tree, where each node—denoted $\mathrm{NODE}\{c\}: t_1, t_2$, with $t_1, t_2 \in \mathcal{T}$—is labeled by a constraint $c \in \mathcal{C}$ over the program variables (where $\mathcal{C}$ is the set of allowed constraints, e.g., linear constraints), and each leaf—denoted $\mathrm{LEAF}: f$— is labeled by a function $f \in \mathcal{F}$ of the program variables (where $\mathcal{F}$ is the set of allowed functions, e.g., affine functions). Nodes recursively partition the space of possible values of the program variables: constraint labeling nodes are satisfied by the left subtrees of the nodes, while the right subtrees satisfy their negation. The leaves of the tree represent the value of the function within each partition.

Algorithm 1 Decision Tree Assignment

1: **function** ASSIGN$_T[\![^l x := \mathsf{e}]\!](t)$
2:    **if** isLeaf(t) **then**
3:        **return** LEAF: ASSIGN$_F[\![^l x := \mathsf{e}]\!](t.f)$
4:    **else if** isNode(t) **then**
5:        $I \leftarrow$ ASSIGN$_C[\![^l x := \mathsf{e}]\!](t.c)$
6:        $J \leftarrow$ ASSIGN$_C[\![^l x := \mathsf{e}]\!](\neg t.c)$
7:        **if** isEmpty$(I) \wedge$ isEmpty(J) **then**
8:            **return** ASSIGN$_T[\![^l x := \mathsf{e}]\!](t.l)$ $\curlyvee_T$ ASSIGN$_T[\![^l x := \mathsf{e}]\!](t.r)$
9:        **else if** isEmpty$(I) \wedge \bot_C \in J$ **then**
10:            **return** ASSIGN$_T[\![^l x := \mathsf{e}]\!](t.l)$
11:        **else if** $\bot_C \in I \wedge$ isEmpty(J) **then**
12:            **return** ASSIGN$_T[\![^l x := \mathsf{e}]\!](t.r)$
13:        **else**
14:            $t_1 \leftarrow$ PRUNE$_T($ASSIGN$_T[\![^l x := \mathsf{e}]\!](t.l), I)$
15:            $t_2 \leftarrow$ PRUNE$_T($ASSIGN$_T[\![^l x := \mathsf{e}]\!](t.r), J)$
16:            **return** $t_1 \curlyvee_T t_2$

Figure 6 shows an example of a decision tree. The leaf with value $\bot_F$ explicitly represents the undefined partition of the (partial) function. Undefined functions can also be represented by the leaf value $\top_F$ in case of an irrecoverable loss of precision of the analysis [8]. The decision tree in Fig. 6 is automatically inferred by our termination resilience static analysis at label 4 of program P_2 in Fig. 3b . It represents a function with con-

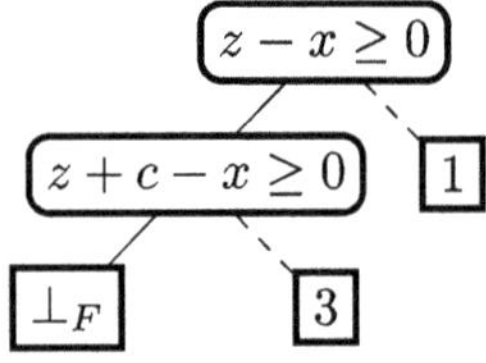

Fig. 6. Decision tree inferred by our analysis at label 4 of program P_2 (cf. Figure 3b).

stant value 1 when $z < x$ (at most one step to termination), with constant value 3 when $z \geq x \wedge z + c < x$ (at most three steps to termination) and undefined otherwise (termination is not proved when $z + c \geq x$), while the concrete termination resilience semantics at the same label 4 is defined when $z < x$ or $c < 0$: the decision tree is thus a sound approximation with respect to $\preceq$ (cf. Eq. 8).

The partitioning induced by decision tree constraints is dynamic: the analysis begins at the end of the program with the decision tree LEAF: 0 (i.e., zero program executions steps to termination) and proceeds *backwards*; during the analysis, constraints are added by boolean conditions or when merging control flows, and are modified by variable assignments.

Algorithm 1 shows the decision tree assignment operator ASSIGN$_T[\![^l x := \mathsf{e}]\!]$ originally defined in [33]. The assignment is performed by recursively descending the given decision tree t (cf. Lines 14 and 15, the left subtree is accessed with $t.l$ and the right one with $t.r$). At the leaves, the assignment is handled by the auxiliary operator ASSIGN$_F[\![^l x := \mathsf{e}]\!]$ operating on functions (accessed by $t.f$, cf. Line 3): after the assignment, the value of defined functions is increased by one to account for one more step to termination. For example, let us consider the

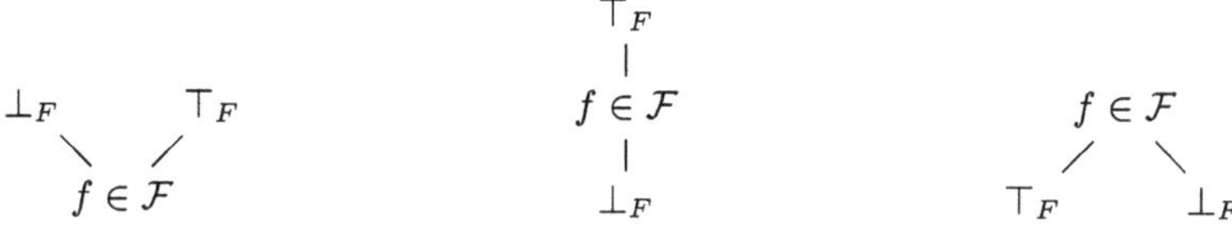

(a) Approximation order $\preceq_F$ (b) Computational order $\sqsubseteq_F$ (c) Resilience order $\leq_F$

Fig. 7. Hasse diagrams for difference orders of decision tree leaves.

decision tree in Fig. 6 and the assignment $c := [-2, 1]$ at label 3 in Fig. 3b. The result of the assignment on LEAF : 3 is LEAF : 4, simply increasing the value of its function by one, while LEAF : $\perp_F$ remains unchanged. Node constraints (accessed with $t.c$) and their negation ($\neg t.c$) are handled by the standard assignment operator $\text{ASSIGN}_C[\![^l x := e]\!]$ of an auxiliary underlying numerical abstract domain used to manage constraints in $\mathcal{C}$, which produces a set $I \in \mathcal{P}(\mathcal{C})$ (cf. Line 5) and a set $J \in \mathcal{P}(\mathcal{C})$ (cf. Line 6) of linear constraint conjuncts resulting from (over-approximating) the assignment to $t.c$ and $\neg t.c$, respectively. For example, considering again the decision tree in Fig. 6, the result of the assignment $c := [-2, 1]$ on the constraint $z + c - x \geq 0$ and its negation $z + c - x < 0$ yields $I = \{z - x \geq -1\}$ and $J = \{z - x < 2\}$. The set of constraints I and J are added (by PRUNE_T) to the subtrees resulting from the recursive calls (cf. Lines 14 and 15). In our example, the pruning of the LEAF : $\perp_F$ with the set of constraints I results in the decision tree shown in Fig. 8a, while the pruning of LEAF : 4 with J results in the tree in Figure. 8b.

(a) Pruning of LEAF: $\perp_F$ with $I = \{z - x \geq -1\}$.

(b) Pruning of LEAF: 4 with $J = \{z - x < 2\}$.

Fig. 8. Decision trees resulting from PRUNE_T.

The special NIL node denotes partitions that are outside the domain of definition of the decision tree. Note that it differs from leaves labeled with $\perp_F$ or $\top_F$, which denote undefined partitions within the domain of definition of the tree. Finally, the resulting trees are joined by the approximation join $\curlyvee_T$ (cf. Line 16), which merges the domains of definitions of the joined trees (essentially removing any NIL nodes) and retains the leaves labeled with an undefined function ($\perp_F$ or $\top_F$) in either of the joined decision trees, cf. the Hasse diagram in Fig. 7a . Continuing the example, joining the trees in Fig. 8 yields the tree in Fig. 9.

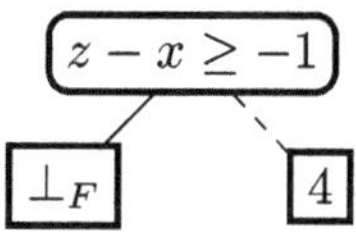

Fig. 9. Approximation join of Fig. 8a and Fig. 8b.

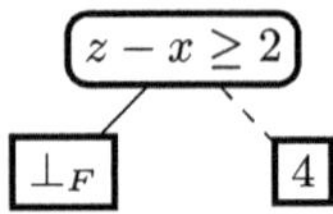

Fig. 10. Resilience join of Fig. 8a and Fig. 8b.

In case both I and J are empty (cf. Line 7), neither the constraint $t.c$ nor its negation $\neg t.c$ exists anymore and thus the subtrees resulting from the recursive calls are joined by the approximation join $\curlyvee_T$. In case I is empty and J is an unsatisfiable set of constraints (i.e., the unsatisfiable constraint $\bot_C$ belongs to J, cf. Line 9), it means that $\neg t.c$ is no longer satisfiable and thus only the left subtree of the decision tree is kept (cf. Line 10). Similarly, in case I is an unsatisfiable set of constraints and J is empty (cf. Line 11), only the right subtree of the decision tree is kept (cf. Line 12).

To minimize the cost of the analysis and to enforce its termination, a widening operator ∇_T limits the height of the decision trees and the number of induced partitions. Abstract fixpoint iterations with widening follow the computational order $\sqsubseteq_T$, which preserves decision tree leaves labeled with a defined function over undefined leaves labeled with $\bot_F$ (i.e., the analysis grows the domain of the inferred ranking function at each iteration), but preserves leaves labeled with $\top_F$ over all other leaves (i.e., the domain of the inferred function shrinks when the analysis loses too much precision), cf. Figure 7b.

5.2   Non-Deterministic Assignments

To leverage the decision tree abstract domain for termination resilience we need to extend it to distinguish between variables assigned with untrusted or (trusted) non-deterministic values. Termination must be ensured for *all* possible values of untrusted variables, while *any* non-deterministic value that ensures termination is enough to satisfy termination resilience. We thus introduce a new *resilience join* operator $\curlyvee_T$ between decision trees that retains the leaves labeled with a *defined* function in either of the joined decision trees, cf. Figure 7c.

We implement $\text{ASSIGN}_T[\![^l x \ := \ [c_1, c_2]]\!]$ for non-deterministic assignments following Algorithm 1, but using the resilience join $\curlyvee_T$ instead of the approximation join $\curlyvee_T$ at Lines 8 and 16. Let us consider again the left subtree in Fig. 6 and the non-deterministic variable assignment $c := [-2, 1]$ at label 3 in Fig. 3b . Joining the trees in Fig. 8 resulting from the PRUNE_T at Lines 14 and 15 yields the decision tree in Fig. 10, which indicates that the termination of program P_2 is possible when $z - x < 2$, for at least *some*

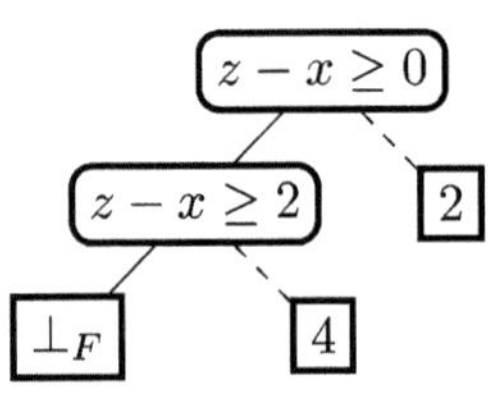

Fig. 11. Decision tree inferred at label 3 of program P_2.

values of c chosen at label 3 (notably, for $c \in \{-2, -1\}$, but the analysis does not explicitly provide these values). The final decision tree at program label 3 is shown in Fig. 11.

Note that this seemingly anodyne modification is actually insidious. Let us consider the decision tree in Fig. 12 and the variable assignment $x := [-\infty, +\infty]$, and assume that the variable y is untrusted. The result of the assignment on the decision tree constraints is the empty set of constraints and, since the variable x is assigned with a non-deterministic value, we can employ the resilience join $\curlyvee_F$ to join all leaves after the assign-

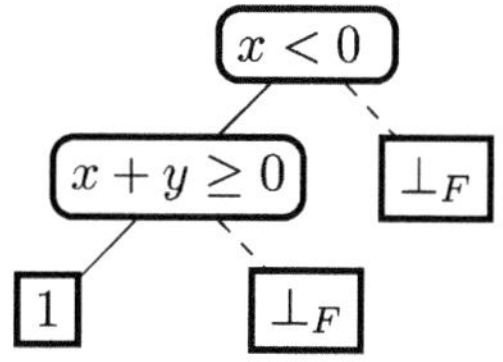

Fig. 12. Unfilled decision tree.

ment is performed on them. This yields the final decision tree LEAF: 2 (obtained from LEAF: 1 after the assignment), which is not sound! In fact, the sufficient precondition for Termination Resilience given by the decision tree in Fig. 12 requires $y > 0$ but this constraint is missing from the decision tree because it is redundant. As a consequence, when the assignment is performed and the other constraints $x < 0$ and $x + y \geq 0$ disappear, the result is an unsound decision tree indicating that termination resilience always holds. To avoid this issue, before non-deterministic variable assignments, decision trees need to be *filled* to explicitly contain *at least* the strongest redundant univariate constraints over the untrusted program variables. We further modify Algorithm 1 to insert a call to $\textrm{FILL}_T(t)$ at the very beginning (*before* Line 2), where $\textrm{FILL}_T$ leverages the underlying numerical abstract domain used to manage $\mathcal{C}$ to find these redundant univariate constraints and adds them to the decision tree t. Additional redundant constraints may be added, though this needlessly increases the analysis time.

The $\textrm{ASSIGN}_T$ operator for non-deterministic variable assignments, implemented by Algorithm 1 modified as discussed above, is a sound over-approximation of the termination resilience semantics (cf. Figure 5):

Proposition 1. $\Lambda_{\mathrm{tr}}[\![^l x := [c_1, c_2]]\!]\gamma(t) \preceq \gamma(\textrm{ASSIGN}_T[\![^l x := [c_1, c_2]]\!]t)$

Resilience Join and Non-Tainted Program Variables. At this point, one may be tempted to extend the use of the resilience join to all variable assignments with a non-tainted right-hand side expression (i.e., an expression that does not depend on untrusted program variables) or, similarly, to `if` and `while` statements with non-tainted boolean conditions. However, this change alone would be unsound. Consider program P in Fig. 1 and let the boolean condition at label 3 be $a * a \geq 0$ where a is assigned with a non-deterministic value before x. Figure 13 shows the most precise decision trees that can be inferred at label 4 and 6.

At program label 3, the boolean condition $a * a \geq 0$ cannot be precisely represented by a (set of) linear constraints. Thus, no further constrains are added before joining the trees. Since the decision tree at label 6 is totally defined (cf. Figure 13b), using the resilience join would also yield a decision tree that is totally defined (cf. Figure 7c), implying that termination resilience always holds. However, program label 6 is not reachable and, indeed, termination resilience does not hold: the program is robustly non-terminating when $x \leq 0$.

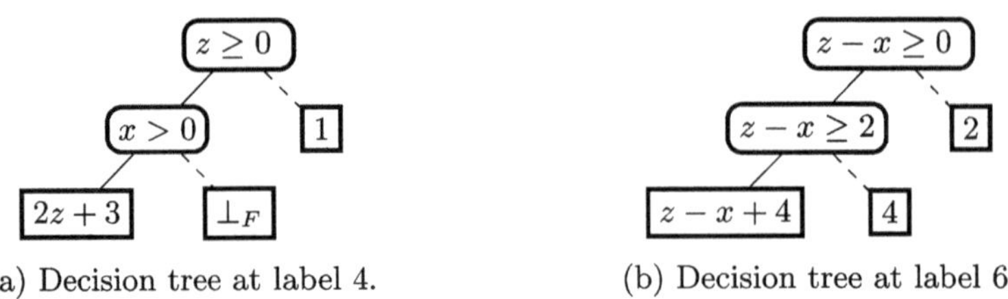

(a) Decision tree at label 4.   (b) Decision tree at label 6.

Fig. 13. Abstract termination resilience semantics for program P.

$$\Lambda_{\mathrm{tr}}^{\natural}[\![^l\, \mathrm{skip}]\!]t \stackrel{\mathrm{def}}{=} \mathrm{SKIP}_T(t)$$

$$\Lambda_{\mathrm{tr}}^{\natural}[\![^l\, x := \mathrm{e}]\!]t \stackrel{\mathrm{def}}{=} \mathrm{ASSIGN}_T[\![^l\, x := \mathrm{e}]\!](t)$$

$$\Lambda_{\mathrm{tr}}^{\natural}[\![\mathrm{if}^l\, \mathrm{a} \bowtie 0\, \{\mathrm{s}\}]\!]t \stackrel{\mathrm{def}}{=} \Theta_{\mathrm{tr}}^{\natural}(t)$$

$$\Theta_{\mathrm{tr}}^{\natural} \stackrel{\mathrm{def}}{=} \lambda X.\mathrm{FILTER}_T[\![\mathrm{a} \bowtie 0]\!](\Lambda_{\mathrm{tr}}^{\natural}[\![\mathrm{s}]\!]X)\; \Upsilon_T\; \mathrm{FILTER}_T[\![\mathrm{a} \not\bowtie 0]\!](t)$$

$$\Lambda_{\mathrm{tr}}^{\natural}[\![\mathrm{while}^l\, \mathrm{a} \bowtie 0\, \{\mathrm{s}\}]\!]t \stackrel{\mathrm{def}}{=} \mathrm{LFP}^{\natural}\, \Theta_{\mathrm{tr}}^{\natural}$$

$$\Lambda_{\mathrm{tr}}^{\natural}[\![\mathrm{s}_1; \mathrm{s}_2]\!]t \stackrel{\mathrm{def}}{=} \Lambda_{\mathrm{tr}}^{\natural}[\![\mathrm{s}_2]\!](\Lambda_{\mathrm{tr}}^{\natural}[\![\mathrm{s}_1]\!]t)$$

Fig. 14. Abstract Program Semantics for Termination Resilience

To ensure soundness in the general case of non-tainted variables and expressions, the resilience join should retain leaves labeled with a defined function only when the domain partition that they represent is actually reachable. In practice, we ensure this by conservatively under-approximating the weakest liberal precondition of program statements by means of the approximation join [33]. We leave finding less conservative approximations for future work.

5.3   Abstract Termination Resilience Semantics

We can finally define in Fig. 14 the *abstract termination resilience semantics* $\Lambda_{\mathrm{tr}}^{\natural}[\![\mathrm{s}]\!] : \mathcal{T} \to \mathcal{T}$ for each program statement s. The SKIP_T operator simply increases by one the value of the functions labeling the leaves of the decision tree. The assignment operator $\mathrm{ASSIGN}_T[\![^l\, x := \mathrm{e}]\!]$ is implemented by Algorithm 1, modified as discussed in Sect. 5.2 for non-deterministic variable assignments. The semantics of conditional if statements $\Theta_{\mathrm{tr}}^{\natural}$ employs the approximation join Υ_T as discussed above, after handling the boolean conditions with the FILTER_T operator (due to space limitations, we refer to [33] for its definition). The semantics for while loops iterates $\Theta_{\mathrm{tr}}^{\natural}$ with widening starting from the totally undefined decision tree $\mathrm{LEAF} : \bot_F$ until an abstract fixpoint is reached.

The *abstract termination resilience semantics* $\Lambda_{\mathrm{tr}}^{\natural}[\![\mathrm{p}]\!] \in \mathcal{T}$ for a $\mathrm{p} \in \mathrm{Prog}$ starts the termination resilience analysis with a decision tree containing a single leaf labeled with the zero function:

Definition 5 (Abstract Termination Resilience Semantics). *Let* $\mathsf{p} \in$ *Prog be a program. Its abstract termination resilience semantics is:*

$$\Lambda^{\natural}_{\mathsf{tr}}[\![\mathsf{p}]\!] = \Lambda^{\natural}_{\mathsf{tr}}[\![\mathsf{s}^l]\!] \overset{def}{=} \Lambda^{\natural}_{\mathsf{tr}}[\![\mathsf{s}]\!](\textsc{leaf} : 0). \tag{9}$$

The termination resilience analysis is sound with respect to the approximation order $\preceq$ (cf. Eq. 8):

Theorem 2 (Soundness). $\Lambda_{\mathsf{tr}}[\![P]\!] \preceq \gamma(\Lambda^{\natural}_{\mathsf{tr}}[\![P]\!])$

Corollary 1 (Soundness). $\Lambda[\![P]\!]I \in \mathcal{TR} \Leftarrow I \subseteq \mathsf{dom}(\gamma(\Lambda^{\natural}_{\mathsf{tr}}[\![P]\!]))$

Example 6. Let us consider again P_1 in Fig. 3a . Its most precise termination resilience semantics at label 3 matches the semantics of program P at label 4 shown in Fig. 13a . The assignment at label 2 – substituting z with 10 and incrementing the value of the functions labeling the defined leaves – yields

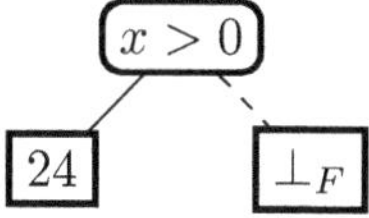

The assignment at label 1 yields $\textsc{leaf} : \bot_F$. Its concretization $\gamma(\textsc{leaf}: \bot_F)$ is the totally undefined function $\dot{\emptyset}$. We are thus unable to prove termination resilience. In this case, this is a true positive since indeed $\Lambda_{\mathsf{tr}}[\![P_1]\!] \notin \mathcal{TR}$.   ◁

Example 7. Let us consider again program P_2 in Fig. 3b . Its abstract termination resilience semantics at program label 3 is shown in Fig. 11. The assignment at label 2 yields

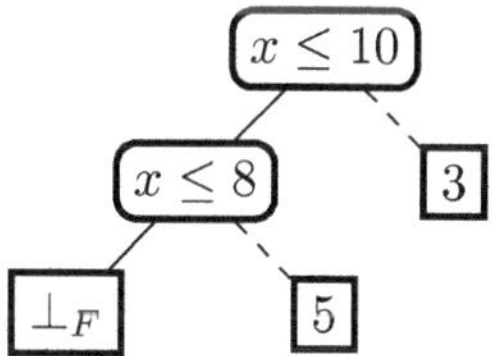

and the assignment at label 1 yields $\textsc{leaf} : \bot_F$. We are again unable to prove termination resilience since $\gamma(\textsc{leaf}: \bot_F)$ is the totally undefined function $\dot{\emptyset}$. This, however, is a false alarm since, in fact, $\Lambda_{\mathsf{tr}}[\![P_2]\!] \in \mathcal{TR}$.

Let us consider instead the most precise abstract termination resilience semantics shown in Fig. 13b . The assignment at label 2 yields

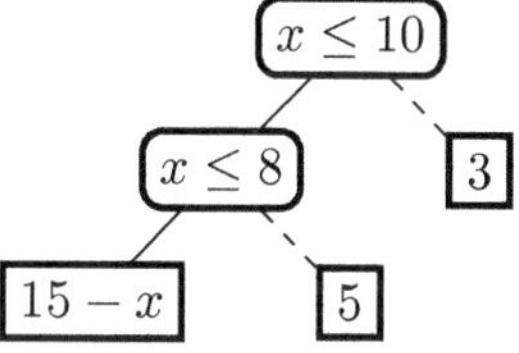

and the assignment at label 1 then yields $\textsc{leaf} : \omega$. In this case, we prove termination resilience since $\gamma(\textsc{leaf}: \omega)$ yields $\lambda\rho.\omega$, which is exactly $\Lambda_{\mathsf{tr}}[\![P_2]\!]$. ◁

6   Implementation and Experimental Evaluation

Implementation. We implemented our termination resilience analysis in the open-source tool FUNCTION-TR [26]. FUNCTION-TR is implemented in OCaml and relies on the APRON numerical abstract domain library [18] to manage decision tree constraints and functions labeling the tree leaves. It is also possible to activate the extension to lexicographic ranking functions [35], and tune the precision of the analysis by adjusting the widening delay. We use a simple implementation of the $FILL_T$ operation (cf. Section 5.2), which adds constraints, obtained from APRON's interval domain, bounding the values of the variables assigned with a non-deterministic value.

The analysis takes as input a numerical program written in a C-like syntax and outputs TRUE if all leaves of the inferred decision tree for the program are labelled with defined functions, i.e., Termination Resilience holds for any initial program state. Otherwise, it outputs UNKNOWN to indicate that Robust Non-Termination may hold (i.e., FUNCTION-TR raises a potential Robust Non-Termination alarm). In this case, the defined partitions of the inferred decision tree are a *sufficient precondition* for Termination Resilience.

Experimental Evaluation. Note that always terminating programs trivially satisfy Termination Resilience. For this reason, in our evaluation, we deliberately focus on programs that admit both terminating and non-terminating executions, thereby targeting the class of benchmarks where termination resilience is a non-trivial property and where our analysis can be meaningfully evaluated.

We selected a total of 62 programs from three sources: SV-COMP 2024[1], the benchmarks of the state-of-the-art non-termination analyzer Pulse [29], and the recent survey on non-termination bugs by Shi et al. [30]. From SV-COMP, we collected 39 programs from the sets *termination-crafted-lit, termination-restricted-15,* and *termination-nla* in the Termination category, all labeled as non-terminating in the associated .yml files (average length of 23 lines of code). From the Pulse benchmarks [29] we retained 10 programs (averaging 20 lines of code), and from the survey benchmarks [30] we retained 13 (averaging 25 lines of code – each program is a simplified extract derived from real-world open-source software that exhibited a non-termination bug). We excluded 16 (resp. 36) always terminating programs from these two latter sources. Furthermore, we discarded 33 (resp. 42) programs making use of pointers, bitwise operations, or recursive functions, which are currently not fully supported by our prototype tool.

We constructed variants of these 62 programs, for all possible combinations of variable initializations (e.g., `__VERIFIER_nondet_int()` for programs in SV-COMP) with untrusted (input) or trusted ($[-\infty, +\infty]$) values. In doing so, we obtained a benchmark of 278 test cases (150 from SV-COMP, 42 from Pulse [29], 86 from Shi et al. [30]). Note that variable initializations also occur within loops for 54 of 278 test cases (28 from SV-COMP, 12 from Pulse [29], 14 from Shi et al. [30]). The experiments were conducted on a 64-bit 8-Core CPU (AMD® Ryzen

---

[1] https://sv-comp.sosy-lab.org/2024/.

Table 1. Evaluation results.

Benchmark	Configuration	Property	Verified	Alarms	TO	Time (s)
SV-COMP	FUNCTION-TR-Boxes	Termination	0	150	0	3.0
		Termination Resilience	58	92	0	3.0
	FUNCTION-TR-Polyhedra	Termination	0	150	0	6.0
		Termination Resilience	99	51	0	14.0
Pulse [29]	FUNCTION-TR-Boxes	Termination	0	42	0	0.5
		Termination Resilience	23	19	0	0.5
	FUNCTION-TR-Polyhedra	Termination	0	42	0	7.0
		Termination Resilience	23	19	0	17.0
Shi et al. [30]	FUNCTION-TR-Boxes	Termination	0	86	0	2.0
		Termination Resilience	58	28	0	2.0
	FUNCTION-TR-Polyhedra	Termination	0	86	0	54.0
		Termination Resilience	58	20	8	257.0

7 pro 5850u) with 16GB of RAM on Ubuntu 20.04. Note that, for the moment, our tool is single-threaded and uses only one CPU core.

Table 1 summarizes the results of the evaluation with FUNCTION-TR configured to use the boxes (FUNCTION-TR-Boxes) or polyhedra (FUNCTION-TR-Polyhedra) abstract domain to manage decision tree constraints. We configured the analysis to perform widening after two iterations, and we left the extension with lexicographic functions disabled. We used a timeout of 120s (wall time) per test case. Activating the extension or increasing the widening delay makes no difference on these benchmarks aside from an increased execution time. As expected, proving Termination Resilience instead of Termination (i.e., Termination Resilience where all variables are considered untrusted) considerably reduces the number of alarms, almost 65% using decision tree with polyhedral constraints. This shows that our termination resilience analysis is an effective way to triage and prioritize alarms related to program non-termination. The execution time reported in the last column of Table 1 corresponds to the total analysis time over all test cases that did not time out.

The full comparison between FUNCTION-TR-Boxes and FUNCTION-TR-Polyhedra is shown in Fig. 15, where the axes denote execution (wall) time in seconds, and green stars ($\star$) label test cases where the configurations have the same precision (i.e., both prove Termination Resilience or raise an alarm), while red crosses ($\times$) and blue circles ($\bullet$) label test cases where FUNCTION-TR-Polyhedra and FUNCTION-TR-Boxes are more precise (i.e., prove Termination Resilience for more test cases), respectively. The presence of blue circles may be surprising: in few test cases, using a more precise numerical domain to manage decision

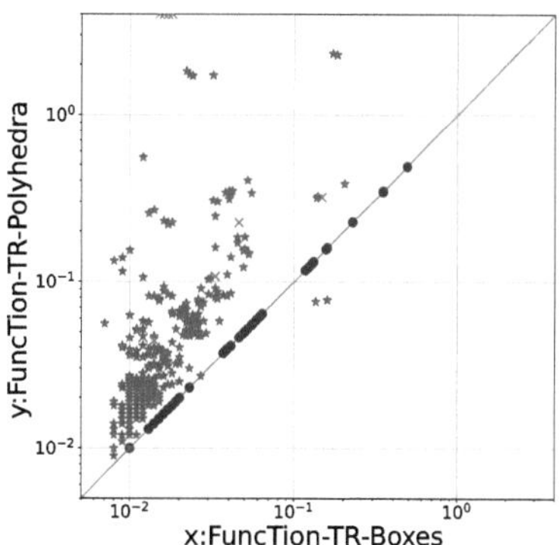

Fig. 15. FUNCTION-TR-Boxes vs FUNCTION-TR-Polyhedra. (Color figure online)

tree constraints actually causes a loss of precision when widening is performed (because the widening heuristic leverages adjacent decision tree partitions, and more partitions are formed with more precise decision tree constraints), while using a less precise numerical domain allows the widening to generalize better. We leave the implementation of better widening heuristics [8] for future work.

Triage of Alarms Found by Non-Termination Analyzers. We evaluated the ability of FUNCTION-TR to filter out non-critical non-termination alarms found by the non-termination analyzers Pulse, Dynamite and Ultimate-Automizer.

We reused the benchmark in which Raad et al. [29] reported non-termination bugs found by Pulse and Dynamite [21]. The programs in this benchmark use non-linear arithmetic and are extracted from the set *termination-nla* in the Termination category of SV-COMP. We excluded 9 out of 40 non-terminating programs that rely on unsigned int, since we assume a mathematical integer semantics. As in the previous evaluation, for each test filename.c, we generated variants for each possible combination of variable initialization which gave us 86 programs to test (with an average of 54 lines of code). In addition to the result of [29], we also ran the non-termination analysis of Ultimate-Automizer [7].

In Table 2, the suffixes of the test names, generated by the regular expression $-[r + i]^*$, describe the sequence of variable initializations. For instance the test `filename-ri.c` is the variant of `filename.c` in which we used a trusted (non-deterministic) value ($[-\infty, +\infty]$) for the first variable initialization and an untrusted (input) value for the second. A non-termination bug found by Pulse (column P in Table 2), Dynamite (column D in Table 2) or Ultimate-Automizer (column U in Table 2) is denoted ⚠, an empty cell indicates that the tool is inconclusive. Note that the tool results should not be compared as they consider different integer semantics. For FUNCTION-TR, ✓ means that it proved Termination Resilience (returns TRUE), while !(FP) or !(TP) denotes the cases where FUNCTION-TR raised a false alarm (upon manual inspection, Termination Resilience, in fact, holds) or a true alarm (upon manual inspection, Termination Resilience indeed does not hold), respectively.

From this evaluation, we conclude that our prototype FUNCTION-TR is able to filter out non-critical non-termination alarms found by non-termination analyzers. More precisely, Termination Resilience is proved for 16 programs where Pulse or Dynamite or Ultimate-Automizer found a non-terminating execution, hence FUNCTION-TR removes 16 alarms. Moreover, Termination Resilience is proved for 4 programs where Pulse, Dynamite and Ultimate-Automizer did not find any non-termination bug. Finally, by manual inspection, we report that 58 out of the 62 alarms raised by FUNCTION-TR are true positives. This result was expected: most of the non-terminating programs from the SV-COMP category *termination-nla* are constructed from terminating ones by making them always diverging, which is a case of Robust Non-Termination where FUNCTION-TR must raise an alarm. Dually, 4 of the 62 alarms are false positives. In our benchmark it is mostly due to the inherited imprecision of the decision tree abstract domain on programs exhibiting non-linear arithmetic operations or, more frequently, when the widening operator does not generalize well.

Table 2. Triage of non-termination alarms by FuncTion-TR.

Test Case	P	D	U	FuncTion-TR
bresenham1-both-nt-ii		⚠		!(TP)
bresenham1-both-nt-ir		⚠		!(TP)
bresenham1-both-nt-ri		⚠		!(TP)
bresenham1-both-nt-rr		⚠		!(FP)
cohencu1-both-nt-i		⚠	⚠	!(TP)
cohencu1-both-nt-r		⚠	⚠	!(TP)
cohencu2-both-nt-i				!(TP)
cohencu2-both-nt-r				!(TP)
cohencu3-both-nt-i				!(TP)
cohencu3-both-nt-r				!(TP)
cohencu4-both-nt-i				!(TP)
cohencu4-both-nt-r				!(TP)
cohencu5-both-nt-i		⚠		!(TP)
cohencu5-both-nt-r		⚠		!(TP)
dijkstra1-both-nt-i	⚠		⚠	!(TP)
dijkstra1-both-nt-r	⚠		⚠	✓
dijkstra2-both-nt-i	⚠		⚠	!(TP)
dijkstra2-both-nt-r	⚠		⚠	✓
dijkstra3-both-nt-i	⚠		⚠	!(TP)
dijkstra3-both-nt-r	⚠		⚠	✓
dijkstra4-both-nt-i	⚠		⚠	!(TP)
dijkstra4-both-nt-r	⚠		⚠	✓
dijkstra5-both-nt-i	⚠		⚠	!(TP)
dijkstra5-both-nt-r	⚠		⚠	✓
dijkstra4-both-nt-i	⚠		⚠	!(TP)
dijkstra6-both-nt-r	⚠		⚠	✓
egcd2-both-nt-ii		⚠	⚠	!(TP)
egcd2-both-nt-ir		⚠	⚠	✓
egcd2-both-nt-ri		⚠	⚠	✓
egcd2-both-nt-rr		⚠	⚠	✓
egcd3-both-nt-ii		⚠	⚠	!(TP)
egcd3-both-nt-ir		⚠	⚠	✓
egcd3-both-nt-ri		⚠	⚠	!(TP)
egcd3-both-nt-rr		⚠	⚠	✓
egcd-both-nt-ii		⚠	⚠	!(TP)
egcd-both-nt-ir		⚠	⚠	✓
egcd-both-nt-ri		⚠	⚠	✓
egcd-both-nt-rr		⚠	⚠	✓
freire1-both-nt-i		⚠		!(TP)
freire1-both-nt-r		⚠		!(TP)
geo1-both-nt-ii		⚠		!(TP)
geo1-both-nt-ir		⚠		!(TP)
geo1-both-nt-ri		⚠		!(TP)
geo1-both-nt-rr		⚠		!(TP)
geo2-both-nt-ii		⚠		!(TP)
geo2-both-nt-ir		⚠		!(TP)
geo2-both-nt-ri		⚠		!(TP)
geo2-both-nt-rr		⚠		!(TP)
geo3-both-nt-ii		⚠		!(TP)
geo3-both-nt-ir		⚠		!(TP)
geo3-both-nt-ri		⚠		!(TP)
geo3-both-nt-rr		⚠		!(TP)
hard2-both-nt-i		⚠		!(TP)
hard2-both-nt-r		⚠		!(TP)
hard-both-nt-ii		⚠	⚠	!(TP)
hard-both-nt-ir		⚠	⚠	✓
hard-both-nt-ri		⚠	⚠	!(TP)
hard-both-nt-rr		⚠	⚠	✓
prod4br-both-nt-ii				!(TP)
prod4br-both-nt-ir				✓
prod4br-both-nt-ri				!(TP)
prod4br-both-nt-rr				✓
prodbin-both-nt-ii				!(TP)
prodbin-both-nt-ir				✓
prodbin-both-nt-ri				!(TP)
prodbin-both-nt-rr				✓
ps2-both-nt-i		⚠		!(TP)
ps2-both-nt-r		⚠		!(TP)
ps3-both-nt-i				!(TP)
ps3-both-nt-r				!(TP)
ps4-both-nt-i		⚠		!(TP)
ps4-both-nt-r		⚠		!(TP)
ps5-both-nt-i				!(TP)
ps5-both-nt-r				!(TP)
ps6-both-nt-i				!(TP)
ps6-both-nt-r				!(TP)
sqrt1-both-nt-i		⚠		!(TP)
sqrt1-both-nt-r		⚠		!(TP)
sqrt2-both-nt-ii				!(TP)
sqrt2-both-nt-ir				!(FP)
sqrt2-both-nt-ri				!(FP)
sqrt2-both-nt-rr				!(FP)

Data Availability. The instrumented programs used for the experimental evaluation are provided (alongside with the tool) in our artifact on Zenodo [26].

7 Related Work

Several approaches have been proposed in the literature to prove program termination [2–4,6,14,19,23,24, etc.], or the existence of diverging executions (potential non-termination) in programs [5,17,19–22,29,37, etc.]. However, to the best of our knowledge, none of the other works studies termination or non-termination in the presence of untrusted inputs and trusted variables. We believe an interesting avenue for future research is to extend these approaches, with the aim of developing new methods to prove Termination Resilience or the existence of Robust Non-Termination bugs.

This work is inspired by the work of Girol et al. [15,16], where they introduce the notion of Robust Reachability of a *safety* bug. They additionally propose symbolic execution and bounded model checking techniques to find robustly reachable bugs. Instead, we introduce the notion of Robust Non-Termination, the robust occurrence of a *liveness* bug, and its negation Termination Resilience. We propose a static analysis approach to verify Termination Resilience and detect potential Robust Non-Termination bugs.

In a similar spirit as Girol et al., Parolini and Miné have introduced the notion of *safety* Non-Exploitability [27,28] and proposed a static analysis, combining taint and reachable values information, to prove that the untrusted inputs (called an attacker in their work) cannot trigger or silence a safety bug in a program. In our work, we focus on *liveness* bugs (notably, non-termination bugs) and we are only concerned with the untrusted inputs triggering these bugs, not silencing them. We also observe that the analysis of Parolini and Miné assumes full knowledge of trusted variables values at the time untrusted inputs values are chosen. In contrast, our semantic and static analysis frameworks naturally accommodate models with incomplete knowledge, including blind models in which the values of the untrusted inputs are chosen before the values of the trusted variables are observed (as in the programs in Fig. 3).

Our work builds on the decision tree abstract domain [33,34] and inherits its imprecision (cf. Section 6). It also means that we could benefit from the extensions of the domain. Urban et al. [36] proposed a static analysis of CTL properties using the decision trees abstract domain augmented with the new abstract operators. Later, Moussaoui-Remil et al. [25], proposed an analysis operating on decision trees to infer minimal sets of variables that need to be constrained in order to ensure a CTL property. We could leverage this work to infer minimal sets of untrusted variables that need to be strained to ensure termination resilience.

Termination Resilience can be formulated within the Alternating-Time Temporal Logic (ATL) [1] whose formulas are the same as those of CTL (in which termination is expressible), except that the universal (A) and existential (E) operators are generalized with a selective quantifier $\langle\langle . \rangle\rangle$ over paths that are possible outcomes of games. Given n players $p_1, \ldots, p_n$, a program satisfies $\langle\langle p_1, \ldots, p_n \rangle\rangle \phi$ if and only if there exists a strategy for the n players to force the program to satisfy ϕ. Termination Resilience is expressed as the existence of a strategy for the trusted variables that ensures program termination. Hence, an abstract interpretation of ATL properties seems to be a natural generalization of our work toward an analysis of resilience CTL properties.

8   Conclusion and Future Work

We have proposed a novel program property, called Termination Resilience, and an abstract interpretation-based static analysis to infer sufficient preconditions ensuring it. To this end, we have enhanced the decision tree abstract domain of piecewise-defined functions [33,34] with non-trivial transformer operators. In our evaluation, we have shown that our static analysis is able to reduce by almost 65% the alarms related to program non-termination.

For future work, an interesting direction is a static analysis for ATL properties with an initial focus on the resilience of CTL properties. We also plan to extend the analysis to data structure-manipulating programs.

References

1. Alur, R., Henzinger, T.A., Kupferman, O.: Alternating-time temporal logic. Journal of the ACM (JACM), 49(5):672–713 (2002)
2. Beyer, D., Dangl, M., Dietsch, D., Heizmann, M.: Correctness witnesses: exchanging verification results between verifiers. In: Zimmermann, T., Cleland-Huang, J., Su, Z., (eds) Proceedings of the 24th ACM SIGSOFT International Symposium on Foundations of Software Engineering, FSE 2016, Seattle, WA, USA, November 13-18, 2016, pp. 326–337. ACM (2016)
3. Beyer, D., Dangl, M., Dietsch, D., Heizmann, M., Stahlbauer, A.: Witness validation and stepwise testification across software verifiers. In: Nitto, E.D., Harman, M., Heymans, P., (eds) Proceedings of the 2015 10th Joint Meeting on Foundations of Software Engineering, ESEC/FSE 2015, Bergamo, Italy, August 30 - September 4, 2015, pp. 721–733. ACM (2015)
4. Brain, M., Joshi, S., Kroening, D., Schrammel, P.: Safety verification and refutation by k-invariants and k-induction. In: Blazy, S., Jensen, T.P., (eds) Static Analysis - 22nd International Symposium, SAS 2015, Saint-Malo, France, September 9-11, 2015, Proceedings, volume 9291 of Lecture Notes in Computer Science, pp. 145–161. Springer (2015)
5. Chatterjee, K., Goharshady, E.K., Novotný, P., Zikelic, D.: Proving non-termination by program reversal. In: Freund, S,N., Yahav, E., (eds) PLDI '21: 42nd ACM SIGPLAN International Conference on Programming Language Design and Implementation, Virtual Event, Canada, June 20-25, pp. 1033–1048. ACM (2021)
6. Chen, H., David, C., Kroening, D., Schrammel, P., Wachter, B.: Bit-precise procedure-modular termination analysis. ACM Trans. Program. Lang. Syst., 40(1):1:1–1:38 (2018)
7. Chen, Y., et al.: Advanced automata-based algorithms for program termination checking. In: Proceedings of the 39th ACM SIGPLAN Conference on Programming Language Design and Implementation, pp. 135–150 (2018)
8. Courant, N., Urban, C.: Precise widening operators for proving termination by abstract interpretation. In: Legay, A., Margaria, T., (eds) Tools and Algorithms for the Construction and Analysis of Systems - 23rd International Conference, TACAS 2017, Held as Part of the European Joint Conferences on Theory and Practice of Software, ETAPS 2017, Uppsala, Sweden, April 22-29, 2017, Proceedings, Part I, volume 10205 of Lecture Notes in Computer Science, pp. 136–152 (2017)
9. Cousot, P.: Constructive design of a hierarchy of semantics of a transition system by abstract interpretation. Theoret. Comput. Sci. **277**(1–2), 47–103 (2002)
10. Cousot, P., Cousot, R.: Abstract interpretation: a unified lattice model for static analysis of programs by construction or approximation of fixpoints. In: Graham, R.M., Harrison, M,A., Sethi, R., (eds) Conference Record of the Fourth ACM Symposium on Principles of Programming Languages, Los Angeles, California, USA, January 1977, pp. 238–252. ACM (1977)

11. Cousot, P., Cousot, R.: Abstract interpretation frameworks. J. Log. Comput. **2**(4), 511–547 (1992)

12. Cousot, P., Cousot, R.: An abstract interpretation framework for termination. In: Field, J., Hicks, M., (eds) Proceedings of the 39th ACM SIGPLAN-SIGACT Symposium on Principles of Programming Languages, POPL 2012, Philadelphia, Pennsylvania, USA, January 22-28, 2012, pp. 245–258. ACM (2012)

13. Floyd, R.W.: Assigning meanings to programs. Proc. Symp. Appl. Math. **19**, 19–32 (1967)

14. Giacobbe, M., Kroening, D., Parsert, J.: Neural termination analysis. In: Roychoudhury, A., Cadar, C., Kim, M., (eds) Proceedings of the 30th ACM Joint European Software Engineering Conference and Symposium on the Foundations of Software Engineering, ESEC/FSE 2022, Singapore, Singapore, November 14-18, 2022, pp. 633–645. ACM (2022)

15. Girol, G., Farinier, B., Bardin, S.: Not all bugs are created equal, but robust reachability can tell the difference. In: Silva, A., Rustan, K., Leino, M., (eds) Computer Aided Verification - 33rd International Conference, CAV 2021, Virtual Event, July 20-23, 2021, Proceedings, Part I, volume 12759 of Lecture Notes in Computer Science, pp. 669–693. Springer (2021)

16. Girol, G., Farinier, B., Bardin, S.: Introducing robust reachability. Formal Meth. Syst. Des. **63**(1), 206–234 (2024)

17. Hensel, J., Mensendiek, C., Giesl, J.: Aprove: non-termination witnesses for C programs - (competition contribution). In: Fisman, D., Rosu, G., (eds) Tools and Algorithms for the Construction and Analysis of Systems - 28th International Conference, TACAS 2022, Held as Part of the European Joint Conferences on Theory and Practice of Software, ETAPS 2022, Munich, Germany, April 2-7, 2022, Proceedings, Part II, volume 13244 of Lecture Notes in Computer Science, pp. 403–407. Springer (2022)

18. Jeannet, B., Miné, A.: Apron: a library of numerical abstract domains for static analysis. In: Bouajjani, A., Maler, O., (eds) Computer Aided Verification, 21st International Conference, CAV 2009, Grenoble, France, June 26 - July 2, 2009. Proceedings, volume 5643 of Lecture Notes in Computer Science, pp. 661–667. Springer (2009)

19. Kobayashi, N., Tanahashi, K., Sato, R., Tsukada, T.: HFL(Z) validity checking for automated program verification. Proc. ACM Program. Lang., 7(POPL):154–184 (2023)

20. Larraz, D., Nimkar, K., Oliveras, A., Rodríguez-Carbonell, E., Rubio, A.: Proving non-termination using max-smt. In: Biere, A., Bloem, R., (eds) Computer Aided Verification - 26th International Conference, CAV 2014, Held as Part of the Vienna Summer of Logic, VSL 2014, Vienna, Austria, July 18-22, 2014. Proceedings, volume 8559 of Lecture Notes in Computer Science, pp. 779–796. Springer (2014)

21. Le, T.C., Antonopoulos,T., Fathololumi, P., Koskinen, E., Nguyen, T.: Dynamite: dynamic termination and non-termination proofs. Proc. ACM Program. Lang., 4(OOPSLA):189:1–189:30 (2020)

22. Le, T.C., Qin, S., Chin, W.: Termination and non-termination specification inference. In: Grove, D., Blackburn, S M., (eds) Proceedings of the 36th ACM SIGPLAN Conference on Programming Language Design and Implementation, Portland, OR, USA, June 15-17, 2015, pp. 489–498. ACM (2015)

23. Malík, V., Necas, F., Schrammel, P., Vojnar, T.: 2ls: arrays and loop unwinding - (competition contribution). In: Sankaranarayanan, S., Sharygina, N., (eds) Tools and Algorithms for the Construction and Analysis of Systems - 29th International

Conference, TACAS 2023, Held as Part of the European Joint Conferences on Theory and Practice of Software, ETAPS 2022, Paris, France, April 22-27, 2023, Proceedings, Part II, volume 13994 of Lecture Notes in Computer Science, pp. 529–534. Springer (2023)

24. Metta, R., Karmarkar, H., Madhukar, K., Venkatesh, R., Chakraborty, S.: PRO-TON: probes for termination or not (competition contribution). In: Finkbeiner, B., Kovács, L., (eds) Tools and Algorithms for the Construction and Analysis of Systems - 30th International Conference, TACAS 2024, Held as Part of the European Joint Conferences on Theory and Practice of Software, ETAPS 2024, Luxembourg City, Luxembourg, April 6-11, 2024, Proceedings, Part III, volume 14572 of Lecture Notes in Computer Science, pp. 393–398. Springer (2024)

25. Remil, N. M., Urban, C., Miné, A.: Automatic detection of vulnerable variables for CTL properties of programs. In: Bjørner, N.S., Heule, M., Voronkov, A., (eds) LPAR 2024: Proceedings of 25th Conference on Logic for Programming, Artificial Intelligence and Reasoning, Port Louis, Mauritius, May 26-31, 2024, volume 100 of EPiC Series in Computing, pp. 116–126. EasyChair (2024)

26. Remil,N.M., Urban, C.: Termination resilience static analysis (artifact) (2025). https://zenodo.org/records/17176433

27. Parolini, F.: Static analysis for security properties of software by abstract interpretation. (Analyse statique des propriétés de sécurité des logiciels par interprétation abstraite) PhD thesis, Sorbonne University, Paris, France (2024)

28. Parolini, F., Miné, A.: Sound abstract nonexploitability analysis. In: Dimitrova, R., Lahav, O., Wolff, S., (eds) Verification, Model Checking, and Abstract Interpretation - 25th International Conference, VMCAI 2024, London, United Kingdom, January 15-16, 2024, Proceedings, Part II, volume 14500 of Lecture Notes in Computer Science, pp. 314–337. Springer (2024)

29. Raad, A., Vanegue, J., O'Hearn, P.W.: Non-termination proving at scale. Proc. ACM Program. Lang. **8**(OOPSLA2), 246–274 (2024)

30. Shi, X., et al.: Large-scale analysis of non-termination bugs in real-world OSS projects. In: Roychoudhury, A., Cadar, C., Kim, M., eds, Proceedings of the 30th ACM Joint European Software Engineering Conference and Symposium on the Foundations of Software Engineering, ESEC/FSE 2022, Singapore, Singapore, November 14-18, 2022, pp. 256–268. ACM (2022)

31. Turing, A.: Checking a large routine. In: Report of a Conference on High Speed Automatic Calculating Machines, pp. 67–69 (1949)

32. Urban, C.: FuncTion: an abstract domain functor for termination - (competition contribution). In: Baier, C., Tinelli, C., (eds) Tools and Algorithms for the Construction and Analysis of Systems - 21st International Conference, TACAS 2015, Held as Part of the European Joint Conferences on Theory and Practice of Software, ETAPS 2015, London, UK, April 11-18, 2015. Proceedings, volume 9035 of Lecture Notes in Computer Science, pp. 464–466. Springer (2015)

33. Urban, C.: Static Analysis by Abstract Interpretation of Functional Temporal Properties of Programs (Analyse Statique par Interprétation Abstraite de Propriétés Temporelles Fonctionnelles des Programmes). PhD thesis, École Normale Supérieure, Paris, France (2015)

34. Urban, C., Miné, A.: A decision tree abstract domain for proving conditional termination. In: Müller-Olm, M., Seidl, H., (eds)Static Analysis - 21st International Symposium, SAS 2014, Munich, Germany, September 11-13, 2014. Proceedings, volume 8723 of Lecture Notes In Computer Science, pp. 302–318. Springer (2014)

35. Urban, C., Miné, A.: An abstract domain to infer ordinal-valued ranking functions. In: Shao, Z., (eds) Programming Languages and Systems - 23rd European Symposium on Programming, ESOP 2014, Held as Part of the European Joint Conferences on Theory and Practice of Software, ETAPS 2014, Grenoble, France, April 5-13, 2014, Proceedings, volume 8410 of Lecture Notes in Computer Science, pp. 412–431. Springer (2014)
36. Urban, C., Ueltschi, S., Müller, P.: Abstract interpretation of CTL properties. In: Podelski, A., (eds) Static Analysis - 25th International Symposium, SAS 2018, Freiburg, Germany, August 29-31, 2018, Proceedings, volume 11002 of Lecture Notes in Computer Science, pp. 402–422. Springer (2018)
37. Zhang, Y., et al.: Endwatch: a practical method for detecting non-termination in real-world software. In: 2023 38th IEEE/ACM International Conference on Automated Software Engineering (ASE), pp. 686–697. IEEE (2023)

Input-Based Three-Valued Abstraction Refinement

Jan Onderka[1,2,3]($\boxtimes$) and Stefan Ratschan[2]

[1] Faculty of Engineering, University of Freiburg,
Freiburg, Germany
`onderka@cs.uni-freiburg.de`

[2] Institute of Computer Science, The Czech Academy
of Sciences, Prague, Czech Republic
`stefan.ratschan@cs.cas.cz`

[3] Faculty of Information Technology, Czech Technical University in Prague,
Prague, Czech Republic

Abstract. Unlike Counterexample-Guided Abstraction Refinement (CEGAR), Three-Valued Abstraction Refinement (TVAR) is able to verify all properties of the µ-calculus. We present a novel algorithmic framework for TVAR that employs a simulator-like approach to build and refine the abstract state space with input-based splitting. This leads to a state space formalism that is much simpler than in previous TVAR frameworks, which use modal transitions. We implemented the framework in our open-source tool **machine-check** and verified properties of machine-code systems for the AVR architecture, showing the ability to verify systems and µ-calculus properties not verifiable by naïve model checking or CEGAR, respectively. This is the first practical use of TVAR for machine-code verification.

Keywords: Model checking · Abstraction · Partial Kripke Structure · µ-calculus

1 Introduction

Abstraction-refinement methodologies are ubiquitous in formal verification, the foremost being Counterexample-guided Abstraction Refinement (CEGAR) [8,10]. Unfortunately, CEGAR does not support the whole propositional µ-calculus, and indeed not even Computation Tree Logic (CTL). This leaves a large class of potentially crucial non-linear-time properties unverifiable. Three-valued Abstraction Refinement (TVAR) is able to verify full µ-calculus. However, previous TVAR frameworks refined abstract states, necessitating state space formalisms based on modal transitions. Frameworks based on simple formalisms [19,47] are not monotone: previously provable properties may no longer be provable after refinement. Intricate monotone formalisms [28,48,52] were devised, their specialised semantics complicating the use of standard model checking algorithms.

Y.-F. Chen et al. (Eds.): VMCAI 2026, LNCS 16417, pp. 237–262, 2026.
https://doi.org/10.1007/978-3-032-15700-3_12

Another problem common to model-checking with abstraction is that the abstract state space cannot be built using a system simulator, which is possible for naïve explicit-state model checking [45]. This is a problem especially with complex systems such as processors executing machine code.

Contribution. Combining the ideas of TVAR and system simulators, we present a novel TVAR framework based on simulator-like generation of the abstract state space and performing refinements by splitting abstract inputs. Our framework does not use modal transitions, leading to simpler and more intuitive reasoning compared to the previous frameworks. Furthermore, it allows simple building and refinement of the abstract state space based on abstract simulators. We prove that the introduced framework is sound, monotone, and complete for μ-calculus properties and existential abstraction domains, provided simple requirements are met. Arbitrary digital systems formalised as automata can be verified, and the choices of abstraction domains and refinement strategy can be tailored to the specific use-case. Unlike the previous abstract-simulator approaches, our framework can be used for the full μ-calculus.

We implemented an instance of our framework in our formal verification tool **machine-check**[1], where the systems are described as simulable finite-state machines in a subset of the Rust language, translated to abstract and refinement analogues used for generating and refining the state space [38]. The ability to refine removes the need for hints such as where to use abstraction [25].

We designed **machine-check** especially for machine-code verification, and were able to verify non-toy programs for the AVR architecture and find a bug in a simplified version of a real-life program using a non-linear-time property not verifiable by CEGAR. To our knowledge, this is the first use of TVAR and verification of μ-calculus properties for machine-code systems.

2   Previous Work

In this section, we list previous relevant work on TVAR in roughly chronological order, with additional information available in summarising papers [15, 18]. After that, we discuss relevant work on abstract simulator approaches to model checking.

Example 1. Consider the finite-state machine in Fig. 1, representing e.g. a controller of aircraft landing gear retraction: if the most significant bit (msb) of the state is 0, the landing gear is extended; if 1, it is retracted. The system is required to follow a single-bit input from the gear lever, with some slack for responses. There is a critical bug, occurring if the aircraft loses power in flight when the landing gear is retracted and the controller restarts in the state 000 after regaining power: as the input is set to retraction (1), the controller proceeds

---

[1] Free and open-source, official website https://machine-check.org/. In this paper, we discuss version 0.6.1 published at https://crates.io/crates/machine-check/0.6.1.

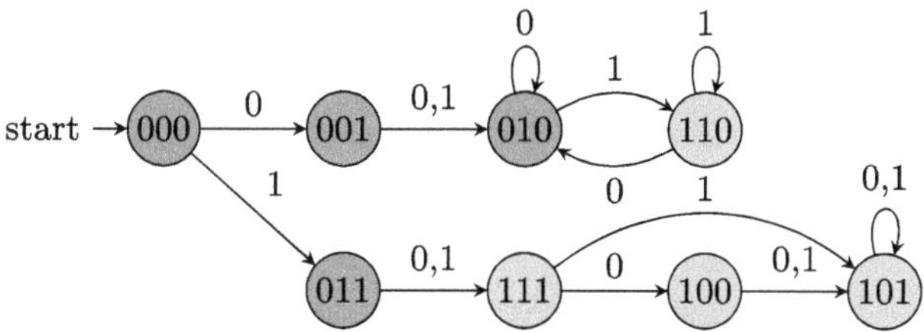

Fig. 1. Example system expressed as a finite-state machine. The states where $\neg msb$ holds are drawn green while the states where msb holds are drawn orange. (Color figure online)

through 011 to states where msb remains 1 forever, a **total loss of capability** to extend the gear again unless the controller is turned off and on again.

The bug is not just dangerous, but also sneaky, as it does not occur during a normal start with lever input 0. To protect ourselves against it, we can verify the property "from every reachable system state, it should be possible to reach a state where the landing gear is extended" holds in the system. This is formalised by a *recovery property* $\mathbf{AG}[\mathbf{EF}[\neg msb]]$ in CTL, not possible to check using CEGAR.

Since the system is buggy, the property should be disproved[2]. For clarity, we will instead reason about proving a dual property, $\mathbf{EF}[\mathbf{AG}[msb]]$, i.e. from the start state, there exists a path ($\mathbf{E}$) where from some state on the path ($\mathbf{F}$), all paths ($\mathbf{A}$) have every state ($\mathbf{G}$) fulfilling msb. We will prove this on Fig. 1 bottom-up as usual for CTL. Clearly, msb holds in the state 101. Since 101 just loops on itself, $\mathbf{AG}[msb]$ holds in it. As 101 is reachable from 000 by the path (000, 011, 111, 100, 101), we conclude $\mathbf{EF}[\mathbf{AG}[msb]]$ holds, finding the bug. While such reasoning is easy for simple systems, in real life, the controller may have billions of possible states, requiring us to abstract some information away[3].

Partial Kripke Structures (PKS). In verification on *partial state spaces* [1], some information is disregarded to produce a smaller state space. PKS enrich standard Kripke structures (KS) by allowing unknown state labellings.

Definition 1. *A partial Kripke structure (PKS) is a tuple* (S, S_0, R, L) *over a set of atomic propositions* $\mathbb{A}$ *with the elements*

- S *(the* set *of states),*
- $S_0 \subseteq S$ *(the* set *of initial states),*
- $R \subseteq S \times S$ *(the* transition relation*),*
- $L : S \times \mathbb{A} \to \{0, 1, \bot\}$ *indicating for each atomic proposition whether it holds, does not hold, or its truth value is unknown (the* labelling function*).*

A Kripke Structure (KS) is a PKS with L *restricted to* $S \times \mathbb{A} \to \{0, 1\}$.

---

[2] In our terminology, *proving* the property determines it holds in the system. *Disproving* it determines it does not. *Verification* aims to either prove or disprove it.

[3] The example is directly inspired by a bug we found, discussed in Sect. 5.

Example 2. While Fig. 1 shows a finite-state machine, it can be converted to a Kripke structure by discarding the inputs, with $S = \{000, 001, \ldots, 111\}$, $S_0 = \{000\}$, and R given by the transitions in Fig. 1. L labels *msb* in states $\{000, 001, 010, 011\}$ as 0, and in $\{100, 101, 110, 111\}$ as 1.

For proving **EF**[**AG**[*msb*]], such a KS is unnecessarily detailed. Using PKS, we could e.g. combine 010 and 110 into a single *abstract* state where it is unknown what the value of the most significant bit is, and the labelling of *msb* is $\bot$.

Existential abstraction. In TVAR, *existential abstraction* [9] is used, where the abstract states in set $\hat{S}$ are related to the original concrete states in S by a function $\gamma : \hat{S} \to 2^S$, the abstract state $\hat{s} \in \hat{S}$ representing some (not fixed) concrete state in $s \in \gamma(\hat{s})$ in each system execution instant. This is a generalisation of domains usable for Abstract Interpretation [11, 12], also allowing non-lattice domains such as wrap-around intervals [16].

Example 3. In the examples, we will use the three-valued bit-vector domain, where each element is a tuple of three-valued bits, each with value '0' (definitely 0), '1' (definitely 1), or 'X' (possibly 0, possibly 1). Except for figures, we write three-valued bit-vectors in quotes, e.g. $\gamma(\text{``0X1''}) = \{001, 011\}$. The bits can also refer to a predicate rather than a specific value. For example, '1' could mean that $v > 5$ holds, '0' that its negation holds, and 'X' that we do not know.

2.1   Previous TVAR Frameworks

Building on the work of Bruns & Godefroid [1–3], Godefroid et al. [19] introduced TVAR by refining the abstract state set, using a state space formalism based on modal transitions. Early TVAR approaches [19, 20, 22, 23, 47] were based on Kripke Modal Transition Structures (KMTS) and did not guarantee previously provable properties stay provable after refinement, i.e. were not monotone.

Definition 2. *A Kripke Modal Transition Structure (KMTS) is a five-tuple* $(S, S_0, R^{may}, R^{must}, L)$ *where* S, S_0, *and* L *follow Definition 1, and*

- $R^{may} \subseteq S \times S$ *is the set of transitions which may be present,*
- $R^{must} \subseteq R^{may}$ *is the set of transitions which are definitely present.*

Intuitively, KMTS allow for transitions with unknown presence ($R^{may} \setminus R^{must}$). PKS can be trivially converted to KMTS by setting $R^{may} = R^{must} = R$. While it is possible to convert a KMTS to an equally expressive PKS by moving the transition presence into the states [21], this requires the set of states to be modified.

Monotone Frameworks. Godefroid et al. recognised non-monotonicity as a problem and suggested keeping previous states when refining [19, p. 3–4]. However, Shoham & Grumberg showed the approach was not sufficient, since in certain cases, it prevents verification of additional properties after refinement.

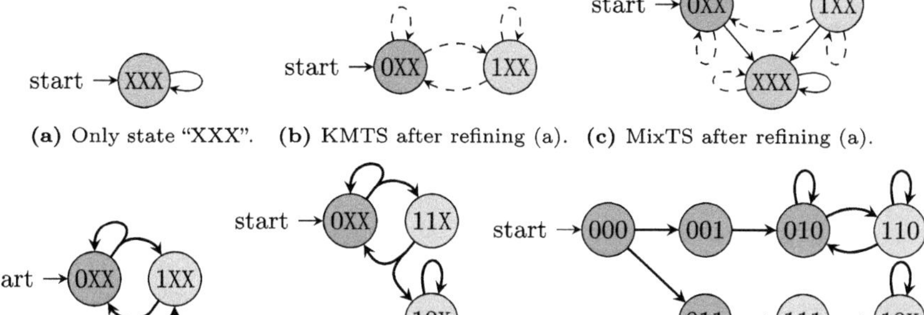

(a) Only state "XXX".   **(b)** KMTS after refining (a).   **(c)** MixTS after refining (a).

(d) GKMTS after refin-**(e)** GKMTS after refining **(f)** GKMTS after refining "0XX" in (e) by
ing (a).   "1XX" in (d).   splitting to "000", "001", "010", and "011".

Fig. 2. State-based refinement with hyper-transitions based on Generalised KMTS. Implied may-transitions, present in all sub-figures except for (c), are not drawn. The states where it is unknown whether msb or $\neg msb$ holds are drawn grey.

As a remedy, they introduced another monotone TVAR framework using Generalized KMTS for CTL [48], later extended to μ-calculus [49]. Gurfinkel & Chechik introduced a framework for verification of CTL properties on Boolean programs using Mixed Transition Systems [28], later extended to lattice-based domains [30][4]. Wei et al. introduced a TVAR framework using Reduced Inductive Semantics for μ-calculus under which the results of model-checking GKMTS and MixTS are equivalent [52].

Definition 3. *A Generalized KMTS (GKMTS) is a tuple* $(S, S_0, R^{may}, R^{must}, L)$ *where* S, S_0, R^{may}, *and* L *follow Definition 2 and* $R^{must} : S \times 2^S$ *is the set of* hyper-transitions, *where* $\forall (a, B) \in R^{must} . \forall b \in B . (a, b) \in R^{may}$.

Definition 4. *A Mixed Transition System (MixTS) is a tuple* $(S, S_0, R^{may}, R^{must}, L)$ *where* S, S_0, R^{may}, *and* L *follow Definition 2 and* $R^{must} \subseteq S \times S$.

Example 4. We will prove **EF[AG[**msb**]]** using the previous state-based TVAR frameworks over the system from Fig. 1, starting with abstract state set {"XXX"}. Clearly, we both **may** and **must** transition from "XXX" to "XXX", visualised in Fig. 2a. Since msb is unknown in "XXX", the model-checking result is unknown and we refine.

Suppose we decide to refine by splitting the abstract state set to {"0XX", "1XX"}, starting in "0XX". From "0XX", we **may** transition either to "0XX" (e.g. by 000 → 001 or 010 → 010) or "1XX" (e.g. by 010 → 110), but cannot

---

[4] An instance of the framework is implemented in the tool Yasm [29], available online at the time of writing [26]. However, it does not support common programming language elements such as bitwise-operation statements (for example, $y - x$ & 1) nor full μ-calculus, which our tool can handle without problems.

conclude that e.g. a transition from "0XX" to itself **must** exist: $011 \in \gamma(\text{"0XX"})$ only transitions to $111 \notin \gamma(\text{"0XX"})$.

PKS cannot be used as they cannot describe unknown-presence transitions. KMTS allow this, producing Fig. 2b. However, it is not possible to prove e.g. **EX[true]**, which was possible in Fig. 2a, i.e. the refinement is not monotone. Using MixTS, we retain "XXX" and the must-transitions to it, producing a *forced choice* in Fig. 2c. Using GKMTS, we obtain Fig. 2d instead. In both, it is possible to prove **EX[true]**, but not **EF[AG[**msb**]]**. Refining further using GKMTS, we obtain Fig. 2e, where it is still not possible to prove **EF[AG[**msb**]]**: the hyper-transitions do not imply that the path ("0XX","11X","10X") corresponds to a concrete path. The property is only proven after additional refinement to Fig. 2f. While the GKMTS in Fig. 2f trivially corresponds to a KMTS or a PKS, fewer refinements and final states may be needed in general when using GKMTS or MixTS as they may guide the refinement better due to monotonicity.

Model Checking. μ-calculus properties can be model-checked on PKS and KMTS by a simple conversion to two KS, applying standard model-checking algorithms, and combining the results [2, 20]. Similar conversions are also possible for multi-valued logics [27, 31]. It is also possible to model-check directly without conversion. A multi-valued model-checker was previously used for the MixTS approach [28]. Game-based model-checking was used for full μ-calculus [22, 23].

Discussion of Previous TVAR Frameworks. It was recognised early on that using KMTS with non-monotone refinement is problematic [19, 47]. GKMTS seem more susceptible to exponential explosion as MixTS can make use of abstract domains. However, specialised algorithms must be used for MixTS to obtain GKMTS-equivalent results [52]. A drawback of all mentioned approaches is their conceptual complexity which, in our opinion, is the main reason for the dearth of available TVAR tools and test sets, compared to CEGAR. This has also made the analysis of these methods difficult, as illustrated by the subtle differences in definitions of expressiveness identified by Gazda & Willemse [17].

2.2   Abstract Simulator Approaches

Naïve explicit-state model checking generates a next state from every state and input combination, with the ability to use a system simulator to perform each step. This simulator can be written in an imperative language such as C. Unfortunately, for machine-code systems where state sizes are in kilobytes even for simple microcontrollers and a single port read can produce e.g. 2^8 next states, this results in exponential explosion for all but the simplest toy programs. We will now discuss approaches where the state space is abstract, but still built by generating the next states using an *abstract simulator*.

Trajectory Evaluation. Bryant used three-valued logic simulators for formal verification of hardware circuits [4, 6]. He showed linear-time properties (expressed by specification machines or circuit assertions) can be proven using a set of three-valued input sequences that together cover all concrete inputs [4, p. 320] by generating permissible state sequences (*trajectories*). *Symbolic trajectory evaluation (STE)* is an extension that allows parametrisation of the introduced *trajectory formulas* [5]. However, the STE formalism drops the distinction between inputs and states. We refer to Melham [32] for a discussion of STE and extensions. Notably, Generalized STE [53] allows verification of properties corresponding to linear-time μ-calculus [14]. While manual refinement was originally needed, automatic refinement was proposed for both STE [51] and GSTE [7].

Delayed Nondeterminism. Noll & Schlich [36] verified machine-code programs by model-checking an abstract state space generated by a simulation-based approach. Each input bit was read as 'X' and split to '0' and '1' only when it was decided to in a subsequent step (e.g. if it was an argument of a branch instruction). This allowed e.g. splitting only one bit of a read 8-bit port if the other bits were masked out by a constant first, allowing soundly proving (but not disproving) properties in path-universal logics such as LTL and ACTL.

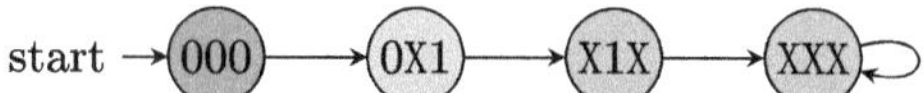

(**a**) Abstract state space corresponding to the system in Figure 1, computed by simulation with unconstrained inputs ('X')

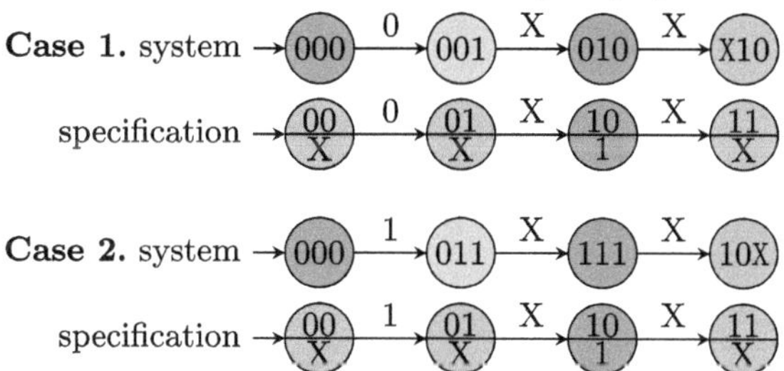

(**b**) Trajectory evaluation: The verification is split into cases, ensuring the abstract input sequences together cover all possible concrete input sequences. Unlike Bryant, we use initial states for consistency with other approaches.

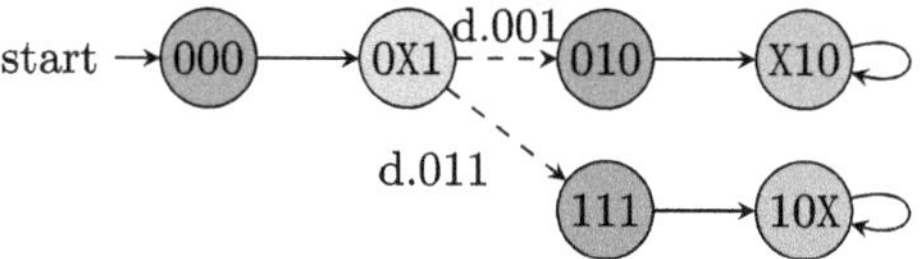

(**c**) Delayed nondeterminism augmented with must-transitions: the 'X' in state "0X1" is split to '0' and '1' before computing the successor

Fig. 3. Simulation-based approaches proving $\mathbf{A}[\mathbf{X}[\mathbf{X}[msb \Leftrightarrow lsb]]]$. (a) can be considered PKS or KMTS, and (c) KMTS. (b) contains four trajectories. Specially in this figure, system states are drawn green if $msb \Leftrightarrow lsb$ holds, orange if it does not, and grey if it is unknown. (Color figure online)

244   J. Onderka and S. Ratschan

Example 5. Due to the restrictions of the approaches, we will illustrate proving the property "in two steps from the initial state, the most significant bit corresponds to the least significant bit", i.e. $\mathbf{A}[\mathbf{X}[\mathbf{X}[msb \Leftrightarrow lsb]]]$. Simulating without splitting, we produce Fig. 3a, unable to prove the property.

To visualise Bryant's trajectory evaluation approach with explicitly considered inputs [4], we encode the specification as a finite-state machine with two bits containing an initially-zero saturating counter. The system output function is $msb \Leftrightarrow lsb$. The specification outputs '1' iff the counter is 10 and 'X' otherwise. To prove the property, we split verification into two cases based on the value of the first input, and obtain simulated trajectories of both machines in Fig. 3b. The property is proven as the trajectories are long enough (at least 3 for the given property) and the specification output always covers the system output.

To better understand Delayed Nondeterminism, we augment with must-transitions where possible. Splitting "0X1" from Fig. 3a, we obtain Fig. 3c, where "010" is obtained as a direct successor of "001", and "111" as a direct successor of "011". We cannot augment during the split as 'X' might not generally correspond to a unique input, potentially e.g. being copied before splitting.

3   Input-Based Three-Valued Abstraction Refinement

We propose a framework that eliminates the need for modal transitions in TVAR by combining ideas from the discussed approaches: using TVAR, build the abstract state space using an abstract simulator and split **inputs** instead of states.

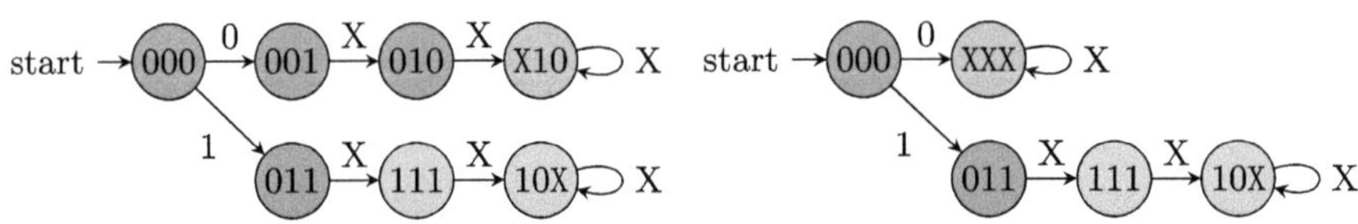

(a) Refining the input after "000" while keeping all others 'X'.

(b) A smaller reachable state space using a *decayed* step function.

Fig. 4. Input-based Three-Valued Abstraction Refinement.

Example 6. We return to the original problem of proving $\mathbf{EF}[\mathbf{AG}[msb]]$. The simulation-based approach initially builds the abstract state space as shown in Fig. 3a. After that, we decide (using e.g. a heuristic, machine learning or human guidance) that the input after "000" should be split. We regenerate the abstract state space as shown in Fig. 4a. We are immediately able to prove $\mathbf{EF}[\mathbf{AG}[msb]]$ holds, meaning the system from Fig. 1 contains a bug.

The part of the abstract state space in Fig. 4a starting with "001" is unnecessarily large for proving the property, potentially causing exponential explosion problems. To prevent them, we also introduce a way to soundly and precisely regulate the outgoing states of transitions, allowing us to e.g. replace "001" by "XXX" as in Fig. 4b when generating the abstract state space, *decaying* to less information. Only one refinement was necessary compared to multiple in Example 4, with the final state space in Fig. 4b smaller than in Fig. 2f. However, this depends[5] on the correct choice to decay "001" and not "011".

The generated state spaces are PKS, which allows us to use previous work on PKS, KMTS, GKMTS, and MixTS, as PKS are trivially convertible to all. This notably includes model-checking using standard formalisms [2,20] and TVAR refinement guidance [22,23,47], with the caveat that we need to select an input instead of a state to refine. Unlike (G)STE [5,53] and Delayed Nondeterminism [36] which were limited to linear-time or path-universal properties, our approach can be used for the full μ-calculus. Verification can be fully automatic or manually guided, and it is also possible to precisely control the number of reachable abstract states and transitions: we can split inputs up to one by one, and decay any newly reachable states before refining the applied decay.

We will now give the framework formalism and simple requirements for its instances to be sound, monotone, and complete. In Sect. 4, we will then discuss how reasonable choices of refinements can be made. Finally, in Sect. 5, we will evaluate an implementation of an instance of our framework in **machine-check**.

3.1   Framework Formalism

We assume that the original Kripke Structure has only one initial state[6], i.e. $K = (S, \{s_0\}, R, L)$. We write the result of model-checking a property ϕ against K as $[\![\phi]\!](K)$, which returns 0 or 1. For a PKS $\hat{K}$, $[\![\phi]\!](\hat{K})$ returns 0, 1, or $\bot$.

We consider the original (concrete) system to be an automaton and will also use automata for abstracting the system, introducing the formalism of *generating automata* that can generate partial Kripke structures.

Definition 5. *A* generating automaton *(GA) is a tuple* $G = (S, s_0, I, q, f, L)$ *with the elements*

- S *(the* set of automaton states*),*
- $s_0 \in S$ *(the* initial state*),*
- I *(the* set of all step inputs*),*
- $q : S \rightarrow 2^I \setminus \{\emptyset\}$ *(the* input qualification function*),*
- $f : S \times I \rightarrow S$ *(the* step function*),*
- $L : S \times \mathbb{A} \rightarrow \{0, 1, \bot\}$ *(the* labelling function*).*

---

[5] As with other frameworks, verification performance depends drastically on abstraction, refinement, and implementation choices, further discussed in Sects. 4 and 5.

[6] This is merely a formal choice. For multiple initial states, a dummy initial state can be introduced before them and the verified property ϕ converted to **AX**[ϕ].

Definition 6. *For a generating automaton* (S, s_0, I, q, f, L), *we define the PKS-generating function* Γ *as*

$$\Gamma((S, s_0, I, q, f, L)) \stackrel{def}{=} (S, \{s_0\}, R, L) \tag{1}$$

$$\textit{where } R = \{(s, f(s, i)) \mid s \in S, i \in q(s)\}. \tag{2}$$

We call a generating automaton $G = (S, s_0, I, q, f, L)$ *concrete* if the labelling function $L : S \times \mathbb{A} \to \{0, 1\}$ (disallowing the value $\bot$) and for all $s \in S$, $q(s) = I$. A concrete GA corresponds to a Moore machine with the output of each state mapping each atomic proposition from $\mathbb{A}$ to either 0 or 1.

Algorithm 1. Input-based Three-Valued Abstraction Refinement Framework

---

Require: a concrete generating automaton (S, s_0, I, q, f, L), a μ-calculus property ϕ
Ensure: return $[\![\phi]\!]((S, s_0, I, q, f, L))$ ▷ If requirements are fulfilled, see Corollary 1

$(\hat{S}, \hat{s}_0, \hat{I}, \hat{q}, \hat{f}, \hat{L}) \leftarrow \text{ABSTRACT}(S, s_0, I, q, f, L)$
while $(r \leftarrow [\![\phi]\!](\Gamma((\hat{S}, \hat{s}_0, \hat{I}, \hat{q}, \hat{f}, \hat{L}))) = \bot$ **do**
    $(\hat{q}, \hat{f}) \leftarrow \text{REFINE}(\hat{S}, \hat{s}_0, \hat{I}, \hat{q}, \hat{f}, \hat{L})$
end while
return r

Algorithm 1 describes our framework. Given a concrete GA, it abstracts it to an *abstract* generating automaton $(\hat{S}, \hat{s}_0, \hat{I}, \hat{q}, \hat{f}, \hat{L})$, successively refining the input qualification function $\hat{q}$ and step function $\hat{f}$ until the result of model-checking is non-$\bot$. $\hat{S}$ and $\hat{I}$ are related to S and I by a state concretization function[7] $\gamma : \hat{S} \to 2^S \setminus \{\emptyset\}$ and an input concretization function $\zeta : \hat{I} \to 2^I \setminus \{\emptyset\}$.

Unlike state-based TVAR, the set of abstract states $\hat{S}$ does not change during refinement. The number of states to be considered is limited by the codomain of $\hat{f}$, allowing structures such as Binary Decision Diagrams to be used. The abstract state space can be built quickly by forward simulation. For backward simulation, care must be taken to pair the states according to inputs.

Example 7. In **machine-check**, states and inputs are composed of bit-vector and bit-vector-array variables, formally represented by flattened $S = \{0, 1\}^w$ and $I = \{0, 1\}^y$ for finite state width w and finite input width y. A dummy s_0 precedes the actual initial system states, f is a function written in an imperative programming language, and L computes relational operations on state variables.

For the abstract GA, we use three-valued bit-vector abstraction [36, 45] with fast abstract operations [40], abstracting as

$$\gamma^{\text{bit}}(\hat{a}) = \{v \in \mathbb{B} \mid (v = 0 \Rightarrow a \neq \text{`1'}) \wedge (v = 1 \Rightarrow a \neq \text{`0'})\}, \tag{3a}$$

$$\hat{S} = \{\text{`0'}, \text{`1'}, \text{`X'}\}^w, \gamma(\hat{s}) = \{s \in S \mid \forall k \in [0, w-1] . s_k \in \gamma^{\text{bit}}(\hat{s}_k)\}, \tag{3b}$$

$$\hat{I} = \{\text{`0'}, \text{`1'}, \text{`X'}\}^y, \zeta(\hat{i}) = \{i \in I \mid \forall k \in [0, y-1] . i_k \in \gamma^{\text{bit}}(\hat{i}_k)\}. \tag{3c}$$

---

[7] We forbid abstract elements with no concretizations as they do not represent any concrete element. Practically speaking, this does not disqualify abstract domains with such elements; we just require such elements are not produced by $\hat{s}_0$, $\hat{q}$, or $\hat{f}$.

Again, $\hat{s}_0$ is a dummy state with $\gamma(\hat{s}_0) = \{s_0\}$. We rewrite the step function f into an abstract function $\hat{f}^{\text{basic}} : \hat{S} \times \hat{I} \to \hat{S}$. To formalise the manipulation in Fig. 4, we use an *input precision function* $\hat{p}_{\hat{q}} : \hat{S} \to \{0,1\}^y$ and a *step precision function* $\hat{p}_{\hat{f}} : \hat{S} \to \{0,1\}^w$, defining the result of $\hat{q}$ and $\hat{f}$ in each bit k by

$$(\hat{p}_{\hat{q}}(\hat{s})_k = 0 \Rightarrow \hat{q}(\hat{s})_k = \{\text{`X'}\}) \wedge (\hat{p}_{\hat{q}}(\hat{s})_k = 1 \Rightarrow \hat{q}(\hat{s})_k = \{\text{`0'}, \text{`1'}\}), \tag{4a}$$

$$(\hat{p}_{\hat{f}}(\hat{s})_k = 0 \Rightarrow \hat{f}(\hat{s},\hat{i})_k = \text{`X'}) \wedge (\hat{p}_{\hat{f}}(\hat{s})_k = 1 \Rightarrow \hat{f}(\hat{s},\hat{i})_k = \hat{f}^{\text{basic}}(\hat{s},\hat{i})_k). \tag{4b}$$

This allows precise control of the size of the reachable abstract state space. For each $\hat{s} \in \hat{S}$, if $\hat{p}_{\hat{q}}(\hat{s}) = (0)^y$, there is exactly one outgoing transition. Each bit set to 1 increases that up to a factor of 2. If $\hat{p}_{\hat{f}}(\hat{s}) = (0)^w$, there is exactly one outgoing transition to the "most-decayed" state $(\text{`X'})^w$.

3.2   Soundness, Monotonicity, and Completeness

In this subsection, we state the requirements sufficient to ensure soundness (the algorithm returns the correct result if it terminates), monotonicity (refinements never lose any information), and completeness (the algorithm always terminates). For reasons of space, we only sketch the proofs in this version of the paper[8].

To intuitively describe the requirements, we formalise the concept of *coverage*. An abstract state $\hat{s}$ or input $\hat{i}$ *covers* a concrete $s \in S$ or $i \in I$ exactly when $s \in \gamma(\hat{s})$ or $i \in \zeta(\hat{i})$, respectively, and it *covers* another abstract state $\hat{s}^* \in \hat{S}$ or input $\hat{i}^* \in \hat{I}$ exactly when $\gamma(\hat{s}^*) \subseteq \gamma(\hat{s})$ or $\zeta(\hat{i}^*) \subseteq \zeta(\hat{i})$, respectively.

We want abstraction to preserve the truth value of μ-calculus properties in the following sense:

Definition 7. *A partial Kripke structure $K^{\uparrow}$ is sound with respect to a partial Kripke structure $K^{\downarrow}$ if, for every property ϕ of μ-calculus over the set of atomic propositions $\mathbb{A}$, it holds that*

$$[\![\phi]\!](K^{\uparrow}) \neq \perp \Rightarrow [\![\phi]\!](K^{\downarrow}) = [\![\phi]\!](K^{\uparrow}). \tag{5}$$

Intuitively, $K^{\uparrow}$ can contain less information than $K^{\downarrow}$, turning some non-$\perp$ proposition results to $\perp$. No other differences are possible.

To ensure the soundness of Algorithm 1, we use the following requirements. Soundness is ensured with any refinement heuristic as long as they are met.

Definition 8. *A GA $\hat{G} = (\hat{S}, \hat{s}_0, \hat{I}, \hat{q}, \hat{f}, \hat{L})$ is a soundness-guaranteeing (γ,ζ)-abstraction of a concrete GA $G = (S, s_0, I, q, f, L)$ iff*

$$\gamma(\hat{s}_0) = \{s_0\}, \tag{6a}$$

$$\forall \hat{s} \in \hat{S} \,.\, \forall s \in \gamma(\hat{s}) \,.\, \forall a \in \mathcal{A} \,.\, (\hat{L}(\hat{s},a) \neq \perp \Rightarrow \hat{L}(\hat{s},a) = L(s,a)), \tag{6b}$$

$$\forall (\hat{s},i) \in \hat{S} \times I \,.\, \exists \hat{i} \in \hat{q}(\hat{s}) \,.\, i \in \zeta(\hat{i}), \tag{6c}$$

$$\forall (\hat{s},\hat{i}) \in \hat{S} \times \hat{I} \,.\, \forall (s,i) \in \gamma(\hat{s}) \times \zeta(\hat{i}) \,.\, f(s,i) \in \gamma(\hat{f}(\hat{s},\hat{i})). \tag{6d}$$

---

[8] The full proofs are given in Appendix A in a version of this paper available at https://arxiv.org/abs/2408.12668.

Informally, the four requirements express the following:

(a) **Initial state concretization.** The abstract initial state has exactly the concrete initial state in its concretization.
(b) **Labelling soundness.** Each abstract state labelling must either correspond to the labelling of all concrete states it covers or be unknown.
(c) **Full input coverage.** In every abstract state, each concrete input must be covered by some qualified abstract input.
(d) **Step soundness.** Each result of the abstract step function must cover all results of the concrete step function where its arguments are covered by the abstract step function arguments.

The requirements ensure the soundness of the used abstractions as follows.

Theorem 1 (Soundness). *For every generating automaton $\hat{G}$ and concrete generating automaton G, state concretization function γ, and input concretization function ζ such that $\hat{G}$ is a soundness-guaranteeing (γ, ζ)-abstraction of G, the partial Kripke structure $\Gamma(\hat{G})$ is sound with respect to $\Gamma(G)$.*

Proof sketch. Show that $\{(s, \hat{s}) \mid \hat{s} \in \hat{S} \wedge s \in \gamma(\hat{s})\}$ is a modal simulation [15, p. 408] from $\Gamma(G)$ to $\Gamma(\hat{G})$ due to (6). Then, $\hat{G}$ is sound wrt. G due to a previous theorem on preservation of μ-calculus formulas [15, p. 410].

Corollary 1. *Assume that the functions* ABSTRACT *and* REFINE *ensure that the generating automaton $(\hat{S}, \hat{s}_0, \hat{I}, \hat{q}, \hat{f}, \hat{L})$ in Algorithm 1 is always a soundness-guaranteeing (γ, ζ)-abstraction of (S, s_0, I, q, f, L). Then, if the algorithm terminates, its result is correct.*

Example 8. Continuing from Example 7, (6a) is fulfilled trivially. (6c) is fulfilled due to (3c) and (4a). From (4b), it is apparent that

$$\forall(\hat{s}, \hat{i}) \in (\hat{S}, \hat{I}) \; . \; \gamma(\hat{f}^{\mathrm{basic}}(\hat{s}, \hat{i})) \subseteq \gamma(\hat{f}(\hat{s}, \hat{i})), \tag{7}$$

i.e. results of $\hat{f}$ cover results of $\hat{f}^{\mathrm{basic}}$ that abstracts f. We carefully implemented the translation of f to $\hat{f}^{\mathrm{basic}}$ and $\hat{L}$ so that (6b) and (6d) hold.

Next, we turn to monotonicity, which ensures no algorithm loop iteration loses information. We give the requirements for the refinement to guarantee it.

Definition 9. *A generating automaton $(\hat{S}, \hat{s}_0, \hat{I}, \hat{q}, \hat{f}, \hat{L})$ is monotone wrt. (γ, ζ)-coverage iff*

$$\forall(\hat{s}, \hat{s}', a) \in \hat{S} \times \hat{S} \times \mathcal{A} \, . \tag{8a}$$
$$((\gamma(\hat{s}') \subseteq \gamma(\hat{s}) \wedge \hat{L}(\hat{s}, a) \neq \bot) \Rightarrow \hat{L}(\hat{s}', a) = \hat{L}(\hat{s}, a)),$$
$$\forall(\hat{s}, \hat{s}', \hat{i}, \hat{i}') \in \hat{S} \times \hat{S} \times \hat{I} \times \hat{I} \, . \tag{8b}$$
$$((\gamma(\hat{s}') \times \zeta(\hat{i}') \subseteq \gamma(\hat{s}) \times \zeta(\hat{i})) \Rightarrow \gamma(\hat{f}(\hat{s}', \hat{i}')) \subseteq \gamma(\hat{f}(\hat{s}, \hat{i}))).$$

Informally, we require that each abstract state covered by an abstract state with non-$\bot$ labelling has the same labelling, and when abstract step function arguments are covered by some other arguments, its result is also covered by the result computed using the other arguments.

Definition 10. *The generating automaton $(\hat{S}, \hat{s}_0, \hat{I}, \hat{q}', \hat{f}', \hat{L})$ is a (γ, ζ)-monotone refinement of the generating automaton $(\hat{S}, \hat{s}_0, \hat{I}, \hat{q}, \hat{f}, \hat{L})$ iff it is monotone wrt. (γ, ζ)-coverage and*

$$\forall \hat{s} \in \hat{S} . \ \forall \hat{i}' \in \hat{q}'(\hat{s}) . \ \exists \hat{i} \in \hat{q}(\hat{s}) . \ \zeta(\hat{i}') \subseteq \zeta(\hat{i}), \tag{9a}$$

$$\forall \hat{s} \in \hat{S} . \ \forall \hat{i} \in \hat{q}(\hat{s}) . \ \exists \hat{i}' \in \hat{q}'(\hat{s}) . \ \zeta(\hat{i}') \subseteq \zeta(\hat{i}), \tag{9b}$$

$$\forall (\hat{s}, \hat{i}) \in \hat{S} \times \hat{I} . \ \gamma(\hat{f}'(\hat{s}, \hat{i})) \subseteq \gamma(\hat{f}(\hat{s}, \hat{i})). \tag{9c}$$

Informally, in addition to the monotonicity wrt. coverage, we also require:

(a) **New qualified inputs are not spurious.** Each new qualified input is covered by at least one old qualified input.
(b) **Old qualified inputs are not lost.** Each old qualified input covers at least one new qualified input.
(c) **New step function covered by old.** The result of the new step function is always covered by the result of the old step function.

The need for both quantifier combinations in the first two requirements in Eq. 9 may be surprising. Their violations correspond to transition addition and removal, respectively, which could make a previously non-$\bot$ property $\bot$.

Theorem 2. (Monotonicity). *If the generating automaton $\hat{G}'$ is a (γ, ζ)-monotone refinement of the generating automaton $\hat{G}$, then for every μ-calculus property ψ for which $[\![\psi]\!](\Gamma(\hat{G})) \neq \bot$, it also holds $[\![\psi]\!](\Gamma(\hat{G}')) \neq \bot$.*

Proof sketch. Show that $\{(\hat{s}', \hat{s}) \mid \hat{s}' \in \hat{S}' \wedge s \in S)\}$ is a modal simulation [15, p. 408] from $\Gamma(\hat{G}')$ to $\Gamma(\hat{G})$ due to (8) and (9). Then, $\hat{G}$ is sound wrt. $\hat{G}'$ due to a previous theorem on preservation of μ-calculus formulas [15, p. 410].

Corollary 2. *If the update in the loop of Algorithm 1 performs a (γ, ζ)-monotone refinement of the generating automaton $(\hat{S}, \hat{s}_0, \hat{q}, \hat{f}, \hat{L})$ and for a μ-calculus property ψ, it held that $[\![\psi]\!](\Gamma((\hat{S}, \hat{s}_0, \hat{q}, \hat{f}, \hat{L}))) \neq \bot$ before the loop iteration, then this is also the case after the iteration.*

Now we turn to termination. We use the following requirements to ensure that the algorithm makes progress.

Definition 11. *The generating automaton $(\hat{S}, \hat{s}_0, \hat{I}, \hat{q}', \hat{f}', \hat{L})$ is a strictly (γ, ζ)-monotone refinement of the generating automaton $(\hat{S}, \hat{s}_0, \hat{I}, \hat{q}, \hat{f}, \hat{L})$ if it is a monotone refinement and either*

$$\exists \hat{s} \in \hat{S} . \ \exists \hat{i} \in \hat{q}(\hat{s}) . \ \forall \hat{i}' \in \hat{q}'(\hat{s}) . \ \exists i \in \zeta(\hat{i}) . \ i \notin \zeta(\hat{i}'), \tag{10}$$

$$or \ \exists (\hat{s}, \hat{i}) \in \hat{S} \times \hat{I} . \ \exists s \in \gamma(\hat{f}(\hat{s}, \hat{i})) . \ s \notin \gamma(\hat{f}'(\hat{s}, \hat{i})). \tag{11}$$

Informally, progress in monotone refinements is ensured either by an abstract state where some old qualified input is not fully covered by any new qualified input or by an abstract state-input combination where the old step function result has at least one concretization absent in the new result concretizations.

Example 9. Continuing from Example 8, during each refinement, we set at least one bit of $\hat{p}_{\hat{q}}$ and/or $\hat{p}_{\hat{f}}$ to 1, and prohibit resetting them to 0, fulfilling (9). We implemented $\hat{L}$ so that (8a) holds. (8b) holds due to (4b). As setting bits in $\hat{p}_{\hat{f}}$ does not necessarily mean $\hat{f}$ changes, we set them until R in $\Gamma(\hat{G})$ changes. This means $\hat{q}$ or $\hat{f}$ change as per (10) or (11), and the refinement is strictly monotone.

Theorem 3 (Completeness). *If $\hat{S}$ and $\hat{I}$ are finite, there is no infinite sequence of generating automata that are soundness-guaranteeing (γ, ζ)-abstractions of some G such that all subsequent pairs in the sequence are strictly (γ, ζ)-monotone refinements.*

Proof sketch. Assume such a sequence exists. For each sequence element $\hat{G}$, define a function $M_{\hat{G}}(\hat{s}) = (\{\zeta(\hat{i}) \mid \hat{i} \in \hat{q}_{\hat{G}}(\hat{s})\}, \{(\hat{i} \in \hat{I}, \gamma(\hat{f}_{\hat{G}}(\hat{s}, \hat{i})))\})$. Show that $M_{\hat{G}}$ must be different for different sequence elements. As there is only a finite number of different functions $M_{\hat{G}}$, the sequence cannot be infinite.

Corollary 3. *If $\hat{S}, \hat{I}$ are finite, the functions* ABSTRACT *and* REFINE *in Algorithm 1 ensure that $(\hat{S}, \hat{s}_0, \hat{I}, \hat{q}, \hat{f}, \hat{L})$ always is a soundness-guaranteeing (γ, ζ)-abstraction of (S, s_0, I, q, f, L), and calls of* REFINE *perform strict (γ, ζ)-monotone refinements, then the algorithm returns the correct result in finite time.*

Fulfilling the requirements of Corollary 3 is not trivial. We must exclude the situation when no strict (γ, ζ)-monotone refinement is possible any more while a non-$\perp$ result has not yet been reached. We propose Lemma 1 for easier reasoning.

Definition 12. *A generating automaton $\hat{G} = (\hat{S}, \hat{s}_0, \hat{I}, \hat{q}, \hat{f}, \hat{L})$ is (γ, ζ)-terminating wrt. a concrete generating automaton $G = (S, s_0, I, q, f, L)$ if it is a (γ, ζ)-abstraction of G monotone wrt. (γ, ζ)-coverage and*

$$\forall (s, \hat{s}) \in S \times \hat{S} \,.\, (\gamma(\hat{s}) = \{s\} \Rightarrow \hat{L}(\hat{s}) = L(s)), \tag{12a}$$

$$\forall \hat{s} \in \hat{S} \,.\, \forall \hat{i} \in \hat{q}(\hat{s}) \,.\, \exists i \in I \,.\, \zeta(\hat{i}) = \{i\}, \tag{12b}$$

$$\forall \hat{s} \in \hat{S} \,.\, \forall i \in I \,.\, \exists \hat{i} \in \hat{q}(\hat{s}) \,.\, \zeta(\hat{i}) = \{i\}, \tag{12c}$$

$$\forall (\hat{s}, \hat{i}, s, i)\hat{S} \times \hat{I} \times S \times I.((\gamma(\hat{s}), \zeta(\hat{i}))(\{s\}, \{i\}) \Rightarrow \gamma(\hat{f}(\hat{s}, \hat{i}))\{f(s, i)\}). \tag{12d}$$

Informally, (12a) requires that each abstract state with a single concretization has the same labelling as the corresponding concrete state. (12b) requires that every qualified abstract input has a single concretization, while (12c) requires that every concrete input corresponds to some qualified abstract input. (12d) requires that the step function applied on single-concretization abstract state-input combinations produces the correct single-concretization result.

Lemma 1. *If $\hat{G}$ is (γ, ζ)-terminating wrt. G, then for every μ-calculus property ψ, $[\![\psi]\!](\Gamma(\hat{G})) = [\![\psi]\!](\Gamma(G))$. Furthermore, if $\hat{G}$ is a soundness-guaranteeing (γ, ζ)-abstraction of G for which some (γ, ζ)-monotone refinement is (γ, ζ)-terminating wrt. G, then $\hat{G}$ itself is (γ, ζ)-terminating wrt. G or the refinement is strict.*

Proof sketch. Show that $\{(\hat{s}, s) \mid \hat{s} \in \hat{S} \land s \in \gamma(\hat{s})\}$ is a modal simulation from $\Gamma(\hat{G})$ to $\Gamma(G)$ due to (6) and (12). As $[\![\phi]\!](\Gamma(G)) \neq \bot$, $[\![\psi]\!](\Gamma(\hat{G})) = [\![\psi]\!](\Gamma(G))$. Then, consider that if the $\hat{G}$ in the second part is not (γ, ζ)-terminating, at least one of (12a), (12b), (12c), (12d) does not hold for $\hat{G}$. Show on a case-by-case basis this is a contradiction for (12a) and forces the refinement to be strict otherwise.

Corollary 4. *If every generating automaton constructed in Algorithm 1 fulfils the conditions of Lemma 1, calls to REFINE can always perform a strict (γ, ζ)-monotone refinement to a soundness-guaranteeing (γ, ζ)-abstraction.*

Example 10. Continuing from Example 9, we implemented $\hat{L}, \hat{f}^{\mathrm{basic}}$ so that they fulfil (12a) and (12d). Since (12b) and (12c) hold due to (4a), $\hat{G}^*$ obtained by $\hat{p}_{\hat{q}} = (1)^y, \hat{p}_{\hat{f}} = (1)^w$ is (γ, ζ)-terminating. Inspecting (9), $\hat{G}^*$ is monotone wrt. all other applicable GA, and Corollary 4 holds. As soundness was already ensured in Example 8 and $\hat{S}, \hat{I}$ are finite by definition, Corollary 3 holds.

4   Making Reasonable Refinement Choices

A sound, monotone, and complete instantiation of the framework can e.g. refine randomly while fulfilling the requirements from Subsect. 3.2. However, for fast verification, refinements must be chosen carefully. We will discuss how reasonable refinements can be made, inspired by the `FindFailure` algorithms to find the cause of an unknown result of verification of CTL [48, p. 551–552] and game-based verification of μ calculus [23, p. 1145].

The goal is to find the root cause of an unknown verification result, an unknown atomic labelling in an abstract state reached through the abstract state space, which we will also call the *culprit*. For this, we extend the semantics of three-valued μ-calculus on PKS, a simplification of three-valued μ-calculus on KMTS [23,49], so that it returns the culprit instead of the unknown value $\bot$. For simplicity, we base this definition on the standard syntax of μ-calculus in positive normal form. Assuming, in addition to the set of atomic labellings $\mathbb{A}$, a set of μ-calculus variables $\mathbb{V}$, a formula then has the form

$$\phi ::= a \mid \neg a \mid Z \mid \phi \land \phi \mid \phi \lor \phi \mid \langle \phi \rangle \mid [\phi] \mid \mu Z \,.\, \phi \mid \nu Z \,.\, \phi \tag{13}$$

where $a \in \mathbb{A}$, $Z \in \mathbb{V}$. The language of μ-calculus then consists of all closed-form formulas given by this definition. We also assume variables are well-bound, i.e. every variable must be bound by μ or ν exactly once.

Definition 13. *Given a GA $\hat{G} = (\hat{S}, \hat{s}_0, \hat{I}, \hat{q}, \hat{f}, \hat{L})$, an* environment *is a function* $\rho : \mathbb{V} \to (\hat{S} \to (\{0, 1\} \cup C))$, *where C is the set of pairs $(\mathcal{S}, a)$ with $\mathcal{S}$ being a non-empty sequence of states from $\hat{S}$ and $a \in \mathbb{A}$ an atomic labelling.*

Now let $\hat{R}$ be the transition relation of the PKS $\Gamma(\hat{G})$. We define the extended semantics *of μ-calculus recursively according to its syntax such that for an environment ρ and a state $\hat{s} \in \hat{S}$,*

$$\llbracket a \rrbracket_{\hat{s}}^{\rho} :: = \begin{cases} \hat{L}(\hat{s}, a) & \text{if } \hat{L}(\hat{s}, a) \neq \bot, \\ ((\hat{s}), a) & \text{otherwise,} \end{cases} \qquad \llbracket \neg a \rrbracket_{\hat{s}}^{\rho} :: = \begin{cases} 1 - \hat{L}(\hat{s}, a) & \text{if } \hat{L}(\hat{s}, a) \neq \bot, \\ ((\hat{s}), a) & \text{otherwise,} \end{cases}$$

$$\llbracket \phi \wedge \psi \rrbracket_{\hat{s}}^{\rho} :: = \begin{cases} 1 & \text{if } \llbracket \phi \rrbracket_{\hat{s}}^{\rho} = 1 \text{ and } \llbracket \psi \rrbracket_{\hat{s}}^{\rho} = 1, \\ 0 & \text{if } \llbracket \phi \rrbracket_{\hat{s}}^{\rho} = 0 \text{ or } \llbracket \psi \rrbracket_{\hat{s}}^{\rho} = 0, \\ c \in C \text{ s.t. } c = \llbracket \phi \rrbracket_{\hat{s}}^{\rho} \text{ or } c = \llbracket \psi \rrbracket_{\hat{s}}^{\rho} & \text{otherwise,} \end{cases}$$

$$\llbracket \langle \phi \rangle \rrbracket_{\hat{s}}^{\rho} :: = \begin{cases} 1, & \text{if there is a } \hat{s}^+ \text{ with } (\hat{s}, \hat{s}^+) \in \hat{R} \text{ and } \llbracket \phi \rrbracket_{\hat{s}^+}^{\rho} = 1, \\ 0, & \text{if for all } \hat{s}^+, (\hat{s}, \hat{s}^+) \in \hat{R} \text{ implies } \llbracket \phi \rrbracket_{\hat{s}^+}^{\rho} = 0, \\ ((\hat{s}, \hat{s}_1, \dots), a) \text{ s.t. } (\hat{s}, \hat{s}_1) \in \hat{R} \text{ and } \llbracket \phi \rrbracket_{\hat{s}_1}^{\rho} = ((\hat{s}_1, \dots), a), \text{otherwise,} \end{cases}$$

$$\llbracket Z \rrbracket_{\hat{s}}^{\rho} :: = \rho(Z)(\hat{s}), \qquad \llbracket \mu Z \, . \, \phi \rrbracket_{\hat{s}}^{\rho} :: = \text{lfp}(\lambda g \, . \, \llbracket \phi \rrbracket_{\hat{s}}^{\rho[Z \mapsto g]}).$$

dummy

Definition 14. *The least fixed point lfp is defined as usual for the ordering $0 < \bot < 1$, but retains the element of C previously obtained when the environmental value remains $\bot$ between subsequent iterations of fixed-point computation. The operators $\phi \vee \phi$, $[\phi]$, and $\nu Z \, . \phi$ are defined correspondingly to their duals.*

As already mentioned, in the case where the result is an element of C, we will call it the *culprit*. The definition is not unique, allowing implementation choice of which culprit to use for refinement. With any choice, the culprit describes a path through the state space ending with a labelling that contributes to the verification result $\llbracket \phi \rrbracket_{\hat{s}_0}$ being unknown.

Using the Culprit for Refinement. If the instance of our framework fulfils the conditions from Corollary 3 then there clearly is at least one application of $\hat{f}$ on the culprit's path that makes the subsequent state cover more than one state, since $\gamma(\hat{s}_0) = \{s_0\}$ and (12a) forbids spuriously unknown labellings. As such, we can decide to only refine $\hat{f}$ from the preceding $\hat{s}$ of such applications. This avoids refinements that definitely would not impact any culprit, unnecessarily increasing the size of the abstract state space. This idea can be extended to the instance details, not refining e.g. inputs that provably do not have an impact on the culprit labelling. The choice of culprit and actual refinement can be fine-tuned according to the typical systems under verification in order to achieve good performance.

Example 11. In **machine-check**, we model-check and obtain the culprit in the spirit of the above discussion. To ensure deterministic behaviour, we prefer the left culprit for operators $\phi \wedge \phi, \phi \vee \phi$ and the culprit continuing with the state with the smallest unique identification number for next-state operators $\langle \phi \rangle, [\phi]$.

For better performance, we use incremental model-checking where we construct a history of environment updates and only update a part of the history after refinement, if possible. To avoid storing the elements of C, where paths can be problematically long, we instead store the evaluation choices made to obtain the given history point from previous history points and atomic labellings. This allows us to reconstruct the culprit if necessary by walking back through history.

After obtaining the culprit $((\hat{s}_0, \hat{s}_1, \ldots, \hat{s}_{n-1}, \hat{s}_n), a)$, we compute a cone of influence on a being unknown in $\hat{s}_n$. For each state $\hat{s} \in \{\hat{s}_0, \hat{s}_1, \ldots, \hat{s}_{n-1}\}$, we determine candidate bit positions $k \in [0, y - 1]$ where $\hat{p}_{\hat{q}}(\hat{s})_k = 0$ and it is possible this has contributed to a being unknown in $\hat{s}_n$, and likewise for $k \in [0, w - 1]$ where $\hat{p}_{\hat{f}}(\hat{s})_k = 0$. We then choose the state in which to refine $\hat{p}_f$ or $\hat{p}_q$ deterministically based on the following heuristic:

- If there are candidate bits for refinement of $\hat{p}_f$, we choose to refine $\hat{p}_f$ in the last state on the culprit path where there is some candidate.
- Otherwise, we pre-select only the states containing candidates that have the highest level of indirection (indirection rises for indices to array reads and writes that impact the culprit labelling). We then choose to refine $\hat{p}_q$ using the candidates on the last pre-selected state on the culprit path.

From multiple candidate bits for the refinement in the chosen state, we choose the input variable deterministically. From multiple bit candidates in a bit-vector variable, we refine the most significant bit.

5   Implementation and Experimental Evaluation

We implemented an instance of our proposed framework in our free and open-source tool **machine-check**. In this section, we discuss our implementation choices and show that our framework is able to mitigate exponential explosion.

5.1   Implementation

Our focus is verification of machine-code systems, where the system is composed of a processor and a machine-code program it executes. However, there is no special handling for typical features of machine-code systems such as the Program Counter, call stack, etc., allowing **machine-check** to support verification of arbitrary finite-state digital systems against properties in propositional μ-calculus.

Systems. Systems to be verified are described in a subset of the Rust programming language. The systems can be comprised of bit-vectors that support standard arithmetical, logical, shift, and extension operations and bit-vector arrays that support indexing.

Properties. Properties are described in a Rust-style format that allows μ-calculus and Computation Tree Logic properties to be expressed. Atomic propositions are formed by equality and comparison operations between variables of the system. CTL and μ-calculus can be freely mixed, and CTL operators are converted to μ-calculus before model-checking. For example, the CTL property $\mathbf{AG}[\texttt{value} = 0]$ is translated to $\nu Z.(\texttt{value} = 0) \wedge [Z]$.

Translating the Systems. The system description, corresponding to the concrete GA under verification, is compiled together with **machine-check** core to form the system verifier, and the description is automatically translated [38] to representations that serve as an abstract simulator and a cone-of-influence calculator. No other interaction between the system and framework is needed.

Abstraction. Three-valued bit-vector abstraction is used for bit-vectors, with fast abstract operations [40]. Arrays are abstracted using a version of sparse representation where elements that are not stored have the value of the nearest stored element with a lesser index.

Building the State Space. The state space is built incrementally using the abstract analogue of the system as the abstract simulator. Abstract state data and a sparse transition graph are retained throughout verification, with garbage collection of abstract states that are no longer reachable.

Model-Checking. As noted in Example 11, we use incremental model-checking of μ-calculus. For reassurance, once the final verification result is obtained, it is double-checked non-incrementally.

Verification Settings. With default verification settings, inputs are initially unknown but decay is not used, i.e. $\hat{p}_{\hat{q}}(\hat{s}) = (0)_{k=0}^{y-1}$, $\hat{p}_{\hat{f}}(\hat{s}) = (1)_{k=0}^{w-1}$ from all states $\hat{s} \in \hat{S}$. It is possible to enable decay ubiquitously, i.e. $\hat{p}_{\hat{f}}(\hat{s}) = (0)_{k=0}^{w-1}$, but we found this to be slower for machine-code systems due to the need to refine each generated state. A naïve strategy with $\hat{p}_{\hat{q}}(\hat{s}) = (1)_{k=0}^{y-1}$ corresponding to explicit-state verification without abstraction would immediately result in hopeless explosion of the state space due to the amount of initially uninitialised memory and read inputs in machine code systems.

5.2   Evaluation Setup

We evaluated verification of μ-calculus properties on machine-code systems for the AVR ATmega328P microcontroller that we described for use in **machine-check** according to the datasheet [33] and instruction set reference [34].

Benchmark Set. As we are not aware of any publicly available and appropriately licensed benchmark sets for formal verification of embedded 8-bit microcontrollers, we used our own set of benchmarks based on machine-code programs for ATmega328P. Our set of benchmarks currently contains 16 programs: 6 programs where we expect a violation of some system-inherent property (e.g. forbidden instruction), 5 simple programs, and 5 more complicated programs, four of which are variations of a simplified version of a program for calibration of a Voltage-Controlled Oscillator used in real life. For C programs, we benchmark the machine code obtained with debug and release builds separately.

Setup. The evaluation was performed on a Linux machine with an Intel Core i9-12900 processor with 128 GB of RAM. The tool was built with Rust 1.83.0 in release configuration. The default verification strategy was used.

Evaluated Properties. We evaluated up to 8 properties on each program[9]. Using w for a propositional formula on atomic properties that varies depending on the program, the properties included, among others:

- Inherent property defined in the processor description, determining if a system with given machine-code is permissible, containing no calls to unimplemented instructions, peripherals, etc. $\mathbf{AG}[w]$, i.e. a safety property.
- Recovery property (as discussed in Example 1), checking whether the system can be recovered from every reachable state to the program loop start with initial output values. $\mathbf{AG}[\mathbf{EF}[w]]$ in CTL. Not a linear-time property.
- Program counter (PC) remains within the main program loop in the main function infinitely often on all paths. $\mathbf{FG}[w]$ in LTL, $\mu X \, . \, \nu Y \, . \, [X] \vee (w \wedge [Y])$ in μ-calculus [13, p. 3133]. Not present in CTL.
- PC is never at the start of the main loop during even times (wrt. instructions). $\nu X \, . \, w \wedge [[X]]$ in μ-calculus, not present in CTL, LTL, nor CTL*.
- The stack pointer stays above a given value, preventing stack overflow problems. $\mathbf{AG}[w]$, i.e. a safety property.

5.3   Evaluation Results

We performed 126 measurements in total on the evaluated machine-code systems and properties, all of which resulted in the property being verified. All results matched our expectations. We will discuss a limited number of the more interesting measurements in detail, demonstrating the applicability of the framework.

---

[9] To ensure the implementation is not faulty, we also test the properties on simple non-machine-code systems in our test suite, e.g. whether the verifier distinguishes between $\mathbf{AFG}[w]$ and $\mathbf{AF}[\mathbf{AG}[w]]$ using the standard example [41, p. 65].

Calibration. We used a simplified version of a program for calibration of a Voltage-Controlled Oscillator (VCO) using binary search. We previously used the program in a real device, where the VCO frequency was adjusted by a digital potentiometer based on the measured frequency of the produced wave. For verification, we replaced the interactions with the digital potentiometer and frequency measure by I/O writes and reads, leaving the core algorithm unchanged.

Verifying using **machine-check**, we found a bug in the program using the recovery property: the output value could never recover to zero after an iteration concludes, since the least significant bit was never set to zero during the calibration, causing a calibration precision loss. This bug would have been tricky to find using source-code verification, as the problem occurred in peripheral manipulation. Linear-time properties of forms $\mathbf{F}[w]$ or $\mathbf{FG}[w]$ would not find the bug, as the calibration command may never be given. After fixing the bug, the recovery property holds, as seen in Table 1.

Despite only the simple three-valued bit-vector abstraction used, the final sizes of abstract state spaces are reasonable. The refinement guidance is remarkably well-behaved, arriving at the same state space for the properties where a fast decision was not possible, despite the non-inherent properties having no knowledge about e.g. illegal instructions. Verification using the infinitely-often

Table 1. Selected measurements of verification of the calibration program compiled in debug configuration. Asterisks mark measurements where the inherent property was verified first and verification of the given property progressed from that state space.

Variant	Property name	Holds	Refin.	States	Transitions	CPU t. [s]	Mem. [MB]
Original	Inherent	✓	513	13059	13573	17.09	87.39
Original	Recovery	✗	513	13059	13573	19.88	111.34
Original	Infinitely often	✓	513	13059	13573	672.05	160.58
Original	Infinitely often*	✓	513	13059	13573	17.98	115.34
Original	Even non-starts	✗	0	17	18	<0.01	3.78
Original	Stack above 0x08FD	✓	513	13059	13573	17.08	87.30
Original	Stack above 0x08FE	✗	0	17	18	<0.01	3.94
Fixed	Inherent	✓	513	13059	13573	17.21	87.60
Fixed	Recovery	✓	513	13059	13573	29.16	112.02
Fixed	Infinitely often	✓	513	13059	13573	672.19	161.65
Fixed	Infinitely often*	✓	513	13059	13573	18.07	115.66
Fixed	Even non-starts	✗	0	17	18	<0.01	3.72
Fixed	Stack above 0x08FD	✓	513	13059	13573	17.2	87.56
Fixed	Stack above 0x08FE	✗	0	17	18	<0.01	3.98
Comp. orig.	Inherent	✓	771	17330	18102	54.43	125.83
Comp. orig.	Recovery	✗	770	17333	18104	65.28	159.20
Comp. orig.	Infinitely often*	✓	771	17330	18102	73.29	173.08
Comp. fixed	Inherent	✓	771	17330	18102	51.63	125.71
Comp. fixed	Recovery	✓	771	17330	18102	70.74	159.52
Comp. fixed	Infinitely often*	✓	771	17330	18102	64.2	173.04

property is notably slow due to the need to model-check with non-trivial nested quantifiers, but this slowdown is effectively mitigated by using the inherent property first.

To show Input-based TVAR holds up where explicit-state verification would be completely infeasible, we created complicated calibration variants where a 64-bit value is read during initialisation, and unrelated 8-bit volatile reads are performed while doing the calibration, ensuring there are more than 2^{80} concrete states even if not considering uninitialised memory. As seen in Table 1, **machine-check** verifies with at most approx. factor-of-4 slowdown.

Factorial. We implemented a program in C which computes the factorial of an input recursively. In the compiled machine code, the computation remained recursive in the debug build but was optimised to an iterative version in the release build. We verified properties including maximal stack sizes for both versions, as seen in Table 2. Regarding the state space size, we noticed that the call return locations remain known even after the stack is popped and they are no longer relevant, unnecessarily growing the state space. We believe a smarter, selective decay of step precision could alleviate these problems.

Comparison with Other Tools. While there has been a multitude of previous machine-code verifiers [39, p. 29–37], we are not aware of any that support full µ-calculus. Our work has been inspired by the research on the tool [mc]square / Arcade.µ C [24,36,37,42,45,46], which could use strategies including Delayed Nondeterminism [36]. However, the tool has been discontinued and is not publicly available. It only supported proving LTL/ACTL properties when

Table 2. Selected measurements of verification of the factorial program. Everything is as expected. The recovery property does not hold as the factorial program never outputs zero after the first computation. The infinitely-often property does not hold due to calls to another function from the main function affecting the Program Counter.

Program	Property name	Holds	Refin.	States	Transitions	CPU t. [s]	Mem. [MB]
Debug	Inherent	✓	576	51595	52940	50.4	359.79
Debug	Recovery	✗	576	51595	52940	76.48	469.88
Debug	Infinitely often	✗	6	1396	1411	5.55	259.89
Debug	Even non-starts	✓	576	51595	52940	46.96	360.40
Debug	Stack above 0x08DD	✓	576	51595	52940	52.77	359.74
Debug	Stack above 0x08DE	✗	3	748	756	0.16	31.58
Release	Inherent	✓	45	4272	4378	0.78	46.84
Release	Recovery	✗	45	4272	4378	1.11	54.58
Release	Infinitely often	✗	6	1006	1021	2.93	148.15
Release	Even non-starts	✗	3	550	558	0.04	17.92
Release	Stack above 0x08FB	✓	45	4272	4378	0.78	46.79
Release	Stack above 0x08FC	✗	0	20	21	< 0.01	3.72

abstraction was used, and needed system-based hints for useful abstraction. As for more recent work, verifiers based on theorem proving have been introduced, with the Serval [35,50] tool supporting automatic verification but not programs with loops, and Islaris [43,44] requiring manual proof support but supporting loops. Both only support verification of safety properties. In contrast, **machine-check** supports fully automatic verification of μ-calculus properties on machine-code programs that include loops thanks to the introduced framework.

6   Conclusion

We presented a novel input-based Three-valued Abstraction Refinement (TVAR) framework for formal verification of μ-calculus properties using abstraction refinement and gave requirements for soundness, monotonicity, and completeness. Our framework can verify properties not verifiable using Counterexample-guided Abstraction Refinement or (Generalized) Symbolic Trajectory Evaluation, eliminating verification blind spots. Compared to previous TVAR frameworks, our framework does not use modal transitions, allowing monotonicity without formalism complications. We implemented the framework in our tool **machine-check**, which can verify machine-code systems while mitigating exponential explosion.

Acknowledgments. The work of Jan Onderka reported in this paper was supported by an Amazon Research Award (Fall 2023) and Czech Technical University in Prague grant SGS23/208/OHK3/3T/18. Development of **machine-check** was supported by the NGI Zero Core fund established by NLnet Foundation. The work of Stefan Ratschan was supported by the research programme of the Strategy AV21 AI: Artificial Intelligence for Science and Society and by institutional support RVO:67985807l.

Data Availability Statement. The artefact containing detailed results summarised in Sect. 5 and reproduction data (including the benchmark set) is available at https://doi.org/10.5281/zenodo.17167265.

References

1. Bruns, G., Godefroid, P.: Model Checking Partial State Spaces with 3-Valued Temporal Logics. In: Halbwachs, N., Peled, D. (eds.) CAV 1999. LNCS, vol. 1633, pp. 274–287. Springer, Heidelberg (1999). https://doi.org/10.1007/3-540-48683-6_25

2. Bruns, G., Godefroid, P.: Generalized Model Checking: Reasoning about Partial State Spaces. In: Palamidessi, C. (ed.) CONCUR 2000. LNCS, vol. 1877, pp. 168–182. Springer, Heidelberg (2000). https://doi.org/10.1007/3-540-44618-4_14

3. Bruns, G., Godefroid, P.: Temporal logic query checking. In: 16th Annual IEEE Symposium on Logic in Computer Science, Boston, Massachusetts, USA, June 16-19, 2001, Proceedings, pp. 409–417. IEEE Computer Society (2001). https://doi.org/10.1109/LICS.2001.932516

4. Bryant, R.E.: A methodology for hardware verification based on logic simulation. J. ACM **38**(2), 299–328 (1991). https://doi.org/10.1145/103516.103519

5. Bryant, R.E., Seger, C.-J.H.: Formal verification of digital circuits using symbolic ternary system models. In: Clarke, E.M., Kurshan, R.P. (eds.) CAV 1990. LNCS, vol. 531, pp. 33–43. Springer, Heidelberg (1991). https://doi.org/10.1007/BFb0023717

6. Bryant, R.: Formal verification of memory circuits by switch-level simulation. IEEE Trans. Comput. Aided Des. Integr. Circuits Syst. **10**(1), 94–102 (1991). https://doi.org/10.1109/43.62795

7. Chen, Y., He, Y., Xie, F., Yang, J.: Automatic abstraction refinement for generalized symbolic trajectory evaluation. In: Formal Methods in Computer-Aided Design, 7th International Conference, FMCAD 2007, Austin, Texas, USA, November 11-14, 2007, Proceedings, IEEE Comput. Soc. 111–118 (2007). https://doi.org/10.1109/FAMCAD.2007.11

8. Clarke, E., Grumberg, O., Jha, S., Lu, Y., Veith, H.: Counterexample-Guided Abstraction Refinement. In: Emerson, E.A., Sistla, A.P. (eds.) CAV 2000. LNCS, vol. 1855, pp. 154–169. Springer, Heidelberg (2000). https://doi.org/10.1007/10722167_15

9. Clarke, E.M., Grumberg, O., Long, D.E.: Model checking and abstraction. ACM Trans. Program. Lang. Syst. **16**(5), 1512–1542 (1994). https://doi.org/10.1145/186025.186051

10. Clarke, E., Gupta, A., Strichman, O.: SAT-based counterexample-guided abstraction refinement. IEEE Trans. Comput. Aided Des. Integr. Circuits Syst. **23**(7), 1113–1123 (2004). https://doi.org/10.1109/TCAD.2004.829807

11. Cousot, P., Cousot, R.: Abstract interpretation: a unified lattice model for static analysis of programs by construction or approximation of fixpoints. In: Proceedings of the 4th ACM SIGACT-SIGPLAN Symposium on Principles of Programming Languages, pp. 238–252. POPL '77, Association for Computing Machinery, New York, NY, USA (1977). https://doi.org/10.1145/512950.512973

12. Cousot, P., Cousot, R.: Systematic design of program analysis frameworks. In: Proceedings of the 6th ACM SIGACT-SIGPLAN Symposium on Principles of Programming Languages, POPL 1979, p. 269–282. Association for Computing Machinery, New York, NY, USA (1979). https://doi.org/10.1145/567752.567778

13. Cranen, S., Groote, J.F., Reniers, M.A.: A linear translation from CTL* to the first-order modal μ-calculus. Theor. Comput. Sci. **412**(28), 3129–3139 (2011). https://doi.org/10.1016/J.TCS.2011.02.034

14. Dam, M.: Fixed points of Büchi automata. In: Shyamasundar, R. (ed.) FSTTCS 1992. LNCS, vol. 652, pp. 39–50. Springer, Heidelberg (1992). https://doi.org/10.1007/3-540-56287-7_93

15. Dams, D., Grumberg, O.: Abstraction and Abstraction Refinement. In: FSTTCS 1992. LNCS, vol. 652, pp. 385–419. Springer, Cham (2018). https://doi.org/10.1007/978-3-319-10575-8_13

16. Gange, G., Navas, J.A., Schachte, P., Søndergaard, H., Stuckey, P.J.: Interval analysis and machine arithmetic: Why signedness ignorance is bliss. ACM Trans. Program. Lang. Syst. **37**(1),(2015). https://doi.org/10.1145/2651360

17. Gazda, M., Willemse, T.A.: Expressiveness and completeness in abstraction. Electr. Proc. Theor. Comput. Sci. **89**, 49–64 (2012). https://doi.org/10.4204/eptcs.89.5

18. Godefroid, P.: May/must abstraction-based software model checking for sound verification and falsification. In: Grumberg, O., Seidl, H., Irlbeck, M. (eds.) Software Systems Safety, NATO Science for Peace and Security Series, D: Information and Communication Security, vol. 36, pp. 1–16. IOS Press (2014). https://doi.org/10.3233/978-1-61499-385-8-1

19. Godefroid, P., Huth, M., Jagadeesan, R.: Abstraction-Based Model Checking Using Modal Transition Systems. In: Larsen, K.G., Nielsen, M. (eds.) CONCUR 2001. LNCS, vol. 2154, pp. 426–440. Springer, Heidelberg (2001). https://doi.org/10.1007/3-540-44685-0_29

20. Godefroid, P., Jagadeesan, R.: Automatic Abstraction Using Generalized Model Checking. In: Brinksma, E., Larsen, K.G. (eds.) CAV 2002. LNCS, vol. 2404, pp. 137–151. Springer, Heidelberg (2002). https://doi.org/10.1007/3-540-45657-0_11

21. Godefroid, P., Jagadeesan, R.: On the Expressiveness of 3-Valued Models. In: Zuck, L.D., Attie, P.C., Cortesi, A., Mukhopadhyay, S. (eds.) VMCAI 2003. LNCS, vol. 2575, pp. 206–222. Springer, Heidelberg (2003). https://doi.org/10.1007/3-540-36384-X_18

22. Grumberg, O., Lange, M., Leucker, M., Shoham, S.: Don't Know in the μ-Calculus. In: Cousot, R. (ed.) VMCAI 2005. LNCS, vol. 3385, pp. 233–249. Springer, Heidelberg (2005). https://doi.org/10.1007/978-3-540-30579-8_16

23. Grumberg, O., Lange, M., Leucker, M., Shoham, S.: When not losing is better than winning: Abstraction and refinement for the full μ-calculus. Inf. Comput. **205**(8), 1130–1148 (2007). https://doi.org/10.1016/j.ic.2006.10.009

24. Gückel, D.: Synthesis of State Space Generators for Model Checking Microcontroller Code. Dissertation thesis, RWTH Aachen (2014). http://aib.informatik.rwth-aachen.de/2014/2014-15.pdf

25. Gückel, D., Kowalewski, S.: Automatic Derivation of Abstract Semantics From Instruction Set Descriptions. In: Brauer, J., Roveri, M., Tews, H. (eds.) 6th International Workshop on Systems Software Verification. Open Access Series in Informatics (OASIcs), vol. 24, pp. 71–83. Schloss Dagstuhl – Leibniz-Zentrum für Informatik, Dagstuhl, Germany (2012). https://doi.org/10.4230/OASIcs.SSV.2011.71

26. Gurfinkel, A.: Yasm: Software model-checker. https://www.cs.toronto.edu/~arie/yasm/

27. Gurfinkel, A., Chechik, M.: Multi-valued model checking via classical model checking. In: Amadio, R., Lugiez, D. (eds.) CONCUR 2003. LNCS, vol. 2761, pp. 266–280. Springer, Heidelberg (2003). https://doi.org/10.1007/978-3-540-45187-7_18

28. Gurfinkel, A., Chechik, M.: Why waste a perfectly good abstraction? In: Hermanns, H., Palsberg, J. (eds.) TACAS 2006. LNCS, vol. 3920, pp. 212–226. Springer, Heidelberg (2006). https://doi.org/10.1007/11691372_14

29. Gurfinkel, A., Wei, O., Chechik, M.: YASM: A software model-checker for verification and refutation. In: Ball, T., Jones, R.B. (eds.) CAV 2006. LNCS, vol. 4144, pp. 170–174. Springer, Heidelberg (2006). https://doi.org/10.1007/11817963_18

30. Gurfinkel, A., Wei, O., Chechik, M.: Model checking recursive programs with exact predicate abstraction. In: Cha, S.S., Choi, J.-Y., Kim, M., Lee, I., Viswanathan, M. (eds.) ATVA 2008. LNCS, vol. 5311, pp. 95–110. Springer, Heidelberg (2008). https://doi.org/10.1007/978-3-540-88387-6_9

31. Konikowska, B., Penczek, W.: Reducing model checking from multi-valued CTL* to CTL*. In: Brim, L., Křetínský, M., Kučera, A., Jančar, P. (eds.) CONCUR 2002. LNCS, vol. 2421, pp. 226–239. Springer, Heidelberg (2002). https://doi.org/10.1007/3-540-45694-5_16

32. Melham, T.: Symbolic trajectory evaluation. In: CONCUR 2002. LNCS, vol. 2421, pp. 831–870. Springer, Cham (2018). https://doi.org/10.1007/978-3-319-10575-8_25

33. Microchip Technology Inc.: ATmega48A/PA/88A/PA/168A/PA/328/P Data Sheet (2018). http://ww1.microchip.com/downloads/en/DeviceDoc/ATmega48A-PA-88A-PA-168A-PA-328-P-DS-DS40002061A.pdf. dS40002061A

34. Microchip Technology Inc.: AVR Instruction Set Manual (2021). https://ww1.microchip.com/downloads/en/DeviceDoc/AVR-InstructionSet-Manual-DS40002198.pdf. dS40002198B
35. Nelson, L., Bornholt, J., Gu, R., Baumann, A., Torlak, E., Wang, X.: Scaling symbolic evaluation for automated verification of systems code with Serval. In: Proceedings of the 27th ACM Symposium on Operating Systems Principles, pp. 225–242. SOSP '19, Association for Computing Machinery, New York, NY, USA (2019). https://doi.org/10.1145/3341301.3359641
36. Noll, T., Schlich, B.: Delayed nondeterminism in model checking embedded systems assembly code. In: Yorav, K. (ed.) Hardware and Software: Verification and Testing, pp. 185–201. Springer, Berlin Heidelberg, Berlin, Heidelberg (2008). https://doi.org/10.1007/978-3-540-77966-7_16
37. Noll, T., Schlich, B.: Delayed nondeterminism in model checking embedded systems assembly code. In: Yorav, K. (ed.) HVC 2007. LNCS, vol. 4899, pp. 185–201. Springer, Heidelberg (2008). https://doi.org/10.1007/978-3-540-77966-7_16
38. Onderka, J.: Formal verification of machine-code systems by translation of simulable descriptions. In: Proceedings of the 13th Mediterranean Conference on Embedded Computing, MECO 2024, pp. 1–4 (2024). https://doi.org/10.1109/MECO62516.2024.10577942
39. Onderka, J.: Abstraction-Based Machine-Code Program Verification. Doctoral thesis, Czech Technical University in Prague (2025). http://hdl.handle.net/10467/122640
40. Onderka, J., Ratschan, S.: Fast three-valued abstract bit-vector arithmetic. In: Finkbeiner, B., Wies, T. (eds.) VMCAI 2022. LNCS, vol. 13182, pp. 242–262. Springer, Cham (2022). https://doi.org/10.1007/978-3-030-94583-1_12
41. Piterman, N., Pnueli, A.: Temporal logic and fair discrete systems. In: Clarke, E.M., Henzinger, T.A., Veith, H., Bloem, R. (eds.) Handbook of Model Checking, pp. 27–73. Springer International Publishing, Cham (2018). https://doi.org/10.1007/978-3-319-10575-8_2
42. Reinbacher, T., Brauer, J., Horauer, M., Schlich, B.: Refining assembly code static analysis for the Intel MCS-51 microcontroller. In: Proceedings of the Fourth IEEE International Symposium on Industrial Embedded Systems, SIES 2009, pp. 161–170 (2009). https://doi.org/10.1109/SIES.2009.5196212
43. Rigorous Engineering of Mainstream Systems Project: Islaris: verification of machine code against authoritative ISA semantics (2023). https://github.com/rems-project/islaris
44. Sammler, M., et al.: In: Islaris: Verification of Machine Code Against Authoritative ISA Semantics, pp. 825–840. , ACM, New York, NY, USA (2022). https://doi.org/10.1145/3519939.3523434
45. Schlich, B., Kowalewski, S.: [mc]square: A model checker for microcontroller code. In: Proceedings of the Second International Symposium on Leveraging Applications of Formal Methods, Verification and Validation, ISOLA 2006, pp. 466–473 (2006). https://doi.org/10.1109/ISoLA.2006.62
46. Schlich, B.: Model checking of software for microcontrollers. ACM Trans. Embedded Comput. Syst. 9(4) (2010). https://doi.org/10/bm83n7
47. Shoham, S., Grumberg, O.: A game-based framework for CTL counterexamples and 3-valued abstraction-refinement. In: Hunt, W.A., Somenzi, F. (eds.) CAV 2003. LNCS, vol. 2725, pp. 275–287. Springer, Heidelberg (2003). https://doi.org/10.1007/978-3-540-45069-6_28

48. Shoham, S., Grumberg, O.: Monotonic abstraction-refinement for CTL. In: Jensen, K., Podelski, A. (eds.) TACAS 2004. LNCS, vol. 2988, pp. 546–560. Springer, Heidelberg (2004). https://doi.org/10.1007/978-3-540-24730-2_40
49. Shoham, S., Grumberg, O.: 3-valued abstraction: more precision at less cost. Inf. Comput. **206**(11), 1313–1333 (2008). https://doi.org/10.1016/j.ic.2008.07.004
50. The UNSAT group: Serval (2019). https://unsat.cs.washington.edu/projects/serval/
51. Tzoref, R., Grumberg, O.: Automatic refinement and vacuity detection for symbolic trajectory evaluation. In: Ball, T., Jones, R.B. (eds.) CAV 2006. LNCS, vol. 4144, pp. 190–204. Springer, Heidelberg (2006). https://doi.org/10.1007/11817963_20
52. Wei, O., Gurfinkel, A., Chechik, M.: On the consistency, expressiveness, and precision of partial modeling formalisms. Inf. Comput. **209**(1), 20–47 (2011). https://doi.org/10.1016/j.ic.2010.08.001
53. Yang, J., Seger, C.J.: Introduction to generalized symbolic trajectory evaluation. In: Proceedings 2001 IEEE International Conference on Computer Design: VLSI in Computers and Processors. ICCD 2001, pp. 360–365 (2001). https://doi.org/10.1109/ICCD.2001.955052

Efficient Discovery of Actual Causality in Stochastic Systems

Arshia Rafieioskouei⬤, Kenneth Rogale⬤, and Borzoo Bonakdarpour[✉]⬤

Michigan State University, East Lansing, MI, USA
`{rafieios,rogaleke,borzoo}@msu.edu`

Abstract. Identifying the actual cause of events in engineered systems is a fundamental challenge in systems. It provides a principled framework for capturing logical dependencies between events, offering insights into the underlying dynamics of a system. Finding such causes becomes more challenging in real-world systems. Finding such causes becomes more challenging in real-world systems. In this paper, we adopt the notion of *probabilistic actual causality* by Fenton-Glynn, which is a probabilistic extension of Halpern and Pearl's actual causality, and propose a novel method to formally reason about causal effect of events in stochastic systems. We (1) formulate the discovery of probabilistic actual causes in computing systems as an SMT problem, and (2) address the scalability challenges by introducing an *abstraction-refinement* technique that improves efficiency by up to 95%. We demonstrate the effectiveness of our approach through three case studies, identifying probabilistic actual causes of safety violations in (1) the Mountain Car problem, (2) the Lunar Lander benchmark, and (3) MPC controller for an F-16 autopilot simulator.

1 Introduction

Modern computing systems often operate in open-ended stochastic environments. For instance, online *cyber-physical systems* (CPS) operations are subject to various disturbances and noise and incur catastrophic risks with their potential misbehavior. Such stochastic behavior directly contributes to challenges in ensuring safety and developing intelligent behavior. Thus, to prevent failures in complex or anomalous scenarios, it is imperative to reason about *causes* of future misbehavior rather than their *correlated* symptoms. Engineers generally build causal systems in which outputs depend only on past and present inputs. Hence, causal inference is the natural way to explain the root causes of potential failures.

Numerous research efforts have focused on the critical challenge of identifying and explaining faults within complex systems. Prior work has effectively employed causal frameworks across various domains, including embedded systems [10, 18–22, 38], and automated bug localization through software trace

A. Rafieioskouei and K. Rogale—Equal contribution.

Y.-F. Chen et al. (Eds.): VMCAI 2026, LNCS 16417, pp. 263–286, 2026.
https://doi.org/10.1007/978-3-032-15700-3_13

analysis [8,11,12,16,17], among several other contexts [13,36]. However, these works do not support models with stochastic behavior. When considering the probabilistic setting, analysis requires expanded notions of causality that can accommodate this stochastic behavior. Relatively few works explore probabilistic causality; many that do, such as [4,6,41], use different causal frameworks that lack counterfactual reasoning. By evaluating how the outcome would change in a counterfactual world where the candidate cause does not occur, while keeping the other variables in the model fixed at the values they have in the actual world, we isolate that cause's contribution to the effect and avoid mistaking correlation for causation. In their absence, methods are liable to mistake correlation for causation by crediting variables that merely co-vary with the outcome or masking the causal contribution when alternative routes to the effect are active.

While the work in [31] also applies a probabilistic causal framework to robotics, its focus is fundamentally on task repair, identifying an effective intervention after a failure. This objective differs from our focus on causal discovery. An approach geared towards finding a sufficient repair may not need to pinpoint the precise root cause. Consequently, their methodology is not designed for the fine-grained analysis required to identify actual causes of the effects in stochastic systems, which is the central challenge we address.

In this paper, we are motivated by the notion of *actual causality* (AC) by Halpern and Pearl (HP) [24], which focuses on the causal effect of particular events, rather than type-level causality, which attempts to make general statements about scientific and natural phenomena (e.g., smoking causes cancer). AC is a formalism to deal with token-level causality, which aims to find the causal effect of individual events. The original definition of AC is for deterministic systems and, hence, can only reason about causation in deterministic systems. For probabilistic systems, Halpern [24] suggests "pulling out" all stochasticity in the exogenous variables whose values are determined by factors outside the model and defining a probability distribution over them. This interpretation of stochastic behavior limits reasoning about the causal effect of events, where the controllable actions of the system have a probabilistic nature (e.g., due to noise or imprecise measurements). Thus, we turn to *probabilistic actual causality* (PAC) introduced by Fenton-Glynn [15] based on the concept of AC. Similar to the original definition of AC, PAC is based on causal settings in terms of structural equations and requires the following: (1) there exists a scenario with non-zero probability in which both the cause and its subsequent effect occur in the actual world; and (2) for any counterfactual scenario whose contingencies are identical to those of the actual world and in which the cause does not occur, the probability of the effect occurring is less than in the actual world.

Our main contribution in this paper is as follows. We begin with the premise that the behavior of the system under scrutiny is given as input by an acyclic discrete-time Markov chain (DTMC). Acyclic DTMCs reflect causal models in Halpern and Pearl's definition of actual causality, which does not permit cycles. For instance, it is not admissible for event A to cause event B while event B causes event A. Fenton-Glynn's causal model is derived from Bayesian networks, so the relationships between individual endogenous variables (whose values are ultimately determined by the exogenous variables) can be made explicitly probabilistic. At the level of structural equations, noise and stochasticity are captured by exogenous variables. For example, suppose we want to model a setting where event A produces outcome B with probability 0.5, that is, $P(B \mid A) = 0.5$. We can write the structural equation for B as $B = A \wedge U$, where U is an exogenous variable. By setting $P(U) = 0.5 = P(B \mid A)$, we capture the noise or stochasticity in this situation. Our high-level goal is to design algorithms that identify the probabilistic actual cause of an effect in a DTMC. More specifically, we aim to design algorithms that take as input (1) a DTMC $\mathcal{M}$, and (2) a state predicate φ^e representing the effect, and synthesize as output a state predicate φ^c that is the probabilistic actual cause of φ^e in $\mathcal{M}$. We propose two techniques:

1. Our first algorithm is based on solving an SMT instance that encodes $\mathcal{M}$, φ^e, and the constraints for PAC. We also encode φ^c as an uninterpreted function f and, hence, the SMT instance is satisfiable if and only if the witness interpretation of f is the actual probabilistic cause of φ^e (soundness and completeness).
2. Since SMT solving does not always scale to handle large models, we also develop a technique based on *abstraction-refinement*. In this approach, the model is first abstracted using a set of predicates. If a cause is found using SMT solving on the abstract model, it is indeed an actual cause. Otherwise, we refine the model to a less coarse abstraction, and the process is repeated. This iterative refinement continues until a valid cause is discovered.

We demonstrate the effectiveness of our approach through three case studies in the context of CPS with stochastic behavior to identify the probabilistic actual causes of safety violations in (1) the Mountain Car problem [9], (2) the Lunar Lander benchmark [9], and (3) an MPC controller for an F-16 autopilot simulator [25].

Organization. The rest of the paper is organized as follows. Section 2 presents the preliminary concepts while Sect. 3 formalizes the notion of PAC. The formal statement of our problem is introduced in Sect. 4. Our SMT-based solution and abstraction-refinement algorithm are presented in Sect. 5 and 6, respectively. We evaluate our algorithms in Sect. 7. Related work is discussed in Sect. 8 and we conclude in Sect. 9. All proofs of correctness are available in the appendix [35].

2   Preliminaries

In this section, we present the preliminary concepts.

Definition 1. *A discrete-time Markov chain (DTMC) is a tuple $\mathcal{M}=(S, \mathbf{P}, AP, L)$ with the following components:*

- *S is a nonempty finite set of states;*

- $\mathbf{P} : S \times S \to [0, 1]$ *is a transition probability function with*

$$\sum_{s' \in S} \mathbf{P}(s, s') = 1$$

for all $s \in S$;
- *AP is a finite set of atomic propositions, and*
- $L : S \to 2^{AP}$ *is a labeling function.*   □

An *(infinite) path of* $\mathcal{M}$ is an infinite sequence $\pi = s_0 s_1 s_2 \ldots \in S^{\omega}$ of states with $\mathbf{P}(s_i, s_{i+1}) > 0$, for all $i \geq 0$; we write $\pi[i]$ for s_i. Let $Paths_s^{\mathcal{M}}$ denote the set of all (infinite) paths of $\mathcal{M}$ starting in s, and $fPaths_s^{\mathcal{M}}$ denote the set of all non-empty finite prefixes of paths from $Paths_s^{\mathcal{M}}$, which we call *finite paths*. For a finite path $\pi = s_0 \ldots s_k \in fPaths_{s_0}^{\mathcal{M}}$, $k \geq 0$, we define $|\pi| = k + 1$.

We will also use the notations $Paths^{\mathcal{M}} = \cup_{s \in S} Paths_s^{\mathcal{M}}$ and $fPaths^{\mathcal{M}} = \cup_{s \in S} fPaths_s^{\mathcal{M}}$. A state $t \in S$ is *reachable* from a state $s \in S$ in $\mathcal{M}$ if there exists a finite path in $fPaths_s^{\mathcal{M}}$ with last state t; we use $fPaths_{s,T}^{\mathcal{M}}$ to denote the set of all finite paths from $fPaths_s^{\mathcal{M}}$ with last state in $T \subseteq S$. A state $s \in S$ is *absorbing* if and only if $\mathbf{P}(s, s) = 1$, and $\mathbf{P}(s, t) = 0$ for all states $t \neq s$.

We label all the absorbing states with halt. Also, let $\mathcal{H}$ denote the set of states labeled with halt.

Example 1. Consider a car located in a valley and aiming to reach the top of a mountain (see Fig. 1a). At each time step, the controller of the car determines whether to apply positive or negative acceleration to guide the car towards the mountain top. We model the behavior of this mountain car in the presence of uncertainty (e.g., noise) with a DTMC (see Fig. 1b). The tuple of three variables (pos, vel, act) defines the set of states in the DTMC. The domain of act is $\{-1, 1, 0\}$, denoting negative, positive, and zero acceleration. The stationary state s_0 is where $pos = 0$, $vel = 0$, and the controller decides to apply positive acceleration, i.e., $act = 1$. Due to uncertainty, with probability 0.5, the successor states are either s_1, where $pos = 0.3$, $vel = 0.01$, and the controller decides to apply positive acceleration, or s_2, where $pos = 0.4$, $vel = 0.03$, and the controller decides not to accelerate. We define the safety specification of mountain car as the absorbing state that does not reach the flag, where $pos = 0.6$, by the following predicate: $\varphi^{\mathsf{fail}} \triangleq \left(pos < 0.6 \wedge \mathsf{halt} \right)$. Thus, two (red) states s_7, s_9 in Fig. 1b are labeled by φ^{fail}.   □

Markov decision processes extend DTMCs with non-deterministic choices.

Definition 2. *A* Markov decision process *(MDP) is a tuple* $\mathbb{M} = (S, Act, \mathcal{P}, AP, L)$ *with the following components:*

- *S is a nonempty finite set of states;*
- *Act is a nonempty finite set of actions;*

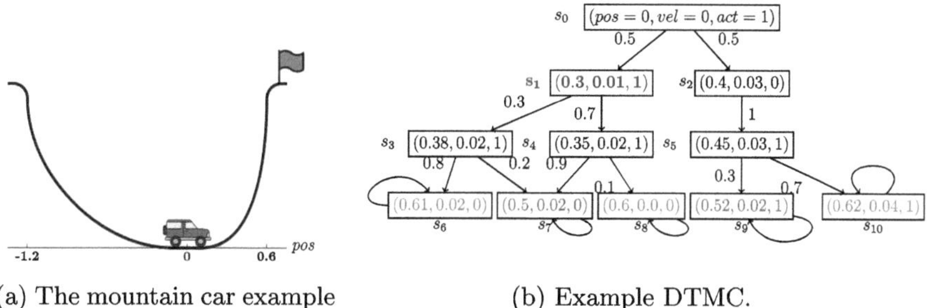

(a) The mountain car example   (b) Example DTMC.

Fig. 1. The mountain car example.

- $\mathcal{P} : S \times Act \times S \to [0,1]$ *is a* transition probability function *such that for all $s \in S$ the* set of enabled actions *in s*

$$Act(s) = \{\alpha \in Act \mid \sum_{s' \in S} \mathcal{P}(s, \alpha, s') = 1\}$$

- *AP is a finite set of* atomic propositions, *and*
- $L : S \to 2^{AP}$ *is a* labeling function.   □

Definition 3. *A memoryless scheduler for an MDP* $\mathbb{M} = (S, Act, \mathcal{P}, AP, L)$ *is a function* $\mathfrak{s} : S \to Act$, *where* $\mathfrak{s}(s) \in Act(s)$ *for all $s \in S$.*   □

Definition 4. *For an MDP* $\mathbb{M} = (S, Act, \mathcal{P}, AP, L)$ *and a scheduler* $\mathfrak{s}$ *for* $\mathbb{M}$, *the DTMC induced by* $\mathbb{M}$ *and* $\mathfrak{s}$ *is defined as* $\mathbb{M}^{\mathfrak{s}} = (S, \mathcal{P}^{\mathfrak{s}}, AP, L)$, *where* $\mathcal{P}^{\mathfrak{s}}(s, s') = \mathcal{P}(s, \mathfrak{s}(s), s')$.   □

A state s' is *reachable* from $s \in S$ in MDP $\mathbb{M}$ if there exists a scheduler $\mathfrak{s}$ for $\mathbb{M}$ such that s' is reachable from s in $\mathbb{M}^{\mathfrak{s}}$. A state $s \in S$ is *absorbing* in $\mathbb{M}$ if s is absorbing in $\mathbb{M}^{\mathfrak{s}}$ for all schedulers $\mathfrak{s}$ for $\mathbb{M}$. We sometimes omit the MDP index $\mathbb{M}$ in the notations when it is clear from the context.

Finally, for a set $B \subseteq S$ of target states, the measure of interest is the maximum, or dually, the minimum probability of reaching a state in B when starting in state $s \in S$ among the set of all (memoryless) schedulers $\mathfrak{S}$:

$$\mathbb{P}^{\max}(s \models \Diamond B) = \sup_{\mathfrak{s} \in \mathfrak{S}} \mathbb{P}^{\mathfrak{s}}(s \models \Diamond B) \qquad \mathbb{P}^{\min}(s \models \Diamond B) = \inf_{\mathfrak{s} \in \mathfrak{S}} \mathbb{P}^{\mathfrak{s}}(s \models \Diamond B).$$

3   Probabilistic Actual Causality

As mentioned in Sect. 1, we use the definition of *probabilistic actual causality* (PAC) due to Fenton-Glynn [15], which is a probabilistic variation of the original definition of actual causality (AC) by Halpern and Pearl [24]. Similar to the original definition of AC [24], the definition of PAC in [15] is based on causal settings in terms of structural equations. Roughly speaking, the definition of PAC requires the following:

- **(PC1)** There exists a scenario with non-zero probability that the cause φ^c and the subsequent effect φ^e both occur in the actual world.
- **(PC2)** For any counterfactual scenario whose contingencies W are identical to those of the actual world, and in which φ^c does not hold, the probability that φ^e becomes true is strictly less than the probability of φ^e occurring in the actual world.

In this paper, since our focus is on Markov models, we do not present the details of PAC in terms of structural equations. Rather, we introduce an interpretation of PAC using the temporal logic HyperPCTL with DTMC semantics [1]. We also do not present the full syntax and semantics of HyperPCTL, as this is the only formula that we will be dealing with throughout the paper. Let $\mathcal{M}=(S,\mathbf{P},\mathsf{AP},L)$ be a DTMC. The following formula expresses PAC due to Fenton-Glynn:

$$\varphi_{\mathsf{pac}} \triangleq \exists\sigma.\forall\sigma'.\underbrace{\mathbb{P}\Big(\neg\varphi^e_\sigma \;\mathcal{U}\; (\varphi^c_\sigma \wedge \mathbb{P}_{>0}(\Diamond\varphi^e_\sigma))\Big)}_{\substack{\psi_{\mathsf{AW}}:\ \text{Probability of effect }\varphi^e \\ \text{occurringafter cause }\varphi^c \\ \text{in actual world }\sigma}} > \overbrace{\mathbb{P}\Big(\neg\varphi^c_{\sigma'} \;\mathcal{U}\; \varphi^e_{\sigma'}\Big)}^{\substack{\psi_{\mathsf{CW}}:\ \text{Probability of cause }\varphi^c \\ \text{not occurringbefore effect }\varphi^e \\ \text{in counterfactual world }\sigma'}} \wedge$$

$$\underbrace{\mathbb{P}_{=1}\Big(\bigwedge_{a\in W} \Box(a_\sigma \leftrightarrow a_{\sigma'})\Big)}_{\substack{\psi_{\mathsf{SE}}:\ \text{Actual and counterfactual worlds} \\ \text{agree with each other with respect} \\ \text{to all propositions in }W}} \tag{1}$$

where:

- σ,σ' are two state variables that range over S, designating the root states of the *actual* and *counterfactual* world computation trees in $\mathcal{M}$, respectively.
- φ^e is a predicate (i.e., a Boolean combination of the propositions in AP) expressing the effect, and φ^c is a predicate expressing the cause. The meaning of φ^c_σ is evaluation of formula φ^c in the computation tree of $\mathcal{M}$ rooted at state σ. Similar interpretation holds for formulas $\varphi^c_{\sigma'}$, φ^e_σ, and $\varphi^e_{\sigma'}$, and
- $W \subseteq \mathsf{AP}$ is a subset of propositions describing all contingencies.

The meaning of formula φ_{pac} is as follows. There exists an actual world (semantically, a computation tree of $\mathcal{M}$ rooted at a state σ), such that for all counterfactual worlds (all computation trees rooted at a state σ') that agree with σ as far as propositions in W are concerned (i.e., subformula ψ_{SE}), and the probability of reaching the effect φ^e after reaching the cause φ^c in the actual world σ (i.e., subformula ψ_{AW}) is strictly greater than the probability of reaching the effect φ^e without reaching the cause φ^c in the counterfactual world σ' (i.e., subformula ψ_{CW}).

The formal semantics of φ_{pac} is based on self-composition of DTMCs.

Definition 5. *The* n*-ary self-composition of a DTMC* $\mathcal{M} = (S,\mathbf{P},\mathit{AP},L)$ *is a DTMC* $\mathcal{M}^n = (S^n,\mathbf{P}^n,\mathit{AP}^n,L^n)$ *with*

- $S^n = S \times \ldots \times S$ *is the n-ary Cartesian product of* S,
- $\mathbf{P}^n(s, s') = \mathbf{P}(s_1, s'_1) \cdot \ldots \cdot \mathbf{P}(s_n, s'_n)$ *for all* $s = (s_1, \ldots, s_n) \in S^n$ *and* $s' = (s'_1, \ldots, s'_n) \in S^n$,
- $AP^n = \cup_{i=1}^n AP_i$, *where* $AP_i = \{a_i \mid a \in AP\}$ *for* $i \in [1, n]$, *and*
- $L^n(s) = \cup_{i=1}^n L_i(s_i)$ *for all* $s = (s_1, \ldots, s_n) \in S^n$ *with* $L_i(s_i) = \{a_i \mid a \in L(s_i)\}$ *for* $i \in [1, n]$. $\qquad\qquad\qquad\square$

The semantics judgment rules to evaluate formula φ_{pac} for a DTMC $\mathcal{M} = (S, \mathbf{P}, AP, L)$ and an n-tuple $s = (s_1, \ldots, s_n) \in S^n$ of states are the following:

$$
\begin{aligned}
\mathcal{M}, s &\models \exists\sigma.\psi & \text{iff}\quad & \exists s_{n+1} \in S.\ \mathcal{M}, (s_1, \ldots, s_n, s_{n+1}) \models \psi[AP_{n+1}/AP_\sigma] \\
\mathcal{M}, s &\models \forall\sigma.\psi & \text{iff}\quad & \forall s_{n+1} \in S.\ \mathcal{M}, (s_1, \ldots, s_n, s_{n+1}) \models \psi[AP_{n+1}/AP_\sigma] \\
\mathcal{M}, s &\models a_i & \text{iff}\quad & a \in L(s_i) \\
\mathcal{M}, s &\models \psi_1 \wedge \psi_2 & \text{iff}\quad & \mathcal{M}, s \models \psi_1 \ \text{and}\ \mathcal{M}, s \models \psi_2 \\
[\![\mathbb{P}(\varphi)]\!]_{\mathcal{M},s} & & =\quad & \Pr\{\pi \in Paths_s^{\mathcal{M}^n} \mid \mathcal{M}, \pi \models \varphi\} \\
\mathcal{M}, s &\models p_1 > p_2 & \text{iff}\quad & [\![p_1]\!]_{\mathcal{M},s} > [\![p_2]\!]_{\mathcal{M},s}
\end{aligned}
$$

where $a \in AP$ is an atomic proposition, σ is a *state variable* from a countably infinite supply of variables $\mathcal{V} = \{\sigma_1, \sigma_2, \ldots\}$, p is a *probability expression*.

The satisfaction relation for HyperPCTL path formulas is defined as follows, where π is a path of $\mathcal{M}^n$ for some $n \in \mathbb{N}_{>0}$; ψ, ψ_1, and ψ_2 are HyperPCTL state formulas:

$$
\mathcal{M}, \pi \models \psi_1 \,\mathcal{U}\, \psi_2 \quad \text{iff}\quad \exists j \geq 0.\Big(\mathcal{M}, \pi[j] \models \psi_2 \wedge \forall i \in [0, j).\mathcal{M}, \pi[i] \models \psi_1\Big)
$$

The satisfaction of formula φ_{pac} by a DTMC $\mathcal{M}=(S, \mathbf{P}, AP, L)$ is defined by:

$$
\mathcal{M} \models \varphi_{\mathsf{pac}} \qquad \text{iff} \qquad \mathcal{M}, () \models \varphi_{\mathsf{pac}}
$$

where $()$ is the empty sequence of states.

4   Problem Statement

We begin with the premise that a DTMC can be obtained by using various methods (e.g., based on learning or statistical experimental design), where each path is an experiment of the system under scrutiny, and probabilities are computed by the frequency of occurrence of states in the experiments. Thus, this paper is not concerned with how one obtains DTMCs. Our goal in this paper is to design algorithms that identify the probabilistic actual causes of an effect in a DTMC. More specifically, our algorithms take as input (1) a DTMC $\mathcal{M}$, and (2) a predicate φ^e and generate as output a predicate φ^c that is the probabilistic actual cause of φ^e in $\mathcal{M}$.

> **Decision Problem**
>
> Given (1) a DTMC $\mathcal{M}$ representing a causal model, and (2) a predicate φ_e, does there exist another predicate φ_c, such that $\mathcal{M} \models \varphi_{\mathsf{pac}}$?

Example 2. Continuing with Example 1, we consider the DTMC shown in Fig. 1b, along with the formula φ_{pac} defined in Equation (1) and the failure condition φ^{fail} introduced in Sect. 2. Suppose in Equation (1), we replace φ^e by φ^{fail} (i.e., the effect is failing to reach the flag). Observe that in this example $W = \{\}$. Now, our goal is answer the above decision problem; i.e., identifying φ^c. In this example, the answer is $\varphi^c \triangleq pos = 0.3 \wedge vel = 0.01 \wedge act = 1$:

- First observe that $\mathbb{P}(\neg\varphi^e_\sigma \; \mathcal{U} \; (\varphi^c_\sigma \wedge \mathbb{P}_{>0}(\Diamond\varphi^e_\sigma))) = 0.5 \times 0.7 \times 0.9 + 0.5 \times 0.3 \times 0.2 = 0.345$, that is the probability of reaching state s_7 through state s_1.
- Now, notice that $\mathbb{P}(\neg\varphi^c_{\sigma'} \, \mathcal{U} \, \varphi^e_{\sigma'}) = 0.5 \times 1 \times 0.3 = 0.15$.
- Since $W = \{\}$, then $(\wedge_{a\in W} \Box(a_\sigma \leftrightarrow a_{\sigma'}))$ trivially holds.

Consequently, since $0.345 > 0.15$, formula Equation (1) is satisfied: predicate φ^c is the cause for φ^{fail}. $\qquad\square$

5   SMT-Based Discovery of PAC

An SMT decision problem consists of two main components: (1) an SMT instance, including variables, their domains, and associated symbolic structures, and (2) a set of constraints that encode the logical relationship among these variables. In our SMT formulation, the objective is to symbolically identify a predicate φ^c, interpreted as the cause of the occurrence of the effect represented by the predicate φ^e. We introduce an uninterpreted function $f : 2^S \to \{\mathtt{True}, \mathtt{False}\}$ which represents the cause. For a set of states $C \subseteq S$, we say $f(C) = \mathtt{True}$ if the predicates associated with the states in C only hold in the actual world but do not hold in the counterfactual world; otherwise, $f(C) = \mathtt{False}$.

Throughout this section, we use p to denote real-valued SMT variables ranging over the interval $[0, 1]$ in the SMT encodings. We proceed by presenting the SMT encoding of the three subformulas introduced in Equation (1), namely ψ_{SE}, ψ_{AW}, and ψ_{CW}.

Encoding Equivalence of Contingencies (ψ_{SE}). Subformula ψ_{SE} requires that the actual and counterfactual computation trees agree on a set of all propositions W. However, due to potential differences in sampling frequencies, a direct state-wise comparison between paths is not meaningful. Instead, we require stutter-trace equivalence with respect to W. Two computation trees are stutter-trace equivalent with respect to a set of variables W if all paths from the root states to the absorbing states in these trees are stutter-equivalent with respect to W [5].

We begin with introducing EqW, which verifies whether two states agree on W, and define it as $\mathsf{EqW}_{s,s'} \triangleq \bigwedge_{a\in W}(a_s \leftrightarrow a_{s'})$.

We define $\mathsf{Cuts}_m(n)$, which returns the set of all strictly increasing tuples of indices $\langle x_0, \ldots, x_m \rangle$ taken from $\{0, \ldots, n\}$ with $x_0 = 0$ and $x_m = n$. Then, we define ST, which verifies whether two paths are stutter-equivalent with respect to the variables in W, as:

$$\mathsf{ST}_{\pi,\pi'} \triangleq \exists m \geq 1 \,.\, \exists x \in \mathsf{Cuts}_m(|\pi|) \,.\, \exists y \in \mathsf{Cuts}_m(|\pi'|) \,.$$

$$\bigwedge_{r=0}^{m-1} \left(\mathsf{EqW}_{\pi[x_r],\pi'[y_r]} \;\wedge\; \bigwedge_{i=x_r}^{x_{r+1}-1} \mathsf{EqW}_{\pi[i],\pi[x_r]} \;\wedge\; \bigwedge_{j=y_r}^{y_{r+1}-1} \mathsf{EqW}_{\pi'[j],\pi'[y_r]} \right)$$

The specification above requires selecting the number of blocks m such that the states in each block agree on W. There exists a way to partition the paths π and π' into m blocks, where x_r and y_r mark the starting indices of block r in π and π', respectively. For each block, we verify that the states at the same block index in both paths agree on W, and that within each block the states of each path also agree on W. We now define SE, which captures whether two computation trees rooted in s and s' are stutter-equivalent with respect to the variables in W, as follows:

$$\mathsf{SE}_{s,s'} \triangleq \forall \pi \in \mathit{fPaths}^{\mathcal{M}}_{s,\mathcal{H}} \,.\, \forall \pi' \in \mathit{fPaths}^{\mathcal{M}}_{s',\mathcal{H}} \,.\, \mathsf{ST}_{\pi,\pi'}$$

Here $\mathit{fPaths}^{\mathcal{M}}_{s,\mathcal{H}}$ and $\mathit{fPaths}^{\mathcal{M}}_{s',\mathcal{H}}$ are non-empty sets of paths from s and s', respectively, to states labeled by halt.

Encoding Actual World Computations (ψ_{AW}). We begin by computing p_{RE}, which corresponds to evaluating $\mathbb{P}(\Diamond \varphi^e)$, that is, the probability of reaching the effect. If $s \models \varphi^e$, then $p_{\mathsf{RE}_s} = 1$; if $s \models \neg\varphi^e \wedge \mathsf{halt}$, then $p_{\mathsf{RE}_s} = 0$. Otherwise:

$$p_{\mathsf{RE}_s} = \sum_{s' \in S} \mathbf{P}(s,s') \cdot p_{\mathsf{RE}_{s'}}$$

Next, we compute p_{AW}, which corresponds to evaluating ψ_{AW}. We set $p_{\mathsf{AW}_s} = 1$ if $(s \models \varphi^c) \wedge (p_{\mathsf{RE}_s} > 0)$. If $\left(s \models \neg\varphi^c \wedge (\mathsf{halt} \vee \varphi^e)\right) \vee (p_{\mathsf{RE}_s} = 0)$, we set $p_{\mathsf{AW}_s} = 0$. Otherwise:

$$p_{\mathsf{AW}_s} = \sum_{s' \in S \setminus \varphi^e} \mathbf{P}(s,s') \cdot p_{\mathsf{AW}_{s'}}$$

Encoding Counterfactual World Computations (ψ_{CW}). To reason about ψ_{CW}, we introduce p_{CW_s} and define it as follows: if $s \models \varphi^e$, then $p_{\mathsf{CW}_s} = 1$. If $s \models \neg\varphi^e \wedge (\mathsf{halt} \vee \varphi^c)$, then $p_{\mathsf{CW}_s} = 0$. Otherwise:

$$p_{\mathsf{CW}_s} = \sum_{s' \in S \setminus \varphi^c} \mathbf{P}(s,s') \cdot p_{\mathsf{CW}_{s'}}$$

Putting Everything Together. Finally, we add the top level SMT quantification:

$$\exists s \in S. \; \forall s' \in S. \; (p_{\mathsf{AW}_s} > p_{\mathsf{CW}_{s'}}) \wedge \mathsf{SE}_{s,s'}$$

If the SMT instance is satisfiable, then the interpretation witness of f returned by the SMT solver identifies the probabilistic actual cause of φ^c. In this formulation, we can have disjunctive causes, meaning that either φ^{c_1} is a cause of φ^e or φ^{c_2}

is a cause of φ^e. This is different from stating that $\varphi^{c_1} \vee \varphi^{c_2}$ is itself the cause of φ^e.

It is apparent that the SMT encoding presented in this section is correct by construction, as they directly mirror the formal PAC conditions in Equation (1). In addition, solving the equations in this section does not require fixed-point reasoning, as our DTMCs are acyclic and have bounded depth, the recursions always terminate in absorbing states. The complexity of HyperPCTL model checking is polynomial in the size of $\mathcal{M}$ while being PSPACE-hard with respect to the number of quantifiers appearing in the formula [1]. The complexity of reasoning the uninterpreted function f matches that of the corresponding SMT theory.

6 Abstraction-Refinement for Probabilistic Causal Models

The overall idea of our algorithm is the following steps:

1. **Predicate abstraction**: Start with the concrete DTMC, $\mathcal{M}$, and apply predicate abstraction to derive the initial abstract model, $\hat{\mathcal{M}}$ as an MDP.
2. **SMT-Based discovery**: Identify potential causal relation within the abstract model $\hat{\mathcal{M}}$. If a cause is discovered, it is confirmed as the explanation and the process terminates.
3. **Refinement**: If no causal relationship is identified, refine $\hat{\mathcal{M}}$ by splitting the relevant abstract state and revisit step 2.

We explain the details of Steps 1–3 in Sects. 6.1 to 6.3, respectively.

6.1   Predicate Abstraction

Predicates are Boolean expressions defined over the variables, and for any such expression ψ, its valuation is a function $[\![\psi]\!] : S \rightarrow \{0, 1\}$, where $[\![\psi]\!]_s = 1$ means state s satisfies predicate ψ. Following the construction in [26], let $\mathfrak{P} = \{\psi_1, \psi_2, \ldots, \psi_n\}$ denote a set of predicates. The set $\mathfrak{P}$ induces a partitioning of the state space into disjoint equivalence classes based on which predicates hold. Each equivalence class is represented as an n-bit vector, where the i-th bit indicates whether the corresponding predicate ψ_i is satisfied in that class. These bit vectors represent abstract states, which we denote by $\hat{S}$. For a given abstract state $\hat{s} \in \hat{S}$, we refer to the corresponding equivalence class as the concretization of $\hat{s}$, denoted by $\gamma(\hat{s})$. We define an abstraction function as $\hat{h}(s) = \left([\![\psi_1]\!]_s, \ldots, [\![\psi_n]\!]_s\right)$. This abstraction function induces an MDP $\hat{\mathcal{M}} = \left(\hat{S}, \mathcal{P}, \mathsf{AP}, \hat{L}, Act\right)$, where

- $Act(\hat{s}) = \{\alpha \in \gamma(\hat{s}) \mid \sum_{\hat{s}' \in \hat{S}} \mathcal{P}(\hat{s}, \alpha, \hat{s}') = 1\}$;
- $\mathcal{P}(\hat{s}, \alpha, \hat{s}') = \sum_{s' \in \gamma(\hat{s}')} \mathbf{P}(\alpha, s')$, and
- $\hat{L}(\hat{s}) = \bigcup_{s \in \gamma(\hat{s})} L(s)$.

The process of state abstraction transforms a DTMC into an MDP. Nondeterminism arises when an abstract state groups together multiple concrete states with different transition probabilities. This ambiguity is resolved by interpreting the distinct transition distributions inherited from the concrete states as the set of available actions in the corresponding abstract state of the MDP.

Lemma 1. *Let $\hat{\mathcal{M}}$ be the abstract model obtained through predicate abstraction. If $\hat{\mathcal{M}}$ satisfies a reachability property, then the concrete model $\mathcal{M}$ also satisfies that property.* □

The soundness of this lemma is guaranteed by the fact that the abstract model $\hat{\mathcal{M}}$ simulates the concrete model $\mathcal{M}$ [26].

Example 3. Continuing Example 1 and the DTMC in Fig. 1b, for the set of predicates $\mathfrak{P} = \{vel \geq 0.03, pos \geq 0.6, pos \geq 0.4, pos \geq 0.3\}$, the states in the DTMC are partitioned into six abstract states. Using the abstraction function $\hat{h}$, we construct the MDP in Fig. 2a. For example, concrete state s_7 is mapped to abstract state $\hat{s}_3$ since $\hat{h}(s_7) = (0, 0, 1, 1)$. The available actions of an abstract state are induced by the concrete states grouped into that abstract state based on $\mathfrak{P}$. For instance, in Fig. 2a, $Act(\hat{s}_1) = \{s_1, s_3, s_4\}$.

Before delving into the SMT-based discovery of probabilistic actual causes in the abstract model, we first address ψ_{SE} in Equation (1). In the abstraction-refinement technique, handling equivalence for W is challenging and we propose two techniques:

1. **Enumerating Subgraphs:** We begin by decomposing the original DTMC $\mathcal{M}$ into subgraphs $\{\mathcal{M}^1, \ldots, \mathcal{M}^n\}$, where each $\mathcal{M}^i$ is a subgraph in which all paths are mutually stutter-equivalent with respect to W. Within each subgraph $\mathcal{M}^i$, the objective is to identify a set of states representing the predicate φ^c. Further details are provided in the appendix [35].
2. **W-Preserving Abstraction:** In this approach, we restrict the abstraction function $\hat{h}$ to preserve the equality of computation trees with respect to variables in W. That is, for two concrete transitions $\mathbf{P}_{>0}(s_\sigma, s'_\sigma)$ and $\mathbf{P}_{>0}(s_{\sigma'}, s'_{\sigma'})$, if (1) $\bigwedge_{a \in W}(a_{s_\sigma} \leftrightarrow a_{s_{\sigma'}})$ and (2) $\bigwedge_{a \in W}(a_{s'_\sigma} \not\leftrightarrow a_{s'_{\sigma'}})$, then we require that (1) $\bigwedge_{a \in W}(a_{s_\sigma} \leftrightarrow a_{\hat{h}(s_{\sigma'})})$ and (2) $\bigwedge_{a \in W}(a_{s'_\sigma} \not\leftrightarrow a_{\hat{h}(s'_{\sigma'})})$, where σ and σ' are state variables. Otherwise, we will not be able to prove the soundness of the abstraction-refinement algorithm with respect to the contingencies defined by W.

6.2   Discovery of Probabilistic Causes in Abstract Model

To identify probabilistic actual causes in the abstract MDP, we first need to adapt the SMT-based encoding originally designed for DTMCs, as discussed in Sect. 5. Since we are dealing with an MDP, we compute the range of reachability

274   A. Rafieioskouei et al.

probabilities, by finding the schedulers that render minimum and maximum of ψ_{AW} and φ_{CW}, respectively:

$$\mathbb{P}^{\min}\left(\neg\varphi^e_{\hat{\sigma}}\ \mathcal{U}\ (\varphi^c_{\hat{\sigma}} \wedge \mathbb{P}^{\min}_{>0}(\Diamond\,\varphi^e_{\hat{\sigma}}))\right) > \mathbb{P}^{\max}\left(\neg\varphi^c_{\hat{\sigma}'}\,\mathcal{U}\,\varphi^e_{\hat{\sigma}'}\right)$$

The formula above indicates that the minimum reachability to the effect states in the actual world ($\psi_{\widehat{\mathsf{AW}}}$) must be greater than the maximum reachability in the counterfactual world ($\psi_{\widehat{\mathsf{CW}}}$). In addition, we reason about contingencies in the abstract model using ($\psi_{\widehat{\mathsf{SE}}}$). **Encoding actual world computations ($\psi_{\widehat{\mathsf{AW}}}$).** We begin by computing $p_{\widehat{\mathsf{RE}}}$, which corresponds to evaluating $\mathbb{P}^{\min}(\Diamond\,\varphi^e)$, that is, the minimum probability of reaching the effect in the abstract model. If $\hat{s} \models \varphi^e$, then $p_{\widehat{\mathsf{RE}}_{\hat{s}}} = 1$; if $\hat{s} \models \neg\varphi^e \wedge \mathsf{halt}$, then $p_{\widehat{\mathsf{RE}}_{\hat{s}}} = 0$. Otherwise:

$$p_{\widehat{\mathsf{RE}}_{\hat{s}}} = \min_{\alpha \in Act(\hat{s})} \sum_{\hat{s}' \in \hat{S}} \mathcal{P}(\hat{s}, \alpha, \hat{s}') \cdot p_{\widehat{\mathsf{RE}}_{\hat{s}'}}$$

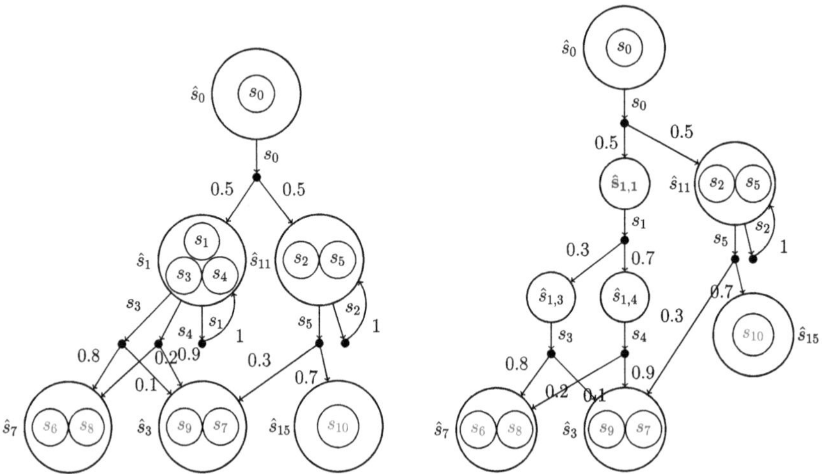

(a) Initial abstraction based on predicate set $\mathfrak{P}$.

(b) Abstract model after one step of refinement.

Fig. 2. Concrete and Abstract model.

Next, we compute $p_{\widehat{\mathsf{AW}}}$, which corresponds to evaluating $\psi_{\widehat{\mathsf{AW}}}$. We set $p_{\widehat{\mathsf{AW}}_{\hat{s}}} = 1$ if $(\hat{s} \models \varphi^c) \wedge (p_{\widehat{\mathsf{RE}}_{\hat{s}}} > 0)$. If $(\hat{s} \models \neg\varphi^c \wedge (\mathsf{halt} \vee \varphi^e)) \vee (p_{\widehat{\mathsf{RE}}_{\hat{s}}} = 0)$, we set $p_{\widehat{\mathsf{AW}}_{\hat{s}}} = 0$. Otherwise:

$$p_{\widehat{\mathsf{AW}}_{\hat{s}}} = \min_{\alpha \in Act(\hat{s})} \sum_{\hat{s}' \in \hat{S} \setminus \varphi^e} \mathcal{P}(\hat{s}, \alpha, \hat{s}') \cdot p_{\widehat{\mathsf{AW}}_{\hat{s}'}}$$

Encoding Counterfactual World Computations ($\psi_{\widehat{\mathsf{CW}}}$). We introduce a variable $p_{\widehat{\mathsf{CW}}_{\hat{s}}}$, such that if $\hat{s} \models \varphi^e$, then $p_{\widehat{\mathsf{CW}}_{\hat{s}}} = 1$. If $\hat{s} \models \neg\varphi^e \wedge (\mathsf{halt} \vee \varphi^c)$, then $p_{\widehat{\mathsf{CW}}_{\hat{s}}} = 0$. Otherwise:

$$p_{\widehat{\mathsf{CW}}_{\hat{s}}} = \max_{\alpha \in Act(\hat{s})} \sum_{\hat{s}' \in \hat{S} \setminus \varphi^c} \mathcal{P}(\hat{s}, \alpha, \hat{s}') \cdot p_{\widehat{\mathsf{CW}}_{\hat{s}'}}$$

Encoding Equivalence of Contingencies ($\psi_{\widehat{\mathsf{SE}}}$). To reason about $\psi_{\widehat{\mathsf{SE}}}$, we consider two cases. If we use the enumerating subgraph strategy, then $\psi_{\widehat{\mathsf{SE}}}$ is trivially satisfied and does not need to be encoded. However, if we use a W-preserving abstraction, we must encode $\psi_{\widehat{\mathsf{SE}}}$ explicitly. To encode $\psi_{\widehat{\mathsf{SE}}}$, we modify the SE defined in Sect. 5 and introduce $\widehat{\mathsf{SE}}$, defined as:

$$\widehat{\mathsf{SE}}_{\hat{s},\hat{s}'} \triangleq \forall \mathfrak{s}, \mathfrak{s}' \in \mathfrak{S} \; . \; \forall \pi \in \mathit{fPaths}_{\hat{s},\mathcal{H}}^{\hat{\mathcal{M}}^{\mathfrak{s}}} \; . \; \forall \pi' \in \mathit{fPaths}_{\hat{s}',\mathcal{H}}^{\hat{\mathcal{M}}^{\mathfrak{s}'}} \; . \; \mathsf{ST}_{\pi,\pi'}$$

where $\mathfrak{S}$ denotes the finite set of memoryless schedulers available in the abstract model $\hat{\mathcal{M}}$.

The rest of the SMT formulation, including the uninterpreted function representing the actual world and the SMT variables φ^c and φ^e, remains consistent with the details presented in Sect. 5.

To proceed with abstraction-refinement algorithm, we search for a cause using the SMT-based approach described in this section. If we find a cause, it is returned as the probabilistic actual cause of the effect. Otherwise, the absence of a cause suggests that the current abstract model is too coarse to capture the underlying causality and must therefore be refined.

Example 4. Continuing with Example 3, consider the abstract model shown in Fig. 2a and the formula φ^{fail}. Our goal is to solve the decision problem of identifying φ^c.

- First, it is clear that states $\hat{s}_7$ and $\hat{s}_{15}$ are not candidates, since their minimum and maximum reachability probabilities to effect are both 0. Similarly, $\hat{s}_3$ is excluded because it is labeled with φ^{fail}. State $\hat{s}_0$ is also not a valid candidate, since it is always included in both actual and counterfactual computation trees.
- Consider $\hat{s}_1$ as a candidate. We find that the minimum probability of reaching $\hat{s}_3$ through $\hat{s}_1$ is $\mathbb{P}^{\min}(\neg\varphi_{\hat{\sigma}}^e \; \mathcal{U} \; (\varphi_{\hat{\sigma}}^c \wedge \mathbb{P}_{>0}^{\min}(\Diamond \varphi_{\hat{\sigma}}^e))) = 0.5 \times 0.2 = 0.1$. On the other hand, in the counterfactual world, the maximum probability is $\mathbb{P}^{\max} = (\neg\varphi_{\hat{\sigma}'}^c \mathcal{U} \, \varphi_{\hat{\sigma}'}^e) = 0.5 \times 0.3 = 0.15$. Since $0.1 \not> 0.15$, $\hat{s}_1$ does not satisfy the condition.
- Next, consider $\hat{s}_{11}$. Using the same reasoning, we find that the minimum probability of reaching the effect in the actual world is 0.15, while the maximum in the counterfactual world is 0.45. Again, $0.15 \not> 0.45$, so this state also fails the condition.

Consequently, none of the abstract states satisfy the specification. Since we couldn't find a valid cause, we conclude that the current abstraction is too coarse and proceed with refinement, described next. $\qquad\square$

6.3   Refinement

As discussed in Sect. 6.2, if a cause cannot be found, then the abstract MDP is possibly too coarse and must be refined to a lower level of abstraction. In the refinement process, the coarser model $\hat{\mathcal{M}}_i = (\hat{S}_i, \mathcal{P}_i, \mathsf{AP}, \hat{L}_i, Act_i)$ is refined to a less coarse model $\hat{\mathcal{M}}_{i+1}$ by identifying the abstract state $\hat{s}_\Delta$. We employ two heuristic methods to identify $\hat{s}_\Delta$:

1. **Number of Available Actions**: The state with the maximum number of available actions in $\hat{\mathcal{M}}_i$.

$$\hat{s}_{\Delta_i} = \underset{\hat{s} \in \hat{S}_i}{\mathrm{argmax}} \; \left| Act(\hat{s}) \right|$$

2. **Range of Reachability:** The state with the maximum range between its minimum ($\mathbb{P}^{\min}(\Diamond\,\varphi^e)$) and maximum ($\mathbb{P}^{\max}(\Diamond\,\varphi^e)$) reachability to the effect states in the $\hat{\mathcal{M}}_i$.

$$\hat{s}_{\Delta_i} = \underset{\hat{s} \in \hat{S}_i}{\mathrm{argmax}} \; \left(\mathbb{P}^{\max}(\hat{s} \models \Diamond\,\varphi^e) - \mathbb{P}^{\min}(\hat{s} \models \Diamond\,\varphi^e) \right)$$

Once $\hat{s}_{\Delta_i}$ is identified, we refine it by splitting the underlying concrete states it represents, thereby constructing a less coarse abstract model $\hat{\mathcal{M}}_{i+1}$ for the next iteration. This refinement step induces a new MDP $\mathcal{M}_{i+1} = (\hat{S}_{i+1}, \mathcal{P}_{i+1}, \mathsf{AP}, \hat{L}_{i+1}, Act_{i+1})$, where $\hat{S}_{i+1} = (\hat{S}_i \setminus \hat{s}_{\Delta_i}) \cup \gamma(\hat{s}_{\Delta_i})$, $Act_{i+1}(\hat{s}) = \{\alpha \in S \mid \sum_{\hat{s}' \in \hat{S}_{i+1}} \mathcal{P}_{i+1}(\hat{s}, \alpha, \hat{s}') = 1\}$, $\mathcal{P}_{i+1}(\hat{s}, \alpha, \hat{s}') = \sum_{s' \in \gamma(\hat{s}')} \mathbf{P}(\alpha, s')$, and $\hat{L}_{i+1}(\hat{s}) = \bigcup_{s \in \gamma(\hat{s})} L(s)$.

Example 5. Continuing from Example 4, we observe that the initial abstract model shown in Fig. 2a is too coarse and requires refinement. In this model, $\hat{s}_\Delta$ corresponds to $\hat{s}_1$ according to both heuristics, as it exhibits the widest reachability range (i.e., $[0.2, 0.9]$) and possesses the highest number of available actions (i.e., 3). To refine the abstraction, we split $\hat{s}_1$ into three more precise abstract states: $\hat{s}_{1,1}$, $\hat{s}_{1,3}$, and $\hat{s}_{1,4}$. The refined abstract model is shown in Fig. 2b. In this refined model, if we consider $\hat{s}_{1,1}$ as a candidate, we compute the minimum probability of reaching $\hat{s}_3$ through $\hat{s}_{1,1}$ in the actual world as, $\mathbb{P}^{\min}(\neg\varphi^e_{\hat{\sigma}} \; \mathcal{U} \; (\varphi^c_{\hat{\sigma}} \wedge \mathbb{P}^{\min}_{>0}(\Diamond\,\varphi^e_{\hat{\sigma}}))) = 0.5 \times 0.7 \times 0.9 + 0.5 \times 0.3 \times 0.2 = 0.345$. On the other hand, in the counterfactual world, the maximum probability of reaching the effect is $\mathbb{P}^{\max} = (\neg\varphi^c_{\hat{\sigma}'} \; \mathcal{U} \; \varphi^e_{\hat{\sigma}'}) = 0.5 \times 0.3 = 0.15$. Since $0.345 > 0.15$, $\hat{s}_{1,1}$ satisfies the specifications.   □

Theorem 1. *Let $\mathcal{M}$ be a concrete causal model and φ_c and φ_e be two predicates. If φ_c is an actual cause of φ_e identified by our abstraction-refinement technique, then φ_c is an actual cause of φ_e in $\mathcal{M}$.*

7   Experimental Evaluation

In this section, we evaluate our approach by applying the techniques discussed in Sects. 5 and 6 to three case studies: (1) the Mountain Car and (2) Lunar Lander environments from OpenAI Gym [9], and (3) an F-16 autopilot simulator [25] that employs an MPC controller.[1]

7.1   Implementation

DTMC Generation. The environments used in our case studies are originally deterministic. To introduce stochastic behavior (e.g., noise), we synthesize DTMCs for each environment. To generate a DTMC, we begin by constructing random probability vectors whose components sum to one. For a state with k outgoing transitions, we sample k values $u_i \sim U(0,1)$ and normalize them as $p_i = \frac{u_i}{\sum_{j=1}^{k} u_j}$. The value of k is randomly chosen, allowing control over the degree of branching in the DTMC. Each distribution defines transitions to successor states, generated by adding noise to the original successor states. We continue this process up to a fixed number of steps to control the overall size of the resulting DTMC.

Algorithm. We used the Python programming language along with the Python API of the Z3 SMT solver [14]. In general, Z3 is incomplete for non-linear arithmetic; however, it successfully handles all our benchmarks, as the considered DTMC models are bounded and acyclic. Our implementation involves two key hyperparameters: the initial set of predicates and the splitting strategy for refining an abstract state $\hat{s}_\Delta$. In order to control the splitting strategy, we introduced a tunable hyperparameter $\beta \in (0,1]$ which controls the fraction of an abstract state to be concretized when refining the abstract state. This fraction determines the trade-off between the number of iterations required to identify a cause and the size of the model in the subsequent iteration. It is important to note that the performance of the abstraction-refinement technique is highly sensitive to these hyperparameters. For example, an unsuitable initial predicate set may result in a model that is too coarse, requiring several rounds of refinement to discover the cause. On the other hand, using a finer abstraction generates a large number of abstract states, thereby diminishing the performance advantages of abstraction and resulting in solving times comparable to those of the concrete model. Moreover, a poor choice of predicates may result in abstract states that conflate concrete states where the effect holds with those where it does not. This ambiguity can lead to difficulties in computing reachability probabilities, as it becomes unclear whether such an abstract state should be considered as satisfying the effect. Therefore, it is crucial to select predicates that distinguish states based on key properties, including the effect. Further details on splitting strategies are available in [35].

---

[1] All implementation artifacts are available at https://github.com/rogaleke/Prob-HP.

7.2   Experimental Settings

All of our experiments were conducted on a single core of the Intel i9-12900K CPU, which features a 16-core architecture and operates @5.2GHz.

7.3   Case Study 1: Mountain Car

Our first case study extends the running example. As illustrated in Fig. 1ba, the car begins in a valley between two mountains with the objective of reaching the peak of the right hand mountain before a specified time-bound. The action controller in the Mountain Car scenario uses a Deep Q-Network (DQN) based on the model provided in [33], and is trained using the hyperparameters from [32]. The objective of this case study is to discover the cause of failure, defined as the car failing to reach the target position within a fixed number of steps. To construct DTMCs, we execute the pre-trained controller under multiple initial valuations, following the procedure detailed in Sect. 7.1.

7.4   Case Study 2: Lunar Lander

In this case study, a lunar lander begins at a certain altitude with the goal of landing on a designated landing pad. The Lunar Lander environment includes eight variables (including x and y coordinates, linear and angular velocities, angle, etc.). Failure in this case study is defined as failing to land safely on the landing pad within a designated number of steps. The action controller is implemented using Proximal Policy Optimization (PPO), following the model architecture from [33] and trained with hyperparameters from [32]. The controller operates over four discrete actions available in the environment, defined as $act = \{0, 1, 2, 3\}$, where 0 corresponds to doing nothing, 1 fires the left orientation engine, 2 fires the main engine, and 3 fires the right orientation engine. We construct DTMCs of varying sizes by executing the pre-trained controller under multiple initial valuations and varying parameters such as the number of simulation steps, while introducing uncertainty through noise injection.

7.5   Case Study 3: F-16 Autopilot MPC Controller [25]

This benchmark models the outer-loop controller of the F-16 fighter jet. We examine two scenarios: the first focuses on reaching a specified speed, and the second on achieving a target altitude.

First Scenario. The first scenario investigates how the engine responds under control inputs. The simulation models two variables: airspeed Vt and engine power lag state pow, with all remaining variables held constant. Control is applied to adjust Vt toward a setpoint value using the throttle input δ_t. The scenario starts with initial values for Vt and alt and aims to verify whether the system can reach the desired Vt setpoint within a specified time. In this scenario, we investigate the causes of failure, which are categorized as either failing to reach the target Vt within the designated time or violating the safety constraints of the aircraft.

Table 1. Mountain Car: Comparing Abstraction-Refinement (Two Heuristics) and Concrete SMT Solving, Averaged Across 10 Independent Runs.

Case	$|S|$	Conc(s)	Abs (Act Based)			Abs (Range Based)		
			Abs-Ref(s)	SMT(s)	Tot(s)	Abs-Ref(s)	SMT(s)	Tot(s)
Mountain Car	203	1.41	0.02	0.64	<u>0.67</u>	0.02	0.31	**0.33**
	359	3.22	0.02	2.52	<u>2.53</u>	0.02	1.82	**1.84**
	1221	14.34	0.06	0.88	**0.94**	0.06	0.87	**0.94**
	1943	4.97	0.06	2.08	**2.14**	0.06	2.12	<u>2.18</u>
	2261	9.85	0.09	4.73	**4.82**	0.09	4.93	<u>5.02</u>
	2603	12.41	0.09	6.38	**6.46**	0.08	6.46	<u>6.54</u>
	2888	1.78	0.08	1.15	**1.23**	0.08	1.15	**1.23**
	4558	40.19	0.19	12.69	<u>12.89</u>	0.19	12.59	**12.79**
	27547	18.37	0.51	0.70	**1.21**	0.51	0.71	<u>1.22</u>

Second Scenario. In this scenario, the aircraft is expected to reach a specific target altitude alt, and it is flying without any roll or yaw. The inputs to the system are the engine throttle δ_t and the elevator δ_e. The model includes seven state variables: airspeed Vt, angle of attack α, pitch angle θ, pitch rate Q, altitude alt, engine power lag pow, and upward acceleration (G-force) Nz. In this scenario, failures are defined as either not reaching the required altitude alt within the given time frame or violating safety parameters, such as limits on G-force and angle of attack.

Table 2. Lunar Lander: Comparing Abstraction-Refinement (Two Heuristics) and Concrete SMT Solving, Averaged Across 10 Independent Runs.

Case	$|S|$	Conc(s)	Abs (Act Based)			Abs (Range Based)		
			Abs-Ref(s)	SMT(s)	Tot(s)	Abs-Ref(s)	SMT(s)	Tot(s)
Lunar Lander	115	**0.07**	0.01	0.07	0.08	0.01	0.07	0.08
	192	**0.09**	0.01	0.19	0.20	0.01	0.20	0.21
	1381	0.59	0.04	0.32	<u>0.36</u>	0.04	0.29	**0.33**
	2458	3.56	0.07	0.62	<u>0.68</u>	0.06	0.53	**0.59**
	3431	9.71	0.09	1.33	**1.41**	0.09	2.65	<u>2.74</u>
	5670	<u>6.63</u>	0.32	18.31	18.63	0.57	3.44	**4.01**
	9282	51.36	0.23	1.16	**1.38**	0.22	1.32	<u>1.54</u>
	11653	24.96	0.30	0.43	<u>0.74</u>	0.30	0.43	**0.73**
	14200	37.20	0.35	3.05	**3.39**	0.34	4.73	<u>5.07</u>
	47915	39.15	1.20	9.89	<u>11.09</u>	1.20	8.73	**9.92**

7.6   Performance Analysis

Tables 1 and 2 summarize the results for the Mountain Car and Lunar Lander environments, and Tables 3 and 4 present the outcomes for the two F-16 scenarios. Bold numbers indicate the best results, and underlined numbers denote the second-best results. As shown in the tables, the abstraction-refinement algorithm (employing both refinement heuristics) achieves significantly better performance in discovering probabilistic actual causes compared to concrete SMT solving especially in larger DTMCs. In most experiments, both refinement heuristics achieve comparable performance, with the range-based heuristic performing slightly better in the Mountain Car, Lunar Lander, and second F-16 scenario, while the action-based heuristic shows an advantage in the first F-16 scenario. In addition, in some smaller DTMCs, the overhead of constructing abstract models and performing refinement steps may outweigh the benefits, making direct SMT solving on the concrete model more efficient for identifying causes. However, for larger DTMCs especially in the F-16 first scenario and second scenario, the performance gains from abstraction-refinement are substantial.

Table 3. F-16 1st Scenario: Comparing Abstraction-Refinement (Two Heuristics) and Concrete SMT Solving, Averaged Across 10 Independent Runs.

Case	$|S|$	**Conc(s)**	**Abs (Act Based)**			**Abs (Range Based)**		
			Abs-Ref(s)	SMT(s)	**Tot(s)**	Abs-Ref(s)	SMT(s)	**Tot(s)**
	84	1.47	0.01	0.10	**0.11**	0.04	0.40	<u>0.44</u>
	130	1.64	0.01	0.13	**0.14**	0.02	0.31	<u>0.32</u>
	371	18.10	0.05	4.31	<u>4.36</u>	0.03	1.99	**2.02**
	401	15.24	0.02	0.53	**0.55**	0.03	0.83	<u>0.87</u>
F-16 1st scenario	599	32.16	0.08	8.30	<u>8.38</u>	0.04	3.84	**3.88**
	630	50.64	0.05	3.48	**3.52**	0.08	4.12	<u>4.20</u>
	2511	883.68	0.19	41.15	**41.33**	0.38	72.47	<u>72.85</u>
	3405	1287.35	0.30	137.66	**137.95**	0.21	145.86	<u>146.07</u>
	4175	1028.07	0.19	33.90	**34.09**	0.37	56.58	<u>56.96</u>
	13842	9582.02	1.00	633.01	<u>634.01</u>	0.50	395.62	**396.12**

Additionally, we observe a meaningful correlation between execution time and the size of the DTMC. Notably, performance is also influenced by the depth-width ratio of the DTMC structure. In our experiments, deeper DTMCs with lower branching factors (as observed in F-16 scenarios) make cause discovery more challenging for both the abstraction-refinement and concrete SMT approaches. For instance, comparing the F-16 first scenario with 13k states in Table 3 to the Lunar Lander with 14k states in Table 2 reveals a significant performance difference. This discrepancy is due to differences in the DTMC structures: the F-16 model is deeper with less branching, while the Lunar Lander DTMC is shallower with more branching. Consequently, the execution time for causal discovery depends on both the number of states and the structure of the DTMC.

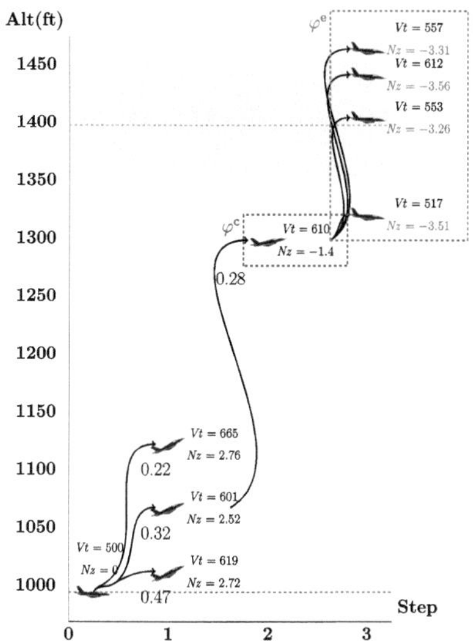

Fig. 3. The F-16 actual world computation tree.

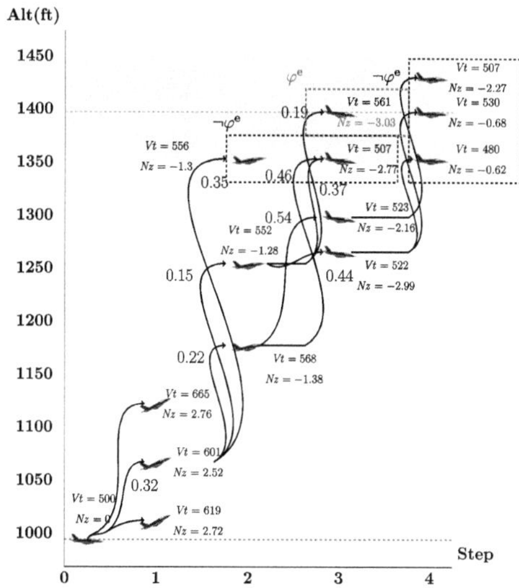

Fig. 4. The F-16 counterfactual world computation tree.

7.7 Causality Analysis

In this section, we apply the causal discovery method to the DTMCs generated from an F-16 simulation scenario by counterfactual computation trees where the probability of failure is lower than in the actual world. Causal analysis for the Lunar Lander case study is provided in appendix [35].

Figure 3 illustrates a scenario from the F-16 Autopilot Simulation where the aircraft starts at an altitude of 1000ft and is expected to reach $1400 \pm \epsilon$ ft within a specific time frame. Failure occurs either when the aircraft does not reach the target altitude within the time limit or violates one of the safety constraints such as G-force (Nz) or angle of attack (α). Our SMT solving technique identifies $\varphi^c \triangleq (alt = 1293 \wedge Vt = 610 \wedge \alpha = -1.1 \wedge \theta = 13 \wedge Nz = -1.4)$ and the computation tree observable in Fig. 3 is the actual world, as a result, Fig. 4 is the counterfactual world. In this simulation, the aircraft starts to gain altitude by adjusting the elevators and throttles. However, due to noise (e.g., turbulence), it may reach one of the following three states: $s_1 \triangleq (alt = 1087 \wedge Vt = 601 \wedge \alpha = 10.3 \wedge \theta = 24 \wedge Nz = 2.52)$ with probability 0.32, $s_2 \triangleq (alt = 1132 \wedge Vt = 665 \wedge \alpha = 9.7 \wedge \theta = 29 \wedge Nz = 2.76)$ with probability 0.22, and $s_3 \triangleq (alt = 1021 \wedge Vt = 619 \wedge \alpha = 9.1 \wedge \theta = 28 \wedge Nz = 2.72)$ with probability 0.47.

When the aircraft reaches state s_1, it continues to gain altitude. At this point, due to noise, the aircraft may enter one of four possible states. Among these, one state is identified as a probabilistic actual cause of failure $\varphi^c \triangleq (alt = 1293, Vt = 610, \alpha = -1.1, \theta = 13, \text{and } Nz = -1.4)$, where the probability of reaching failure (specifically due to a violation of the G-force constraint) is 1.

Table 4. F-16 2nd Scenario: Comparing Abstraction-Refinement (Two Heuristics) and Concrete SMT Solving, Averaged Across 10 Independent Runs.

Case	$|S|$	Conc(s)	Abs (Act Based)			Abs (Range Based)		
			Abs-Ref(s)	SMT(s)	Tot(s)	Abs-Ref(s)	SMT(s)	Tot(s)
F-16 2nd scenario	101	**1.03**	0.06	4.62	4.67	0.03	1.32	<u>1.35</u>
	187	**3.57**	0.04	6.71	6.76	0.04	3.72	<u>3.76</u>
	403	<u>13.19</u>	0.05	39.59	39.63	0.03	6.13	**6.15**
	417	15.54	0.03	3.70	**3.73**	0.03	5.90	<u>5.93</u>
	425	41.99	0.03	2.08	**2.11**	0.04	2.39	<u>2.43</u>
	621	33.83	0.04	22.24	<u>22.28</u>	0.04	19.10	**19.14**
	879	195.42	0.05	7.87	**7.92**	0.07	9.97	<u>10.04</u>
	1285	153.02	0.09	141.37	<u>141.45</u>	0.17	100.49	**100.66**
	1925	427.00	0.10	88.66	<u>88.76</u>	0.10	29.66	**29.75**
	6535	7463.79	0.17	490.09	<u>490.26</u>	0.16	319.36	**319.52**

In contrast, the counterfactual world produces three alternative successor states. One of them, $s_4 \triangleq (alt = 1349 \wedge Vt = 556 \wedge \alpha = -1.1 \wedge \theta = 12 \wedge Nz = -1.3)$ with a probability of 0.35, successfully reaches the target altitude without violating safety constraints. Another, $s_5 \triangleq (alt = 1167 \wedge Vt = 568 \wedge \alpha = -1.1 \wedge \theta = 12 \wedge Nz = -1.38)$, has a probability of zero for reaching the effect (i.e., failure). The third, $s_6 \triangleq (alt = 1241 \wedge Vt = 552 \wedge \alpha = -1.1 \wedge \theta = 12 \wedge Nz = -1.28)$, leads to failure with a probability of 0.19.

A closer examination of the dynamics reveals that s_6 and the identified causal state φ^c are similar in terms of system variables. In both s_6 and φ^c, the aircraft gains a significant amount of altitude, prompting the simulator to issue a strong correction using the elevator. This abrupt correction induces a negative G-force condition, which is more physiologically demanding for pilots than positive G-force. In comparison with s_6 and φ^c, s_4 successfully reaches the target altitude without safety violations, and s_5 ascends more gradually, maintaining compliance with all safety thresholds and achieving the target with a probability of 1. This experiment highlights how environmental noise, such as turbulence, can drive the system into failure-inducing states, demonstrating the importance of accounting for such uncertainties in the analysis of safety failures within CPS.

8   Related Work

Causal analysis is of growing importance in formal verification in hopes of answering the question of "why?" when it comes to faults in complex systems. A survey written by Baier et al. [3] reviews numerous approaches that are inspired by the HP causality framework to explain the behavior of the system. Recent research applies temporal logic to model and explain causality and bugs [8,11,12,17]. In the CPS domain and robotics, causality has been explored to repair AI-enabled controllers via HP models, search algorithms, and constraint verification [2,30,31,40]. The studies [10,18–22,38] present alternative formal frameworks to the HP causal model, whereas our approach uses causal analysis specifically to pinpoint the cause of a given effect. Yet, these works often address only the modeling aspects or overlook the scalability challenges in automated, counterfactual reasoning. A recent approach presented in [34] offers a more efficient method for directly identifying failure inducing causes from system execution traces. Furthermore, causal analysis is also applied in fields like medicine, where it helps uncover the causes of biological phenomena [23,36]. Although methods have been proposed to explain counterexamples in model checking [7,11], our work specifically targets efficient failure cause identification in embedded systems.

Expanding from deterministic to probabilistic systems, such as probabilistic programs and Markov chains, demands adaptations of causality notions to account for event likelihoods. In [41], Ziemek et al. use a different notion of causality extended to Markov Chains where the cause, which is composed of a set of executions, must cover all instances of the effect. In [4], Baier et al. introduce the notion of probability-raising causation in MDPs, and extended this idea to explore quality measures of predictors and the notion of probability-raising policies as schedulers in MDPs [6]. While their framework allows for cycles, our work explores a different dimension of causality that preserves contingencies between the actual and counterfactual worlds, rather than focusing solely on probability-raising causes. In parallel to the above formal verification-oriented approach, Kleinberg et al. introduce actual causality in Markov chains from a data-driven perspective [28]. This perspective was later extended to token causality in [29] and further extended in [39]. In [31] the authors explore the probabilistic setting, but do so from the perspective of robot task execution. They derive probability distributions from a simulation in which a robot is tasked with pouring one container of marbles into another container. Their goal is to identify causes of unwanted behavior and to evaluate corrective actions the robotic agent can take. However we aim to propose an efficient method to discover causes, specifically in DTMCs.

9   Conclusion and Future Work

In this paper, we proposed two algorithms for efficient discovery of probabilistic actual causes due to Fenton-Glynn in systems characterized by stochastic behavior and noise. We (1) formulated the discovery of probabilistic actual causality in computing systems as an SMT problem, and (2) addressed the scalability challenges by introducing an abstraction-refinement technique that significantly improves efficiency. We demonstrated the effectiveness of our approach through three case studies, identifying probabilistic causes of safety violations in (1) the Mountain Car problem, (2) the Lunar Lander benchmark, and (3) MPC controller for an F-16 autopilot simulator.

This paper is the first step in formal analysis of actual causality in probabilistic systems. There are two immediate extensions with significant practical implications: actual causal inference in input models that allow nondeterministic decision making (MDPs) and those where full observability is not possible

(POMDPs). In this paper, we formalize PAC in the hyperproperty setting. In [27], reinforcement learning is used to generate policies that maximize the satisfaction of hyperproperties. Combining these two studies could enable training agents that learn to identify counterfactual scenarios or achieve counterfactual realizability [37].

References

1. Ábrahám, E., Bonakdarpour, B.: HyperPCTL: a temporal logic for probabilistic hyperproperties. In: Proceedings of the 15th International Conference on Quantitative Evaluation of Systems (QEST), pp. 20–35 (2018)
2. Araujo, H., et al.: Kaspar causally explains. In: Cavallo, F., et al. (eds.) Social Robotics, pp. 85–99. Springer, Cham (2022). https://doi.org/10.1007/978-3-031-24670-8_9
3. Baier, C., et al.: From verification to causality-based explications (2021)
4. Baier, C., Funke, F., Piribauer, J., Ziemek, R.: On probability-raising causality in Markov decision processes. In: FoSSaCS 2022. LNCS, vol. 13242, pp. 40–60. Springer, Cham (2022). https://doi.org/10.1007/978-3-030-99253-8_3
5. Baier, C., Katoen, J.-P.: Principles of Model Checking. The MIT Press, Cambridge (2008)
6. Baier, C., Klüppelholz, S., Piribauer, J., Ziemek, R.: Formal quality measures for predictors in Markov decision processes (2024)
7. Beer, I., Ben-David, S., Chockler, H., Orni, A., Trefler, R.: Explaining counterexamples using causality. In: Bouajjani, A., Maler, O. (eds.) CAV 2009. LNCS, vol. 5643, pp. 94–108. Springer, Heidelberg (2009). https://doi.org/10.1007/978-3-642-02658-4_11
8. Beutner, R., Finkbeiner, B., Frenkel, H., Siber, J.: Checking and sketching causes on temporal sequences. In: Proceedings of the 21st International Symposium Automated Technology for Verification and Analysis (ATVA), pp. 314–327 (2023)
9. Brockman, G., et al.: OpenAI Gym (2016)
10. Broy, M.: Time, causality, and realizability: engineering interactive, distributed software systems. J. Syst. Softw. **210**, 111940 (2024)
11. Coenen, N., et al.: Explaining hyperproperty violations. In: Proceedings of the 34th International Conference on Computer Aided Verification(CAV), Part I, pp. 407–429 (2022)
12. Coenen, N., Finkbeiner, B., Frenkel, H., Hahn, C., Metzger, N., Siber, J.: Temporal causality in reactive systems. In: Proceedings of the 20th International Symposium on Automated Technology for Verification and Analysis (ATVA), pp. 208–224. Springer, Heidelberg (2022). https://doi.org/10.1007/978-3-031-19992-9_13
13. Datta, A., Garg, D., Kaynar, D., Sharma, D., Sinha, A.: Program actions as actual causes: a building block for accountability. In: 2015 IEEE 28th Computer Security Foundations Symposium, pp. 261–275 (2015)
14. de Moura, L.M., Bjørner, N.: Z3: an efficient SMT solver. In: Tools and Algorithms for the Construction and Analysis of Systems (TACAS), pp. 337–340 (2008)
15. Fenton-Glynn, L.: A proposed probabilistic extension of the halpern and pearl definition of 'Actual Cause'. Br. J. Phil. Sci. **68**(4), 1061–1124 (2017)
16. Finkbeiner, B., Jahn, F., Siber, J.: Counterfactual explanations for MITL violations. In: Barman, S., Lasota, S. (eds.) 44th IARCS Annual Conference on Foundations of Software Technology and Theoretical Computer Science (FSTTCS 2024),

vol. 323 of Leibniz International Proceedings in Informatics (LIPIcs), Dagstuhl, Germany, pp. 22:1–22:25. Schloss Dagstuhl – Leibniz-Zentrum für Informatik (2024)

17. Finkbeiner, B., Kupriyanov, A.: Causality-based model checking. In: Proceedings 2nd International Workshop on Causal Reasoning for Embedded and safety-critical Systems Technologies (CREST), vol. 259, pp. 31–38 (2017)

18. Goessler, G., Astefanoaei, L.: Blaming in component-based real-time systems. In: Proceedings of the International Conference on Embedded Software (EMSOFT), pp. 7:1–7:10. ACM (2014)

19. Gößler, G., Métayer, D.L.: A general trace-based framework of logical causality. In: Fiadeiro, J.L., Liu, Z., Xue, J. (eds.) Proceedings of the10th International Symposium on Formal Aspects of Component Software (FACS), pp. 157–173 (2013)

20. Gößler, G., Métayer, D.L., Raclet, J.: Causality analysis in contract violation. In Proceedings of the First International Conference on Runtime Verification (RV), pp. 270–284 (2010)

21. Gößler, G., Sokolsky, O., Stefani, J.: Counterfactual causality from first principles? In: Proceedings 2nd International Workshop on Causal Reasoning for Embedded and safety-critical Systems Technologies (CREST), vol. 259, pp. 47–53 (2017)

22. Gössler, G., Stefani, J.: Causality analysis and fault ascription in component-based systems. Theor. Comput. Sci. **837**, 158–180 (2020)

23. Guha, A., et al.: Ai-driven prediction of cardio-oncology biomarkers through protein corona analysis. Chem. Eng. J. **509**, 161134 (2025)

24. Halpern, J.Y.: Actual Causality. MIT Press, Cambridge (2016)

25. Heidlauf, P., Collins, A., Bolender, M., Bak, S.: Verification challenges in F-16 ground collision avoidance and other automated maneuvers. In: ARCH18. 5th International Workshop on Applied Verification of Continuous and Hybrid Systems, vol. 54, pp. 208–217 (2018)

26. Hermanns, H., Wachter, B., Zhang, L.: Probabilistic CEGAR. In: Gupta, A., Malik, S. (eds.) CAV 2008. LNCS, vol. 5123, pp. 162–175. Springer, Heidelberg (2008). https://doi.org/10.1007/978-3-540-70545-1_16

27. Hsu, T.-H., Rafieioskouei, A., Bonakdarpour, B.: HYPRL: reinforcement learning of control policies for hyperproperties. In: The Thirty-ninth Annual Conference on Neural Information Processing Systems (2025)

28. Kleinberg, S., Mishra, B.: The temporal logic of causal structures. In: Proceedings of the Twenty-Fifth Conference on Uncertainty in Artificial Intelligence, UAI '09, Arlington, Virginia, USA, pp. 303–312. AUAI Press (2009)

29. Kleinberg, S., Mishra, B.: The temporal logic of token causes. In: Proceedings of the Twelfth International Conference on Principles of Knowledge Representation and Reasoning, KR'10, pp. 575–577. AAAI Press (2010)

30. Lu, P., Ruchkin, I., Cleaveland, M., Sokolsky, O., Lee, I.: Causal repair of learning-enabled cyber-physical systems. In: Proceedings of the IEEE International Conference on Assured Autonomy (ICAA), pp. 1–10 (2023)

31. Maldonado, J., Krumme, J., Zetzsche, C., Didelez, V., Schill, K.: Robot pouring: identifying causes of spillage and selecting alternative action parameters using probabilistic actual causation (2025)

32. Raffin, A.: Rl baselines3 zoo (2020). https://github.com/DLR-RM/rl-baselines3-zoo

33. Raffin, A., Hill, A., Gleave, A., Kanervisto, A., Ernestus, M., Dormann, N.: Stable-baselines3: Reliable reinforcement learning implementations. J. Mach. Learn. Res. **22**(268), 1–8 (2021)

34. Rafieioskouei, A., Bonakdarpour, B.: Efficient discovery of actual causality using abstraction refinement. IEEE Trans. Comput. Aided Des. Integr. Circuits Syst. **43**(11), 4274–4285 (2024)
35. Rafieioskouei, A., Rogale, K., Bonakdarpour, B.: Efficient discovery of actual causality in stochastic systems (2025)
36. Rafieioskouei, A., Rogale, K., Saei, A.A., Mahmoudi, M., Bonakdarpour, B.: Beyond correlation: establishing causality in protein corona formation for nanomedicine. Molec. Pharmaceut. (2025)
37. Raghavan, A., Bareinboim, E.: Counterfactual realizability. In: The Thirteenth International Conference on Learning Representations (2025)
38. Wang, S., Geoffroy, Y., Gössler, G., Sokolsky, O., Lee, I.: A hybrid approach to causality analysis. In: Bartocci, E., Majumdar, R. (eds.) RV 2015. LNCS, vol. 9333, pp. 250–265. Springer, Cham (2015). https://doi.org/10.1007/978-3-319-23820-3_16
39. Zheng, M., Kleinberg, S.: A method for automating token causal explanation and discovery. In: Rus, V., Markov, Z. (eds.) Proceedings of the Thirtieth International Florida Artificial Intelligence Research Society Conference, FLAIRS 2017, Marco Island, Florida, USA, 22–24 May 2017, pp. 176–181. AAAI Press (2017)
40. Zibaei, E., Borth, R.: Building causal models for finding actual causes of unmanned aerial vehicle failures. Front. Rob. AI **11**, 1123762 (2024)
41. Ziemek, R., Piribauer, J., Funke, F., Jantsch, S., Baier, C.: Probabilistic causes in Markov chains. Innov. Syst. Softw. Eng. **18**(3), 347–367 (2022)

Verification of Generic VHDL Designs and Their Translation to Rocq

Ocan Sankur[✉], Benoît Boyer, and Florian Faissole

Mitsubishi Electric R&D Centre Europe, Rennes, France
o.sankur@fr.merce.mee.com

Abstract. We present our methodology to formally prove properties of VHDL designs by first translating them to Rocq. Because our translation keeps all parameters (a.k.a. generics) uninstantiated, we develop algorithms that check the correctness of given VHDL designs under all parameter valuations. These checks detect whether there are combinatorial loops, missing or multiple signal assignments, wrong integer assignments with respect to specified ranges, array access and assignment errors, and integer overflows for some valuation of the parameters. Once these checks pass, we show how to compute a topological ordering of the signal assignments that is valid for all parameter valuations, and which allows us to translate to simple Rocq functions that capture the functional behaviors of the VHDL designs given as input. We further show to address pipelined circuits and present an application on the verification of a FPU.

1 Introduction

Hardware Verification. Hardware designs were historically one of the first applications of formal verification techniques [10], and these are routinely used today to verify the functional behaviors of such designs. Various techniques are applied including symbolic simulation [5], equivalence checking [12], model checking [9], and the use of proof assistants [11].

Proof assistants are particularly powerful since they allow one to formalize the system under study, and to state and prove their properties using very detailed theorems. Moreover, the produced proofs can be independently checked, so the correctness of the verification results do not rely on complex software. Accordingly, the research community has developed methodologies to formally verify hardware designs using theorem provers. Proving properties of hardware designs require formalizing the semantics of the hardware description language [33]. One typically writes a high-level specification and the description of the implementation within the proof system and proves that the implementation conforms to the specification. Various proof assistants have been used to prove hardware designs including PVS [1,7], ACL2 [13], HOL [29], and Rocq [3].

Y.-F. Chen et al. (Eds.): VMCAI 2026, LNCS 16417, pp. 287–308, 2026.
https://doi.org/10.1007/978-3-032-15700-3_14

Translation Between RTL and Proof Systems. There are two approaches to link the actual RTL design to its formal representation in the proof system. In some works, RTL designs were automatically translated into an equivalent representation within the proof system often relying on a formalization of a simpler RTL language [13], while some other works focus on providing complete hardware development methodologies where the hardware design is written within the proof system using an abstract language, and RTL code or netlist is later extracted automatically [6,20].

While designing and verifying hardware within the proof system is appealing, in an industrial setting, it is not always possible to require hardware engineers to actually design hardware using proof systems or alternative languages. So it is important to be able to prove properties of existing hardware designs written in standard languages such as VHDL or Verilog. At Mitsubishi Electric R&D Centre Europe, we have been using the Rocq proof assistant [32] for verifying critical hardware designs developed independently in VHDL. To verify such designs, we systematically translate VHDL designs to Rocq's language Gallina relying on a formalization of VHDL [33]. Because performing such translations manually would be error-prone, we have automatized this translation, following other work *e.g.* [24,31]. One difficulty we encountered in this automatization was the handling of generics (a.k.a. parameterized) designs: while we needed to translate such designs by keeping the generics as inputs, because the VHDL compiler cannot check the correctness for all possible valuations of the generics, the translation could be erroneous for some generic valuations. The goal of this paper is to present our full methodology including parametrically checking and translating generic VHDL designs, thus ensuring the correctness of the translations to Rocq, and the application of our approach beyond combinational designs, namely, for the formal verification of pipelined designs.

Parameterized Translation and Verification. The originality of our approach lies in *parameterized* checks we have developed and which apply to VHDL code to ensure the correctness of the designs and their translations to Rocq under *all* parameter valuations. Parameters are called *generics* in VHDL, and are often used to specify the sizes of vectors in VHDL entities. For example, a typical floating-point unit (FPU) can be instantiated for single precision (32-bit) or double precision (64-bit) by appropriately instantiating its parameters. Our checks detect the existence of combinatorial loops, missing or multiple signal assignments, wrong integer assignments with respect to specified ranges, array access and assignment errors, and integer overflows that can occur for *some* valuation of the parameters.

While similar checks are routinely performed by VHDL simulators such as GHDL either in compile time or during simulation when parameters are *instantiated*, some errors can only appear for particular values of the parameters; so a successful compilation or simulation for selected values of the parameters cannot guarantee the correctness for all values. Parameterized checks are absent in existing translation or simulation tools. In fact, simulators require all parameters to be instantiated to proceed to simulation, so it is natural that they check the code only once parameters are instantiated. Existing formal verification tools also translate RTL to C code after instantiating all parameters, and then apply model checking (e.g. [22]).

One advantage of instantiating parameters first is that translation to sequential code becomes simpler. For example, when parameter values are known, signal assignments can be ordered topologically in a combinational circuit so that each assignment can be evaluated only once to compute the values of the outputs at a given cycle. This results in a simplified translation that is very easy to read and reason with. Therefore, proving the properties of the design in Rocq also becomes easier. However, this is nontrivial when parameter values are unknown since such an order may not exist (see Example 1).

We develop these parameterized checks because we translate designs to Rocq keeping all parameters as arguments in the target function, so these checks ensure that the produced function is correct and corresponds to the instantiation of the original design for all possible valuations of the parameters. Keeping the parameters as arguments has the advantage of enabling general theorem statements about the design in which parameters are quantified universally. For instance, we prefer proving general theorems about a single parameterized FPU design under all parameter valuations rather than separate theorems and proofs for each instantiation (32-bit, 64-bit, etc.).

In this work, we first focus on combinational designs (i.e. without state holding elements) with the objective of producing simple Rocq translations. We show how to perform the parameterized checks mentioned above by analyzing given VHDL designs and using satisfiability modulo theory (SMT) solvers to deal with unknown parameters. Once these checks pass, our translator attempts to produce a sequential functional program in Rocq by computing a topological ordering of assignments that is valid for all parameter valuations, if such an order exists. The translated programs are thus always "simple" in the sense that each signal is assigned only once. We advocate such an approach whenever possible since this has the advantage of simplifying the proof effort in Rocq.

One could establish some of these correctness properties by proving them on the produced Rocq function after translation, provided a more detailed and heavier formalization of VHDL in Rocq. Our approach does not exclude such efforts and can be seen as a set of completely automatic, thus easier preliminary checks applied during development by hardware engineers rather than formal verification experts. We found that using these checks during development time brings formal verification closer to the development phase by fostering discussions with developers without requiring them to understand the details of formal verification. This was also helpful in avoiding a possibly large number of small modification requests to the source code to make it compatible with formal proofs.

All these features are implemented in a translator called VHDL2Rocq which we have used in internal projects. This tool can also be seen as a linter, of independent interest, which forces the designers to write designs that are valid for all parameter valuations, *e.g.* by restricting the ranges of the parameters so that all resulting instantiations are valid.

While our tool is restricted to combinational designs, we have combined it with equivalence checking methods to prove properties of pipelined designs. Note that equivalence checking have been used before in combination with proof assistants, see *e.g.* [26,31]. We detail how we apply our method in combination with combinational equivalence checking to formally verify a pipelined floating-point unit (FPU).

We thus advocate the simplicity of our methodology when proving properties of combinational hardware designs in Rocq. Our translation simplifies the proof effort while we can still deal with pipelined circuits by combining formal proofs with equivalence checking.

Outline. We start by explaining the targeted translation objective in more detail in Sect. 2, where we briefly present supported VHDL types and statements on examples, and introduce several notations. Section 3 presents the set of parameterized checks we perform along with the computation of the topological ordering of statements. We discuss more specifically the verification of errors related to integer signals in Sect. 4. Section 5 presents the translation from combinational VHDL to Rocq through examples. We discuss the use of our translation on pipelined designs in Sect. 7. Section 8 contains additional discussion on related works and conclusions.

2   Preliminaries

Parameterized Translation. We are interested in this work in a parameterized translation in Rocq in the following sense. Figure 1 shows a typical example of an entity[1] which takes as input two vectors of size n, and outputs a vector sum of size n along with a single bit carry. Other architectures[2] in the VHDL design can instantiate this component for any positive n.

To state the properties of `IntegerAdder`, we might want to prove a theorem stating that "**for all** positive n, for all vectors a, b of size n, the output sum contains the n first bits of the sum of the numbers represented by a, b", rather than restating such a theorem with a separate proof for all values of n. Accordingly, parameterized translation consists in automatically producing a Rocq function with the following type.

```
Definition integerAdder (n : N) (a b : bit_vector) (c_in : bit) :
    bit_vector * bit := ...
```

This is a function that takes the argument n of type natural number (N), a, b and c of type `bit_vector` and `bit` respectively, and returns a tuple corresponding to the types of sum and carry. Thus the component is translated as a single function in which n appears as an argument.

---

[1] A VHDL entity defines the interface (inputs and outputs) of the circuit.

[2] In VHDL, an architecture corresponds to an implementation of an entity.

```
1    library IEEE;
2    use IEEE.std_logic_1164.all;
3
4    entity IntegerAdder is
5      generic(n: positive);
6      port (a, b: in std_logic_vector(n-1 downto 0);
7           c_in: in std_logic;
8           sum: out std_logic_vector(n-1 downto 0);
9           carry: out std_logic);
10   end entity IntegerAdder;
11
```

Fig. 1. An entity for an integer adder computing the sum of two vectors of size n.

Note that the function type does not contain any information about the range of n nor the sizes of a, b. Our tool nevertheless symbolically checks that any component instantiation in VHDL has a port map that assigns arguments of the right size and range (that is, if the generic n is instantiated as an expression e, then the size of the vector passed as argument to a must be equal to e); see Sect. 3. This implies that all function calls in Rocq provide arguments satisfying the associated preconditions.

Bits versus IEEE std_logic *Types.* In VHDL, the IEEE standard library provides the std_logic which models signals which can have nine different values. These are the standard '0', '1' values but also the weak values 'W', 'L', 'H', the undefined value 'U', the don't care value '–', the high impedance value 'Z', and the undetermined value 'X'. Because only combinational circuits are considered in our work, we do not deal with uninitialized register states ('U'), and only deal with the case where all inputs are '0' or '1'. Because each signal is assigned exactly once at all cases as guaranteed by our checks (see Sect. 3.3), this means that each signal has a unique driver, so the undetermined value 'X' or high impedance value 'Z' never occur by construction. We also forbid the don't care value '–' syntactically. Thus, our checks guarantee that all signal values are '0' or '1' at all times.

We thus map the std_logic type to the type bit which is an alias of the Boolean type bool, and their vectors to bit vectors in Rocq (see Sect. 5).

Definitions. We consider VHDL designs made of possibly several entities and architectures but with no process statement, which ensures that the considered designs are combinational. We allow the following types: std_logic, ranged integer types including integer, natural, positive, and custom ranges of these, and arrays of aforementioned types. We allow generics of integer types only. The ports of entities can contain in or out signals. We consider five types of statements: simple concurrent assigment, conditional concurrent assignment, component instantiation, if-generate and for-generate statements.

An example architecture of entity IntegerAdder is shown in Fig. 2. Line 17 is a simple concurrent assignment of target signal sc_in(0). The component FullAdder declared in line 2 (and defined in another file) is instantiated in line 9. For-generate and if-generate statements can be seen in lines 7, 8; the for-generate statement contains a list of statements that it *generates* for each value

of the index in the given range, and the if-generate statement generates its list
of statements only if the condition holds.

```
1    architecture structural of IntegerAdder is
2      component FullAdder is
3        port (a, c_in, b: in std_logic; sum, carry: out std_logic);
4      end component FullAdder;
5      signal sc_in : std_logic_vector(n downto 0);
6    begin
7      st1: for i in 0 to n-1 generate
8        FIRST_IT: if i < n-1 generate
9          FA: FullAdder port map
10            (a=>a(i),b=>b(i),c_in=>sc_in(i),carry=>sc_in(i+1),sum=>sum(i));
11         end generate;
12         LAST_IT: if i = n-1 generate
13           LAST_FA: FullAdder port map
14             (a=>a(n-1),b=>b(n-1),c_in=>sc_in(n-1),carry=>carry,sum=>sum(n-1));
15         end generate;
16       end generate;
17       st2: sc_in(0)<=c_in;
18     end structural;
19
```

Fig. 2. An architecture for IntegerAdder entity defined in Fig. 1

We assume that a unique architecture is given for each entity. Consider an
entity E. Let I denote the set of loop indices, G the set of generics, and $\mathsf{V} = \mathsf{I} \cup \mathsf{G}$.
We assume that all loop indices in a given design are unique; if not, these can
be rewritten by renaming.

Let $\mathsf{Stmts}(E)$ denote the list of statements of E. For a for-generate or if-
generate statement st, let $\mathsf{Stmts}(\mathtt{st})$ denote the list of statements in the body of
these statements. For example, in Fig. 2, $\mathsf{Stmts}(E) = (\mathtt{st1}, \mathtt{st2})$, $\mathsf{Stmts}(\mathtt{st1}) =$
$(\mathtt{FIRST\_IT})$, and $\mathsf{Stmts}(\mathtt{FIRST\_IT}) = (\mathtt{FA})$.

For each statement st, let $\mathsf{reads}(\mathtt{st})$ denote the set of signals that it reads,
and $\mathsf{writes}(\mathtt{st})$ the set of signals that st writes. For instance, for statement
st2 in Fig. 2, $\mathsf{reads}(\mathtt{st2}) = \{\mathtt{c\_in}\}$, and $\mathsf{writes}(\mathtt{st2}) = \{\mathtt{sc\_in(0)}\}$. Furthermore,
$\mathsf{reads}(\mathtt{FA}) = \{\mathtt{a(i)}, \mathtt{b(i)}, \mathtt{sc\_in(i)}\}$, and $\mathsf{writes}(\mathtt{FA}) = \{\mathtt{sc\_in(i+1)}, \mathtt{sum(i)}\}$.
Note that we have $\mathsf{writes}(\mathtt{st1}) = \mathsf{writes}(\mathtt{FIRST\_IT}) = \mathsf{writes}(\mathtt{FA})$ so high-level
statements are also considered to be reading and writing the signals read or
written by their internal statements.

We also define the *context* of a statement as a set of constraints containing the
range constraints of all generic parameters, the range constraints of loop indices
and the conditions of if-generate statements at the current scope. For example,
$\mathsf{context}(\mathtt{FA}) = \{1 \leq n < N_{\max}, 0 \leq i \leq n - 1, i < n - 1\}$ where $N_{\max} = 2^{31} - 1$
which is the maximal positive integer in the VHDL 1993 standard [15]. Here the
first constraint comes from the fact that n is a generic of type **positive**; the
second one comes from the range of loop index i in line 7, and the third one from
the condition of line 8. We have $\mathsf{context}(\mathtt{st2}) = \{1 \leq n < N_{\max}\}$ since the
range of the generic is the only constraint at this statement.

A *valuation for generics of* E is a map of type $v_{\mathsf{G}} : \mathsf{G} \to \mathbb{Z}$ assigning each
generic to an integer that conforms to its range as specified in E. Let $\mathsf{Val}(\mathsf{G})$
define the set of these valuations. A valuation for generics must be specified

to give a semantics to the circuit since the sizes of input and output vectors and ranges of integer signals typically depend on the generics. For an integer expression e whose variables belong to G, we denote by $e[v_G]$ its value uniquely determined by v_G[3]. A *valuation* for a signal of type `std_logic` is a value in $\{0, 1\}$. Given v_G, a valuation for a signal of type `std_logic_vector` of length e is a value in $\{0, 1\}^{e[v_G]}$, where e only depends on G. An *integer range* is a pair of expressions $[l, r]$ whose variables belong to G, and we denote by $[l, r]v_G$ the interval $[l[v_G], r[v_G]]$.

Let In denote the set of *inputs*, and Out the set of its *outputs* as specified in the entity definition. Given a valuation v_G for the generics of E, an *input valuation* v_{In} maps each input signal to a valuation, and an *output valuation* v_{Out} maps each output signal to a valuation. The sets of input and output valuations are denoted, respectively, by $\mathsf{Val}_{v_G}(\mathsf{In})$ and $\mathsf{Val}_{v_G}(\mathsf{Out})$. Given v_G, a *valuation for loop indices* is a map $v_{\mathsf{I}} : \mathsf{I} \to \mathbb{Z}$ where for each loop index i, $v_{\mathsf{I}}(i) \in r_i[v_G]$ where r_i is the range of the loop index specified in its for-generate loop. Loop index valuations are denoted $\mathsf{Val}_{v_G}(\mathsf{I})$.

In considered designs, all integer expressions e will have unknowns in $V = G \cup I$. So given valuations v_G, v_{I}, we will denote by $e[v_G, v_{\mathsf{I}}]$ the value of e given by these valuations.

Given v_G and v_{In} for an entity E, let $E[v_G, v_{\mathsf{In}}]$ denote a valuation of all local and output signals as computed by the circuit. This is well defined since the design is combinational, thus functional.

3 Parameterized Checks and Topological Ordering

In this section, we detail all checks and operations we perform prior to translation to ensure the correctness of the generated Rocq functions.

3.1 Type Annotations

We first annotate all expressions of the given design with types. This consists in visiting the syntax tree and assigning each expression a type among all supported types. This is fairly standard and easy for VHDL, and is done by compilers such as GHDL. In fact, because all generics and signals are declared in each entity and architecture with precise types in VHDL, one can easily determine the type of all expressions.

Our type annotations additionally contain the sizes of the vectors symbolically, where generics appear as unknowns. Consider the code below.

```
1    signal s, t : std_logic_vector(n-1 downto 0);
2    ...
3    s(n/2 downto 1) <= "1" & t(n-2 downto n/2);
```

---

[3] Integer expressions are built by combining integer variables with arithmetic operations. For example, if $G = \{n\}$, and $v_G(n) = 3$, then $(3 * n + 1)[v_G] = 10$.

Because $\mathtt{s}$, $\mathtt{t}$ are declared as vectors of size n, in line 3, we can easily determine that $\mathtt{s}$ is a vector of size n, $\mathtt{s(n/2\ downto\ 1)}$ is a vector of size $n/2$, $\mathtt{t(n\text{-}2\ downto\ n/2)}$ is of size $n/2 - 1$, and "1" is of size 1.

One of the uses of these type annotations is to detect assignments between vectors of incompatible sizes such as $\mathtt{s\ <=\ t(n\text{-}1\ downto\ 1)}$. For all assignments between vectors of lengths given by expressions e and e' (this includes vector arguments in component instantiations), we check the validity of $e = e'$ using an SMT solver. If this equality is not valid, then we reject the design and report a valuation of generics under which the equality fails.

3.2   Array Accesses

We further check the absence of array access errors to avoid situations as the following one. Consider the following listing where $\mathtt{n}$, $\mathtt{m}$ are positive generics.

```
1   ...
2     generic(n, m: positive);
3   ...
4
5   signal s, t : std_logic_vector(n-1 downto 0);
6   ...
7   FOR1: for i in 0 to m-1 generate
8     s(i) <= t(i);
9   end generate;
```

If tested for values $n = m$, then the design could compile and pass all simulations or formal verification attempts successfully. However, the translated Rocq function will not be correct whenever called with $n \neq m$.

Thus, for all array accesses $\mathtt{s}(e)$, we check the validity of the following formula.

$$\forall v \in \mathsf{Val}(\mathsf{V}), e[v] \subseteq r_s[v]. \tag{1}$$

Note that e is either an expression of type integer, or an expression of the form $e'\ \mathtt{downto}\ e''$; in the former case, we interpret $e[v]$ as a singleton interval, and in the latter case as the interval with endpoints $e'[v]$ and $e''[v]$.

Combined with the checks described in Sect. 3.1, this ensures that array accesses and assignments are correct.

3.3   Unique Assignments to Signals

Our mapping of $\mathtt{std\_logic}$ values to a simple Boolean type requires verifying that only the '0' and '1' values can be produced by considered VHDL designs. This is the case if all signals are assigned exactly once under all valuations for generics since the don't-care value and high impedance values are forbidden syntactically, and circuits are assumed to receive only '0' and '1' values as inputs. Let us show how to perform this check.

A signal $\mathtt{s}$ is multiply assigned if there are two statements which write to $\mathtt{s}$, which are both "generated" under some valuation for generics. Consider the following formula.

$$\exists \mathsf{st1} \neq \mathsf{st2} \in \mathsf{Stmts}(E), \exists e_1 \in \mathsf{writes}(\mathsf{st1}), \exists e_2 \in \mathsf{writes}(\mathsf{st2}),$$
$$\exists v_\mathsf{G} \in \mathsf{Val}(\mathsf{G}), \exists u_1, u_2 \in \mathsf{Val}_{v_\mathsf{G}}(\mathsf{I}), (e_1, u_1) \neq (e_2, u_2),$$
$$\mathsf{context}(\mathsf{st1})[v_\mathsf{G}, u_1] \wedge \mathsf{context}(\mathsf{st2})[v_\mathsf{G}, u_2] \wedge \mathsf{intersects}(e_1[v_\mathsf{G}, u_1], e_2[v_\mathsf{G}, u_2]).$$
$$(2)$$

Intuitively, we are looking here for two statements $\mathsf{st1}$, $\mathsf{st2}$ that write to expressions e_1, e_2 which "intersect". For now, assume that $\mathsf{intersects}$ checks whether e_1 and e_2 refer to the exact same signal. Because several copies of a statement is created within for-generate loops, the formula also guesses loop index valuations u_1, u_2, and requires that $(e_1, u_1) \neq (e_2, u_2)$ (which means $e_1 \neq e_2$ or $u_1 \neq u_2$). This ensures that we indeed guess two expressions which are either syntactically distinct, or are identical but correspond to different "iterations" of the for-generate loops within their scope. The constraints $\mathsf{context}(\mathsf{st1})[v_\mathsf{G}, u_1] \wedge \mathsf{context}(\mathsf{st2})[v_\mathsf{G}, u_2]$ ensure that copies of the statements are indeed generated for the guessed loop indices (the loop indices satisfy the if-generate conditionals).

We now detail and explain the $\mathsf{intersects}$ function. VHDL allows one to assign to vector components either individually or by slices. For example the following code contains a double assignment to signal $\mathsf{s(0)}$:

```
signal s, t : std_logic_vector (n-1 downto 0);
...
s(0)  <= '0';
s(n-1 downto 0) <= t;
```

In (2), we would detect this by choosing $e_1 = \mathsf{s(0)}$ and $e_2 = \mathsf{s(n\text{-}1\ downto\ 0)}$ and, say, for $v_\mathsf{G}(n) = 16$. However, $\mathsf{s(0)}[v_\mathsf{G}] = \mathsf{s(0)}$, and $\mathsf{s(n\text{-}1\ downto\ 0)}[v_\mathsf{G}] = \mathsf{s(15\ downto\ 0)}$ and these are not equal (Here v_I is irrelevant since $\mathsf{I} = \emptyset$). They do however "intersect" since they both contain the signal $\mathsf{s(0)}$. Thus, given valuations $v_\mathsf{G}, v_\mathsf{I}$, the $\mathsf{intersects}$function returns true iff the set of signals contained in the two given expressions intersect. The detailed definition is given in Algorithm 1. Notice that the intervals r_i, r_i' that appear in the expressions passed to this function are integer intervals since $v_\mathsf{G}, v_\mathsf{I}$ are applied to expressions before calling $\mathsf{intersects}$.

> **Function** $\mathsf{intersects}(s(r_1) \ldots (r_k),\ t(r_1') \ldots (r_l'))$:
> > **if** $s \neq t$ **then   return** false ;
> > **else if** $k = 0$ *or* $l = 0$ **then   return** true ;
> > **else   return** $r_1 \cap r_1' \neq \emptyset \wedge \mathsf{intersects}(s(r_2) \ldots (r_k),\ t(r_2') \ldots (r_l'))$;

Algorithm 1: The $\mathsf{intersects}$function checking whether two expressions of the form $\mathsf{intersects}(s(r_1) \ldots (r_k),\ t(r_1') \ldots (r_l'))$ refer to the same signal. Here s, t are names of the signals of type $\mathsf{std_logic}$or $\mathsf{std_logic_vector}$given with a possibly empty list of indices where all r_i, r_i' are integer intervals. We assume that a simple index i as in $\mathsf{s(i)}$ is seen as the interval $[i, i]$.

The query (2) can be solved using an SMT solver, for example, using the linear theory of integers when all expressions are linear. Notice that the existential quantifiers in the first line are simply disjunctions and can be implemented in a loop since they range over the finite set of statements and expressions, so the formulas sent to the solver are often quite small.

We now explain how we check that each signal is indeed assigned at least once. We provide the query to be solved for a single-dimensional vector s but the generalization to simpler signals or multi-dimensional arrays is straightforward.

$$\exists v_G \in \mathsf{Val}(G), \exists j_0 \in r_{\mathsf{s}}[v_G], \forall e \in \mathsf{l-exps}(\mathsf{s}), \forall v_I \in \mathsf{Val}(I),$$
$$\mathsf{context}(e)[v_G, v_I] \rightarrow \neg\mathsf{intersects}(\mathsf{s}(j_0), e[v_G, v_I]). \tag{3}$$

Here, $\mathsf{l-exps}(\mathsf{s})$ is the set of left-expressions of s, that is expressions involving s which are targets of an assignment. We abusively use here $\mathsf{context}(e)$ to refer to the context of the statement that contains e. The formula checks whether for some index j_0 that belongs to the range of the vector s (denoted $r_{\mathsf{s}}[v_G]$), $\mathsf{s}(j_0)$ is not assigned. This is the case if for each assignment to signal e, for all possible loop index valuations v_I, $\mathsf{s}(j_0)$ is distinct from all signals assigned in $e[v_G, v_I]$, that is $\neg\mathsf{intersects}(\mathsf{s}(j_0), e[v_G, v_I])$.

Here, the quantification on e is encoded as a conjunction, but the quantification over v_I uses a universal quantifier. This query can be solved using an SMT solver supporting quantifiers.

3.4   Dead Code

It is straightforward to check whether some statement is never generated in a generic design. This can be the case due to unsatisfying constraints as a result of generic range specifications and if-generate conditionals. For each statement st, one can check the satisfiability of

$$\exists v \in \mathsf{Val}(V), \mathsf{context}(\mathsf{st})[v]. \tag{4}$$

Any unsatisfiable formula indicates that st is never generated, or executed when translated to sequential code.

3.5   Topological Ordering of Statements

We first focus on ordering the statements $\mathsf{Stmts}(E)$ of the considered architecture. The statements inside for-generate and if-generate bodies will be ordered recursively.

Let us define an ordering between statements. We write $\mathsf{st} < \mathsf{st}'$ iff st writes to a signal that is read by st' for some valuation of the generics and loop indices. Intuitively, $\mathsf{st} < \mathsf{st}'$ iff under some valuations of the generics, statement st must come before st' if we were to execute all statements sequentially. Formally,

$$\mathsf{st} < \mathsf{st}' \iff \exists e \in \mathsf{writes}(\mathsf{st}), \exists e' \in \mathsf{reads}(\mathsf{st}'), \exists v \in \mathsf{Val}(V),$$
$$\mathsf{context}(e)[v] \wedge \mathsf{context}(e')[v] \wedge \mathsf{intersects}(e[v], e'[v]). \tag{5}$$

Note that we guess a single valuation for I here because loop indices are distinct.

The partial order $<$ is computed as follows. Given $\mathsf{st}, \mathsf{st}'$, one iterates over expressions $e' \in \mathsf{reads}(\mathsf{st}')$, $e \in \mathsf{writes}(\mathsf{st})$, and solves the SMT query $\exists v \in$

$\mathsf{Val}(V), \mathsf{context}(e)[v] \wedge \mathsf{context}(e')[v] \wedge \mathsf{intersects}(e[v], e'[v])$. Now, to reorder the statements, we simply run a topological ordering algorithm [16] on a graph in which the statements are the nodes, and there is an edge $\mathtt{st}$ to $\mathtt{st'}$ iff $\mathtt{st} < \mathtt{st'}$.

Once the statements $\mathsf{Stmts}(E)$ are reordered, we visit each if-generate or for-generate statement in this list, and reorder recursively the statements in their bodies.

3.6 Combinatorial Loops and Absence of Generic Topological Orders

The topological sorting algorithm can fail to find a topological ordering. There can be two reasons for such a failure.

A combinatorial loop occurs in a combinational circuit if signal updates contain a cycle as in the following case.

```
1  ]
2     signal  a,  b,  c  :  std_logic;
3     ...
4     a  <=  b;
5     b  <=  c;
6     c  <=  a;
```

Combinatorial loops must be avoided because they do not have a clear semantics, and hardware simulators run into infinite loops attempting to evaluate them. Because in generic designs, signal assignments are generated conditionally on the values of the generics, we also need to check for the absence of such loops for all valuations for the generics.

In the presence of such loops, there does not exist a topological order of the statements. These loops are detected by the topological ordering algorithm and the sequence of statements involved in such a loop can be printed.

Second, notice that our procedure attempts to find an ordering of the statements which is a topological order under all valuations of the generics. It is possible that there is no such a generic order for all valuations, even though no instantiation contains a combinatorial loop, as in the following example.

Example 1. Consider the design below.

```
1     entity  unsortable  is
2        generic(p,q,n  :  positive);
3        port(  a  :  in  std_logic_vector(n  downto  0)  );
4     end  entity;
5
6     architecture  rtl  of  unsortable  is
7        signal  t1  :  std_logic_vector(q  downto  0);
8        signal  t2  :  std_logic_vector(p  downto  0);
9     begin
10       t1  <=  a(q  downto  0)  when  p  <=  q  else  (others  =>  t2(0));
11       t2  <=  a(p  downto  0)  when  p  >  q  else  (others  =>  t1(0));
12    end  architecture;
13
```

Both signals $\mathtt{t1}$, $\mathtt{t2}$ are assigned conditionally on the values of generics $\mathtt{p}$, $\mathtt{q}$. When $\mathtt{p}\ \mathtt{<=}\ \mathtt{q}$, then $\mathtt{t1}$ must be defined first, because all elements of $\mathtt{t2}$ are assigned to $\mathtt{p1(0)}$. When $\mathtt{p}\ \mathtt{>}\ \mathtt{q}$, then $\mathtt{t2}$ must be defined first, because all elements of $\mathtt{t1}$ are assigned to $\mathtt{t2(0)}$. It follows that there is no ordering of these statements that is correct for all valuations of the generics.

In our approach, we reject such designs, and the translation fails. We have not encountered such a case in real VHDL designs we have worked with.

4 Integer Signal Range and Overflow Errors

VHDL allows one to work with integer signals which are limited to 32 bits in VHDL 1993 and to 64 bits in VHDL 2008, and their ranges can be further restricted. However, existing VHDL simulators and synthesizers have different behaviors when these signals are assigned values outside their ranges. Consider the following listing.

```
1   signal a : integer range 0 to 63;
2   signal s : std_logic_vector(6 downto 0);
3   signal o : std_logic;
4   ...
5   s <= "1000000"; -- 64
6   a <= to_integer(unsigned(s)); -- attempting to assign 64 to a
7   o <= '1' when a >= unsigned(s) else '0'; -- is o '0' or '1'?
```

The integer signal a is restricted to the range $[0, 63]$ but is then assigned the value 64. What should be its value? Some simulators such as GHDL throw a runtime error during simulation. However, when synthesizing a netlist, GHDL only allocates 6 bits for the signal a, thus 64 is truncated to 0 during the assignment to a. Xilinx Vivado has a different behavior: it always allocates 32 bits ignoring the integer ranges during RTL synthesis, and silently assigns 64 to a, violating the specified range. It follows that signal o has the value '0' when using GHDL, and '1' when using Xilinx Vivado.

Another type of error is, as in any program or hardware design working with integers, overflows can occur in intermediate computations:

```
1   generic(n : integer);
2   ...
3   signal a, b : integer;
4   ...
5   a <= n;
6   b <= (a+1)/2;
```

If n is instantiated to the maximal integer, then an overflow occurs in the assignment to b. While overflows can sometimes be detected during simulation, one must ensure that they can never so that the synthesized circuit is correct.

Such situations must thus be avoided in a rigorous development framework. Moreover, we map integer values in VHDL to the (infinite precision) natural type N in Rocq. This map is only correct if we can formally verify that integer values are within legal bounds, and that there is no overflow.

We address the first type of errors by generating Rocq theorems stating that all integer assignments are within ranges declared for the target signals, relying on the user to prove them. Although we initially attempted implementing a static analysis to do this check, this was nontrivial due to arbitrary conversions between bit vectors and integers, so determining possible integer values required tracking bit-level operations. While a model checking approach could be used for this purpose, the presence of arrays of integers of unknown sizes (due to generics) made this nontrivial as well. We thus opted for automatically generating Rocq theorems which happen to be easy to prove in most cases. An example of such a theorem is given in Appendix A.

We did develop a basic static analysis to detect possible overflows in intermediate computations using abstract interpretation using the interval domain [25]. This was useful in our applications since overflows occurred rarely, and their absence was easy to verify with a simple static analysis. In this analysis, all generics were assumed to take arbitrary values within their ranges (they were abstracted by their ranges), and arithmetic operations were applied on these abstractions as usual. A conversion of the form `to_integer(unsigned(s))` from a vector `s` of size n to an unsigned integer results in the interval $[0, 2^{n_{\max}} - 1]$ where $n_{\max}$ is the maximal possible value for the generic n.

5   Translation to Rocq

We provide here a glance of our Rocq library for combinational VHDL designs, and illustrate how various statements are translated. Thanks to the checks of Sect. 3.3, we can map the `std_logic` type of the standard IEEE library to simple Boolean values. Moreover, we define the following type to represent `std_logic_vector`, thus a bit vector.

```
1    Record vector := make_vector { repr :> N; size : N}.
```

A bit vector is thus represented by its `size` and the natural number `repr` represented by the bit vector (where the most significant bit is on the left). Representing bit vectors with a natural number representation allows immediate conversion to and from the `signed`, `unsigned`, and `integer` types.

The notation $[\![v]\!]$ provides the size of the vector v, and $\langle x|n \rangle$ creates a vector of size n, representing the natural number x. Moreover, for a vector v, the notation $v[i] \leftarrow$ true is used to build a fresh vector from v in which the component i is set to true (and others have values as in v). This is extended to $v[i \text{ downto } j] \leftarrow u$ where u is a vector of size $j - i + 1$, which creates a fresh vector copying v and assigning all $v[i + k] = u[k]$ for all $0 \leq k \leq j - i$.

All integer types are mapped to the natural number type N of Rocq. While the typing information does not contain ranges, these are checked as explained in Sect. 4.

We now illustrate how different VHDL constructs are translated. Let us assume here that all parameterized checks of Sect. 3 have passed, and that the statements were ordered topologically according to Sect. 3.5.

Entities and Component Instantiations. Each entity is mapped to a function taking as argument all generics and all inputs signals, and returning a tuple of values for the output signals. An example was shown in Sect. 2.

All component instantiations are translated as function calls. Consider the following entity and architecture.

```
1  entity FullAdder is
2    port (a, c_in, b: in std_logic;
3        sum, carry: out std_logic);
4  end entity FullAdder;
```

```
 5
 6  architecture structural of FullAdder is
 7    component HalfAdder is
 8      port (a, b: in std_logic;
 9           sum, carry: out std_logic);
10    end component HalfAdder;
11    signal s1, s2, s3: std_logic;
12  begin
13    H1: HalfAdder port map (b=>b, a=>a, sum=>s1, carry=>s3);
14    H2: HalfAdder port map (a=>s1, b=>c_in, sum=>sum, carry=>s2);
15    carry<=s2 or s3;
16  end structural;
```

This is translated as

```
1  Definition fulladder (a : bool) (c_in : bool) (b : bool) : bool * bool :=
2    let '(s1, s3) := halfadder a b in
3    let '(sum, s2) := halfadder s1 c_in in
4    let carry := (s2 || s3) in
5    (sum, carry).
```

where **halfadder** is a function that was previously translated.

Simple and Conditional Concurrent Assignments. The previous example also illustrates how simple concurrent assignments are translated as let expressions. Hence **carry <= s2 or s3;** is translated as **let carry := (s2 || s3)**.

In VHDL, each component of a vector is seen as a separate signal, and one can assign values to these separately. In contrast, a vector in Rocq is an object as defined above, so assignment to a component i of vector v is simulated by redefining the symbol v with a let-expression, and assigning it to a fresh vector obtained by the $v[i] \leftarrow \dots$ notation:

```
1    let v := v[0] <- true in ...
```

It must be clear that while v on the right hand side refers to the previously defined symbol v, the **let v := ...** expression redefines the symbol v, thus overriding the previous symbol. Such updates thus require v to be initialized; otherwise, v would be unknown in the first such assignment.

We therefore initialize all vectors to sizes given in their VHDL type declarations. We arbitrarily set their initial values to the 0 vector. These initial values do not impact the semantics of the function. In fact, thanks to Sect. 3.3, we know that every signal is assigned exactly once; and moreover, Sect. 3.5 ensures that each signal that is read was assigned in previous statements. Therefore the initial values are never read and thus their values are irrelevant.

Conditional concurrent assignments are also translated as let expressions using the if-then-else construct:

```
1    v[0] <= '0' when t[0] = '1' else t[1];
```

becomes

```
1    let v := v[0] <- if t[0] then false else t[1] in ...
```

If-Generate Statements. An if-generate statement is translated using an if-then-else expression:

```
1    IFG: if n >= 1 generate
2      v[0] <= '1';
3      s <= '0';
4    end generate;
```

becomes

```
1    let (v, s) :=
2      if n >= 1 then
3        let v := v[0] <- true in
4        let s := false in
5        (v, s)
6      else
7        (v,s)
8    in ...
```

Here the else branch simply returns the two signals `v`, `s` unmodified. Notice that this assumes that both signals were defined previously. While this is always the case for vectors, we also initialize `std_logic` signals such as `s` that are assigned inside conditional assignments so that the above code is correct. The initial value set for `s` is irrelevant thanks to our parameterized checks as explained for vectors.

For-Generate Statements. For-generate statements are translated using the iterator `N.iter` which, given a natural number n, a function $f : X \to X$, and initial value x, computes $f^n(x)$. Consider a for-generate statement of the following form.

```
1    FG: for i in 0 to n-1 generate
2      -- signal assignments to s1, ..., sn
3    end generate;
```

Let $(s_1, \ldots, s_n)$ denote the set of signals that are modified inside the loop body. This is translated as follows:

```
1  let (s1, ..., sn,_) := N.iter n
2    (fun (s1, ..., sn, i) ->
3      let s1 := ... in
4      ...
5      let sn := ... in
6      (s1,...,sn,i+1)
7    ) (s1, ..., sn, 0)
8  in ...
```

Thus, the iteration starts with the tuple `(s1, ..., sn, 0)` made of the current values of the signals (which are initialized as seen above), and the first index 0. The given function is repeated n times. At each iteration, the function returns a tuple of the values at the end of this iteration, along with the next index $i + 1$. Thus, the iterations mimick the generated iterations in the for-generate loop.

While the VHDL's for-generate statement "generates" concurrent statements for each i in the given range, the Rocq translation "executes" these iterations for increasing values of i. In VHDL, the i-th "iteration" can read signals defined at later iterations, while this would not be correct in Rocq. In fact, this translation is correct under the condition that for each i, all signals read in the i-th iteration were assigned previously. Intuitively, this ensures that if we unroll the loop, then the statements would be in a topological ordering. It is therefore possible to compute the signal assignments by "executing" the loop once.

Let $S = \{\mathsf{st}_1, \ldots, \mathsf{st}_n\}$ denote the statements inside a loop. The loop condition is violated if the following formula is satisfiable.

$$\exists e \in \bigsqcup_{\mathsf{st} \subset S} \mathsf{reads}(\mathsf{st}), \exists e' \in \bigsqcup_{\mathsf{st} \in S} \mathsf{writes}(\mathsf{st}), \exists v_\mathsf{G} \in \mathsf{Val}(\mathsf{G}), v_\mathsf{I}, v_\mathsf{I}' \in \mathsf{Val}(\mathsf{I}), v_\mathsf{I} \prec v_\mathsf{I}',$$

$$\mathsf{context}(e[v_\mathsf{G}, v_\mathsf{I}]) \wedge \mathsf{context}(e'[v_\mathsf{G}, v_\mathsf{I}']) \wedge \mathsf{intersects}(e[v_\mathsf{G}, v_\mathsf{I}], e'[v_\mathsf{G}, v_\mathsf{I}']),$$

where $v_\mathsf{l} \prec v_\mathsf{l}'$ means that the iteration for the loop index values v_l is executed before that for v_l'. For a single loop with a range `i in 0 to n-1`, this simply means $v_\mathsf{l}(i) < v_\mathsf{l}'(i)$ (and for a range `i in n-1 downto 0`, this means $v_\mathsf{l}(i) > v_\mathsf{l}'(i)$). If there are two nested loops, say, of range `j in 0 to m-1` contained in the above loop, then this is the lexical ordering, *i.e.*

$$v_\mathsf{l}(i) < v_\mathsf{l}'(i) \vee v_\mathsf{l}(i) = v_\mathsf{l}'(i) \wedge v_\mathsf{l}(j) < v_\mathsf{l}'(j),$$

etc. The query precisely checks the existence of a signal that appears anywhere in the loop body, that is read at iteration v_l and written later in iteration v_l'. Designs not conforming to this condition are rejected.

6   The VHDL2Rocq Tool

Our tool is written in Python 3 and uses the parser and the document object model of the pyGHDL library[4]. It uses the Z3 SMT solver [8] to solve all queries necessary for the parameterized checks. The tool has a verification mode where it only parametrically checks the given VHDL code, which non-experts can use as a linter. When specified, in addition to these checks, it outputs Rocq code.

The Rocq code is produced for the given design along with a library that formalizes the IEEE library providing operations such as concatenation, conversion to signed, unsigned and integer and back to vectors etc. Some details of this library were described in Sect. 5. Our formalization contains all necessary operations along with their formally proven properties referring to the VHDL standard [15].

One of the applications of our translator was the verification and translation of a combinational floating-point unit design of over 1000 lines. Our FPU design is based on a publicly available open-source design[5], with several modifications done by ourselves to adapt the design to our needs and to prepare it for verification. Typical modifications included restricting the ranges of the generics, and rewriting some modules to avoid the use of the process statement. All parameterized checks were efficient, and they could be done within a second for each translated file. In fact, the SMT queries are often small since the number of generics and the set of constraints in the context of each statement is also small. The only possibly hard query is done by (3) in Sect. 3.3 which involves quantifiers. This check always ran in less than a second in our experiments since our designs do not contain architectures with large numbers of loops (the maximum was two nested loops).

We limit the scope of the present paper to the parameterized checks and to the translation, and we plan to publish separately a detailed case study on the formal verification of the FPU, including the VHDL source code, its translation, and the Rocq proofs of its compliance with the IEEE-754 standard.

We cannot publish our tool at this point since it is proprietary due to company policy.

---

[4] https://ghdl.github.io/ghdl/.
[5] https://gitlab.inria.fr/taran/FPU.

7   Methodology for Pipelined Circuits

While our checks and translation were designed for combinational designs, we applied our methodology for the verification of a pipelined version of the FPU mentioned above combining formal proofs in Rocq with hardware equivalence checking methods [18]. Note that several previous works also report on the combination of two types of techniques [23,26]. We focus here on restricting the application of the Rocq proof assistant to combinational designs, and transferring the verification results to pipelined circuits using equivalence checking.

In our project, we first developed a generic combinational FPU design in VHDL, verified and translated the design to Rocq using our tool VHDL2Rocq. As advocated before, the combinational nature of the design was extremely useful because the translation of each VHDL entity/architecture yielded simple functions made of sequences of assignments, possibly within loops. Moreover, the proofs were easier to carry out since the model was purely functional, and there was no need to track the current state of the pipeline registers. The goal of this phase was to prove in Rocq that the design conformed to the IEEE Standard for Floating-Point Arithmetic (IEEE 754).

Once this was established, we built an alternative pipelined version of the FPU design. This phase was carried out by an independent team. We then established the equivalence of the two designs using the equivalence checking algorithms of the tool ABC [4]. More precisely, we verified each operation (addition, subtraction, multiplication, division) of the FPU separately. For the subcircuit corresponding to each operation, we first synthesized an RTL netlist from the VHDL code using ABC (after translating VHDL to Verilog using GHDL). We unrolled the pipelined design P for K clock cycles where K is the number of pipeline stages, and obtained a combinational circuit P^K which, given inputs at the first stage, produces outputs after K clock cycles. We then applied the combinational equivalence check algorithm (the `cec` command of ABC) to check P^K against the netlist for the combinational design.

The overall process is summarized in Fig. 3 and allows us to formally verify that all proven properties in Rocq carry over the pipelined design which is to be deployed in a product.

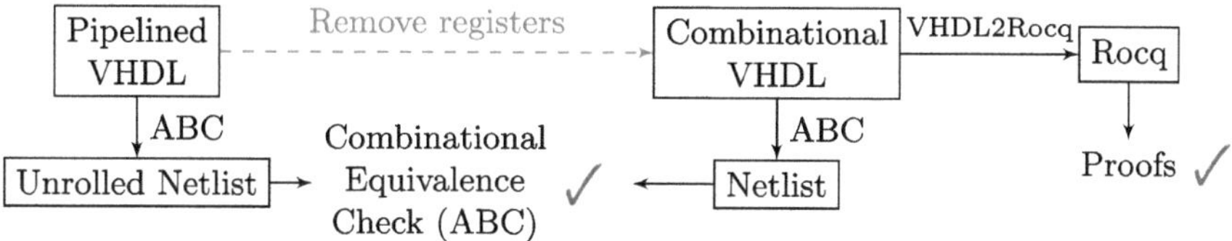

Fig. 3. Our methodology for verifying pipelined designs in Rocq using VHDL2Rocq.

It is not desirable to build two separate versions of a design, one to be used in production and one for formal verification. While Fig. 3 presents a general methodology, one can actually obtain the combinational design automatically. Assume that the pipelined design was built in which delays were introduced by instantiating a unique component `Reg` which models a sequence of D-flip-flops, taking as input a set of signals, and an integer k, and outputting the same signals delayed by k clock cycles (the component also takes as input the clock, and the standard enable, and reset signals that control the flip-flop). In our case, this was the unique VHDL file that contained an architecture with memory. For a given design with K pipeline stages, a combinational version can then be obtained automatically by replacing the VHDL source code for `Reg` by an alternative file `DummyReg` which has the same inputs and outputs as `Reg` but simply connects the inputs to its outputs without delay. Note that this transformation has the same effect as unrolling the pipelined circuit for K steps, but it is done at the source code level. Such transformations to obtain equivalent combinational circuits from pipelines have been used before on the abstract model [17,28]. We transform the design directly at the RTL source code level, and moreover apply equivalence checking as shown in Fig. 3 which ensures that both versions are indeed equivalent.

Thus, by carefully isolating the registers, it was possible to design pipelined circuits, and extract simpler combinational designs automatically to carry out formal proofs in Rocq, and transfer back the verification results.

8   Conclusion

Related Works. There are several translators from VHDL or Verilog to programming languages, namely, to C, C++, and SystemC. Some of these available translators are actually hardware simulators which encode the simulation semantics of the given designs into these languages, and compile these into a simulation program. This is the case of Verilator [30] which translates to C++. The tools vhdl2systemc and verilog2systemc from EDAUtils[6] translate hardware designs to SystemC. Although these tools produce highly efficient code, because they have support for all the simulation features of the languages, the translated code is difficult to read. In particular, it is not possible to map the translated code directly to Rocq or a language of a similar proof assistant. The translator V2C [22] was designed to apply model checking to Verilog designs through translation to C so the translated code is simpler to interpret. However, all aforementioned tools require all parameters (or generics) to be instantiated, and the translation is performed for the instantiated designs. Therefore, these do not fulfill our needs both in terms of the target language and in terms of the genericity of the translations.

Many projects focus on translating from the proof systems to register transfer languages (RTL); see e.g. for PVS [2]. In [20], designs written in HOL are translated to Verilog along with an equivalence proof within a formalization of a fragment of Verilog. An extension of this work even allows one to synthesize technology-mapped netlists using a verified compiler [19]. Kami [6] allows one to

---

[6] https://edautils.com.

write designs in a domain-specific language defined within Rocq and to compile it to Verilog via Bluespec.

There have been automatic translations from a fragment of the Bluespec SystemVerilog (BSV) designs to PVS [24]. BSV is a high-level language based on Haskell and requires the computation of a scheduler by the compiler to execute specified actions. Thus, the translation to PVS must also contain the description of the scheduler [21]. In contrast, we consider simpler combinational designs in the older but more widely used language VHDL, and the unique topological order that is computed means that there is no need to compute a scheduler.

Other works target a fragment of simpler languages such as AMD RTL and develop a formalization which is then used to translate RTL descriptions to proof assistants' languages (ACL2) [27]. The latter work also focuses only on combinational circuits and present an extension to pipelined circuits by a method similar to the unrolling we use here.

Perspectives. We have presented a verified translation scheme for translating combinational VHDL designs to Rocq to prove properties of generic designs. Although the support for combinational designs were sufficient in our work thanks to the approach detailed in Sect. 4, we have plans for extending this work to sequential designs which would broaden its applicability beyond pipelined circuits. We are planning to publish the details of the formal verification of a floating-point unit design with respect to the IEEE-754 standard [14] using the Rocq proof assistant and the approach presented here. Currently, we also support Verilog by translating to VHDL using Icarus Verilog[7] but a stronger guarantee could be obtained via a direct support.

Acknowledgments.

A   Examples for Integer Range Verification

Consider the following VHDL design.

```vhdl
1    library IEEE;
2    use IEEE.std_logic_1164.all;
3    USE ieee.numeric_std.ALL;
4
5    entity RangedIntegerEx is
6       generic(n : positive range 1 to 15);
7          port (a, b: in std_logic_vector(n-1 downto 0)
8             );
9    end entity;
10
11   architecture structural of RangedIntegerEx is
12      component Submodule is
13         generic(n : positive);
14         port (i: in integer range 0 to 101;
15               y: out integer range 0 to 1000);
16      end component;
17
18      signal x : integer range 100 downto 1;
19      signal y : natural;
20      signal z : natural range 1 to 10;
21   begin
```

---

[7] https://steveicarus.github.io/iverilog/.

```
22    x <= 100 when unsigned(a)+unsigned(b)>=100 else 1 when unsigned(a)+
      unsigned(b)<= 0 else to_integer(unsigned(a)+unsigned(b));
23    CI: Submodule generic map (n => n/2+1) port map (i => x, y => y);
24    z <= 1 + x/10;
25  end structural;
```

The model has a generic **n**, and two input vectors **a, b**. Furthermore, it defines internal integer signals **x,y,z**, and it instantiates **Submodule** which takes as argument an integer, and outputs an integer. We need to prove that all involved integer signals are within their declared ranges.

Our translator first generates a a Rocq function which returns the tuple of output *and* internal signals which must be proven correct:

```
1   Definition signals_of_rangedintegerex (n : N) (a : Std.Logic.vector) (b
      : Std.Logic.vector) :=
2   let x :=
3     (if (Unsigned.ge_integer (unsigned_adder (n) (Unsigned.of a) (
      Unsigned.of b)) 100) then
4       100
5     else if (Unsigned.le_integer (unsigned_adder (n) (Unsigned.of a) (
      Unsigned.of b)) 0) then
6       1
7     else
8       (Unsigned.to_integer (unsigned_adder (n) (Unsigned.of a) (Unsigned
      .of b)))) in
9   let y := submodule ((n / 2) + 1) x in
10  let z := (1 + (x / 10)) in
11  (x, y, z).
```

The translator also generates a correctness theorem which assumes that the inputs are correct, and states that the outputs and internal signals are correct, in the following sense.

```
1   Lemma lm_no_overflow_rangedintegerex:
2   let '(x, y, z) := signals_of_rangedintegerex n a b in
3     (1 <= n /\ n <= 15)
4     /\
5     ( |a| = n)
6     /\
7     ( |b| = n)
8     ->
9     (0 <= y /\ y <= 1000)
10    /\
11    (0 <= x /\ x <= 101)
12    /\
13    (1 <= ((n / 2) + 1) /\ ((n / 2) + 1) <= 2147483647)
14    /\
15    (1 <= x /\ x <= 100)
16    /\
17    (0 <= y /\ y <= 2147483647)
18    /\
19    (1 <= z /\ z <= 10).
20
```

This lemma assumes that the inputs of the entity **RangedIntegerEx** is correct: that **n** is within its declared range, and vectors **a,b** have size **n**, and states that internal signals **x,y,z** are within their ranges, but also all values passed to component instantiations such as the arguments to the instantiation on line 23. Note that signal **y** is the output of **Submodule**, so the proof of this lemma requires that **Submodule** was already proven correct. This is always possible since component instantiations are acyclic, so there is always an order in which these lemmas can be proved.

References

1. Berg, C., Jacobi, C.: Formal verification of the VAMP floating point unit. In: Margaria, T., Melham, T. (eds.) CHARME 2001. LNCS, vol. 2144, pp. 325–339. Springer, Heidelberg (2001). https://doi.org/10.1007/3-540-44798-9_26
2. Beyer, S., Jacobi, C., Kroening, D., Leinenbach, D.: Correct hardware by synthesis from pvs. In: Submitted for Publication (2002)
3. Braibant, T.: Coquet: a coq library for verifying hardware. In: Jouannaud, J.-P., Shao, Z. (eds.) CPP 2011. LNCS, vol. 7086, pp. 330–345. Springer, Heidelberg (2011). https://doi.org/10.1007/978-3-642-25379-9_24
4. Brayton, R., Mishchenko, A.: ABC: an academic industrial-strength verification tool. In: Touili, T., Cook, B., Jackson, P. (eds.) CAV 2010. LNCS, vol. 6174, pp. 24–40. Springer, Heidelberg (2010). https://doi.org/10.1007/978-3-642-14295-6_5
5. Bryant, R.E.: Symbolic simulation—techniques and applications. In: Proceedings of the 27th ACM/IEEE Design Automation Conference, pp. 517–521 (1991)
6. Choi, J., Vijayaraghavan, M., Sherman, B., Chlipala, A., Arvind: Kami: a platform for high-level parametric hardware specification and its modular verification. Proc. ACM Program. Lang. **1**(ICFP), 1–30 (2017)
7. Cyrluk, D., Rajan, S., Shankar, N., Srivas, M.K.: Effective theorem proving for hardware verification. In: Kumar, R., Kropf, T. (eds.) TPCD 1994. LNCS, vol. 901, pp. 203–222. Springer, Heidelberg (1995). https://doi.org/10.1007/3-540-59047-1_50
8. de Moura, L., Bjørner, N.: Z3: an efficient SMT solver. In: Ramakrishnan, C.R., Rehof, J. (eds.) TACAS 2008. LNCS, vol. 4963, pp. 337–340. Springer, Heidelberg (2008). https://doi.org/10.1007/978-3-540-78800-3_24
9. Grumberg, O., Veith, H. (eds.): 25 Years of Model Checking. LNCS, vol. 5000. Springer, Heidelberg (2008). https://doi.org/10.1007/978-3-540-69850-0
10. Gupta, A.: Formal hardware verification methods: a survey. Formal Methods Syst. Des. **1**, 151–238 (1992)
11. Harrison, J.: Formal verification at intel. In: 18th Annual IEEE Symposium of Logic in Computer Science, 2003. Proceedings, pp. 45–54. IEEE (2003)
12. Huang, S.Y., Cheng, K.T.T.: Formal equivalence checking and design debugging, vol. 12. Springer, Heidelberg (2012)
13. Hunt, W.A., Jr., Kaufmann, M., Moore, J.S., Slobodova, A.: Industrial hardware and software verification with ACL2. Phil. Trans. R. Soc. A: Math. Phys. Eng. Sci. **375**(2104), 20150399 (2017)
14. IEEE. IEEE standard for floating-point arithmetic. IEEE Std 754-2019 (Revision of IEEE 754-2008), pp. 1–84 (2019)
15. IEEE. IEEE/IEC international standard – behavioural languages – part 1-1: VHDL language reference manual. IEC 61691-1-1:2023-10 (IEEE Std 1076-2019), pp. 1–678 (2023)
16. Kahn, A.B.: Topological sorting of large networks. Commun. ACM **5**(11), 558–562 (1962)
17. Kaufmann, M., Russino, D.: Verification of pipeline circuits. In: ACL2 Workshop (2000)
18. Kuehlmann, A., Somenzi, F., Hsu, C.J., Bustan, D.: Equivalence checking. In: Electronic Design Automation for IC Implementation, Circuit Design, and Process Technology, pp. 99–130. CRC Press (2017)
19. Lööw, A.: Lutsig: a verified verilog compiler for verified circuit development. In: Proceedings of the 10th ACM SIGPLAN International Conference on Certified Programs and Proofs, pp. 46–60 (2021)

20. Lööw, A., Myreen, M.O.: A proof-producing translator for verilog development in hol. In: 2019 IEEE/ACM 7th International Conference on Formal Methods in Software Engineering (FormaliSE), pp. 99–108. IEEE (2019)
21. Moore, N., Lawford, M.: A case study in the automated translation of bsv hardware to pvs formal logic with subsequent verification. In: International Symposium on Theoretical Aspects of Software Engineering, pp. 65–72. Springer, Heidelberg (2022). https://doi.org/10.1007/978-3-031-10363-6_5
22. Mukherjee, R., Tautschnig, M., Kroening, D.: v2c – a verilog to C translator. In: Chechik, M., Raskin, J.-F. (eds.) TACAS 2016. LNCS, vol. 9636, pp. 580–586. Springer, Heidelberg (2016). https://doi.org/10.1007/978-3-662-49674-9_38
23. Pisini, V.K., Tahar, S., Curzon, P., Ait-Mohamed, O., Song, X.: Formal hardware verification by integrating hol and mdg. In: Proceedings of the 10th Great Lakes Symposium on VLSI, pp. 23–28 (2000)
24. Richards, D., Lester, D.: A monadic approach to automated reasoning for bluespec systemverilog. Innov. Syst. Softw. Eng. **7**, 85–95 (2011)
25. Rival, X., Yi, K.: Introduction to Static Analysis. MIT Press, Boston (2020)
26. Russinoff, D., Bruguera, J., Chau, C., Manjrekar, M., Pfister, N., Valsaraju, H.: Formal verification of a chained multiply-add design: combining theorem proving and equivalence checking. In: 2022 IEEE 29th Symposium on Computer Arithmetic (ARITH), pp. 120–126. IEEE (2022)
27. Russinoff, D., Kaufmann, M., Smith, E., Sumners, R.: Formal verification of floating-point rtl at amd using the acl2 theorem prover. In: Proceedings of the 17th IMACS World Congress on Scientific Computation, Applied Mathematics and Simulation, Paris, France (2005)
28. Russinoff, D.M., Flatau, A.: Rtl verification: a floating-point multiplier. In: Computer-Aided Reasoning: ACL2 Case Studies, pp. 201–231. Springer, Heidelberg (2000)
29. Schneider, K., Kumar, R., Kropf, T.: Automating most parts of hardware proofs in HOL. In: Larsen, K.G., Skou, A. (eds.) CAV 1991. LNCS, vol. 575, pp. 365–375. Springer, Heidelberg (1992). https://doi.org/10.1007/3-540-55179-4_35
30. Snyder, W.: Verilator 4.0: open simulation goes multithreaded. In: Open Source Digital Design Conference (ORConf) (2018)
31. Swords, S., Chau, C.: Robust, end-to-end correctness proofs of industrial divide and square root rtl designs. In: 2025 IEEE 32nd Symposium on Computer Arithmetic (ARITH), pp. 149–156 (2025)
32. The Coq Development Team. The coq proof assistant (2024)
33. Umamageswaran, K., Pandey, S.L., Wilsey, P.A.: Formal Semantics and Proof Techniques for Optimizing VHDL Models. Springer, Cham (1999)

Data Race Detection by Digest-Driven Abstract Interpretation

Michael Schwarz[1][(✉)] and Julian Erhard[2,3]

[1] National University of Singapore, Singapore, Singapore
m.schwarz@nus.edu.sg
[2] Technische Universität München, Garching, Germany
julian.erhard@tum.de
[3] Ludwig-Maximilians-Universität München, Munich, Germany

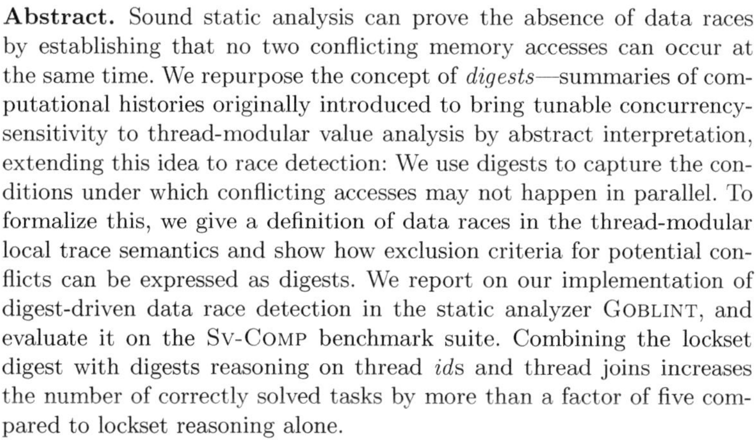

Abstract. Sound static analysis can prove the absence of data races by establishing that no two conflicting memory accesses can occur at the same time. We repurpose the concept of *digests*—summaries of computational histories originally introduced to bring tunable concurrency-sensitivity to thread-modular value analysis by abstract interpretation, extending this idea to race detection: We use digests to capture the conditions under which conflicting accesses may not happen in parallel. To formalize this, we give a definition of data races in the thread-modular local trace semantics and show how exclusion criteria for potential conflicts can be expressed as digests. We report on our implementation of digest-driven data race detection in the static analyzer GOBLINT, and evaluate it on the SV-COMP benchmark suite. Combining the lockset digest with digests reasoning on thread *id*s and thread joins increases the number of correctly solved tasks by more than a factor of five compared to lockset reasoning alone.

Keywords: data races · concurrency · static program analysis · software verification · abstract interpretation

1 Introduction

Data races happen when the same data is accessed by two different threads, the accesses are unordered, and at least one access is a write. As data races and any bugs they may trigger do not manifest deterministically, finding them by testing is exceedingly difficult. For this reason, many dynamic [19,32,37,39,41,49] and static [18,20,26,30,38,50] analyses have been proposed, with some of the latter [12,13,16,28,34,51] building on abstract interpretation [14].

Showing data race freedom amounts to proving at least one of two properties for each pair of accesses where at least one is a write. Either

(a) a separation in space (accesses go to different data); or
(b) a separation in time (accesses are not concurrent)

is to be established. Here, we propose a novel framework for property (b) and a principled way of designing new abstractions of it.

```
1    main:                    8    t1:
2        init(a);             9        lock(a);
3        g = 0;              10        g = 12;
4        create(t1);         11        unlock(a);
5        lock(a);
6        g = 5;
7        unlock(a);
```

Fig. 1. Multi-threaded program with thread prototypes main and $t1$.

Example 1. Consider the program in Fig. 2 with thread prototypes main and $t1$. It uses the mutex a for synchronization. The mutex is first initialized by main. Later, main creates a thread from prototype $t1$. There are two write-accesses to the global variable g in main, and one in $t1$. While the access to g in line 3 is unprotected, i.e., a is not held, it occurs before $t1$ is created. Thus, the program is still single-threaded at that time and the access does not race with any other access. The accesses in lines 6 and 10 do not race as both are protected by a. □

Example 1 illustrates that one has to rely on different forms of reasoning to establish race freedom. Sometimes, the program is still in single-threaded mode for an access, sometimes the set of mutexes overlaps. In other cases, two writes from the same thread prototype can be shown to not race because only one unique instance of that thread prototype is created. Here, we introduce a framework that captures all these different types of arguments. Conceptually,

(a) we associate with each access an abstraction of the execution history leading up to the access from some set $\mathcal{A}$ of *digests*. Locksets, thread *ids*, and whether the program is single-threaded all are (rather coarse) history abstractions.
(b) we introduce, for each global g, a predicate $(\|_g^?) : \mathcal{A} \to \mathcal{A} \to \{\mathsf{false}, \top\}$ which checks whether two accesses with respective digests may happen in parallel. While false indicates that the accesses may definitely not happen in parallel, $\top$ indicates that either such accesses may happen in parallel, or that the analyzer cannot exclude that they may based on digests alone. Whenever the definition does not depend on g, we write $\|^?$, omitting the index.

Example 2. For a digest tracking locksets, one can, irrespective of the global, set

$$A_a \parallel^? A_b = \begin{cases} \mathsf{false} & \text{if } A_a \cap A_b \neq \emptyset \\ \top & \text{otherwise} \end{cases}$$

to capture that accesses with overlapping locksets do not happen in parallel. □

As a semantic foundation for our framework, we build on the *local trace semantics* [21,43,46,47] which is a concrete semantics which, for each thread, tracks only *relevant* history information. While commonly some sort of interleaving semantics [29] is used to characterize multi-threaded programs, the local trace semantics as an equivalent, but thread-modular concrete semantics, is better-suited to our approach, as digests also take a thread-modular view of the execution history.

Previous work [47] introduced the notion of *digests* as a way to track relevant history information in a thread-modular way to make value analyses by abstract interpretation more precise. In this way, one obtains *families* of abstract interpretations parametrized by digests which offer a configurable degree of what was later dubbed *concurrency-sensitivity* [43]. However, neither of these works addresses data races. We take on this gap and repurpose digests for data race detection. The existing digest-driven abstract interpretation framework provides a notion of *admissibility* for instances of digests from which the soundness of analyses refined by such an instance directly follows. Here, we enhance this framework with a predicate $\|_g^?$ as discussed above and show how a first definition of such a predicate can automatically be derived from an admissible digest.

Our proposed framework does not only drastically simplify reasoning about race freedom by decoupling the base analysis from the admissibility of the digests, but also provides the conceptual foundation for our existing implementation of data race detection in the successful static analyzer GOBLINT.

The rest of the paper is structured as follows: Sect. 2 recalls the local trace semantics and its abstract interpretation, whereas Sect. 3 recalls digests and introduces the notion of *access stability*. Section 4 proposes two definitions of data races, one closely aligned with intuition and one more directly amenable to abstraction, and establishes their equivalence. Section 5 formally introduces the predicate $\|_g^?$ and its soundness requirements, before showing how a definition can automatically be derived and how bespoke definitions can sometimes be more precise. Section 6 presents our digest-driven race detection algorithm, whereas Sect. 7 reports on experiments within the static analyzer GOBLINT. Section 8 outlines related work, with Sect. 9 describing future work and concluding.

2 Local Trace Semantics and Its Abstract Interpretation

Thread-modular abstract interpretation scales as it avoids considering all possible interleavings. Traditional concrete semantics, on the other hand, distinguish all interleavings. This disconnect renders soundness proofs of thread-modular analyses cumbersome. The local trace semantics [21,43,46,47], which has been shown equivalent to the interleaving semantics w.r.t. safety properties [21], embraces thread-modularity already at the level of the concrete semantics:

– Instead of considering interleavings and hence total orders, the local trace semantics considers a *partial order* of actions. Other than actions within one thread, only relevant actions are ordered, namely those where information flows between threads as is, e.g., the case for locking and unlocking mutexes.

312   M. Schwarz and J. Erhard

- A local trace corresponds to the *local perspective* of one thread. It only contains those parts of the computational history that the thread can have knowledge about: This includes its own past as well as those parts of the past of other threads that it has learned about by communicating. Such communication happens via pairs of *observable* and *observing actions*, such as unlocking and locking some specific mutex.

		MT-dig.	Lockset-dig.
1	*main* :	ST_{main}	$\emptyset$
2	init(a);	ST_{main}	$\emptyset$
3	init(m_g);	ST_{main}	$\emptyset$
4	lock(m_g);	ST_{main}	$\{m_g\}$
5	g = 0;	ST_{main}	$\{m_g\}$
6	unlock(m_g);	ST_{main}	$\emptyset$
7	create(t1);	MT_{main}	$\emptyset$
8	lock(a);	MT_{main}	$\{a\}$
9	lock(m_g);	MT_{main}	$\{a, m_g\}$
10	g = 5;	MT_{main}	$\{a, m_g\}$
11	unlock(m_g);	MT_{main}	$\{a\}$
12	unlock(a);	MT_{main}	$\emptyset$

		MT-dig.	Lockset-dig.
13	*t1* :	MT	$\emptyset$
14	lock(a);	MT	$\{a\}$
15	lock(m_g);	MT	$\{a, m_g\}$
16	g = 12;	MT	$\{a, m_g\}$
17	unlock(m_g);	MT	$\{a\}$
18	unlock(a);	MT	$\emptyset$

Fig. 2. Multi-threaded program from Fig. 1 enhanced with the atomicity mutex m_g required by the local trace semantics. All program locations are annotated using the MT-digest (see Fig. 4) and the lockset-digest (see Fig. 6).

At an observing action (e.g., locking a mutex), the observed local trace (e.g., one ending in an unlock of the same mutex) is incorporated into the local trace of the observing thread and the result prolonged to record the observing action having taken place. At this step, only *compatible* local traces can be incorporated. Abstract interpretations of the local trace semantics track values for program points and for observable actions and thus share the rough structure with the concrete semantics [46, 47] allowing for scalable analyses that enjoy modular soundness proofs. Abstractions of the aforementioned compatibility of local traces are a key ingredient for making thread-modular abstract interpretation and, as we show in this paper, data race detection more precise.

Local Traces. More concretely, a local trace consists of several swimlanes, one for each thread appearing in the trace, which records a sequence of thread-local configurations attained by this thread. Such thread-local configurations record, e.g., the values of *local* variables and the current program point of the thread, but crucially no information about the global state, such as values of global variables. The order of configurations within one swimlane is the program order of the corresponding thread. Additionally, some dependencies between thread-local configurations are recorded. Such dependencies may, e.g., be between the action creating a thread[1] and the first configuration of the resulting thread, or between unlock and lock operations of the same mutex. We call the union of all such dependencies and the program orders the causality order, and demand that it is acyclic and has a unique maximal element. This unique maximal configuration belongs to one thread—the *ego thread*—whose perspective is taken.

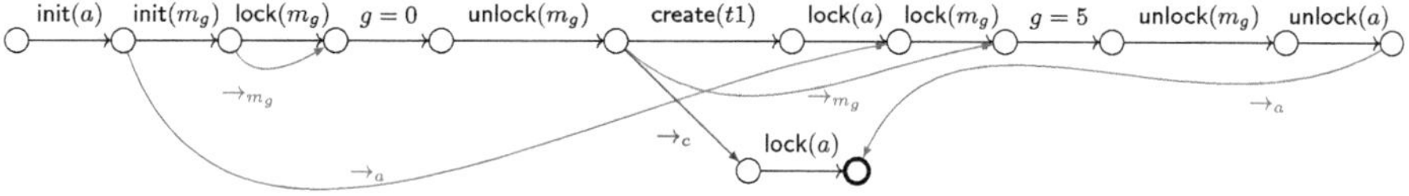

Fig. 3. Stylized local trace of the program in Fig. 2, where the ego thread is the single instance of $t1$, which completes its first action lock(a). The black arrows indicate the order of configurations within a thread. The red arrows indicate an ordering between two configurations that is mediated via a unlock/init and lock of a certain mutex. The blue arrow indicates the order between a configuration of a thread creating another thread and the first configuration of the newly created thread.

Example 3. Consider the program in Fig. 2, for now ignoring any annotations. Figure 3 shows an example local trace for this program, with the last appearing configuration of the ego thread highlighted by a thicker border.   □

The program in Fig. 2 is a variant of the one from Fig. 1 where all accesses to the global variable g are surrounded by an additional mutex m_g, its so-called *atomicity mutex*. In fact, this is how the local trace semantics enforces a sequential-consistency-like property for accesses to global variables. Let $\mathcal{G}$ be the set of global variables. Then, any access χ to a global $g \in \mathcal{G}$ (we allow accesses of the form x = g and g = x to copy the value of a global g from/to a local variable x), is immediately preceded by locking the atomicity mutex for g and immediately followed by unlocking it, i.e., it occurs as the sequence lock(m_g); χ; unlock(m_g). This allows conveniently treating accesses to global variables as local actions—with all communication happening via locking and unlocking atomicity mutexes. We will outline in Sect. 4 how this assumption interacts with the goal of detecting data races.

More formally, let $\mathcal{N}$ denote the set of all program points. We consider disjoint, uniquely labeled, control-flow graphs where each control-flow graph corresponds to the prototype of a thread. Control-flow graphs consist of nodes corresponding to program points and edges $(u, \mathsf{act}, v) \in \mathcal{E}$ where $u, v \in \mathcal{N}$ and act is from some set $\mathcal{A}ct$ of actions. This set of actions is given as the disjoint union of the sets of observing, observable, local, and creating actions. Each control-flow graph has exactly one start node which has no incoming edges. We identify the

---

[1] For technical reasons relating to the use of creation histories as concrete thread *ids* in prior work, the dependency is from the last configuration of the creating thread before the create action to the first configuration of the created thread.

label of a control-flow graph with its start node. One of the control-flow graphs is labeled main. It represents the main thread with which execution starts.

Let us call the set of all local traces $\mathcal{T}$. Within $\mathcal{T}$, there is a subset $\mathsf{init} \subset \mathcal{T}$ of local traces consisting only of an initial configuration of the main thread at its initial program point u_0.

An instantiation of the local trace semantics to some set of actions then provides for each edge (u, act, v) a function $[\![(u, \mathsf{act}, v)]\!]$ prolonging a local trace with the corresponding action, where possible. For non-observing actions, $[\![(u, \mathsf{act}, v)]\!]$ has type $\mathcal{T} \to 2^{\mathcal{T}}$ where the returned set is either a singleton or empty.[2] This corresponds to extending the local trace with the action act when possible. For example, extending a local trace t along an edge originating at some program point u is only possible if t ends in u. For an observing action act, $[\![(u, \mathsf{act}, v)]\!]$ has type $\mathcal{T} \to \mathcal{T} \to 2^{\mathcal{T}}$, where the returned set is again either empty or a singleton. When $[\![(u, \mathsf{act}, v)]\!]\, t_0\, t_1$ returns a non-empty set, the local traces t_0 and t_1 are said to be *compatible* w.r.t. the edge (u, act, v) and each other. In particular, to be compatible, t_1 must end in an observable action whose type matches the ones observed by act. Depending on the type of observing action, additional requirements have to be fulfilled. For instance, the operations corresponding to locking and unlocking some mutex a need to be totally ordered. For edges corresponding to the creation of new threads, which are of the form $(u_1, \mathsf{create}\, u', u_2)$, we additionally require a function $\mathsf{new} : \mathcal{T} \to \mathcal{N} \to \mathcal{N} \to 2^{\mathcal{T}}$; $\mathsf{new}\, t\, u_1\, u'$ computes from a local trace t ending in u_1 the local trace of the newly created thread starting its execution at u', provided such a thread creation is possible at this edge.

Instantiation. While we avoid fixing the set of observable, observing, and local actions to remain general, we will fix subsets to give examples. We then have, in addition to thread creates as described above, the following actions:

- The observing action $\mathsf{lock}(a)$, and the observable actions $\mathsf{unlock}(a)$ and $\mathsf{init}(a)$, for each mutex a from the set of mutexes $\mathcal{S}$. We require $\{m_g \mid g \in \mathcal{G}\} \subseteq \mathcal{S}$, i.e., atomicity mutexes for all global variables are included in the set of mutexes.
- As local actions, we have at least $x = g$ and $g = x$ to copy values between local variables and globals. We call either an *access* to g.

Consider again the local trace in Fig. 3 which originates from the program in Fig. 2. This local trace belongs to thread $t1$ as evidenced by the maximal element being a configuration of that thread. Of particular interest are the locking orders $\to_a$ and $\to_{m_g}$ for mutexes a and m_g, which illustrate additional conditions that need to hold in local traces: In the lock order $\to_a$ for each mutex a, every lock must be preceded by exactly one unlock of a or be the first lock of a in which case it is preceded by $\mathsf{init}(a)$. Each unlock and init operation of a must be directly followed by at most one lock operation. Some further requirements on these orders are detailed in previous work [42].

---

[2] This is purely for technical reasons. Constructs such as non-deterministic assignment remain possible, as CFGs can have several outgoing edges per node.

Fixpoint Formulation. While the set $\mathcal{T}$ of local traces can be characterized as the least fixpoint obtained when, starting from init, applying the functions $[\![\cdot]\!]$ and new to all (pairs of) local traces and taking the union with init, a different—but equivalent [42, 43]—view is more conducive to static analysis: Here, one considers the least solution of a constraint system with several constraint system variables, called *unknowns*. One such unknown is introduced for each program point, and for each observable action. An unknown $[u]$ associated with a program point u collects all local traces ending in u. For each observable action act an unknown $[\mathsf{act}]$ collects all local traces ending with this action. When lifting $[\![\cdot]\!]$ and new point-wise to sets, the resulting constraint system takes the following form:

$$[u_0] \supseteq \mathsf{init} \tag{1}$$

$$[u'] \supseteq \mathsf{new}([u_1], u_1, u'), \text{for } (u_1, \mathsf{create}\, u', u_2) \in \mathcal{E} \tag{2}$$

$$[u_2] \supseteq [\![(u_1, \mathsf{act}, u_2)]\!]([u_1]), \text{for } (u_1, \mathsf{act}, u_2) \in \mathcal{E}, \mathsf{act} \text{ not observing} \tag{3}$$

$$[\mathsf{act}] \supseteq [\![(u_1, \mathsf{act}, u_2)]\!]([u_1]), \text{for } (u_1, \mathsf{act}, u_2) \in \mathcal{E}, \mathsf{act} \text{ observable} \tag{4}$$

$$[u_2] \supseteq [\![(u_1, \mathsf{act}, u_2)]\!]([u_1], [\mathsf{act}']), \text{for } (u_1, \mathsf{act}, u_2) \in \mathcal{E}, \mathsf{act} \text{ observing,}$$
$$\mathsf{act}' \text{ observed by act} \tag{5}$$

For edges labelled with an observable action, there are two constraints: Constraints (3) propagate sets of resulting local traces to the unknown corresponding to the control-flow successor, whereas constraints (4) ensure that the resulting local traces are recorded at the unknown corresponding to the observable action.

Abstract Interpretation. To set up an abstract interpretation, it suffices to let unknowns range over values from a suitable abstract domain $\mathcal{D}$ instead of over $2^{\mathcal{T}}$ and replace functions $[\![\cdot]\!]$, init, and new with suitable abstract counterparts $[\![\cdot]\!]^{\sharp}$, $\mathsf{init}^{\sharp}$, and $\mathsf{new}^{\sharp}$. Adapting constraints (1) to (5) in this way, one obtains the abstract constraint system. For simplicity, we assume that $\mathcal{D}$ is a complete lattice with least element $\bot$ (denoting unreachability), greatest element $\top$, and that $\sqcup$ is used to denote the join (binary least-upper-bound). Furthermore, we require a monotonic concretization function $\gamma_{\mathcal{D}} : \mathcal{D} \to 2^{\mathcal{T}}$. An analysis is *sound* when $[\![\cdot]\!]^{\sharp}$, $\mathsf{init}^{\sharp}$, and $\mathsf{new}^{\sharp}$ are sound abstractions (w.r.t. $\gamma_{\mathcal{D}}$) of their concrete counterparts. Our considerations in this paper are generic in the base analysis and abstract domain. We assume that it is sound but do not provide details.

3   Digests as Concurrency-Sensitivity

Digests [42, 43, 47] are a way to refine abstract interpretations of multi-threaded programs to yield more precise results. Unknowns are split according to an abstraction of the computational past (the *digest*). Then, the abstract value at each unknown only describes those local traces that agree w.r.t. the digest. This approach can be understood as a generalization of trace partitioning [25, 33, 35, 40] to a multi-threaded setting. It offers a convenient way to specify which aspects of the computational history to consider and what to abstract away.

Digests exploit that, when at observing actions different views about the computational past are combined, some views are known to be incompatible already after considering their abstraction as given in the digest alone. These combinations then do not need to be considered by the analysis. Technically, digests overapproximate the concrete trace compatibility: If two digests are *incompatible*, so are all traces they represent. However, the reverse does not necessarily hold: digests may be compatible when the corresponding traces are, in fact, not.

More formally, let $\mathcal{A}$ be a set and $\alpha_{\mathcal{A}} : \mathcal{T} \to \mathcal{A}$ a function to extract such information from a local trace. Given a local trace t, we refer to $\alpha_{\mathcal{A}} t$ as its *digest*. We require that for each observable action $\mathsf{act} \in \mathcal{A}ct$, there is a function $[\![\mathsf{act}]\!]^{\sharp}_{\mathcal{A}} : \mathcal{A} \to \mathcal{A} \to 2^{\mathcal{A}}$ overapproximating the set of resulting digests of traces obtained by executing act. Similarly, for non-observable actions act and functions $[\![\mathsf{act}]\!]^{\sharp}_{\mathcal{A}} : \mathcal{A} \to 2^{\mathcal{A}}$. More formally, for an edge $e = (u, \mathsf{act}, v)$ we require

$$
\begin{aligned}
\forall t_0, t_1 \in \mathcal{T} : \alpha_{\mathcal{A}}([\![e]\!](t_0, t_1)) \subseteq [\![\mathsf{act}]\!]^{\sharp}_{\mathcal{A}}(\alpha_{\mathcal{A}} t_0, \alpha_{\mathcal{A}} t_1) \quad &\text{(act observable)} \\
\forall t_0 \in \mathcal{T} : \alpha_{\mathcal{A}}([\![e]\!](t_0)) \subseteq [\![\mathsf{act}]\!]^{\sharp}_{\mathcal{A}}(\alpha_{\mathcal{A}} t_0) \quad &\text{(act non-observable)}
\end{aligned}
\tag{6}
$$

where $\alpha_{\mathcal{A}}$ is lifted element-wise to sets. Intuitively, this corresponds to $[\![\cdot]\!]^{\sharp}_{\mathcal{A}}$ soundly overapproximating $[\![\cdot]\!]$. Additionally, we require $[\![\mathsf{act}]\!]^{\sharp}_{\mathcal{A}}$ to be deterministic, i.e., to either yield an empty set of digests or a singleton set:[3]

$$
\begin{aligned}
\forall A_0, A_1 \in \mathcal{A} : |[\![\mathsf{act}]\!]^{\sharp}_{\mathcal{A}}(A_0, A_1)| \leq 1 \quad &\text{(act observable)} \\
\forall A_0 \in \mathcal{A} : |[\![\mathsf{act}]\!]^{\sharp}_{\mathcal{A}}(A_0)| \leq 1 \quad &\text{(act non-observable)}
\end{aligned}
\tag{7}
$$

As an abstract counterpart for new, we require a function $\mathsf{new}^{\sharp}_{\mathcal{A}} : \mathcal{A} \to \mathcal{N} \to 2^{\mathcal{A}}$ that returns for a thread starting execution at program point u_1 the digest of this new trace. We once again require determinism in the sense discussed above.

$$
\forall t_0 \in \mathcal{T} : \alpha_{\mathcal{A}}(\mathsf{new}\, t_0\, u\, u_1) \subseteq \mathsf{new}^{\sharp}_{\mathcal{A}}(\alpha_{\mathcal{A}} t_0)\, u_1 \quad \forall A_0 \in \mathcal{A} : |\mathsf{new}^{\sharp}_{\mathcal{A}} A_0\, u_1| \leq 1 \tag{8}
$$

Furthermore, we require that whenever there is a successor digest for the edge corresponding to creating a new thread, there is also an appropriate digest for the newly created thread, i.e.,

$$
[\![\mathsf{create}(u_1)]\!]^{\sharp}_{\mathcal{A}}(A_0) \neq \emptyset \implies \mathsf{new}^{\sharp}_{\mathcal{A}} A_0\, u_1 \neq \emptyset \tag{9}
$$

For the initial digest at program start, we define:

$$
\mathsf{init}^{\sharp}_{\mathcal{A}} = \{\alpha_{\mathcal{A}} t \mid t \in \mathsf{init}\} \tag{10}
$$

Deviating from earlier work, we here additionally require a property called *access stability*, which intuitively states that executing an access sequence of the form $\mathsf{lock}(m_g); x = g; \mathsf{unlock}(m_g)$ or $\mathsf{lock}(m_g); g = x; \mathsf{unlock}(m_g)$ does not affect the

---

[3] This assumption simplifies reasoning about digests. To lift this restriction, one can instead consider *abstract digests* [43] where digests come from (complete) lattices. As (normal) digests are already quite expressive, we stick with them here for simplicity.

digest, whenever it is defined. More formally, consider a sequence $\bar{\chi}$ consisting of an access χ (read or write) to a global g and locking and unlocking the atomicity mutex m_g. For such *access sequences* of the form $(v_0^i, \mathsf{lock}(m_g), v_1^i) \cdot (v_1^i, \chi, v_2^i) \cdot (v_2^i, \mathsf{unlock}(m_g), v_3^i)$, we define the effect on the digest by

$$[\![\bar{\chi}]\!]_{\mathcal{A}}^{\sharp}(A_0, A_1) = \left([\![\mathsf{unlock}(m_g)]\!]_{\mathcal{A}}^{\sharp} \circ [\![\chi]\!]_{\mathcal{A}}^{\sharp} \circ [\![\mathsf{lock}(m_g)]\!]_{\mathcal{A}}^{\sharp}\right)(A_0, A_1).$$

Here, composition is lifted to return $\emptyset$ when an intermediate computation does so, and we identify singleton sets of resulting digests with their members. We call a digest *access stable*, if for all access sequences $\bar{\chi}$ we have

$$\forall A_0 \in \mathcal{A} : \bigcup_{A_1 \in \mathcal{A}} [\![\bar{\chi}]\!]_{\mathcal{A}}^{\sharp}(A_0, A_1) \subseteq \{A_0\}. \tag{11}$$

Definition 1. *A set $\mathcal{A}$ together with functions $[\![act]\!]_{\mathcal{A}}^{\sharp}$, $new_{\mathcal{A}}^{\sharp}$, and $init_{\mathcal{A}}^{\sharp}$ is called* digest. *If properties* (6) - (10) *hold for a digest, it is called* admissible. *If property* (11) *also holds, the digest is also called* access stable.

We remark that the question of admissibility is independent of the used analysis and the abstract domain, and is tied to the concrete semantics only. We take the freedom to refer to both the structure such as in Definition 1 as well as to an element of its set $\mathcal{A}$ as a *digest*, as the meaning is clear from the context.

A digest then gives rise to a refined abstract constraint system where each of the unknowns is split into several unknowns, corresponding to the different digests associated with the reaching traces: Each unknown $[u]$ for a program point u is replaced with unknowns $[u, A]$ for $A \in \mathcal{A}$, and each unknown $[act]$ for an observable action act is replaced with unknowns $[act, A]$ for $A \in \mathcal{A}$.

The refined constraint system then takes the following form:

$$[u_0, A_0] \sqsupseteq \mathsf{init}^{\sharp} \qquad\qquad (\text{for } A_0 \in \mathsf{init}_{\mathcal{A}}^{\sharp})$$

$$[u_2, A'] \sqsupseteq [\![(u_1, \mathsf{act}, u_2)]\!]^{\sharp}([u_1, A_0]) \qquad (\text{for } (u_1, \mathsf{act}, u_2) \in \mathcal{E}, \mathsf{act} \text{ not observing},$$
$$A' \in [\![\mathsf{act}]\!]_{\mathcal{A}}^{\sharp} A_0)$$

$$[\mathsf{act}, A'] \sqsupseteq [\![(u_1, \mathsf{act}, u_2)]\!]^{\sharp}([u_1, A_0]) \qquad (\text{for } (u_1, \mathsf{act}, u_2) \in \mathcal{E}, \mathsf{act} \text{ observable},$$
$$A' \in [\![\mathsf{act}]\!]_{\mathcal{A}}^{\sharp} A_0)$$

$$[u_2, A'] \sqsupseteq [\![(u_1, \mathsf{act}, u_2)]\!]^{\sharp}([u_1, A_0], [\mathsf{act}', A_1]) \qquad (\text{for } (u_1, \mathsf{act}, u_2) \in \mathcal{E},$$
$$\mathsf{act} \text{ observing}, \mathsf{act}' \text{ observed by } \mathsf{act}, A' \in [\![\mathsf{act}]\!]_{\mathcal{A}}^{\sharp}(A_0, A_1))$$

$$[u', A'] \sqsupseteq \mathsf{new}^{\sharp}([u_1, A_0], u_1, u') \qquad (\text{for } (u_1, \mathsf{create}\, u', u_2) \in \mathcal{E}, A' \in \mathsf{new}_{\mathcal{A}}^{\sharp} A_0 u').$$

We quickly state the main soundness theorem for such refined analyses:

Proposition 1. *Refining a sound analysis with an admissible digest yields a sound refined analysis.*

Proof. (Sketch). The proof proceeds by first proving a refined version of the concrete semantics sound w.r.t. the original semantics and then showing that the refined analysis is sound w.r.t. the refined semantics, provided the original analysis is sound w.r.t. the original semantics (see [42, Chapter 2.3] and [43]). □

Consider the local trace semantics with the actions described in the previous section. We provide a first example digest for this setting.

Example 4. A lightweight thread *id* digest may track for each thread whether

$$\mathcal{A} = \{\mathsf{ST}_{\mathsf{main}}, \mathsf{MT}_{\mathsf{main}}, \mathsf{MT}\}$$

$$\mathsf{init}^{\sharp}_{\mathcal{A}} = \{\mathsf{ST}_{\mathsf{main}}\}$$

$$\mathsf{new}^{\sharp}_{\mathcal{A}} M\, u_1 = \{\mathsf{MT}\}$$

$$[\![\mathsf{create}\, u_0]\!]^{\sharp}_{\mathcal{A}}\, M = \begin{cases} \{\mathsf{MT}\} & \text{if } M = \mathsf{MT} \\ \{\mathsf{MT}_{\mathsf{main}}\} & \text{otherwise} \end{cases}$$

$$[\![\mathsf{act}]\!]^{\sharp}_{\mathcal{A}}\, M = \{M\} \qquad \text{(other non-observing)}$$

$$[\![\mathsf{lock}(a)]\!]^{\sharp}_{\mathcal{A}}\, (M_0, M_1) = \begin{cases} \emptyset & \text{if } M_0 = \mathsf{ST}_{\mathsf{main}} \wedge M_1 \neq \mathsf{ST}_{\mathsf{main}} \\ \{M_0\} & \text{otherwise} \end{cases}$$

Fig. 4. Digest for lightweight thread *ids*.

- it is the main thread and has not started any other threads ($\mathsf{ST}_{\mathsf{main}}$), or
- it is the main thread and has started other threads ($\mathsf{MT}_{\mathsf{main}}$), or
- it is some other thread (MT).

The corresponding definitions are given in Fig. 4. $[\![\mathsf{lock}(a)]\!]^{\sharp}_{\mathcal{A}}$ exploits that if the digest for the ego thread is $\mathsf{ST}_{\mathsf{main}}$, i.e., the ego thread is the main thread and has not created other threads yet, it cannot acquire a mutex from another thread (with the MT digest) or from itself already having created threads (with the $\mathsf{MT}_{\mathsf{main}}$ digest). Thus, in these cases, it yields $\emptyset$. However, the result of $[\![\mathsf{lock}(a)]\!]^{\sharp}_{\mathcal{A}}(\mathsf{MT}_{\mathsf{main}}, \mathsf{ST}_{\mathsf{main}})$ cannot be set to $\emptyset$, as the last unlock of the mutex a can in fact have happened while the main thread was single-threaded, even if further threads were created in the meantime. This digest is *access stable*.   □

Example 5. Consider the program in Fig. 2. The MT-digests reachable after each statement is executed are annotated. Here, for each program point only one digest is reachable. Reasoning about data races using digests, we will later be able to conclude that line 5 does not race with either of the other accesses.   □

Remark 1. By setting $\mathcal{D} = \{\bot, \bullet\}$ with $\gamma_{\mathcal{D}}(\bullet) = \mathcal{T}$ one can obtain a trivial, sound analysis. This way, one can analyze a program using only digests for reasoning.

4 Races in Terms of Local Traces

To arrive at techniques for checking data races that are based on a thread-modular abstract interpretation of the local trace semantics, we first precisely define what exactly constitutes a race in the context of the local trace semantics.

We first make the following observation about the atomicity mutexes we have introduced: In race-free programs, any two conflicting accesses are ordered w.r.t. each other. Thus, the extra edges and dependencies introduced in the local trace semantics for atomicity mutexes only serve to make transitive dependencies between conflicting accesses in race-free executions explicit. Additionally, an ordering between read operations is introduced. However, as both orderings of these reads are considered, and executions only differing in the order in which such reads are performed are indistinguishable, all executions are still captured. In fact, by virtue of having atomicity mutexes, the local trace semantics assigns a meaning even to executions which would not traditionally have one, as they contain races.

To detect whether two accesses are racy, it thus suffices to check whether these appear in any local trace such that they are unordered but for their special atomicity mutex, leading to the following definition:

Definition 2 (Racy Accesses). *Let g be a global variable and χ_a and χ_b be two accesses to g. Accesses χ_a and χ_b are* racy *if at least one is a write and there exists a local trace in which these accesses are unordered w.r.t. each other when not considering the order provided by the m_g mutex.*

A local trace containing a racy access to g is visualized in Fig. 5.

While this definition corresponds closely to the intuition, it does not directly lend itself to abstraction. We thus provide an alternative definition that can be given in terms of the functions $[\![\cdot]\!]$ of the local trace semantics, which is better amenable to abstraction, and prove them equivalent.

The latter definition builds on the following observation: When accesses are unordered when ignoring the m_g mutexes, one can, from a local trace containing both accesses and ending in one of them, construct a local trace where the order of these accesses is flipped.

We first introduce some helpful notation: Consider once again an access sequence $\bar{\chi} \equiv (v_0, \mathsf{lock}(m_g), v_1) \cdot (v_1, \chi, v_2) \cdot (v_2, \mathsf{unlock}(m_g), v_3)$ where χ is an access to the global g (either read or write). We, akin to the definition of $[\![\bar{\chi}]\!]_{\mathcal{A}}^{\sharp}$, denote by $[\![\bar{\chi}]\!](t_0, t_1)$ the function $([\![(v_2, \mathsf{unlock}(m_g), v_3)]\!] \circ [\![(v_1, \chi, v_2)]\!] \circ [\![(v_0, \mathsf{lock}(m_g), v_1)]\!])(t_0, t_1)$ with composition lifted to return $\emptyset$ when an intermediate computation does so.

Definition 3 (Bidirectional Compatibility). *We call two access sequences $\bar{\chi}_a$ and $\bar{\chi}_b$ for a global g bidirectionally (trace) compatible if*

$$\exists t_a, t_b, t_l \in \mathcal{T} : ([\![\bar{\chi}_b]\!](t_b, [\![\bar{\chi}_a]\!](t_a, t_l)) \neq \emptyset) \wedge ([\![\bar{\chi}_a]\!](t_a, [\![\bar{\chi}_b]\!](t_b, t_l) \neq \emptyset)),$$

i.e., both orders of accesses are possible.

Here, t_l corresponds to the local trace ending in the last unlock of m_g before the pair of considered accesses (or the initialization if no accesses have happened yet), whereas t_a corresponds to the local trace ending in the last action of the thread performing $\bar{\chi}_a$ before the pair of considered accesses, and conversely for t_b and $\bar{\chi}_b$. It thus remains to relate the two definitions:

Theorem 1. *Two access sequences for a global variable g—at least one of which corresponds to a write—are bidirectionally compatible if and only if the accesses are racy.*

Proof. We consider the two directions separately:

($\Rightarrow$) Consider two bidirectionally compatible access sequences $\bar{\chi}_a$ and $\bar{\chi}_b$, with at least one a write. Then, $\exists t_a, t_b, t_l \in \mathcal{T} : [\![\bar{\chi}_b]\!](t_b, [\![\bar{\chi}_a]\!](t_a, t_l)) \neq \emptyset$. Let us call the trace in this set t'. By the construction of the actions for locking atomicity mutexes, the two accesses in t' are unordered safe for the m_g mutexes and any program order edges (as these are the only edges introduced). For the accesses to be ordered by program order edges, both access would have to belong to the same thread, as program order edges are thread-local. However, in this case we would have $[\![\bar{\chi}_a]\!](t_a, [\![\bar{\chi}_b]\!](t_b, t_l)) = \emptyset$, as the access $\bar{\chi}_a$ would already appear in t_b. Thus, the accesses are unordered safe for m_g mutexes and thus racy.

($\Leftarrow$) Consider two racy accesses χ_a and χ_b of g. At least one is a write and there exists a local trace t in which these accesses are unordered safe for the m_g mutexes. Consider the subtrace ending in the second access. The accesses in this subtrace remain unordered safe for the m_g mutexes. We prolong this trace by appending the unlock operation of m_g. This yields another trace, say t', in which the accesses remain unordered safe for the m_g mutexes. t' now contains sequences $\bar{\chi}_a, \bar{\chi}_b$ corresponding to accesses to g and locking and unlocking m_g.

Then, let the subtrace of t' ending in the last program point preceding the access sequence of the thread performing the first access be called t_a and conversely for t_b and the second access. Furthermore, let t_l denote the subtrace ending in the last $\mathsf{unlock}(m_g)$ before either of these access sequences or in $\mathsf{init}(m_g)$ if no such unlock exists. Then, $[\![\bar{\chi}_b]\!](t_b, [\![\bar{\chi}_a]\!](t_a, t_l)) = t'$ and $[\![\bar{\chi}_a]\!](t_a, [\![\bar{\chi}_b]\!](t_b, t_l))$ also yields a local trace, as both t_a and t_b originate from the same computation, $[\![\bar{\chi}_b]\!](t_b, t_l)$ ends in $\mathsf{unlock}(m_g)$, and there are no requirements enforcing a particular order of $\bar{\chi}_a$ and $\bar{\chi}_b$. Thus, $\bar{\chi}_b$ and $\bar{\chi}_a$ are bidirectionally compatible. $\qquad\square$

Thus, both definitions coincide.

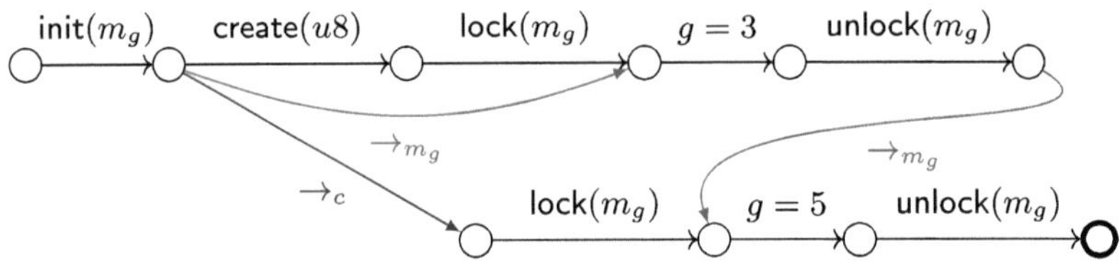

Fig. 5. An example local trace of a racy program.

Remark 2. Unlike in many other definitions, the local trace semantics assigns meaning even to executions past their first racy access, assuming sequential consistency. Thus, the local trace semantics has a notion of traces containing multiple races. However, our notion of how program execution continues *after the first race* need not coincide with other notions: For instance, *garbled* writes are not considered in the local trace semantics. A semantics that allows for such garbled writes may then disagree with the local trace semantics on races that happen after the first race.

5   Races in Terms of Digests

With the previous section having established bidirectional trace compatibility as a key ingredient to a formal definition of data races in the local trace semantics, we now turn to the question of designing sound abstractions of bidirectional trace compatibility that can be used in practical analyses. To this end, we introduce for each digest a family of new (commutative) predicates

$$(\|_g^?) : \mathcal{A} \to \mathcal{A} \to \{\mathsf{false}, \top\}$$

which is meant to express whether two accesses to global g associated with the respective digests can happen in parallel. Here, false means that two accesses with these digests can definitely not happen in parallel, whereas $\top$ means that this cannot be excluded.[4]

We tie this definition directly to the definition of bidirectional trace compatibility (Definition 3). We call $\|_g^?$ *sound* if for all traces t_a, t_b, t_l and sequences $\bar{\chi}_a$ and $\bar{\chi}_b$ of accesses to some global g as above, the following implication holds:

$$\emptyset \neq [\![\bar{\chi}_b]\!](t_b, [\![\bar{\chi}_a]\!](t_a, t_l)) \wedge \emptyset \neq [\![\bar{\chi}_a]\!](t_a, [\![\bar{\chi}_b]\!](t_b, t_l)) \\ \implies (\alpha_\mathcal{A}\, t_a) \,\|_g^?\, (\alpha_\mathcal{A}\, t_b) = \top \tag{12}$$

5.1   Bidirectional Digest Compatibility

A natural question arises: Can we soundly derive a predicate $\|_g^?$ for any admissible, access stable digest? To this end, we first define the notion of *bidirectional digest compatibility*:

Definition 4. *Sequences $\bar{\chi}_a$ and $\bar{\chi}_b$ corresponding to accesses to a global g are called* bidirectionally digest compatible *if*

$$\exists A_a, A_b, A_l \in \mathcal{A} : \\ \left([\![\bar{\chi}_b]\!]_\mathcal{A}^\sharp(A_b, [\![\bar{\chi}_a]\!]_\mathcal{A}^\sharp(A_a, A_l)) \neq \emptyset \right) \wedge \left([\![\bar{\chi}_a]\!]_\mathcal{A}^\sharp(A_a, [\![\bar{\chi}_b]\!]_\mathcal{A}^\sharp(A_b, A_l) \neq \emptyset) \right). \tag{13}$$

Theorem 2. *Bidirectional trace compatibility implies bidirectional digest compatibility.*

Proof. By repeated application of the properties of digests (Eqs. (6) and (7)), and $\alpha_\mathcal{A}$ being total. □

We remark that the opposite direction does not hold: (13) holding for two sequences $\bar{\chi}_a$ and $\bar{\chi}_b$ does not imply that these sequences are in fact bidirectionally trace compatible, as the digests (by design) introduce an overapproximation.

With this insight, we can now give a definition of the predicate $\|_g^?$ for admissible and *access stable* digests.

---

[4] As the digests are an overapproximation, they do not generally carry enough information to determine that two accesses can definitely happen in parallel.

$$\mathcal{A} = 2^{\mathcal{S}}$$
$$\mathsf{init}^{\sharp}_{\mathcal{A}} = \{\emptyset\}$$
$$\mathsf{new}^{\sharp}_{\mathcal{A}} S\, u_1 = \{\emptyset\}$$
$$[\![\mathsf{act}]\!]^{\sharp}_{\mathcal{A}} S = \{S\} \text{ (other non-observing)}$$
$$[\![\mathsf{act}]\!]^{\sharp}_{\mathcal{A}} (S_0, S_1) = \{S_0\} \text{ (other observing)}$$

$$[\![\mathsf{lock}(a)]\!]^{\sharp}_{\mathcal{A}} (S_0, S_1) = \begin{cases} \emptyset & \text{if } a \in S_0 \\ \{S_0 \cup \{a\}\} & \text{otherwise} \end{cases}$$

$$[\![\mathsf{unlock}(a)]\!]^{\sharp}_{\mathcal{A}} S = \begin{cases} \emptyset & \text{if } a \notin S \\ \{S \setminus \{a\}\} & \text{otherwise} \end{cases}$$

Fig. 6. Right-hand sides for lockset digest [43].

Proposition 2. *For an access stable and admissible digest $\mathcal{A}$, the definition*

$$A_a \;||^?_g\; A_b = \begin{cases} \textit{false} & \textit{if } [\![\mathit{lock}(m_g)]\!]^{\sharp}_{\mathcal{A}}(A_a, A_b) = \emptyset \vee [\![\mathit{lock}(m_g)]\!]^{\sharp}_{\mathcal{A}}(A_b, A_a) = \emptyset \\ \top & \textit{otherwise} \end{cases}$$

is sound.

Proof. This amounts to showing Eq. (12). Consider two bidirectionally compatible access sequences $\bar{\chi}_a$ and $\bar{\chi}_b$ and traces t_a, t_b, and t_l such that

$$([\![\bar{\chi}_b]\!](t_b, [\![\bar{\chi}_a]\!](t_a, t_l)) \neq \emptyset) \wedge ([\![\bar{\chi}_a]\!](t_a, [\![\bar{\chi}_b]\!](t_b, t_l)) \neq \emptyset)).$$

Let $A_a = \alpha_{\mathcal{A}}\, t_a$, $A_b = \alpha_{\mathcal{A}}\, t_b$, and $A_l = \alpha_{\mathcal{A}}\, t_l$. By the properties of digests (Eqs. (6) and (7)), also

$$\emptyset \neq [\![\bar{\chi}_b]\!]^{\sharp}_{\mathcal{A}}(A_b, [\![\bar{\chi}_a]\!]^{\sharp}_{\mathcal{A}}(A_a, A_l)) \wedge \emptyset \neq [\![\bar{\chi}_a]\!]^{\sharp}_{\mathcal{A}}(A_a, [\![\bar{\chi}_b]\!]^{\sharp}_{\mathcal{A}}(A_b, A_l)).$$

As the digest is *access stable* (Eq. (11)), this implies $\emptyset \neq [\![\bar{\chi}_b]\!]^{\sharp}_{\mathcal{A}}(A_b, A_a) \wedge \emptyset \neq [\![\bar{\chi}_a]\!]^{\sharp}_{\mathcal{A}}(A_a, A_b)$ which in turn implies $\emptyset \neq [\![\mathsf{lock}(m_g)]\!]^{\sharp}_{\mathcal{A}}(A_b, A_a) \wedge \emptyset \neq [\![\mathsf{lock}(m_g)]\!]^{\sharp}_{\mathcal{A}}(A_a, A_b)$, and thus $A_a \;||^?_g\; A_b = \top$. $\square$

Example 6. Consider again the program in Fig. 2, which is annotated using the digest for lightweight thread *ids* (Fig. 4). For the two access sequences starting after line 3 and line 14, with associated digests $\mathsf{ST}_{\mathsf{main}}$ and MT, it can be excluded that these two accesses race with each other: While $[\![\mathsf{lock}(m_g)]\!]^{\sharp}_{\mathcal{A}} (\mathsf{MT}, \mathsf{ST}_{\mathsf{main}}) = \{\mathsf{MT}\} \neq \emptyset$, $[\![\mathsf{lock}(m_g)]\!]^{\sharp}_{\mathcal{A}} (\mathsf{ST}_{\mathsf{main}}, \mathsf{MT}) = \emptyset$ and the respective sequences thus are known to not be bidirectionally compatible. $\square$

This definition in Proposition 2 is sound and generic—but can be overly conservative:

Example 7. Consider the lockset digest [43] (see Fig. 6) which associates with each program point the set of locks held when it is reached. The program in Fig. 2 is annotated with this information. The digests for lines 8 and line 14, right before the following access sequences, are given by $\{a\}$. $[\![\mathsf{lock}(m_g)]\!]^{\sharp}_{\mathcal{A}}(\{a\}, \{a\})$ yields $\{\{a, m_g\}\}$, i.e., a non-empty set of digests, and thus $\{a\} \;||^?_g\; \{a\} = \top$. Using the generic definition of $||^?_g$ from Proposition 2, we cannot exclude that the accesses in line 10 and line 16 race, despite both happening while holding the mutex a. $\square$

Remark 3. Setting $[\![\mathsf{lock}(m_g)]\!]^{\sharp}_{\mathcal{A}}(S_a, S_b) = \emptyset$ when $S_a \cap S_b \neq \emptyset$ renders the digest inadmissible: Consider a program where all accesses to a global g are protected by some mutex a. Then, any unlock of m_g happens while holding a, as does any lock of m_g. Setting $[\![\mathsf{lock}(m_g)]\!]^{\sharp}_{\mathcal{A}}(\{a\}, \{a\}) = \emptyset$ thus violates Eq. (6).

Nevertheless, two accesses where a common program mutex is held cannot race. In fact, synchronization via mutexes is a common way to prevent races. Thus, this first definition in terms of bidirectional digest compatibility—while sound and generic—does not sufficiently capture the property for all digests.

5.2   Bespoke Definitions

While the definition in Proposition 2 allows using information computed by any access stable digest to exclude races, often better definitions of $||^{?}_{g}$ are possible.

Example 8. Consider again the definition of $||^{?}$ for the lockset digest provided in Example 2.

$$S_a \ ||^{?} \ S_b = \begin{cases} \mathsf{false} & \text{if } S_a \cap S_b \neq \emptyset \\ \top & \text{otherwise} \end{cases}$$

This definition is more precise than the definition in terms of bidirectional digest compatibility and captures the intent behind mutex-based synchronization.   □

Proposition 3. *The definition of $||^{?}$ for the lockset digest given above is sound.*

Proof Sketch 1. *While immediately believable, we still sketch a proof. It is by contraposition: Consider sequences $\bar{\chi}_a$ and $\bar{\chi}_b$ of accesses to a global g and local traces t_a, t_b and t_l where at the end of t_a and t_b, both threads hold some mutex m. Then, $m \in (\alpha_{\mathcal{A}}\, t_a) \cap (\alpha_{\mathcal{A}}\, t_b)$ and thus $(\alpha_{\mathcal{A}}\, t_a) \ ||^{?} \ (\alpha_{\mathcal{A}}\, t_b) = \mathsf{false}$. We show that then the left-hand side of implication (12) is false, i.e., the accesses are not bidirectionally compatible. If both accesses are performed by the same thread, these sequences are not bidirectionally compatible as the program order orders one sequence before the other. Consider the case where the accesses are performed by different threads: Neither $\bar{\chi}_a$ nor $\bar{\chi}_b$ unlocks m, and both t_a and t_b contain operations that lock m. As locks of m are totally ordered, one of the sequences must have happened before the other thread acquired m. Thus, trace compatibility holds at most in one direction. For an in-depth proof, see the extended version [45, Appendix A].*   □

Later, we will usually not flesh out soundness arguments where they are intuitive.

Example 9. For a further example, consider again the digest from Fig. 4. The definition in terms of bidirectional digest compatibility already serves to exclude races where one access happens in single-threaded mode whereas the other one happens in multi-threaded mode. However, two accesses performed by the main

```
1    # From program, digest, and RHS of abstract interpretation, construct constr. sys. (Sect. 3)
2    constraintSystem = constructConSys(program, digest, ([[·]]#, new#, init#, D));
3
4    # Add helper unknowns and constraints (Eq. (14)) and solve using fixpoint engine
5    hConstraintSystem = addHelperConstraints(constraintSystem);
6    solution = solve(hConstraintSystem);
7
8    # Postprocessing
9    for g in globals:
10       accesses = solution[R_g];   # Get set of accesses to g
11
12       # For each pair of accesses, check whether they may race and at least one is a write
13       for (u_0, τ_0, A_0) in accesses:
14           for (u_1, τ_1, A_1) in accesses:
15               if (τ_0 == W or τ_1 == W ) and (A_0||?_g A_1 == T):
16                   report_race(u_0, u_1);
```

Fig. 7. Pseudocode of the digest-driven algorithm for data race detection.

thread can also not race as the main thread is unique, and such accesses are totally ordered by the program order. To exploit this, we can set:

$$M_a \;||^? M_b = \begin{cases} \mathsf{false} & \text{if } (M_a = \mathsf{ST}_{\mathsf{main}} \vee M_b = \mathsf{ST}_{\mathsf{main}}) \vee (M_a = M_b = \mathsf{MT}_{\mathsf{main}}) \\ \top & \text{otherwise.} \end{cases}$$

We define $||^?$ for general thread ids in the extended version [45, Appendix B]. $\square$

5.3   Combination of Digests

Several admissible digests can be combined into an admissible product digest [43] with all operations given pointwise. If all digests are access stable, so is the product digest. To leverage such a product digest for race detection, we consider the range of answers $\{\mathsf{false}, \top\}$ of the predicate $||^?_g$ as a lattice, with false as the least element. The predicate $||^?_g$ for a product $(\Pi_{j=0}^{n-1} \mathcal{A}_j)$ of n *active digests* then is defined as the component-wise meet of predicates $||^?_{g,j}$ for individual digests

$$(A_0^0, \ldots, A_{n-1}^0) \;||^?_g\; (A_0^1, \ldots, A_{n-1}^1) = \sqcap_{j=0}^{n-1}(A_j^0 \;||^?_{g,j}\; A_j^1)$$

As all $||^?_{g,j}$ soundly overapproximate bidirectional compatibility, so does the meet.

Example 10. Consider again the program in Fig. 2 where the main thread first initializes a global g and then starts two threads which each write to g, with only the latter accesses synchronized via some mutex. Neither the lockset digest (Example 8) nor the lightweight thread id digest (Example 9) on their own suffice to show the absence of all data races, but their combination does. $\square$

Alternatively, one can define a custom predicate for the product digest to jointly exploit information from all digests—at the expense of providing a dedicated soundness argument. Such a custom predicate resembles the *reduced* product construction as proposed by Cousot and Cousot [15], while the definition via the component-wise meet is more in the spirit of the *direct* product.

6   The Digest-Driven Race Detection Algorithm

The predicate $\|_g^?$ as a convenient yet expressive overapproximation of bidirectional trace compatibility can serve as a cornerstone of a race detection algorithm for a static analyzer. Such an algorithm works in two phases: During the analysis, for each global g the digest associated with each access and the access type is recorded. In a post-processing phase, the predicate $\|_g^?$ is then used to check whether there are any accesses that may race.

The algorithm is shown in Fig. 7. In line 2 a refined abstract constraint system is constructed that is parameterized by the program, the abstract domain and the digest, which we require to be admissible and access stable. This constraint system is extended with constraints that enforce that for each global variable information about each access to it is accumulated (line 5). For each global g, a dedicated unknown $[\mathcal{R}_g]$ is introduced, with values ranging over $2^{(\mathcal{N} \times \{W,R\} \times (\Pi_i \mathcal{A}_i))}$ where each tuple corresponds to an access. It consists of the location of access, its type (W(rite) or R(ead)), and digest information from all active digests. This is achieved by introducing for each constraint of the form

$$[\mathsf{unlock}(m_g), A'] \sqsupseteq [\![(u_1, \mathsf{unlock}(m_g), u_2)]\!]^{\sharp}[u_1, A_0]$$

an additional constraint

$$[\mathcal{R}_g] \sqsupseteq \left(\lambda\, x \to \begin{cases} \{(u_1, \tau\, u_1, A')\} & \text{if } x \neq \bot \\ \emptyset & \text{else} \end{cases} \right) [u_1, A_0] \tag{14}$$

where $\tau\, u_1 \in \{W, R\}$ is the type of the access which can be determined by checking whether the single (by construction of the programs with atomicity mutexes) incoming edge of u_1 is a write or a read. The case distinction ensures that only the digests associated with potentially reachable accesses are recorded, while the digest A' in this case is guaranteed to be identical to the digest before executing the access sequence due to access stability (Eq. (11)). The constraint system is then solved (line 6). In the post-processing phase (line 9 to 16) for each global the set of accumulated accesses is considered. For any two program points u_0 and u_1, a potential race for a global g is then flagged if

$$\exists \tau_0, A_0, \tau_1, A_1 : (u_0, \tau_0, A_0), (u_1, \tau_1, A_1) \in [\mathcal{R}_g] \wedge (\tau_0{=}W \vee \tau_1{=}W) \wedge (A_0 \,\|_g^?\, A_1) = \top,$$

i.e., at least one of the accesses is a write, and it cannot be excluded that they are unordered w.r.t. each other. We remark that the tuples (u_0, τ_0, A_0) and (u_1, τ_1, A_1) are allowed to coincide, as two accesses with the same digest may still race as they could be performed by different (concrete) threads.

Remark 4. Where an analyzer does not have concrete memory locations, but only abstract memory locations, as may be the case when analyzing programs with dynamic memory allocation with an allocation site abstraction, the check is performed for all accesses to abstract memory locations that potentially overlap.

Table 1. SV-COMP 2025 results for no-data-race [9] of tools with at least 650 correct true verdicts. *Correct true* (*correct false*) is the number of programs where a tool correctly deduced race-freedom (raciness). *Incorrect true* (*incorrect false*) is the number of programs where a tool incorrectly claimed race-freedom (raciness).

	CoOperRace	Goblint	Infer	Deagle	RacerF
Correct true	754	712	747	685	674
Correct false	6	0	0	0	68
Incorrect true	1	0	213	1	0
Incorrect false	0	0	0	0	4

7   Experiments and Benchmark Set

The Goblint static analyzer [51] is based on thread-modular abstract interpretation of the local trace semantics [42,47]. While it has supported data race detection for a long time, digest-driven data race detection proposed here serves as the theoretical underpinning and justification for this feature. Goblint implements various digests, and the digest-driven algorithm we propose in Sect. 6 is what emerged after unifying various older ad-hoc mechanisms.

Goblint participates in the data race track of Sv-Comp since its inception in 2021, receiving the highest score in 2021, 2023, and 2024[5] and the second-highest score in 2025—beaten only by CoOperRace which internally calls Goblint. [6–8,11] Table 1 shows the 2025 results. Among these, Goblint is the only tool producing no incorrect verdicts. We pose the following research question about digest-driven data race detection as implemented in Goblint:

RQ What impact do different digests and combinations of digests have on the number of solved tasks in the Sv-Comp data race benchmark set?

We modify Goblint to allow us to selectively change the predicate $\|_g^?$ for some digests to always return $\top$. Then, we compare for how many tasks Goblint still succeeds in proving data-race freedom depending on which digests are used to exclude races. Goblint implements the following digests:[6]

L   Locksets (c.f. Figure 6 and [43], plus support for Reader/Writer-Locks)
TF   Threadflag for multi- vs. single-threaded mode (c.f. Figure 4)
TID History-based thread *ids* [47] (recording parts of the creation history)
J   Must-joined threads [47] (based on the history-based thread *ids*)
FR   Tracking allocated memory that has not escaped its thread, i.e., is *fresh*[7]

---

[5] Deagle initially scored higher but was disqualified for fingerprinting tasks [8].

[6] Digests marked with a $\star$ go beyond the framework presented here in that they combine time and space aspects. We include them as they nevertheless give rise to a predicate $\|^?$ akin to ours and can thus be employed in the digest-based algorithm.

[7] It may seem surprising that FR can be considered a time-separation property. However, any access where the accessed memory has not escaped its thread yet must be ordered w.r.t. all other accesses to the same memory: Either both accesses happen in the same thread, or the other access happens later after the memory has escaped.

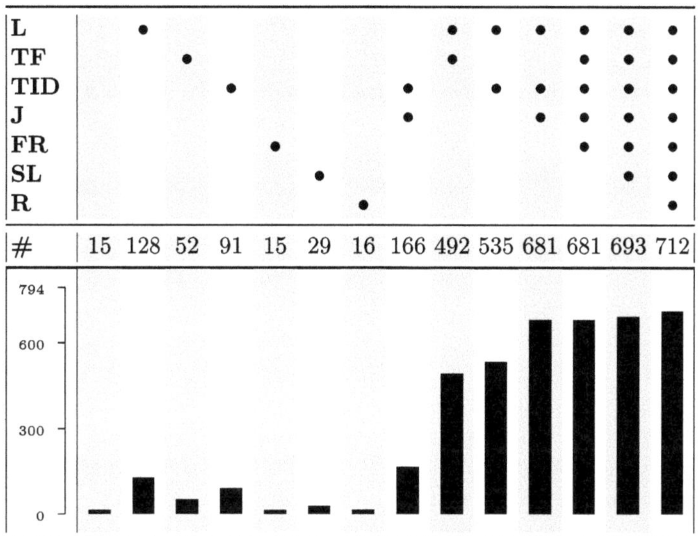

Fig. 8. The number of tasks where each digest succeeded in proving race freedom, out of a total of 794 race-free tasks in the benchmark set. None of the approaches produced any false negatives for the other 235 tasks in the benchmark set, which are racy.

SL Symbolic per-element locksets [51, Sections 5&6]*
R Region analysis [48]*

The benchmarks were conducted on a machine with two Intel Xeon Platinum 8260 @ 2.40GHz processors with 24 cores each and 256GB of RAM running Ubuntu 24.04.1 LTS. BENCHEXEC [10] was used to limit each run to 5 min and 5GB of memory. With these settings, the system ran out of memory for 4 tasks, 3 tasks timed out, and the analyzer crashed for 7 tasks. Where execution terminated, it did so within at most 20 s, where for all but 6 tasks the runtime was below 10 s. As the analysis was always performed with the same settings, and only the definition of the predicate $\|_g^?$ was changed, we do not further detail runtimes.[8] Figure 8 summarizes the evaluation results: Each column corresponds to one configuration, and a • in a row indicates that this configuration uses the given digest to exclude races. The number indicates for how many out of a total of 794 race-free tasks the configuration succeeded in establishing race freedom.

Interestingly, for 15 of the 794 tasks, none of the digests is necessary to prove race freedom. These programs only contain accesses to shared memory locations marked as atomic using the _Atomic specifier from C11 or its GCC equivalent. When it comes to single digests, **L** succeeds in proving race freedom for a sizable chunk of tasks, as do **TF** and **TID**. Here, **TID** subsumes **TF**. The other digests are less successful on their own, succeeding for less than 30 tasks each. While

---

[8] For **TID**, prior experiments [42, Chapter 5.3.2] on large, real-world programs, have established a median slowdown of around 1.2× (with a maximum slowdown of 5.2×) in the context of value analyses.

enabling **J** on top of **TID** almost doubles the number of proven tasks, it is only the combination of thread-based and lock-based techniques that enables race-freedom to be proven for the majority of tasks (**L+TF**: 492, **L+TID**: 535, **L+TID+J**: 681). Enabling **FR** and **TF** on top of **L+TID+J** does not yield an improvement, but enabling **SL** and **R** modestly increases the number of successful tasks to 712. Overall, the experiments show that digest-based data race detection and the *combination* of digests reasoning about mutexes, created threads, and joins underlie the success of GOBLINT. The impact of the more specialized techniques used by GOBLINT on the other hand is rather limited.

Threats to Validity. All digests remained enabled, with only $||_g^?$ modified to always return $\top$. This may lead to underestimating their impact, as a digest may help show program points unreachable. However, we do not expect programs to contain many unreachable accesses. Additionally, digests can constrain other digests with which program points are reached. As our digests do not exploit guards, we expect such improvements (through the value analysis) to be rare.

8 Related Work

Detecting data races is a well-studied program analysis problem and there is a wealth of dynamic [19,32,37,39,41,49] and static race detectors [18,20,22,23,26, 30,38,50–52]. A recent comparison of sound static race detectors [27] concludes that, to this day, open challenges remain—both w.r.t. showing time- and space-separation of accesses. Heeding this call for further research, our work provides a solid foundation for reasoning about time separation of accesses.

For the unsound static race detector RACERD, Gorogiannis et al. [24] state under which (idealized) conditions the tool becomes sound. Their definition of a race is for an interleaving semantics, and they require that there is an interleaving that can be extended by performing either of the racy accesses next. This differs from our setting building on the local trace semantics, where, in general, no local view of the execution can be extended directly by either racy access, rendering their definition inapplicable for local traces. Chopra et al. [13], for an analysis of interrupt-driven kernels, define data races based on overlapping sections of executions which also does not apply to the local trace semantics.

A line of work closely tied to the time-separation aspect of data race detection is *May-Happen-in-Parallel* (MHP) analysis [1–5,17,31,36,53], which attempts to determine sets of statements that may execute in parallel. Approaches there are, once more, often based on flavors of interleaving semantics, rather than a local view. Typically, control flow within threads, i.e., whether given statements are reachable at all, is abstracted away. Approaches may, e.g., extract a graph-based overapproximation of the *inter*-thread control flow in a first step, with an analysis of this graph as a second step. [5,17,31] Our framework, on the other hand, is integrated into a thread-modular abstract interpreter and thus seamlessly combines MHP reasoning for races with value analyses. Furthermore, as opposed to the more monolithic approaches in the MHP literature, our framework can be instantiated with different digests and their admissibility and the soundness of their predicate $||_g^?$ can be established modularly for each digest.

Digests were proposed as a way to refine the static analysis of the *values* of global variables of multi-threaded programs [42,47] with later work [43] proposing further instances of digests and generalizing the setup along various axes— within the scope of value analysis. We do not study value analyses, but instead show how digests can be repurposed to obtain principled static data race checkers in an extensible framework allowing for modular soundness proofs.

9 Conclusion and Future Work

We have provided a framework for reasoning about which accesses cannot happen simultaneously for digest-driven abstract interpretation. Giving two definitions of data races for the local trace semantics and establishing their equivalence allowed us to derive a sound digest-based data race detection algorithm. It relies on a novel predicate $\|_g^?$ to check whether two accesses with given digests may happen in parallel. We have shown how a generic definition of this predicate can be derived in terms of building blocks provided by any digest, and how, for some digests, bespoke definitions can yield more precise results. Digest-Driven Abstract Interpretation underlies our implementation in the static analyzer GOBLINT. To evaluate the impact of digests on the results of GOBLINT, we have turned specific digests on or off, finding that it is the combination of digests that underlies its success in SV-COMP. Combining the lockset digest with digests reasoning on thread *ids* and joins increases the number of successful tasks by more than fourfold when compared to the thread-related digests and by more than fivefold when compared to the lockset digest alone.

There are many directions for future work: Preliminary experiments show that digests can capture further synchronization primitives such as barriers, and that our techniques also apply to the analysis of other concurrency issues such as deadlocks. Going beyond time separation, it is also worth exploring how to generalize digests to also tackle space separation, by, e.g., enhancing pointers with allocation time stamps. Lastly, it would be interesting to investigate how digests can serve as succinct explanations of analysis results to end users that do not require them to understand the inner workings of the analyzer and are thus easily digestible.

Acknowledgments and Data Availability Statement. This research is supported in part by the National Research Foundation, Singapore, and Cyber Security Agency of Singapore under its National Cybersecurity R&D Programme (Fuzz Testing NRF-NCR25-Fuzz-0001). Any opinions, findings and conclusions, or recommendations expressed in this material are those of the author(s) and do not reflect the views of National Research Foundation, Singapore, and Cyber Security Agency of Singapore. This work is also supported in part by Deutsche Forschungsgemeinschaft (DFG)— 378803395/2428 CONVEY. An artifact [44] allowing for the reproduction of our results is available on Zenodo.

References

1. Agarwal, S., Barik, R., Sarkar, V., Shyamasundar, R.K.: May-happen-in-parallel analysis of x10 programs. In: PPoPP '07, pp. 183–193, ACM (2007). https://doi.org/10.1145/1229428.1229471
2. Albert, E., Flores-Montoya, A., Genaim, S.: Analysis of may-happen-in-parallel in concurrent objects. In: Giese, H., Rosu, G. (eds.) Formal Techniques for Distributed Systems - Joint 14th IFIP WG 6.1 International Conference, FMOODS 2012 and 32nd IFIP WG 6.1 International Conference, FORTE 2012, Stockholm, Sweden, June 13-16, 2012. Proceedings, Lecture Notes in Computer Science, vol. 7273, pp. 35–51, Springer (2012). https://doi.org/10.1007/978-3-642-30793-5_3
3. Albert, E., Genaim, S., Gordillo, P.: May-happen-in-parallel analysis for asynchronous programs with inter-procedural synchronization. In: Blazy, S., Jensen, T.P. (eds.) Static Analysis - 22nd International Symposium, SAS 2015, Saint-Malo, France, September 9-11, 2015, Proceedings, Lecture Notes in Computer Science, vol. 9291, pp. 72–89, Springer (2015). https://doi.org/10.1007/978-3-662-48288-9_5
4. Albert, E., Genaim, S., Gordillo, P.: May-happen-in-parallel analysis with returned futures. In: D'Souza, D., Kumar, K.N. (eds.) Automated Technology for Verification and Analysis - 15th International Symposium, ATVA 2017, Pune, India, October 3-6, 2017, Proceedings, Lecture Notes in Computer Science, vol. 10482, pp. 42–58, Springer (2017). https://doi.org/10.1007/978-3-319-68167-2_3
5. Barik, R.: Efficient computation of may-happen-in-parallel information for concurrent Java programs. In: LCPC '06, vol. 4339 LNCS, pp. 152–169, Springer (2006). https://doi.org/10.1007/978-3-540-69330-7_11
6. Beyer, D.: Software verification: 10th comparative evaluation (SV-COMP 2021). In: Groote, J.F., Larsen, K.G. (eds.) Tools and Algorithms for the Construction and Analysis of Systems - 27th International Conference, TACAS 2021, Held as Part of the European Joint Conferences on Theory and Practice of Software, ETAPS 2021, Luxembourg City, Luxembourg, March 27 - April 1, 2021, Proceedings, Part II, Lecture Notes in Computer Science, vol. 12652, pp. 401–422, Springer (2021). https://doi.org/10.1007/978-3-030-72013-1_24
7. Beyer, D.: Competition on software verification and witness validation: SV-COMP 2023. In: Sankaranarayanan, S., Sharygina, N. (eds.) Tools and Algorithms for the Construction and Analysis of Systems - 29th International Conference, TACAS 2023, Held as Part of the European Joint Conferences on Theory and Practice of Software, ETAPS 2022, Paris, France, April 22-27, 2023, Proceedings, Part II, Lecture Notes in Computer Science, vol. 13994, pp. 495–522, Springer (2023). https://doi.org/10.1007/978-3-031-30820-8_29
8. Beyer, D.: State of the art in software verification and witness validation: SV-COMP 2024. In: Finkbeiner, B., Kovács, L. (eds.) Tools and Algorithms for the Construction and Analysis of Systems - 30th International Conference, TACAS 2024, Held as Part of the European Joint Conferences on Theory and Practice of Software, ETAPS 2024, Luxembourg City, Luxembourg, April 6-11, 2024, Proceedings, Part III, Lecture Notes in Computer Science, vol. 14572, pp. 299–329, Springer (2024). https://doi.org/10.1007/978-3-031-57256-2_15
9. Beyer, D.: NoDataRace-Main – BenchExec results (2025). https://sv-comp.sosy-lab.org/2025/results/results-verified/no-data-race.NoDataRace-Main.table.html#/

10. Beyer, D., Löwe, S., Wendler, P.: Reliable benchmarking: requirements and solutions. Int. J. Softw. Tools Technol. Transf. **21**(1), 1–29 (2019). https://doi.org/10.1007/S10009-017-0469-Y

11. Beyer, D., Strejcek, J.: Improvements in software verification and witness validation: SV-COMP 2025. In: Gurfinkel, A., Heule, M. (eds.) Tools and Algorithms for the Construction and Analysis of Systems - 31st International Conference, TACAS 2025, Held as Part of the International Joint Conferences on Theory and Practice of Software, ETAPS 2025, Hamilton, ON, Canada, May 3-8, 2025, Proceedings, Part III, Lecture Notes in Computer Science, vol. 15698, pp. 151–186, Springer (2025). https://doi.org/10.1007/978-3-031-90660-2_9

12. Blackshear, S., Gorogiannis, N., O'Hearn, P.W., Sergey, I.: Racerd: compositional static race detection. Proc. ACM Program. Lang. **2**(OOPSLA),(2018). https://doi.org/10.1145/3276514

13. Chopra, N., Pai, R.R., D'Souza, D.: Data races and static analysis for interrupt-driven kernels. In: Caires, L. (ed.) Programming Languages and Systems - 28th European Symposium on Programming, ESOP 2019, Held as Part of the European Joint Conferences on Theory and Practice of Software, ETAPS 2019, Prague, Czech Republic, April 6-11, 2019, Proceedings, Lecture Notes in Computer Science, vol. 11423, pp. 697–723, Springer (2019). https://doi.org/10.1007/978-3-030-17184-1_25

14. Cousot, P., Cousot, R.: Abstract interpretation: A unified lattice model for static analysis of programs by construction or approximation of fixpoints. In: Graham, R.M., Harrison, M.A., Sethi, R. (eds.) Conference Record of the Fourth ACM Symposium on Principles of Programming Languages, Los Angeles, California, USA, January 1977, pp. 238–252, ACM (1977). https://doi.org/10.1145/512950.512973

15. Cousot, P., Cousot, R.: Systematic design of program analysis frameworks. In: Aho, A.V., Zilles, S.N., Rosen, B.K. (eds.) Conference Record of the Sixth Annual ACM Symposium on Principles of Programming Languages, San Antonio, Texas, USA, January 1979, pp. 269–282, ACM Press (1979). https://doi.org/10.1145/567752.567778

16. Dacík, T., Vojnar, T.: Racerf: Lightweight static data race detection for c code (2025). https://arxiv.org/abs/2502.04905, to appear at COOP 2025

17. Di, P., Sui, Y., Ye, D., Xue, J.: Region-based may-happen-in-parallel analysis for C programs. In: ICPP, pp. 889–898, IEEE (2015), ISBN 978-1-4673-7587-0, https://doi.org/10.1109/ICPP.2015.98

18. Dietsch, D., Heizmann, M., Klumpp, D., Schüssele, F., Podelski, A.: Ultimate taipan and race detection in ultimate - (competition contribution). In: Sankaranarayanan, S., Sharygina, N. (eds.) Tools and Algorithms for the Construction and Analysis of Systems - 29th International Conference, TACAS 2023, Held as Part of the European Joint Conferences on Theory and Practice of Software, ETAPS 2022, Paris, France, April 22-27, 2023, Proceedings, Part II, Lecture Notes in Computer Science, vol. 13994, pp. 582–587, Springer (2023). https://doi.org/10.1007/978-3-031-30820-8_40

19. Effinger-Dean, L., Lucia, B., Ceze, L., Grossman, D., Boehm, H.: Ifrit: interference-free regions for dynamic data-race detection. In: Leavens, G.T., Dwyer, M.B. (eds.) Proceedings of the 27th Annual ACM SIGPLAN Conference on Object-Oriented Programming, Systems, Languages, and Applications, OOPSLA 2012, part of SPLASH 2012, Tucson, AZ, USA, October 21-25, 2012, pp. 467–484, ACM (2012). https://doi.org/10.1145/2384616.2384650

20. Engler, D.R., Ashcraft, K.: RacerX: effective, static detection of race conditions and deadlocks. In: Scott, M.L., Peterson, L.L. (eds.) Proceedings of the 19th ACM Symposium on Operating Systems Principles 2003, SOSP 2003, Bolton Landing, NY, USA, October 19-22, 2003, pp. 237–252, ACM (2003). https://doi.org/10.1145/945445.945468

21. Erhard, J., Bentele, M., Heizmann, M., Klumpp, D., Saan, S., Schüssele, F., Schwarz, M., Seidl, H., Tilscher, S., Vojdani, V.: Correctness witnesses for concurrent programs: Bridging the semantic divide with ghosts. In: Shankaranarayanan, K., Sankaranarayanan, S., Trivedi, A. (eds.) Verification, Model Checking, and Abstract Interpretation, pp. 74–100, Springer Nature Switzerland, Cham (2025), ISBN 978-3-031-82700-6

22. Farzan, A., Klumpp, D., Podelski, A.: Sound sequentialization for concurrent program verification. In: Jhala, R., Dillig, I. (eds.) PLDI '22: 43rd ACM SIGPLAN International Conference on Programming Language Design and Implementation, San Diego, CA, USA, June 13 - 17, 2022, pp. 506–521, ACM (2022). https://doi.org/10.1145/3519939.3523727

23. Gavrilenko, N., de León, H.P., Furbach, F., Heljanko, K., Meyer, R.: BMC for weak memory models: Relation analysis for compact SMT encodings. In: Dillig, I., Tasiran, S. (eds.) Computer Aided Verification - 31st International Conference, CAV 2019, New York City, NY, USA, July 15-18, 2019, Proceedings, Part I, Lecture Notes in Computer Science, vol. 11561, pp. 355–365, Springer (2019). https://doi.org/10.1007/978-3-030-25540-4_19

24. Gorogiannis, N., O'Hearn, P.W., Sergey, I.: A true positives theorem for a static race detector. Proc. ACM Program. Lang. 3(POPL), 57:1–57:29 (2019). https://doi.org/10.1145/3290370

25. Handjieva, M., Tzolovski, S.: Refining static analyses by trace-based partitioning using control flow. In: Levi, G. (ed.) Static Analysis, pp. 200–214, Springer Berlin Heidelberg, Berlin, Heidelberg (1998). ISBN 978-3-540-49727-1

26. He, F., Sun, Z., Fan, H.: Deagle: An SMT-based verifier for multi-threaded programs (competition contribution). In: Fisman, D., Rosu, G. (eds.) Tools and Algorithms for the Construction and Analysis of Systems - 28th International Conference, TACAS 2022, Held as Part of the European Joint Conferences on Theory and Practice of Software, ETAPS 2022, Munich, Germany, April 2-7, 2022, Proceedings, Part II, Lecture Notes in Computer Science, vol. 13244, pp. 424–428, Springer (2022). https://doi.org/10.1007/978-3-030-99527-0_25

27. Holter, K., Saan, S., Lam, P., Vojdani, V.: Sound static data race verification for c: Is the race lost? ACM Trans. Program. Lang. Syst. (2025), ISSN 0164-0925, https://doi.org/10.1145/3732933, just Accepted

28. Kästner, D., et al.: Finding all potential run-time errors and data races in automotive software. In: WCX^{TM} 2017-SAE World Congress Experience, pp. 1–9, SAE International (2017)

29. Lamport: how to make a multiprocessor computer that correctly executes multi-process programs . IEEE Trans. Comput. 28(09), 690–691 (1979), ISSN 1557-9956, https://doi.org/10.1109/TC.1979.1675439

30. Liu, B., Liu, P., Li, Y., Tsai, C., Silva, D.D., Huang, J.: When threads meet events: efficient and precise static race detection with origins. In: Freund, S.N., Yahav, E. (eds.) PLDI '21: 42nd ACM SIGPLAN International Conference on Programming Language Design and Implementation, Virtual Event, Canada, June 20-25, 2021, pp. 725–739, ACM (2021). https://doi.org/10.1145/3453483.3454073

31. Lu, F., Wang, X., Zeng, Q., Yuan, G., Bao, Y.: Segment-based may-happen-in-parallel analysis for c programs. Concurr. Comput.: Pract. Exper. **37**(21–22), e70203 (2025). https://doi.org/10.1002/cpe.70203
32. Marino, D., Musuvathi, M., Narayanasamy, S.: Literace: effective sampling for lightweight data-race detection. In: Hind, M., Diwan, A. (eds.) Proceedings of the 2009 ACM SIGPLAN Conference on Programming Language Design and Implementation, PLDI 2009, Dublin, Ireland, June 15-21, 2009, pp. 134–143, ACM (2009). https://doi.org/10.1145/1542476.1542491
33. Mauborgne, L., Rival, X.: Trace partitioning in abstract interpretation based static analyzers. In: Sagiv, M. (ed.) Programming Languages and Systems, pp. 5–20, Springer Berlin Heidelberg, Berlin, Heidelberg (2005), ISBN 978-3-540-31987-0
34. Miné, A., et al.: Taking static analysis to the next level: proving the absence of run-time errors and data races with astrée. In: 8th European Congress on Embedded Real Time Software and Systems (ERTS 2016) (2016)
35. Montagu, B., Jensen, T.: Trace-based control-flow analysis. In: PLDI '21, p. 482–496, ACM (2021). https://doi.org/10.1145/3453483.3454057
36. Naumovich, G., Avrunin, G.S., Clarke, L.A.: An efficient algorithm for computing mhp information for concurrent Java programs. In: ESEC/FSE '99, vol. 1687 LNCS, pp. 338–354, Springer (1999). https://doi.org/10.1007/3-540-48166-4_21
37. O'Callahan, R., Choi, J.: Hybrid dynamic data race detection. In: Eigenmann, R., Rinard, M.C. (eds.) Proceedings of the ACM SIGPLAN Symposium on Principles and Practice of Parallel Programming, PPOPP 2003, June 11-13, 2003, San Diego, CA, USA, pp. 167–178, ACM (2003). https://doi.org/10.1145/781498.781528
38. Pratikakis, P., Foster, J.S., Hicks, M.W.: LOCKSMITH: context-sensitive correlation analysis for race detection. In: Schwartzbach, M.I., Ball, T. (eds.) Proceedings of the ACM SIGPLAN 2006 Conference on Programming Language Design and Implementation, Ottawa, Ontario, Canada, June 11-14, 2006, pp. 320–331, ACM (2006). https://doi.org/10.1145/1133981.1134019
39. Raman, R., Zhao, J., Sarkar, V., Vechev, M.T., Yahav, E.: Scalable and precise dynamic datarace detection for structured parallelism. In: Vitek, J., Lin, H., Tip, F. (eds.) ACM SIGPLAN Conference on Programming Language Design and Implementation, PLDI '12, Beijing, China - June 11–16, 2012, pp. 531–542, ACM (2012). https://doi.org/10.1145/2254064.2254127
40. Rival, X., Mauborgne, L.: The trace partitioning abstract domain. ACM Trans. Program. Lang. Syst. **29**(5), 26–es (2007), ISSN 0164-0925, https://doi.org/10.1145/1275497.1275501
41. Savage, S., Burrows, M., Nelson, G., Sobalvarro, P., Anderson, T.E.: Eraser: a dynamic data race detector for multithreaded programs. ACM Trans. Comput. Syst. **15**(4), 391–411 (1997). https://doi.org/10.1145/265924.265927
42. Schwarz, M.: Thread-Modular Abstract Interpretation: The Local Perspective. Ph.D. thesis, Technical University of Munich (2025). https://d-nb.info/137113345X/34
43. Schwarz, M., Erhard, J.: The digest framework: concurrency-sensitivity for abstract interpretation. International Journal on Software Tools for Technology Transfer (2024), ISSN 1433-2787, https://doi.org/10.1007/s10009-024-00773-y
44. Schwarz, M., Erhard, J.: Artifact for 'data race detection by digest-driven abstract interpretation' (2025). https://doi.org/10.5281/zenodo.17128591
45. Schwarz, M., Erhard, J.: Data race detection by digest-driven abstract interpretation (extended version) (2025). https://doi.org/10.48550/arXiv.2511.11055

46. Schwarz, M., Saan, S., Seidl, H., Apinis, K., Erhard, J., Vojdani, V.: Improving thread-modular abstract interpretation. In: Dragoi, C., Mukherjee, S., Namjoshi, K.S. (eds.) Static Analysis - 28th International Symposium, SAS 2021, Chicago, IL, USA, October 17-19, 2021, Proceedings, Lecture Notes in Computer Science, vol. 12913, pp. 359–383, Springer (2021). https://doi.org/10.1007/978-3-030-88806-0_18
47. Schwarz, M., Saan, S., Seidl, H., Erhard, J., Vojdani, V.: Clustered relational thread-modular abstract interpretation with local traces. In: European Symposium on Programming, pp. 28–58, Springer (2023). https://doi.org/10.1007/978-3-031-30044-8_2
48. Seidl, H., Vojdani, V.: Region analysis for race detection. In: International Static Analysis Symposium, pp. 171–187, Springer (2009)
49. Serebryany, K., Iskhodzhanov, T.: Threadsanitizer: data race detection in practice. In: Proceedings of the Workshop on Binary Instrumentation and Applications, p. 62–71, WBIA '09, Association for Computing Machinery, New York, NY, USA (2009), ISBN 9781605587936, https://doi.org/10.1145/1791194.1791203
50. Terauchi, T.: Checking race freedom via linear programming. In: Gupta, R., Amarasinghe, S.P. (eds.) Proceedings of the ACM SIGPLAN 2008 Conference on Programming Language Design and Implementation, Tucson, AZ, USA, June 7-13, 2008, pp. 1–10, ACM (2008). https://doi.org/10.1145/1375581.1375583
51. Vojdani, V., Apinis, K., Rõtov, V., Seidl, H., Vene, V., Vogler, R.: Static race detection for device drivers: the goblint approach. In: Lo, D., Apel, S., Khurshid, S. (eds.) Proceedings of the 31st IEEE/ACM International Conference on Automated Software Engineering, ASE 2016, Singapore, September 3–7, 2016, pp. 391–402, ACM (2016). https://doi.org/10.1145/2970276.2970337
52. Voung, J.W., Jhala, R., Lerner, S.: RELAY: static race detection on millions of lines of code. In: Crnkovic, I., Bertolino, A. (eds.) Proceedings of the 6th joint meeting of the European Software Engineering Conference and the ACM SIG-SOFT International Symposium on Foundations of Software Engineering, 2007, Dubrovnik, Croatia, September 3-7, 2007, pp. 205–214, ACM (2007). https://doi.org/10.1145/1287624.1287654
53. Zhou, Q., Li, L., Wang, L., Xue, J., Feng, X.: May-happen-in-parallel analysis with static vector clocks. In: CGO '18, pp. 228–240, ACM (2018). https://doi.org/10.1145/3168813

A Formal Executable Semantics of PROMELA

Byoungho Son and Kyungmin Bae[(✉)]

Pohang University of Science and Technology,
Pohang, South Korea
{byhoson,kmbae}@postech.ac.kr

Abstract. PROMELA, the modeling language of SPIN, is widely used to specify and model check finite-state concurrent systems but lacks support for deductive verification. This paper presents an executable semantics of PROMELA in the $\mathbb{K}$ framework that enables code-level deductive verification. To address the nontrivial interactions between guarded nondeterminism and concurrency, we introduce LOAD-AND-FIRE, an elegant semantic pattern that yields a modular, uniform treatment of guarded nondeterminism, cross-process interference, and atomicity in $\mathbb{K}$. Our semantics enables the full suite of analyses provided by $\mathbb{K}$, including deductive verification of PROMELA programs with infinite state spaces, a capability previously unavailable for PROMELA models. We illustrate the approach with a case study in deductive verification of an infinite-state concurrent system.

Keywords: PROMELA · Mechanized semantics · K framework · Semantic pattern · Deductive verification

1 Introduction

PROMELA (PROcess MEta LAnguage) [17,36], the modeling language of the SPIN model checker, has been widely used for decades to specify concurrent systems. It combines imperative constructs—such as assignments, conditionals, and loops—with first-class nondeterministic choice and message passing via buffered and handshake channels. PROMELA is broadly adopted across academia and industry for teaching, prototyping, and validating concurrency designs in domains including protocols, operating systems, and multithreaded software [8,13,22].

As PROMELA becomes more widely used, many verification tasks increasingly demand *deductive* reasoning: parametric invariants (e.g., "for all buffer capacities N"), proofs for infinite-state systems, compositional arguments, and reusable proof artifacts. SPIN's explicit-state LTL model checking [17] is highly effective for analyzing finite-state concurrent systems, yet it does not directly address these needs. What is missing is a mechanized, executable semantics of PROMELA that provides both a precise, runnable reference for the language and a basis for code-level deductive verification.

Y.-F. Chen et al. (Eds.): VMCAI 2026, LNCS 16417, pp. 335–358, 2026.
https://doi.org/10.1007/978-3-032-15700-3_16

Prior semantic definitions of Promela, such as labeled transition systems [5,27], structural operational semantics [14,42], and denotational semantics [9], provide insight but often lack mechanized formalizations that support deductive verification of Promela programs. Spin remains the reference implementation, yet it is informally specified [26]. To our knowledge, no prior line of work offers, in a single framework, (i) an executable interpreter for Promela; (ii) a proof system that reasons about Promela programs; and (iii) modular extensibility that scales to the language's difficult features.

In this paper, we present an executable semantics of Promela in $\mathbb{K}$ that also supports deductive verification. $\mathbb{K}$ [33] is a rewriting-based semantic framework that enables modular, machine-readable definitions of programming languages and automatically derives tools such as interpreters, model checkers, and deductive verifiers from the semantics. Concretely, our semantics (i) executes Promela models as an interpreter, (ii) provides a proof engine (`kprove`) to establish safety properties for both finite-state and infinite-state models, and (iii) constitutes a precise, machine-readable reference for the language.

Defining an executable semantics of Promela is challenging because the language exhibits nontrivial interactions between guarded nondeterminism and concurrency. For example, Promela permits nondeterministic choice whose options are guarded by arbitrary statements rather than Boolean conditions, and those statements may themselves nest additional choices. A satisfactory semantics must surface enabledness at the frontier before committing to a step and integrate blocking, nondeterminism, and atomicity in a uniform way. Formalizing these interactions in an elegant, executable form is difficult; consequently, prior work on Promela often targets restricted subsets that avoid these complications.

To address these challenges, we propose the Load-and-Fire semantic pattern. *Load* normalizes nested guarded constructs (including statement-guarded options) into a canonical multiset of enabled frontiers and is purely structural and produces no effects. *Fire* then commits atomically to one enabled frontier. A simple lock mechanism enforces atomic blocks, and a lock-tossing rule preserves atomicity across handshake (e.g., a send inside an atomic block transfers the lock to the receiver). This pattern yields modular rules and a uniform treatment of nested guarded nondeterminism, cross-process interference, and atomicity.

Our semantics enables the full suite of analyses provided by the $\mathbb{K}$ framework [21], including deductive verification of Promela programs with infinite state spaces, a capability previously unavailable for Promela models. The executability of our semantics also enables *validation* by differential execution against the reference implementation, Spin [17]. We use benchmark examples—including those from the Spin repository—as a conformance suite, also covering corner cases that exercise the nontrivial features described above.

The contributions of this paper are summarized as follows. (1) We present the first mechanized, executable semantics of Promela in $\mathbb{K}$ that enables code-level deductive verification. (2) We introduce the Load-and-Fire semantic pattern for guarded nondeterminism with arbitrary statement guards, applicable to languages with similar features (e.g., Go's `select`). We demonstrate the approach with a case study in deductive verification of an infinite-state concurrent system. The full $\mathbb{K}$ semantics, the case study, and additional benchmark models are available at https://github.com/postechsv/spink.

The rest of the paper is organized as follows. Section 2 provides background on the $\mathbb{K}$ framework and PROMELA. For readability, we present the semantics incrementally: Sect. 3 considers a core subset without nested guards, atomic blocks, or impure expressions; Sect. 4 extends this subset with nested guards; and Sect. 5 adds atomic blocks and impure expressions. Section 6 presents a case study in deductive verification. Section 7 discusses related work. Finally, Sect. 8 presents some concluding remarks.

2   Preliminaries

2.1   The $\mathbb{K}$ Semantics Framework

$\mathbb{K}$ [33] is a semantic framework for programming languages, based on rewriting logic [23]. It has been widely used to formalize a variety of languages, including C [12], Java [6], JavaScript [28], and so on. $\mathbb{K}$ provides several analysis tools, such as deductive verifiers and symbolic execution engines.

Semantics Definition in $\mathbb{K}$. In $\mathbb{K}$, program states are represented as multisets of nested cells, called *configurations*. Each cell represents a component of a program state, such as computations and stores. Transitions between configurations are specified as (labeled) $\mathbb{K}$ rules, specifying only the relevant parts of these cells.

A computation in $\mathbb{K}$ (called the $\mathbb{K}$ continuation) is defined as a $\curvearrowright$-separated sequence (of sort K) of computational tasks (of sort *KItem*). For example, $t_1 \curvearrowright t_2 \curvearrowright \ldots \curvearrowright t_n$ represents the computation consisting of t_1 followed by t_2, and so on. A task can be decomposed into simpler tasks, and the result of a task is forwarded to the subsequent tasks. E.g., $(1 + x) * 5$ is decomposed into $x \curvearrowright 1 + \square \curvearrowright \square * 5$, where $\square$ is a placeholder for the result of a previous task. If x evaluates to some value, say 3, then $3 \curvearrowright 1 + \square \curvearrowright \square * 5$ becomes $1 + 3 \curvearrowright \square * 5$, which eventually becomes 20. We denote the empty computation by $.K$.

The following shows typical examples of $\mathbb{K}$ rules for variable lookup and assignment, respectively, where lookup and assign are labels, the k cell contains a computation, *env* contains a map from variables to locations, and *store* contains a map from locations to values:

$$\text{lookup:} \quad \frac{\langle X \curvearrowright \ldots \rangle_k}{V} \ \langle \ldots X \mapsto L \ldots \rangle_{env} \ \langle \ldots L \mapsto V \ldots \rangle_{store}$$

$$\text{assign:} \quad \frac{\langle X \ := \ V \curvearrowright \ldots \rangle_k}{.K} \ \langle \ldots X \mapsto L \ldots \rangle_{env} \ \langle \ldots L \mapsto \frac{\cdot}{V} \ldots \rangle_{store}$$

A horizontal line represents a state change, and "..." indicates irrelevant parts. A cell without horizontal lines is not changed by the rule. By the `lookup` rule, if the first item in k is X, then X is replaced by the value V of X in its location

L. Similarly, the assign rule updates the *store* cell at L to V and removes the completed assignment from the k cell (yielding $.K$ at that position).

The following shows an example trace obtained by applying the rules lookup and assign, together with "structural" rules (denoted by $\dashrightarrow^*$), for the program x := y + 1 under $x \mapsto 4$ and $y \mapsto 2$:

$$\langle \text{x := y + 1} \rangle_k \langle \text{x} \mapsto 0; \text{y} \mapsto 1 \rangle_{env} \langle 0 \mapsto 4; 1 \mapsto 2 \rangle_{store} \langle 42 \rangle_{output}$$

$$\dashrightarrow^* \langle \text{y} \curvearrowright \square \text{ + 1} \curvearrowright \text{x := } \square \rangle_k \langle \text{x} \mapsto 0; \text{y} \mapsto 1 \rangle_{env} \langle 0 \mapsto 4; 1 \mapsto 2 \rangle_{store} \langle 42 \rangle_{output}$$

$$\xrightarrow{\text{lookup}} \langle 2 \curvearrowright \square \text{ + 1} \curvearrowright \text{x := } \square \rangle_k \langle \text{x} \mapsto 0; \text{y} \mapsto 1 \rangle_{env} \langle 0 \mapsto 4; 1 \mapsto 2 \rangle_{store} \langle 42 \rangle_{output}$$

$$\dashrightarrow^* \langle \text{x := 3} \rangle_k \langle \text{x} \mapsto 0; \text{y} \mapsto 1 \rangle_{env} \langle 0 \mapsto 4; 1 \mapsto 2 \rangle_{store} \langle 42 \rangle_{output}$$

$$\xrightarrow{\text{assign}} \langle .K \rangle_k \langle \text{x} \mapsto 0; \text{y} \mapsto 1 \rangle_{env} \langle 0 \mapsto 3; 1 \mapsto 2 \rangle_{store} \langle 42 \rangle_{output}$$

In this trace, locations are represented as natural numbers, and the *output* cell remains unchanged since no rule matches it.

$\mathbb{K}$ is effective at defining nontrivial features of real-world languages regarding control flow, concurrency, and nondeterminism. This allowed many features in C (e.g., pointers, goto, malloc) to be defined near completely in $\mathbb{K}$ [12]. Likewise, $\mathbb{K}$'s inherent concurrency and nondeterminism enabled elegant treatments of Java multithreading [6] and JavaScript for-in enumeration [28].

For example, goto statements can be can be defined naturally in $\mathbb{K}$ by treating control as data. A standard approach [12] precomputes a *gotoMap* cell that maps labels to continuations, encoding the local control-flow for each label. The following shows a simple $\mathbb{K}$ rule for goto:

$$\text{goto:} \quad \frac{\langle \text{goto } X \curvearrowright ... \rangle_k \langle ...X \mapsto K... \rangle_{gotoMap}}{K}$$

This rule takes the first item goto X in the k cell, looks up the continuation K bound to X in *gotoMap*, and replaces the current continuation with K.

Deductive Verification in $\mathbb{K}$. The $\mathbb{K}$ framework provides a language-agnostic deductive verifier based on Reachability Logic (RL) [37], which is parameterized by the user-defined $\mathbb{K}$ semantics of a language. It has been widely used to verify real-world programs, such as smart contracts [15,29,30].

RL is a proof system for proving partial correctness of a program, whose properties are specified as (all-path) reachability claims of the form $\phi \Rightarrow \psi$, where ϕ and ψ are Matching Logic [32] *patterns* describing a *set* of states (i.e., $\mathbb{K}$ configurations). Such a claim means that any *terminating* execution starting from a state satisfying ϕ reaches a state satisfying ψ. Skeirik et al. [35] extend the use of RL to verify invariant properties of reactive systems[1] by "freezing" every reachable state to induce a corresponding finite path to which reachability claims apply (see Sect. 6 for a PROMELA case study using this technique).

---

[1] Under its partial-correctness semantics, RL cannot be directly used to reason about *infinite* behaviors of reactive systems.

```
 1  requires "imp.k"                        10       <store>
 2  (...)                                   11       s |-> (
 3  module SPEC                             12         S:Int
 4    imports IMP // imp semantics          13         =>
 5    (...)                                 14         S +Int ((N +Int 1) *Int N /Int 2));
 6    claim                                 15       n |-> (N:Int => 0)
 7      <k> while ( 0 < n )                 16       </store>
 8         { s = s + n; n = n - 1; }        17     requires N >=Int 0
 9      => .K ...</k>                       18  endmodule
```

Fig. 1. An example reachability claim in `spec.k`.

Figure 1 shows a concrete specification of a reachability claim, excerpted from the $\mathbb{K}$ tutorial.[2] $\mathbb{K}$ configurations in the specification are written in an XML-like notation, where each (possibly nested) tag "`<cell>...</cell>`" denotes a $\mathbb{K}$ cell $\langle ... \rangle_{cell}$, and "`=>`" (line 9, 13, 15) is written locally within tags to distinguish the initial (ϕ) and goal (ψ) patterns. Concretely, the figure defines

$$\phi : \langle \texttt{while(0 < n)} \ \{...\} \curvearrowright K \rangle_k \ \langle s \mapsto S; n \mapsto N \rangle_{store} \quad \text{where } N \geq 0$$

$$\psi : \langle .K \curvearrowright K \rangle_k \ \langle s \mapsto S + \tfrac{N(N+1)}{2}; n \mapsto 0 \rangle_{store}$$

where the tail continuation K is written as '`...`' (line 9). This claim asserts that, upon termination (`=> .K`) of the `while` loop (line 9), starting from a state with $N \geq 0$ (line 17), the value of `s` is incremented by $((N + 1) * N/2)$ (line 14).

This claim can be proved automatically via the following process: first, one writes the claim in the file `spec.k`, importing the semantics definition (e.g., `imp.k`). Then, running `kprove` for `spec.k` completes the proof as follows:

```
$ kprove spec.k
(output) #Top // success
```

Typically, `kprove` may require auxiliary claims to prove the main claim $\phi \Rightarrow \psi$. Auxiliary claims can be used to ignore the cycles reachable from ϕ, as they correspond to infinite paths. When such cycles involve an intermediate pattern ϕ' but not ϕ, they should be discharged via the auxiliary claim $\phi' \Rightarrow \psi$.

2.2 The PROMELA Language

PROMELA is a popular high-level modeling language for specifying concurrent and distributed systems. It is used as the input language for the SPIN [17] model checker—received the 2001 ACM Software System Award—for verification and simulation. SPIN/PROMELA has been applied to a wide range of systems, including cryptographic protocols [22], Linux synchronization primitives [13], and flight-guidance systems [8].

A PROMELA program consists of declarations of global data (e.g., integers, booleans, channels) and processes (**proctype**). Like C functions, process declarations provide arguments and a body containing a `;`-separated sequence (*Seq*) of

---

[2] https://kframework.org/k-distribution/k-tutorial/1_basic/22_proofs/.

$$
\begin{array}{lll}
Seq & ::= & Stmt^+ \\
Stmt & ::= & BStmt \mid CStmt \mid Decl \\
Decl & ::= & Type \; Id\,[\,Int\,] \\
 & & \mid \; \texttt{chan} \; Id\,[\,Int\,] \; \texttt{= [}Int\texttt{] of \{} \; Type^+ \; \texttt{\}} \\
Type & ::= & \texttt{int} \mid \texttt{bool} \mid \texttt{chan} \\
BStmt & ::= & Expr \\
 & & \mid \; Id\,[\,Expr\,] \; \texttt{=} \; Expr \\
 & & \mid \; Id\,[\,Expr\,] \; \texttt{!} \; Expr^+ \\
 & & \mid \; Id\,[\,Expr\,] \; \texttt{?} \; Arg^+ \\
Arg & ::= & Id\,[\,Expr\,] \mid Int \mid Bool
\end{array}
$$

$$
\begin{array}{lll}
CStmt & ::= & \texttt{if} \; Option^+ \; \texttt{fi} \\
 & & \mid \; \texttt{do} \; Option^+ \; \texttt{od} \\
 & & \mid \; \texttt{goto} \; Id \mid \texttt{break} \\
 & & \mid \; \texttt{atomic \{} \; Seq \; \texttt{\}} \\
Expr & ::= & \texttt{run} \; Id\,(\,Expr^*\,) \\
 & & \mid \; Id\,[\,Expr\,] \mid Int \mid Bool \\
 & & \mid \; \texttt{nfull}\,(Id\,[\,Expr\,]) \\
 & & \mid \; Id\,[\,Expr\,] \; \texttt{?} \; \texttt{[}Arg^+\texttt{]} \\
 & & \mid \; \ominus Expr \mid Expr \odot Expr \\
Option & ::= & \texttt{::} \; Expr \; \texttt{->} \; Seq
\end{array}
$$

Fig. 2. An abstract syntactic subset of the full PROMELA. $+/*$ denote repetitions in EBNF notation; unary/binary operators are denoted $\ominus/\odot$, resp.

statements, whose syntax considered in this paper is given in Fig. 2. Statements include *basic* statements (*BStmt*) which define one-step atomic action, *control* statements (*CStmt*) which organize the control-flow, and variable declarations (*Decl*). All variables are regarded as arrays; e.g., a variable x is syntactic sugar for x[0] (unlike C). An informal description of PROMELA is available at [36].

Basic statements include condition statements, assignments, and channel operations. A condition statement is a standalone expression that gets skipped when it evaluates to true. It is typically used to guard the subsequent sequence. Channel operations are send/receive operations through a channel.

Similar to related formalisms (e.g., Hoare's CSP [16], Dijkstra's GCL [11]), PROMELA provides features for nondeterminism with if/do statements, whose inner options (*Option*) are selected nondeterministically when enabled. Unlike CSP/GCL, there is no syntactic restriction for the guards in each option sequence; options may start with arbitrary statements, allowing nested options.

A PROMELA program models a concurrent system of processes instantiated from proctype declarations. Processes may be active at startup (by annotating the proctype with active) or spawned by run expressions (e.g., in the initializer process init). Annotating active [N] instantiates N concurrent copies of a proctype. Concurrent processes execute as interleavings of locally executed basic statements. Enclosing a sequence SL as atomic $\{SL\}$ enforces atomic execution. Processes communicate via global variables or channels. Channels are either buffered or handshake: buffered channels provide asynchronous FIFO queues of finite capacity, while handshake channels synchronize sender and receiver. A channel c is declared as chan c = [N] of $\{TL\}$, where N is the capacity ($N = 0$ for handshake) and TL specifies the message format.

In PROMELA, basic statements can only be executed if their *enabledness* (or *executability* in PROMELA terminology) condition holds; otherwise, they block until they become enabled by global state changes. For example, a buffered receive c ? x is enabled iff the channel c is nonempty and the arguments match the message. Likewise, a sequence of statements is enabled iff its first statement is enabled; an atomic statement is enabled iff its inner sequence is enabled; an if/do statement is enabled iff either one of the options is enabled. Accord-

```
1   int x = 0, turn = 0;
2   chan c = [10] of { int };
3   active proctype producer() {
4     do
5       :: turn == 0 -> c ! 42; turn = 1
6       :: turn != 0 -> skip
7     od
8   }
9   active proctype consumer() {
10    do
11      :: turn == 1 -> c ? x; turn = 0
12      :: turn != 1 -> skip
13    od
14  }
```

Fig. 3. A producer–consumer model written in PROMELA.

ingly, `atomic` acquires atomicity only when the inner sequence is enabled; `if`/`do` chooses an option only when it is enabled.

Figure 3 shows a producer–consumer model written in PROMELA. Two `active` processes, `producer` and `consumer`, run concurrently and communicate via the buffered channel c. In its `do` loop, `producer` either (when `turn == 0`) sends 42 on c and sets `turn` to 1 (line 5), or (when `turn != 0`) executes `skip` (line 6). The process `consumer` behaves symmetrically: when `turn == 1` it receives 42 from c into x and sets `turn` to 0 (line 11). Note that "`->`" and `skip` are syntactic sugar for `;` and `true` (as a condition statement), respectively.

3   A Basic Semantics of PROMELA in $\mathbb{K}$

In this section, we give an overview of the semantics of PROMELA in $\mathbb{K}$. Rather than covering the full syntax presented in Fig. 2, we focus on a simplified subset for which it is straightforward to define $\mathbb{K}$ rules. In subsequent sections, we use these rules as a baseline and extend them modularly to cover the full syntax within our scope.

3.1   Syntactic Subset

We impose three syntactic restrictions on the full syntax shown in Fig. 2.

(EXPRESSION-GUARDED OPTIONS) We require that each nondeterministic option in `if`/`do` start with an explicit expression as its guard. Namely, we restrict the syntax of *Option* to be:

$$Option ::= :: \; \textbf{\textit{Expr}} \; \textbf{\textit{->}} \; \textbf{\textit{Seq}}$$

Under this restriction, the enabledness conditions of options are explicitly tied to the outer `if`/`do` wrappers, so that nondeterministic selection is defined in a straightforward way: an option can be selected if its guard evaluates to true. This restriction will be dropped in Sect. 4.

(BASIC GRANULARITY) We assume a simple setting in which basic statements are the sole granularity of atomic execution in PROMELA: interleaving *cannot* occur during the execution of a basic statement, and *can* occur whenever a basic statement completes. This is enforced by excluding `atomic` statements from the syntactic category *CStmt*. This restriction will be dropped in Sect. 5.

$$\langle\langle\langle Id\rangle_{ptName} \langle Decl^*\rangle_{params} \langle Seq\rangle_{code} \langle Id \hookrightarrow K\rangle_{gotoMap}\rangle_{ptype*}\rangle_{ptypes}$$

$$\langle\langle\langle Int\rangle_{pid} \langle Id\rangle_{pName} \langle K\rangle_k \langle Id \times Int \hookrightarrow Ref\rangle_{env}\rangle_{proc*}\rangle_{procs}$$

$$\langle Id \times Int \hookrightarrow Ref\rangle_{genv} \langle Int \hookrightarrow PVal\rangle_{str} \langle Int \hookrightarrow Queue\rangle_{net} \langle Int_\perp\rangle_{lock}$$

$$\langle Int\rangle_{nextPid} \langle Int\rangle_{nextLoc} \langle Int\rangle_{nextBCid} \langle Int\rangle_{nextHCid}$$

$BChan ::= bch(Int)$		$Value ::= PVal \mid CVal$
$HChan ::= hch(Int)$		$Ref \quad ::= loc(Int) \mid CVal$
$CVal \quad ::= uch \mid BChan \mid HChan$		$Queue ::= q(Int, Msg^*)$
$PVal \quad ::= Int \mid Bool$		$Msg \quad ::= m(Value^*)$

Fig. 4. (a) Top-level configuration. (b) Semantic domains.

$$\left\langle \left\langle \begin{array}{c} \left\langle \begin{array}{c} \langle 0\rangle_{pid} \langle \text{producer}\rangle_{pName} \langle \text{turn = 1} \curvearrowright \text{do } \ldots \text{ od}\rangle_k \\ \langle \texttt{x[0]} \mapsto loc(0); \texttt{turn[0]} \mapsto loc(1); \texttt{c[0]} \mapsto bch(0)\rangle_{env} \end{array}\right\rangle_{proc} \\ \left\langle \begin{array}{c} \langle 1\rangle_{pid} \langle \text{consumer}\rangle_{pName} \langle \text{do } \ldots \text{ od}\rangle_k \\ \langle \texttt{x[0]} \mapsto loc(0); \texttt{turn[0]} \mapsto loc(1); \texttt{c[0]} \mapsto bch(0)\rangle_{env} \end{array}\right\rangle_{proc} \end{array}\right\rangle_{procs} \right.$$

$$\langle\langle\cdots\rangle\rangle_{ptypes} \langle \texttt{x[0]} \mapsto loc(0); \texttt{turn[0]} \mapsto loc(1); \texttt{c[0]} \mapsto bch(0)\rangle_{genv} \langle 0 \mapsto 0; 1 \mapsto 0\rangle_{str}$$

$$\langle 0 \mapsto queue(10, m(42))\rangle_{net} \langle\perp\rangle_{lock} \langle 2\rangle_{nextPid} \langle 2\rangle_{nextLoc} \langle 1\rangle_{nextBCid} \langle 0\rangle_{nextHCid}$$

Fig. 5. A $\mathbb{K}$ configuration of the producer–consumer model.

(PURE EXPRESSIONS) We assume that every expression used in PROMELA code is pure: evaluating an expression never produces side-effects. To enforce this, we exclude **run** expressions from the syntactic category *Expr*, where expressions containing **run** as a sub-expression carry the side-effect of spawning new processes. This restriction will be dropped in Sect. 5.

3.2 Semantic Domains and Configuration

We describe the structure of our configuration in Fig. 4a upon which the $\mathbb{K}$ rules are defined. Some core semantic domains are shown in Fig. 4b.

The configuration contains core semantic components as nested cells such as: a set of **proctype** declarations (*ptypes*); a set of active processes (*procs*); a store (*str*) mapping locations to primitive values (*PVal*); a network (*net*) mapping buffered channel id's to the corresponding queues (*Queue*) of messages (*Msg*).

A *ptype* cell contains static information of a process: e.g., its name (*ptName*), parameters (*params*), code (*code*), and the gotomap (*gotoMap*). A *proc* cell contains dynamic information of a process: e.g., its continuation (k), pid (*pid*), and environment (*env*). The *env* (resp., *genv*) cell binds integer-indexed local (resp., global) variables (e.g., $\texttt{x[0]}$) to a reference (*Ref*), either to a store location, or to a channel. The *lock* cell contains a global lock used for enforcing atomic execution of a given process; it may contain a pid to which exclusive execution is granted, or $\perp$ indicating that the lock is free. The cells *nextPid*, *nextLoc*, *nextBCid*, and *nextHCid* are used to supply fresh identifiers.

Figure 5 shows the $\mathbb{K}$ configuration of the producer–consumer model from Sect. 2, after executing two basic statements `turn == 0; c ! 42` in `producer`. Note that the global variables `x` and `turn` point to the values stored in *str*, and the channel `c` points to the queue (capacity 10) with a message 42 in *net*.

Following the notational conventions of $\mathbb{K}$, we omit nested structure when unambiguous: e.g., when k and *env* cells appear in $\mathbb{K}$ rules without mentioning any *proc*, they are implicitly assumed to be contained in the same *proc* cell.

3.3 Basic Statements Under Concurrency

As noted, basic statements should execute atomically. Ideally, we can enforce this in the $\mathbb{K}$ framework by encoding each complete one-step behavior of basic statements as a *single* $\mathbb{K}$ rule. This would immediately enforce atomic execution for each basic statement.

However, because PROMELA is a high-level language, the behavior of basic statements often decomposes into several orthogonal steps that are more modular to specify as separate rules. For example, a buffered receive `c ? x, y` naturally expands into: "enabledness check; dequeue; assign x; assign y", where $\mathbb{K}$ rules are modularly defined for each of these intermediate steps.

Under concurrency, this modular style raises consistency concerns. E.g., if two processes `p1` and `p2` execute `c ? x, y` concurrently when the channel `c` holds a single message, an interleaving such as "enabledness check (`p1`); enabledness check (`p2`); dequeue (`p1`)" will abruptly block `p2` from dequeueing the message.

To prevent such inconsistencies while retaining modularity, we introduce two $\mathbb{K}$ rules, fire and rel, which enforce mutual exclusion between concurrent applications of *execution rules* that perform intermediate computations of basic statements. Intuitively, for a basic statement *BS*, fire initiates its execution by granting a free lock to the owner process, upon checking the enabledness of *BS*; symmetrically, rel completes the execution by releasing the lock.

The following rules formalize these behaviors, where *effect*(*BS*) defines a $\mathbb{K}$ continuation representing intermediate computations for *BS* and *enabled*(*BS*) defines its enabledness condition. The fire rule checks *enabled*(*BS*) as a side-condition,[3] via the evaluation function $\llbracket \cdot \rrbracket_{(\cdot,\cdot,\cdot)}$ parameterized by the local environment, the store, and the network. By BASIC GRANULARITY, the lock must be released immediately after *effect*(*BS*) completes.

$$\text{fire:} \quad \langle P \rangle_{pid} \left\langle \frac{BS}{\mathit{effect}(BS) \curvearrowright \texttt{\#rel}} \curvearrowright \ldots \right\rangle_k \left\langle \frac{\bot}{P} \right\rangle_{lock} \langle ENV \rangle_{env} \langle STR \rangle_{str} \langle NET \rangle_{net}$$

$$\text{if } \llbracket \mathit{enabled}(BS) \rrbracket_{ENV,STR,NET} = \top$$

$$\text{rel:} \quad \left\langle \frac{\texttt{\#rel}}{.K} \curvearrowright \ldots \right\rangle_k \left\langle \frac{\cdot}{\bot} \right\rangle_{lock}$$

Let $\mathit{nfull}(X[E])$ and $X[E]?[AL]$ be the predicates stating that the buffered channel $X[E]$ is sendable (i.e. not full) and receivable (i.e., the oldest message

---

[3] In the paper, we use $\top/\bot$ and `true`/`false` interchangeably for boolean values.

on $X[E]$ is $m(VL)$ with VL *matching* the pattern AL), respectively. We write $VL \sqsubseteq_{pm} AL$ for pattern matching (e.g., $1, 2 \sqsubseteq_{pm}$ x,2 but $1, 2 \not\sqsubseteq_{pm}$ x,1).

The side-effects and the enabledness conditions for the four basic statements—namely, condition/assignment/buffered send/buffered receive—are defined as follows, via the functions $effect : BStmt \to K$ and $enabled : BStmt \to Expr$:

$$effect(E) = .K \qquad\qquad enabled(E) = E$$
$$effect(X\,[E]\ =\ E') = \#\texttt{assign}(X, E, E') \quad enabled(X\,[E]\ =\ E') = \top$$
$$effect(X\,[E]\ !\ EL) = \#\texttt{send}(X, E, EL) \quad enabled(X\,[E]\ !\ EL) = nfull(X[E])$$
$$effect(X\,[E]\ ?\ AL) = \#\texttt{recv}(X, E, AL) \quad enabled(X\,[E]\ ?\ AL) = X[E]?[AL]$$

where the side effects are represented by *effect-markers* (e.g., #assign), which is reduced via the (multi-step) execution rules, *atomically*. Handshake operations are excluded here; they are handled specially via a dedicated firing rule below.

Example (Assignment). We demonstrate concretely how fire and rel help define execution rules for assignments in a modular way. The following rules reduce the effect-marker $\#\texttt{assign}(X, E, E')$ in two steps:

$$\text{assign-eval:} \quad \frac{\langle \#\texttt{assign}(X, \underline{E}, \underline{E'}) \curvearrowright \ldots\rangle_k}{[\![E]\!]_{ENV,STR,NET}\ [\![E']\!]_{ENV,STR,NET}}$$
$$\langle ENV\rangle_{env}\ \langle STR\rangle_{str}\ \langle NET\rangle_{net} \quad \text{if at least one of } E \text{ or } E' \text{ is a non-value}$$

$$\text{assign-prim:} \quad \frac{\langle \#\texttt{assign}(X, I : Int, V : PVal) \curvearrowright \ldots\rangle_k\ \langle \ldots X[I] \mapsto loc(J) \ldots\rangle_{env}\ \langle \ldots J \mapsto \underline{\cdot} \ldots\rangle_{str}}{.K \qquad\qquad\qquad\qquad\qquad\qquad\qquad\qquad \underline{V}}$$

$$\text{assign-chan:} \quad \frac{\langle \#\texttt{assign}(X, I : Int, V : CVal) \curvearrowright \ldots\rangle_k\ \langle \ldots X[I] \mapsto \underline{\cdot} \ldots\rangle_{env}}{.K \qquad\qquad\qquad\qquad\qquad\qquad \underline{V}}$$

The assign-eval rule evaluates the index E to an integer I and E' to a value V. If V is a primitive value, assign-prim updates the store by V at the location pointed by $X[I]$; otherwise, if V is a channel value, assign-chan updates the environment so that $X[I]$ points to the channel V.

Now consider an assignment x = x + y in the process with pid 1, where x = 4 and y = 2. This executes in lock-step via the sequence "fire; assign-eval; assign-prim; rel", as shown by the trace below (mutated portions underlined):

$$\langle 1\rangle_{pid}\ \langle \underline{\texttt{x = x + y}} \curvearrowright K\rangle_k\ \langle \texttt{x}[0] \mapsto loc(0); \texttt{y}[0] \mapsto loc(1)\rangle_{env}\ \langle 0 \mapsto 4; 1 \mapsto 2\rangle_{str}\ \langle \underline{\bot}\rangle_{lock}$$
$$\to\ \langle 1\rangle_{pid}\ \langle \underline{\#\texttt{assign}(\texttt{x, 0, x + y})} \curvearrowright \#\texttt{rel} \curvearrowright K\rangle_k\ \langle \texttt{x}[0] \mapsto loc(0); \texttt{y}[0] \mapsto loc(1)\rangle_{env}$$
$$\langle 0 \mapsto 4; 1 \mapsto 2\rangle_{str}\ \langle \underline{1}\rangle_{lock}$$
$$\to\ \langle 1\rangle_{pid}\ \langle \underline{\#\texttt{assign}(\texttt{x, 0, 6})} \curvearrowright \#\texttt{rel} \curvearrowright K\rangle_k\ \langle \texttt{x}[0] \mapsto loc(0); \texttt{y}[0] \mapsto loc(1)\rangle_{env}$$
$$\langle 0 \mapsto \underline{4}; 1 \mapsto 2\rangle_{str}\ \langle 1\rangle_{lock}$$
$$\to\ \langle 1\rangle_{pid}\ \langle \underline{\#\texttt{rel}} \curvearrowright K\rangle_k\ \langle \texttt{x}[0] \mapsto loc(0); \texttt{y}[0] \mapsto loc(1)\rangle_{env}\ \langle 0 \mapsto 6; 1 \mapsto 2\rangle_{str}\ \langle \underline{1}\rangle_{lock}$$
$$\to\ \langle 1\rangle_{pid}\ \langle K\rangle_k\ \langle \texttt{x}[0] \mapsto loc(0); \texttt{y}[0] \mapsto loc(1)\rangle_{env}\ \langle 0 \mapsto 6; 1 \mapsto 2\rangle_{str}\ \langle \bot\rangle_{lock}$$

3.4 Communication via Channels

Following the approach above, we present the execution rules for channel operations. This concludes our execution rules for all four basic statements; condition statements do not require any execution rules by PURE EXPRESSIONS.

Buffered Channels. Communication via a buffered channel proceeds by processing the markers $\#\mathtt{send}(X,E,EL)$ and $\#\mathtt{recv}(X,E,AL)$, where $EL : Expr^*$ and $AL : Arg^*$ are sender's message payloads and receiver's arguments, respectively. We list the execution rules for buffered channels below.

$$
\text{send-eval:} \quad \frac{\langle\, \#\mathtt{send}(X,\ \underline{E}\ ,\ \underline{EL}\ ) \curvearrowright \ldots\rangle_k}{\llbracket E\rrbracket_{ENV,STR,NET}\ \llbracket EL\rrbracket_{ENV,STR,NET}}
$$

$$
\langle ENV\rangle_{env}\ \langle STR\rangle_{str}\ \langle NET\rangle_{net} \quad \text{if } E \text{ is a non-value or } EL \text{ is not a list of values}
$$

$$
\text{send-msg:} \quad \frac{\langle \#\mathtt{send}(X,I,VL) \curvearrowright \ldots\rangle_k\ \langle \ldots X[I]\mapsto bch(C)\ldots\rangle_{env}\ \langle \ldots C\mapsto q(\cdot, ML;\ \underline{nil}\ )\ldots\rangle_{net}}{.K \qquad\qquad\qquad\qquad\qquad\qquad\qquad\qquad\qquad m(VL)}
$$

$$
\text{recv-eval:} \quad \frac{\langle \#\mathtt{recv}(X,\ \underline{E}\ ,AL) \curvearrowright \ldots\rangle_k\ \langle ENV\rangle_{env}\ \langle STR\rangle_{str}\ \langle NET\rangle_{net}}{\llbracket E\rrbracket_{ENV,STR,NET} \qquad\qquad\qquad\qquad\qquad\qquad \text{if } E \text{ is a non-value}}
$$

$$
\text{recv-msg:} \quad \frac{\langle \#\mathtt{recv}(X,I,AL) \curvearrowright \ldots\rangle_k\ \langle \ldots X[I]\mapsto bch(C)\ldots\rangle_{env}\ \langle \ldots C\mapsto q(I,\underline{m(VL)};\ ML)\ldots\rangle_{net}}{\#\mathtt{asgnMsg}(AL,VL) \qquad\qquad\qquad\qquad\qquad\qquad\qquad\qquad nil}
$$

$$
\text{assign-msg1:} \quad \frac{\langle\ \underline{.K}\ \curvearrowright \#\mathtt{asgnMsg}(X[I],\ AL,\ \underline{V},\ VL) \curvearrowright \ldots\rangle_k}{\#\mathtt{assign}(X,I,V) \qquad\qquad nil \qquad nil}
$$

$$
\text{assign-msg2:} \quad \frac{\langle \#\mathtt{asgnMsg}(\ \underline{V},\ AL,\ \underline{V},\ VL) \curvearrowright \ldots\rangle_k}{nil \qquad nil} \qquad\qquad \text{assign-msg3:} \quad \frac{\langle \#\mathtt{asgnMsg}(nil,nil) \curvearrowright \ldots\rangle_k}{.K}
$$

The rules send-eval and recv-eval work similarly to assign-eval. By slight abuse of notation in send-eval, the evaluation function $\llbracket\cdot\rrbracket_{(\cdot,\cdot,\cdot)}$ is used as a pointwise extension, returning $Value^*$ from $Expr^*$. The send msg rule enqueues the message $m(VL)$ to the queue pointed by $X[I]$ (via a buffered channel $bch(C)$). Similarly, recv-msg dequeues the message from the queue, and then receives the message to the arguments by iterative assignments via the rules assign-msg$i$ for $1 \leq i \leq 3$.

Handshake Channels. As mentioned, the firing rule for handshake is separately defined, as it involves two k cells for inter-process synchronization:

$$
\text{hs-fire:} \quad \frac{\langle\ldots\langle \underline{X[E]\ \mathtt{!}\ EL} \curvearrowright \ldots\rangle_k\ \langle ENV\rangle_{env}\ldots\rangle_{proc}\ \langle STR\rangle_{str}\ \langle NET\rangle_{net}\ \langle\underline{\perp}\rangle_{lock}}{.K \qquad\qquad\qquad\qquad\qquad\qquad\qquad\qquad\qquad\qquad P}
$$

$$
\frac{\langle\ldots\langle P\rangle_{pid}\ \langle\ \underline{X'[E']\ \mathtt{?}\ AL}\ \curvearrowright \ldots\rangle_k\ \langle ENV'\rangle_{env}\ldots\rangle_{proc}}{\#\mathtt{asgnMsg}(AL,EL) \curvearrowright \#\mathtt{rel}}
$$

$$
\text{if } \llbracket X[E]\rrbracket_{ENV,STR,NET} =_{HChan} \llbracket X'[E']\rrbracket_{ENV',STR,NET} \wedge AL \sqsupseteq_{\mathsf{pm}} \llbracket EL\rrbracket_{ENV,STR,NET}
$$

The side condition checks if $X[E]$ and $X'[E']$ denote the same handshake channel, and if the sender's (evaluated) payload EL matches the receiver arguments

AL. After handshake, the control is given to the receiver (P), who continues to receive the message via assign-msg$i$, $1 \leq i \leq 3$, defined above.

3.5   Other $\mathbb{K}$ Rules

Nondeterminism. As EXPRESSION-GUARDED OPTIONS require each option's enabledness condition to appear as its guard, the nondeterministic selection rule (select) for if statements can be defined straightforwardly, which selects the option whose guard evaluates to true and yields its sequence.

$$\text{select:} \quad \frac{\langle \text{if } OL \ (:: E \to SL) \ OL' \text{ fi} \curvearrowright \ldots \rangle_k \ \langle ENV \rangle_{env} \ \langle STR \rangle_{str} \ \langle NET \rangle_{net} \ \langle \bot \rangle_{lock}}{SL}$$
$$\text{if } [\![E]\!]_{ENV,STR,NET} = \top$$

Note that this rule requires the global lock to be free, thereby preventing inconsistent guard evaluation during the execution of a basic statement.

Structural Rules. Below we list the structural rules, which rearrange the configuration or control without changing the program's observable state.

$$\text{seq:} \quad \frac{\langle \ S \ ; \ SL \curvearrowright \ldots \rangle_k}{S \curvearrowright SL} \qquad \text{loop:} \quad \frac{\langle \quad .K \quad \curvearrowright \text{do } OL \text{ od} \curvearrowright \ldots \rangle_k}{\text{if } OL \text{ fi}}$$

$$\text{goto:} \quad \frac{\langle \ldots \langle X \rangle_{pName} \ \langle \text{goto } Y \curvearrowright \ldots \rangle_k \ldots \rangle_{proc} \ \langle \ldots \langle X \rangle_{ptName} \ \langle \ldots Y \mapsto K \ldots \rangle_{gotoMap} \ldots \rangle_{ptype}}{K}$$

$$\text{break1:} \quad \frac{\langle \text{break} \curvearrowright \text{do } OL \text{ od} \curvearrowright \ldots \rangle_k}{.K} \qquad \text{break2:} \quad \frac{\langle \text{break} \curvearrowright KI : KItem \curvearrowright \ldots \rangle_k}{.K} \quad \text{otherwise}$$

The seq rule decomposes a sequence SL into $\mathbb{K}$ continuations. The goto rule performs an unconditional jump to the destination label, as covered in Sect. 2. We assume *gotoMap* is appropriately preprocessed, eliminating all the labels in the code. The loop rule produces if indefinitely, whenever do is encountered. The loop breaks by the rules break1 and break2 (the side-condition "otherwise" means no other rules match).

4   Handling Nondeterminism

In this section, we drop the EXPRESSION-GUARDED OPTIONS restriction introduced in the previous section. This syntactic extension allows options to be guarded by arbitrary statements, as in Fig. 2. Consequently, the previous select rule no longer applies under the extended syntax. Devising a new set of rules for this setting is nontrivial. We begin with motivating examples that lead to our LOAD-AND-FIRE approach, which we develop at the end of this section.

In Fig. 6, the outer if involves nested options: the second option is guarded by a whole do block. Consequently, if is guarded by the three *leading* basic statements A, B, and C, each of which can be selected when enabled.

```
active proctype p1() {
  if
    :: A
    :: do :: B :: C od ; D
  fi
} // A,B,C : any basic statement
```

```
int x = 0, y = 0;
chan c = [0] of { int };
active proctype p1()
{ if :: c ! 1 :: y = 1 fi }
active proctype p2()
{ if :: y = 1 :: c ? x fi }
```

Fig. 6. Nested Options.   **Fig. 7.** Cross-Process Interference.

In Fig. 7, both of the `if` statements in processes `p1` and `p2` are guarded by send/receive operations via a handshake channel `c`. This raises cross-process interference among two nondeterministic choices: selecting the first option in `p1` forces `p2` to take its second (the handshake must occur), whereas selecting the second option in `p1` forces `p2` to take its first (the handshake cannot occur).

As motivated by the examples, `if` statements may involve (i) nested options, (ii) cross-process interference, or (iii) their combinations. Consequently, unlike `select`, their behavior cannot be specified by simply matching a guard (i.e., a leading basic statement) among the options, since guards may appear at *arbitrary nesting depth* within options.

4.1 The Load-and-Fire Approach

We address the challenge above by introducing *loading* rules, which are structural normalization rules that flatten nested `if` options into a canonical form. The canonical form is a *multiset* of $\mathbb{K}$ continuations, each being the flattened computation for each local branch of `if`.

For example, a continuation of the form `if` $\dots$ `fi` $\curvearrowright K_{global}$ is normalized by loading rules, into the canonical form $[BS_1 \curvearrowright K_1 \mid \dots \mid BS_n \curvearrowright K_n] \curvearrowright K_{global}$ for each leading basic statement BS_i appearing in `if` $\dots$ `fi`, where $[\_ \mid \_]$ is the multiset constructor. Intuitively, this canonical form represents a "superposition" of local continuations $BS_i \curvearrowright K_i \curvearrowright K_{global}$, where each BS_i is available for syntactic matching (modulo associativity/commutativity for multisets) at the top-level of the continuation.

Accordingly, we can lift the previous `fire` and `hs-fire` rules to match under this set-lifted context, organized by the loading rules. This allows the firing rules to match against a basic statement under arbitrarily nested options. Below we give concrete definitions of the (lifted) firing rules and the loading rules, and revisit the motivating examples to demonstrate how they work.

Firing Rules. The following rules are the lifted versions of `fire` and `hs-fire`:

$$\text{fire:}\quad \frac{\langle P\rangle_{pid}\,\langle\ \underline{[BS \curvearrowright K_{local}\mid K_{rest}]}\ \curvearrowright \ldots\rangle_k\ \langle ENV\rangle_{env}\ \langle STR\rangle_{str}\ \langle NET\rangle_{net}\ \langle\bot\rangle_{lock}}{\mathit{effect}(BS) \curvearrowright K_{local} \curvearrowright \texttt{\#rel} \qquad\qquad\qquad\qquad\qquad\qquad\qquad\qquad P}$$

$$\text{if } [\![\mathit{enabled}(BS)]\!]_{ENV.STR.NET} = \top$$

$$\text{hs-fire:}\quad \frac{\langle\ldots\langle[X\,\texttt{[}E\texttt{]}\ \texttt{!}\ EL \curvearrowright K_{local}\mid K_{rest}]\ \curvearrowright \ldots\rangle_k\ \langle ENV\rangle_{env\ldots}\rangle_{proc}\ \langle STR\rangle_{str}\ \langle NET\rangle_{net}}{K_{local}}$$

$$\frac{\langle\ldots\langle P\rangle_{pid}\,\langle\ \underline{[X'\,\texttt{[}E'\texttt{]}\ \texttt{?}\ AL \curvearrowright K'_{local}\mid K'_{rest}]}\ \curvearrowright \ldots\rangle_k\ \langle ENV'\rangle_{env\ldots}\rangle_{proc}\ \langle\bot\rangle_{lock}}{\texttt{\#asgnMsg}(AL,EL) \curvearrowright K'_{local} \curvearrowright \texttt{\#rel} \qquad\qquad\qquad\qquad\qquad\qquad\qquad P}$$

$$\text{if } [\![X\,\texttt{[}E\texttt{]}]\!]_{ENV.STR.NET} =_{HChan} [\![X'\,\texttt{[}E'\texttt{]}]\!]_{ENV'.STR.NET}\ \wedge\ AL \sqsupseteq_{\mathsf{pm}} [\![EL]\!]_{ENV.STR.NET}$$

Compared to the previous versions, the changes are found only in the k cells: A basic statement BS is matched under the set-lifted context $[BS \curvearrowright K_{local} \mid K_{rest}]$, where K_{rest} matches the other branches. When BS fires, the set-lifted continuation *collapses* back into a linear form, reflecting a nondeterministic selection.

Loading Rules. Loading rules decompose a continuation of the form `if ... fi` $\curvearrowright K$ into $[BS_1 \curvearrowright K_1 \mid \cdots \mid BS_n \curvearrowright K_n] \curvearrowright K$, so that enabledness becomes visible at the front of each branch. As a special case, $BS \curvearrowright K$ is also lifted to $[BS] \curvearrowright K$, viewing $BS \equiv$ `if :: ` BS ` fi`. The loading rules are defined as follows:

$$\text{load-lift:}\quad \frac{\langle\ \underline{S}\ \curvearrowright \ldots\rangle_k}{\texttt{\#load} \curvearrowright [S]}\quad \text{if } \mathit{Loadable}(S) = \top \qquad \text{load-bs:}\quad \frac{\texttt{\#load} \curvearrowright [BS \curvearrowright K]}{.K}$$

$$\text{load-seq:}\quad \frac{LT \curvearrowright [\ \underline{S;SL} \curvearrowright K]}{S \curvearrowright SL} \qquad \text{load-do:}\quad \frac{LT \curvearrowright [\ \underline{.K}\ \curvearrowright \text{do } OL \text{ od} \curvearrowright K]}{\texttt{if } OL \texttt{ fi}}$$

$$\text{load-if1:}\quad \frac{LT \curvearrowright [\texttt{if } (:: SL)\ OL\ \texttt{fi} \curvearrowright K]}{[LT \curvearrowright [SL \curvearrowright K] \mid LT \curvearrowright [\texttt{if } OL\ \texttt{fi} \curvearrowright K]]}$$

$$\text{load-if2:}\quad \frac{LT \curvearrowright [\texttt{if } (:: SL)\ \texttt{fi} \curvearrowright K]}{[LT \curvearrowright [SL \curvearrowright K]]}$$

$$\text{load-goto:}\quad \frac{LT \curvearrowright [\ \underline{.K} \curvearrowright \text{goto } X \curvearrowright K]}{\mathbf{true}} \qquad \text{load-break:}\quad \frac{LT \curvearrowright [\ \underline{.K} \curvearrowright \text{break} \curvearrowright K]}{\mathbf{true}}$$

Loading begins by the rule load-lift, which lifts *Loadable* statements (i.e. basic and `if` statements) in a k cell into multisets with a special loader token `#load`. In the remaining rules, `#load` (bound to a metavariable LT) guides the local rewrite *within* the multiset, by recursively decomposing each inner option until a leading basic statement is reached.

4.2   Examples

Our LOAD-AND-FIRE semantics handle nested options and cross-process interference in an elegant way, as illustrated by the running examples below.

Example 1. The nested `if` appearing in Fig. 6 is loaded so that the basic statements A, B, and C appear upfront, exposing each branch's enabledness for firing (via fire/hs-fire). We underline the portion reduced by the loading rules:

$$\langle \text{if :: A :: do :: B :: C od fi; D} \rangle_k$$
$$\rightarrow \langle \underline{\text{if :: A :: do :: B :: C od fi}} \curvearrowright \text{D} \rangle_k$$
$$\rightarrow \langle \text{\#load} \curvearrowright \underline{[\text{if :: A :: do :: B :: C od fi}]} \curvearrowright \text{D} \rangle_k$$
$$\rightarrow \langle [\underline{\text{\#load} \curvearrowright [\text{A}]} \,|\, \text{\#load} \curvearrowright [\text{if :: do :: B :: C od fi}]] \curvearrowright \text{D} \rangle_k$$
$$\rightarrow \langle [\text{A} \,|\, \text{\#load} \curvearrowright \underline{[\text{if :: do :: B :: C od fi}]}] \curvearrowright \text{D} \rangle_k$$
$$\rightarrow \langle [\text{A} \,|\, \text{\#load} \curvearrowright \underline{[\text{do :: B :: C od}]}] \curvearrowright \text{D} \rangle_k$$
$$\rightarrow \langle [\text{A} \,|\, \text{\#load} \curvearrowright \underline{[\text{if :: B :: C fi}} \curvearrowright \text{do :: B :: C od}]] \curvearrowright \text{D} \rangle_k$$
$$\rightarrow \langle [\text{A} \,|\, \text{\#load} \curvearrowright [\underline{\text{B} \curvearrowright \text{do :: B :: C od}} \,|\, \text{\#load} \curvearrowright [\text{if :: C fi} \curvearrowright \text{do :: B :: C od}]]] \curvearrowright \text{D} \rangle_k$$
$$\rightarrow \langle [\text{A} \,|\, \text{B} \curvearrowright \text{do :: B :: C od} \,|\, \text{\#load} \curvearrowright \underline{[\text{if :: C fi} \curvearrowright \text{do :: B :: C od}]}] \curvearrowright \text{D} \rangle_k$$
$$\rightarrow \langle [\text{A} \,|\, \text{B} \curvearrowright \text{do :: B :: C od} \,|\, \text{\#load} \curvearrowright \underline{[\text{C} \curvearrowright \text{do :: B :: C od}]}] \curvearrowright \text{D} \rangle_k$$
$$\rightarrow \langle [\text{A} \,|\, \text{B} \curvearrowright \text{do :: B :: C od} \,|\, \text{C} \curvearrowright \text{do :: B :: C od}] \curvearrowright \text{D} \rangle_k$$

Example 2. The cross-process interference in Fig. 7 is resolved in the fully loaded form (intermediate loading steps omitted):

$$\langle \text{if :: c ! 1 :: y = 1 fi} \rangle_{k_1} \, \langle \text{if :: y = 1 :: c ? x fi} \rangle_{k_2}$$
$$\rightarrow^! \langle [\text{c ! 1} \,|\, \text{y = 1}] \rangle_{k_1} \langle [\text{c ? x} \,|\, \text{y = 1}] \rangle_{k_2}$$

where k_1 and k_2 denote (by abuse of notation) the k cells of p1 and p2, respectively. The rule `hs-fire` applies to this loaded form, making a joint nondeterministic choice across the two processes and thus capturing the interference.

5   More Extensions

In this section, we further extend the LOAD-AND-FIRE to incrementally add `atomic` blocks and `run`-expressions to our syntax, in a modular way. We briefly present the added/revised rules.

5.1   Adding Atomic Blocks

We add `atomic` blocks to our syntax (i.e., we drop BASIC GRANULARITY). Sequences in an `atomic` block execute atomically; e.g., `atomic{ x++; x++; }` is equivalent to `x = x + 2`. Intermixing `atomic` with constructs such as goto and handshake introduces subtle control-flow issues (Fig. 8, 9) to be elaborated below. We outline extensions to the relevant $\mathbb{K}$ rules that cover these cases.

Loading Rules. We modularly extend the loading rules to handle atomicity. Since `if`, `do`, and `atomic` can be nested in arbitrary combinations, `atomic` blocks may hide leading basic statements and must therefore be decomposed by loading. We present the extended portion, treating `atomic` blocks as *Loadable*. The old load-lift and load-bs are deprecated; all other rules remain unchanged.

```
chan q = [0] of { bool };
active proctype p1() {
  atomic { A; q!0; B }
}
active proctype p2() {
  atomic { q?0 ; C }
}
```

```
active proctype p() {
  goto L;
  atomic {
    A; L: B
  }
}
```

Fig. 8. Handshake under atomicity.   **Fig. 9.** Goto under atomicity.

$$\text{load-lift:} \quad \langle P \rangle_{pid} \langle \frac{S}{\texttt{\#load}(P = L) \curvearrowright [S]} \curvearrowright ... \rangle_k \langle L \rangle_{lock} \quad \text{if } Loadable(S) = \top$$

$$\text{load-bs1:} \frac{\texttt{\#load}(\top) \curvearrowright [BS \curvearrowright K]}{.K} \qquad \text{load-bs2:} \frac{\texttt{\#load}(\bot) \curvearrowright [BS \curvearrowright .K \curvearrowright K]}{.K \quad \texttt{\#rel}}$$

$$\text{load-at1:} \frac{\texttt{\#load}(\top) \curvearrowright [\texttt{atomic } \{ SL \} \curvearrowright K]}{SL}$$

$$\text{load-at2:} \frac{\texttt{\#load}(\bot) \curvearrowright [\texttt{atomic } \{ SL \} \curvearrowright K]}{\top \qquad SL \curvearrowright \texttt{\#rel}}$$

The purpose of these rules is to insert a single **#rel** at the end of the *outermost* **atomic** block, thereby enforcing atomic execution within it. We refine the loader token with a Boolean flag (written as $\top/\bot$ inside of **#load(·)**) that records whether **#rel** has already been inserted to prevent incorrect early release. Together with the other loading rules, **#load(·)** is propagated along the nested structure, inserting **#rel** at each valid release points.

Firing Rules. We lift the firing rules to operate under atomicity, in accordance with the extended loading rules. The revised firing rules (i) no longer inserts **#rel**, as it is inserted by loading rules; and (ii) also matches a self-acquired lock, since basic statements may fire consecutively within an **atomic** block without releasing the lock.

$$\text{fire:} \quad \frac{\langle P \rangle_{pid} \langle [BS \curvearrowright K_{local} \mid K_{rest}] \curvearrowright ... \rangle_k \langle ENV \rangle_{env} \langle STR \rangle_{str} \langle NET \rangle_{net} \langle L \rangle_{lock}}{effect(BS) \curvearrowright K_{local} \qquad\qquad\qquad\qquad P}$$
$$\text{if } L \in \{\bot, P\} \wedge [\![enabled(BS)]\!]_{ENV,STR,NET} = \top$$

$$\text{hs-fire:} \quad \frac{\langle ... \langle P \rangle_{pid} \langle [X[E] \; ! \; EL \curvearrowright K_{local} \mid K_{rest}] \curvearrowright ... \rangle_k \langle ENV \rangle_{env} ... \rangle_{proc} \langle STR \rangle_{str} \langle NET \rangle_{net}}{[\texttt{true} \curvearrowright K_{local}]}$$
$$\frac{\langle ... \langle P' \rangle_{pid} \langle [X'[E'] \; ? \; AL \curvearrowright K'_{local} \mid K'_{rest}] \curvearrowright ... \rangle_k \langle ENV' \rangle_{env} ... \rangle_{proc}}{(L = P \; ? \; \texttt{\#toss}(P) : .K) \curvearrowright \texttt{\#asgnMsg}(AL,EL) \curvearrowright K'_{local}}$$
$$\frac{\langle L \rangle_{lock}}{P'} \qquad \begin{array}{l} \text{if } [\![X[E]]\!]_{ENV,STR,NET} =_{HChan} [\![X'[E']]\!]_{ENV',STR,NET} \\ \wedge \; L \in \{\bot, P, P'\} \wedge AL \sqsupseteq_{pm} [\![EL]\!]_{ENV,STR,NET} \end{array}$$

Additionally, under atomicity, handshake operations chained across multiple processes induce a global atomic chain of executions [36], as shown in Fig. 8.

If a handshake occurs while p1 is executing atomically, atomicity is transferred to p2, which executes block C; upon completion, atomicity returns to p1. This mechanism is implemented in hs-fire via *lock-tossing*: if the sender (P) already held the lock $(L = P)$, the lock granted to the receiver (P') gets tossed back to the sender upon encountering #rel in the receiver's k cell. Consequently, #rel never gets processed in the middle of the atomic chain, preventing any invalid interleavings. The rules for lock-tossing appear below.

Execution Rules. We list the execution rules updated for atomicity. Irrelevant execution rules from Sect. 3 remain unchanged.

$$
\text{toss1:} \quad \frac{\langle \text{\#toss}(P) \curvearrowright K \curvearrowright \text{\#rel} \curvearrowright \dots \rangle_k}{K \curvearrowright \text{\#toss}(P)} \qquad \text{toss2:} \quad \frac{\langle \text{\#toss}(P) \curvearrowright \text{\#rel} \curvearrowright \dots \rangle_k \; \langle \frac{\cdot}{P} \rangle_{lock}}{.K}
$$

$$
\text{loose:} \quad \langle P \rangle_{pid} \; \langle \dots \curvearrowright \frac{KI : KItem}{(KI = \text{\#toss}(\cdot) \; ? \; .K \; : \; KI)} \curvearrowright \text{\#rel} \curvearrowright \dots \rangle_k \; \langle \frac{P}{\bot} \rangle_{lock} \quad \text{otherwise}
$$

$$
\text{goto:} \quad \langle \dots \langle P \rangle_{pid} \; \langle X \rangle_{pName} \; \langle \frac{\text{goto } Y \curvearrowright \dots}{K} \rangle_k \dots \rangle_{proc} \; \langle \frac{L}{(B \; ? \; L \; : \; \bot)} \rangle_{lock}
$$
$$
\langle \dots \langle X \rangle_{ptName} \; \langle \dots Y \mapsto (B, K) \dots \rangle_{gotoMap} \dots \rangle_{ptype}
$$

The rules toss1 and toss2 implement lock-tossing used in hs-fire. First, the rule toss1 locates the marker #toss(P) (with P the sender's pid) right before #rel. Then, toss2 sets the lock to P instead of releasing it.

When execution blocks inside an atomic block, atomicity is lost temporarily. The loose rule implements this by releasing the lock when no (firing) rules match, via the side-condition "otherwise". Any pending toss-marker, if present, is invalidated. Note that #rel still remains, preserving the original atomicity.

Finally, atomicity may switch abruptly via goto when the atomicity of the source and destination differs (Fig. 9). The revised goto rule reflects the atomicity of the destination Y: if Y is (resp., is not) within an atomic block, indicated by $B = \top$ (resp., $B = \bot$), the lock is preserved (resp., released). Accordingly, the *gotoMap* is augmented with atomicity flags B for each label.

5.2   Adding Impure Expressions

As our final language extension, we add run expressions (i.e., we drop PURE EXPRESSIONS). For simplicity, we only consider run expressions used in condition statements, and in the righthand-side of the assignment.

Evaluating run expressions incurs side-effects of spawning new processes. For example, assuming the *nextPid* cell contains 42, the side-effect of X = run p1() + run p2() is to spawn two processes p1 and p2 and to assign X = 42 + 43. To reflect this, we extend the function *effect* : $BStmt \times Int \to K$ to be parameterized by *nextPid* as the second argument:

$$
\textit{effect}(E, I) = \text{\#run}(E)
$$
$$
\textit{effect}(X\,[E] \; = \; E', I) = \text{\#run}(E') \curvearrowright \text{\#assign}(X, E, E' \upharpoonright_I)
$$
$$
\textit{effect}(E, I) = \textit{effect}(E) \quad \text{(otherwise, define as before)}
$$

```
int n = N, s = 0; // N is parameter
active proctype sum() {
  do
    :: 0 < n -> s = s + n ; n--
    :: 0 >= n -> break
  od
}
```

Fig. 10. Summation in PROMELA.

```
int disp = 0, serv = 0, crit = 0;
active [2] proctype p() { // activates p1, p2
  int tick = 0;
  do
    :: atomic { tick = disp; disp = disp + 1 }
     ; atomic { tick == serv; crit = crit + 1 }
     ; atomic { serv = serv + 1; crit = crit - 1 }
  od
}
active proctype monitor() { freeze } // for mutex
```

Fig. 11. 2-process bakery alg. in PROMELA.

Here, the purification function $(\cdot) \upharpoonright_{(\cdot)}: BStmt \times Int \to BStmt$ substitutes all occurrences of **run**'s in a basic statement by fresh pid's w.r.t. the given $nextPid$.

Accordingly, the fire rule is revised to use the updated effect and enabledness for basic statements that may involve **run** as subexpressions:

$$\text{fire:} \quad \frac{\langle P \rangle_{pid} \langle\ [BS \curvearrowright K_{local} \mid K_{rest}] \curvearrowright ... \rangle_k \langle ENV \rangle_{env} \langle STR \rangle_{str} \langle NET \rangle_{net}}{effect(BS, P') \curvearrowright K_{local}}$$
$$\frac{\langle \bot \rangle_{lock} \langle P' \rangle_{nextPid}}{P} \quad \text{if } L \in \{\bot, P\} \wedge [\![enabled(BS \upharpoonright_{P'})]\!]_{ENV,STR,NET} = \top$$

For execution rules, we add rules that process the **#run(·)** markers. These rules decompose **#run(E)** into **#run(run X(EL))**'s for each **run X(EL)** contained in E. The run rule spawns a new process upon each **#run(run X(EL))**:

$$\text{run:} \quad \frac{\dfrac{\langle \text{\#run(run } X\text{(}EL\text{))} \curvearrowright ... \rangle_k \langle ENV \rangle_{env} \langle GENV \rangle_{genv}}{\text{\#wait}}}{\langle\langle P \rangle_{pid} \langle X \rangle_{pName} \langle \text{\#init(}DL\text{,}EL\text{)} \curvearrowright \text{\#done} \curvearrowright SL \rangle_k \langle GENV \rangle_{env} \rangle_{proc}}$$
$$\frac{\langle ... \langle X \rangle_{ptName} \langle DL \rangle_{params} \langle SL \rangle_{code} ... \rangle_{ptype} \left\langle \dfrac{P}{P+1} \right\rangle_{nextPid}}{.Cell}$$

6 Case Study: Application to Deductive Verification

We implemented a prototype $\mathbb{K}$ semantics of PROMELA with which we validated our semantic definitions via several examples with respect to the SPIN implementation. Using this prototype, we present a case study of using the $\mathbb{K}$ deductive verifier to prove reachability claims for PROMELA programs under our semantics. In our case study, we verify examples for which SPIN—an explicit model checker—cannot prove the property.

As a simple case, we revisit the reachability specification (Fig. 1) discussed in Sect. 2. Following a similar approach there, we can also prove the same claim for the counterpart program written in PROMELA shown in Fig. 10. Note that leaving N as a parameter in the code yields an infinite family of models: each fixed N has a finite reachable state space (verifiable by SPIN), but SPIN cannot verify the property for arbitrary N.

As a nontrivial example, we verify mutual exclusion for Lamport's bakery algorithm [20] with two processes. This instance yields an infinite number of reachable states, so that SPIN cannot verify the property. Figure 11 presents the PROMELA program for the bakery algorithm involving two concurrent processes, say p1 and p2. Analogous to a real bakery (or bank), each process repeatedly: (i) obtains a ticket (`tick`) from the dispenser (`disp`), (ii) enters the critical section when the server (`serv`) calls its ticket, and then (iii) exits, via `LOOP:=do ... od`. We denote each stage by the code fragment (note that `tick` can grow indefinitely, inducing an infinite number of states):

```
WAIT  := atomic { tick = disp; disp = disp + 1 }; ENTER
ENTER := atomic { tick == serv; crit = crit + 1 }; EXIT
EXIT  := atomic { serv = serv + 1; crit = crit - 1 }
```

The mutual exclusion between p1 and p2 can be specified by asserting the invariant `crit` ≤ 1 for all reachable states. To express this property for nonterminating systems in reachability logic, we follow a similar approach proposed in [35]: we introduce a `monitor` process that nondeterministically "freezes" the global state via a pseudo-statement `freeze`, by setting the lock to a special value `#frozen` (also denoted "$\circledast$"). By defining the goal pattern ϕ_{goal} as the set of "frozen" states where `crit` ≤ 1 holds, the main claim $\phi_{init} \Rightarrow \phi_{goal}$ asserts mutual exclusion for all finite prefixes of the original bakery code without `monitor`, from the initial pattern ϕ_{init} with `disp`=`serv`=N for some integer N.

Figure 12 shows the concrete specification for our main claim. We succinctly write this main claim in the following high-level notation:

$$\langle \texttt{LOOP}, \texttt{LOOP}, \texttt{freeze}, \texttt{N}, \texttt{N}, \texttt{0}, \bot \rangle \Rightarrow \langle ?\_, ?\_, .\texttt{K}, ?\_, ?\_, ?\texttt{C} \leq 1, \circledast \rangle \qquad (1)$$

The seven components in the tuple correspond to: (the continuations of) p1, p2, `monitor`, (the values of) `disp` (line 22), `serv` (line 22), `crit` (line 23), and the global lock (line 26). In the goal pattern, "?_" denotes "don't care" variables, and "?C$\leq$1" means `crit` ≤ 1, where "?" indicate newly introduced variables.

During the proof of the main claim, we found that the bakery code induces three cycles that do not contain ϕ_{init}. As noted in Sect. 2, they can be ignored by adding auxiliary claims, listed as follows:

$$\langle \texttt{WAIT!(T)}, \texttt{WAIT!(T)}, \texttt{freeze}, \texttt{N}, \texttt{N}, \texttt{0}, \bot \rangle \Rightarrow \langle ?\_, ?\_, .\texttt{K}, ?\_, ?\_, ?\texttt{C} \leq 1, \circledast \rangle \qquad (2)$$

$$\langle \texttt{ENTER!(N)}, \texttt{WAIT!(·)}, \texttt{freeze}, \texttt{N}+1, \texttt{N}, \texttt{0}, \bot \rangle \Rightarrow \langle ?\_, ?\_, .\texttt{K}, ?\_, ?\_, ?\texttt{C} \leq 1, \circledast \rangle \quad (3)$$

$$\langle \texttt{WAIT!(·)}, \texttt{ENTER!(N)}, \texttt{freeze}, \texttt{N}+1, \texttt{N}, \texttt{0}, \bot \rangle \Rightarrow \langle ?\_, ?\_, .\texttt{K}, ?\_, ?\_, ?\texttt{C} \leq 1, \circledast \rangle \quad (4)$$

where `WAIT!(T)` denotes the fully-loaded continuation for `WAIT`, under the local variable `tick`= T, and likewise for `ENTER!(T)`. Indeed, the initial pattern of Claim 2 is simply the fully-loaded version of the initial pattern of Claim 1. The initial patterns of Claims 3 and 4 are obtained by executing the `atomic`

```
 1  claim:
 2    <procs>
 3      <proc>... // proc p1
 4        <k> LOOP => ?_ </k>
 5        <env>
 6    disp[0] |-> loc(0); serv[0] |-> loc(1)
 7    crit[0] |-> loc(2); tick[0] |-> loc(3)
 8        </env>
 9    ...</proc>
10      <proc>... // proc p2
11        <k> LOOP => ?_ </k>
12        <env>
13    disp[0] |-> loc(0); serv[0] |-> loc(1)
14    crit[0] |-> loc(2); tick[0] |-> loc(4)
15        </env>
16    ...</proc>
17      <proc>... // proc monitor
18        <k> freeze => .K </k>
19    ...</proc>
20    </procs>
21    <str>
22      0 |-> (N:Int => ?_); 1 |-> (N => ?_);
23      2 |-> (0 => ?C);
24      3 |-> (0 => ?_); 4 |-> (0 => ?_)
25    </str>
26    <lock>#none => #frozen </lock>
27  ensures ?C <=Int 1 // MUTEX
```

Fig. 12. The main claim.

```
 1  claim:
 2    <procs>
 3      <proc>... // proc p1
 4        <k> ENTER! => ?_ </k>
 5        <env>
 6    disp[0] |-> loc(0); serv[0] |-> loc(1)
 7    crit[0] |-> loc(2); tick[0] |-> loc(3)
 8        </env>
 9    ...</proc>
10      <proc>... // proc p2
11        <k> WAIT! => ?_ </k>
12        <env>
13    disp[0] |-> loc(0); serv[0] |-> loc(1)
14    crit[0] |-> loc(2); tick[0] |-> loc(4)
15        </env>
16    ...</proc>
17      <proc>... // proc monitor
18        <k> freeze => .K </k>
19    ...</proc>
20    </procs>
21    <str>
22      0 |-> (N +Int 1 => ?_);
23      1 |-> (N:Int => ?_); 2 |-> (0 => ?C);
24      3 |-> (N => ?_); 4 |-> (0 => ?_)
25    </str>
26    <lock>#none => #frozen </lock>
27  ensures ?C <=Int 1 // MUTEX
```

Fig. 13. An auxiliary claim.

block of **p1** and **p2**, respectively, from the initial pattern of Claim 2. Figure 13 shows the concrete specification for Claim 3.

Encoding the four claims (1 main/3 auxiliary) in **spec.k** and running it with **kprove**, the proof has terminated successfully with output **#Top** in approximately 5 min in a laptop computer (i5-1335U 2.50 GHz/16 GB RAM).

7   Related Work

Substantial work exists on establishing formal semantics of PROMELA. In [5,26,27], the semantics are given by low-level labeled transition systems. The semantics in [14,34,42] follow the style of structural operational semantics (SOS) [31], while a denotational semantics is given in [9]. From the analysis perspective, the works [27,42] remain in the scope of SPIN's LTL model checking. Other lines of work extend the analysis to other techniques such as abstract interpretation [9,14], and PROMELA-to-C refinement [34]. The works [4,18] extend the original C implementation of SPIN to support various kinds of analysis, but they do not formalize the semantics of PROMELA. We identified two works that establish mechanized, executable semantics of PROMELA in ACL2 [5] and in Isabelle/HOL [27]. However, both focus on providing reference implementations for PROMELA independent of SPIN, and do not cover code-level deductive verification of user programs.

K [33] is an executable semantic framework for defining programming languages. It emphasizes modularity—addressing a well-known limitation of SOS—so new features can be added without revising unrelated rules. K has proved effective for control-intensive features (e.g., exceptions, call/cc) and has been used to formalize several real-world languages [6,12,28]. The framework was later formalized as Matching Logic [32], yielding a deductive system [37] with commercial applications, notably in smart-contract verification [15,29,30]. This work presents the first executable PROMELA semantics defined in $\mathbb{K}$.

$\mathbb{K}$ is grounded in Rewriting Logic, which is introduced as a general formalism for specifying concurrent systems [23]. Classic process calculi—including CCS [24] and the Pi-calculus [25]—have been studied within Rewriting Logic [7,10,38–41]. Rewriting Logic's inherent nondeterminism and true concurrency make such semantics natural to specify; we exploit this in our LOAD-AND-FIRE semantics via multiset matching.

8 Concluding Remarks

We have presented a faithful, executable semantics of PROMELA in the $\mathbb{K}$ framework. Our semantics enables code-level deductive verification of PROMELA models, including infinite-state systems, a capability previously unavailable for PROMELA. Beyond providing a precise, machine-readable reference, it bridges model checking and deductive reasoning: properties beyond explicit-state model checking can now be proved directly on PROMELA programs.

We have also introduced LOAD-AND-FIRE, an elegant semantic pattern that yields a modular, uniform treatment of guarded nondeterminism, cross-process interference, and atomicity in $\mathbb{K}$. Beyond PROMELA, this constitutes a reusable methodology for $\mathbb{K}$-based language semantics of guarded concurrency, with natural applications to constructs such as Go's `select` and Erlang's `receive`.

Future work includes supporting temporal reasoning for deductive verification (e.g., LTL) strengthening proof automation for PROMELA (e.g., reusable lemma libraries), and conducting additional case studies on complex PROMELA models. In particular, it is interesting to extend the bakery example in Sect. 6 to an arbitrary number of processes, using rewriting-based techniques for parameterized verification [1–3,19]. We also plan to apply LOAD-AND-FIRE to other programming languages with guarded choice and synchronous communication.

Acknowledgments. This work was partially supported by the Institute of Information & Communications Technology Planning & Evaluation (IITP) grant (No. RS-2024-00439856) and by the National Research Foundation of Korea (NRF) grants (No. RS-2021-NR060080 and No. RS-2024-00413202), both funded by the Korea government (MSIT).

Data Availability Statement. The full $\mathbb{K}$ semantics of PROMELA, the case study, and the benchmark models for validation are available at https://doi.org/10.5281/zenodo.17183743.

References

1. Bae, K., Escobar, S., López-Rueda, R., Meseguer, J., Sapiña, J.: Verifying invariants by deductive model checking. In: International Workshop on Rewriting Logic and Its Applications (WRLA). Lecture Notes in Computer Science, vol. 14953, pp. 3–21. Springer, Heidelberg (2024). https://doi.org/10.1007/978-3-031-65941-6_1
2. Bae, K., Escobar, S., Meseguer, J.: Abstract logical model checking of infinite-state systems using narrowing. In: International Conference on Rewriting Techniques and Applications (RTA). LIPIcs, vol. 21, pp. 81–96. Schloss Dagstuhl - Leibniz-Zentrum für Informatik (2013). https://doi.org/10.4230/LIPICS.RTA.2013.81
3. Bae, K., Meseguer, J.: Predicate abstraction of rewrite theories. In: Rewriting and Typed Lambda Calculi - Joint International Conference, (RTA-TLCA). Lecture Notes in Computer Science, vol. 8560, pp. 61–76. Springer, Heidelberg (2014). https://doi.org/10.1007/978-3-319-08918-8_5
4. van der Berg, F.I., Laarman, A.: SpinS: extending LTSmin with PROMELA through SpinJa. In: International Workshop on Parallel and Distributed Methods in Verification (PDMC). Electronic Notes in Theoretical Computer Science, vol. 296, pp. 95–105. Elsevier (2012). https://doi.org/10.1016/J.ENTCS.2013.07.007
5. Bevier, W.: Toward an operational semantics of PROMELA in ACL2. In: Proceedings of the Third SPIN Workshop (1997)
6. Bogdanas, D., Rosu, G.: K-Java: a complete semantics of Java. In: ACM SIGPLAN-SIGACT Symposium on Principles of Programming Languages (POPL), pp. 445–456. ACM (2015). https://doi.org/10.1145/2676726.2676982
7. Braga, C., Meseguer, J.: Modular rewriting semantics in practice. In: International Workshop on Rewriting Logic and its Applications (WRLA). Electronic Notes in Theoretical Computer Science, vol. 117, pp. 393–416. Elsevier (2004). https://doi.org/10.1016/J.ENTCS.2004.06.019
8. Choi, Y.: From NuSMV to SPIN: experiences with model checking flight guidance systems. Formal Methods Syst. Des. **30**(3), 199–216 (2007). https://doi.org/10.1007/S10703-006-0027-9
9. Comini, M., Gallardo, M., Villanueva, A.: A denotational semantics for PROMELA addressing arbitrary jumps. CoRR arxiv:2108.12348 (2021)
10. Degano, P., Gadducci, F., Priami, C.: A causal semantics for CCS via rewriting logic. Theor. Comput. Sci. **275**(1–2), 259–282 (2002). https://doi.org/10.1016/S0304-3975(01)00165-7
11. Dijkstra, E.W.: Guarded commands, nondeterminacy and formal derivation of programs. Commun. ACM **18**(8), 453–457 (1975). https://doi.org/10.1145/360933.360975
12. Ellison, C., Rosu, G.: An executable formal semantics of C with applications. In: ACM SIGPLAN-SIGACT Symposium on Principles of Programming Languages (POPL), pp. 533–544. ACM (2012). https://doi.org/10.1145/2103656.2103719
13. Evrard, H., Donaldson, A.F.: Model checking futexes. In: Model Checking Software (SPIN). Lecture Notes in Computer Science, vol. 13872, pp. 41–58. Springer, Heidelberg (2023). https://doi.org/10.1007/978-3-031-32157-3_3
14. Gallardo, M., Merino, P., Pimentel, E.: A generalized semantics of PROMELA for abstract model checking. Formal Aspects Comput. **16**(3), 166–193 (2004). https://doi.org/10.1007/S00165-004-0040-Y

15. Hildenbrandt, E., et al.: KEVM: a complete formal semantics of the Ethereum virtual machine. In: Computer Security Foundations Symposium (CSF), pp. 204–217. IEEE Computer Society (2018). https://doi.org/10.1109/CSF.2018.00022
16. Hoare, C.A.R.: Communicating sequential processes. In: Theories of Programming: The Life and Works of Tony Hoare, ACM Books, vol. 39, pp. 157–186. ACM/Morgan & Claypool (2021). https://doi.org/10.1145/3477355.3477364
17. Holzmann, G.J.: The model checker SPIN. IEEE Trans. Softw. Eng. **23**(5), 279–295 (1997). https://doi.org/10.1109/32.588521
18. de Jonge, M., Ruys, T.C.: The SpinJa model checker. In: Model Checking Software (SPIN). Lecture Notes in Computer Science, vol. 6349, pp. 124–128. Springer, Heidelberg (2010). https://doi.org/10.1007/978-3-642-16164-3_9
19. Kang, B., Bae, K.: Narrowing and heuristic search for symbolic reachability analysis of concurrent object-oriented systems. Sci. Comput. Program. **235**, 103097 (2024). https://doi.org/10.1016/J.SCICO.2024.103097
20. Lamport, L.: A new solution of Dijkstra's concurrent programming problem. Commun. ACM **17**(8), 453–455 (1974). https://doi.org/10.1145/361082.361093
21. Lazar, D., et al.: Executing formal semantics with the K tool. In: International Symposium on Formal Methods (FM). Lecture Notes in Computer Science, vol. 7436, pp. 267–271. Springer, Heidelberg (2012). https://doi.org/10.1007/978-3-642-32759-9_23
22. Maggi, P., Sisto, R.: Using SPIN to verify security properties of cryptographic protocols. In: Model Checking of Software (SPIN). Lecture Notes in Computer Science, vol. 2318, pp. 187–204. Springer, Heidelberg (2002). https://doi.org/10.1007/3-540-46017-9_14
23. Meseguer, J.: Conditional rewriting logic as a united model of concurrency. Theor. Comput. Sci. **96**(1), 73–155 (1992). https://doi.org/10.1016/0304-3975(92)90182-F
24. Milner, R.: A Calculus of Communicating Systems, Lecture Notes in Computer Science, vol. 92. Springer, Heidelberg (1980). https://doi.org/10.1007/3-540-10235-3
25. Milner, R., Parrow, J., Walker, D.: A calculus of mobile processes. I. Inf. Comput. **100**(1), 1–40 (1992). https://doi.org/10.1016/0890-5401(92)90008-4
26. Natarajan, V., Holzmann, G.: Outline for an operational semantics of PROMELA. In: Proceedings of the Second SPIN Workshop (1996)
27. Neumann, R.: Using PROMELA in a fully verified executable LTL model checker. In: International Conference on Verified Software: Theories, Tools and Experiments (VSTTE). Lecture Notes in Computer Science, vol. 8471, pp. 105–114. Springer, Heidelberg (2014). https://doi.org/10.1007/978-3-319-12154-3_7
28. Park, D., Stefanescu, A., Rosu, G.: KJS: a complete formal semantics of JavaScript. In: ACM SIGPLAN Conference on Programming Language Design and Implementation (PLDI), pp. 346–356. ACM (2015). https://doi.org/10.1145/2737924.2737991
29. Park, D., Zhang, Y., Rosu, G.: End-to-end formal verification of Ethereum 2.0 deposit smart contract. In: International Conference on Computer Aided Verification (CAV). Lecture Notes in Computer Science, vol. 12224, pp. 151–164. Springer, Heidelberg (2020). https://doi.org/10.1007/978-3-030-53288-8_8

30. Park, D., Zhang, Y., Saxena, M., Daian, P., Rosu, G.: A formal verification tool for Ethereum VM bytecode. In: ACM Joint Meeting on European Software Engineering Conference and Symposium on the Foundations of Software Engineering (ESEC/FSE), pp. 912–915. ACM (2018). https://doi.org/10.1145/3236024.3264591
31. Plotkin, G.D.: A structural approach to operational semantics. J. Log. Algebraic Methods Program. **60–61**, 17–139 (2004)
32. Rosu, G.: Matching logic. Log. Methods Comput. Sci. **13**(4) (2017). https://doi.org/10.23638/LMCS-13(4:28)2017
33. Rosu, G., Serbanuta, T.: An overview of the K semantic framework. J. Log. Algebraic Methods Program. **79**(6), 397–434 (2010). https://doi.org/10.1016/J.JLAP.2010.03.012
34. Sharma, A.: A refinement calculus for PROMELA. In: International Conference on Engineering of Complex Computer Systems (ICECCS), pp. 75–84. IEEE Computer Society (2013). https://doi.org/10.1109/ICECCS.2013.20
35. Skeirik, S., Stefanescu, A., Meseguer, J.: A constructor-based reachability logic for rewrite theories. In: International Symposium on Logic-Based Program Synthesis and Transformation (LOPSTR). Lecture Notes in Computer Science, vol. 10855, pp. 201–217. Springer, Heidelberg (2017). https://doi.org/10.1007/978-3-319-94460-9_12
36. Spin Online References: PROMELA manual pages. https://spinroot.com/spin/Man/promela.html. Accessed 20 Nov 2025
37. Stefanescu, A., Ciobâcă, Ş., Mereuta, R., Moore, B.M., Serbanuta, T., Rosu, G.: All-path reachability logic. In: Rewriting and Typed Lambda Calculi - Joint International Conference (RTA-TLCA). Lecture Notes in Computer Science, vol. 8560, pp. 425–440. Springer, Heidelberg (2014). https://doi.org/10.1007/978-3-319-08918-8_29
38. Stehr, M.: CINNI - a generic calculus of explicit substitutions and its application to lambda-, varsigma- and pi- calculi. In: International Workshop on Rewriting Logic and its Applications (WRLA). Electronic Notes in Theoretical Computer Science, vol. 36, pp. 70–92. Elsevier (2000). https://doi.org/10.1016/S1571-0661(05)80125-2
39. Thati, P., Sen, K., Martí-Oliet, N.: An executable specification of asynchronous pi-calculus semantics and may testing in Maude 2.0. In: International Workshop on Rewriting logic and Its Applications (WRLA). Electronic Notes in Theoretical Computer Science, vol. 71, pp. 261–281. Elsevier (2002). https://doi.org/10.1016/S1571-0661(05)82539-3
40. Verdejo, A., Martí-Oliet, N.: Two case studies of semantics execution in Maude: CCS and LOTOS. Formal Methods Syst. Des. **27**(1–2), 113–172 (2005). https://doi.org/10.1007/S10703-005-2254-X
41. Viry, P.: Input/output for ELAN. In: International Workshop on Rewriting Logic and its Applications (WRLA). Electronic Notes in Theoretical Computer Science, vol. 4, pp. 51–64. Elsevier (1996). https://doi.org/10.1016/S1571-0661(04)00033-7
42. Weise, C.: An incremental formal semantics for PROMELA. In: Proceedings of the Third SPIN Workshop (1997)

Multi-variable Quantification of BDDs in External Memory using Nested Sweeping

Steffan Christ Sølvsten[(✉)] [iD] and Jaco van de Pol [iD]

Aarhus University, Aarhus, Denmark
{soelvsten,jaco}@cs.au.dk

Abstract. Previous research on the Adiar BDD package has been successful at designing algorithms capable of handling large Binary Decision Diagrams (BDDs) stored in external memory. To do so, it uses consecutive sweeps through the BDDs to resolve computations. Yet, this approach has kept algorithms for multi-variable quantification, the relational product, and variable reordering out of its scope.

In this work, we address this by introducing the *nested sweeping* framework. Here, multiple concurrent sweeps pass information between each other to compute the result. We have implemented the framework in Adiar and used it to create a new external memory multi-variable quantification algorithm. In practice, this improves Adiar's running time by a factor of 1.7. In turn, this work extends the previous research results on Adiar to also apply to its quantification operation: compared to conventional depth-first implementations, Adiar with nested sweeping is able to solve more problems and/or solve them faster.

Keywords: Time-forward Processing · External Memory Algorithms · Binary Decision Diagrams

1 Introduction

The ability of Binary Decision Diagrams (BDDs) to represent Boolean formulae as small directed acyclic graphs (DAGs) have made them an invaluable tool to solve many complex problems. For example, recently they have been used to check type-and-effect systems [33,34], to generate proofs for SAT and QBF solvers [13–15], for circuit synthesis [21,30], to solve games [35,42,53], and for symbolic model checking [3,18,19,22,24,26,32].

Implementations of decision diagrams conventionally make use of recursive depth-first algorithms and a unique node table [10,20,27,31,38,52]. Both of these introduce random access, which pauses the entire computation while missing data is fetched [28,37,41]. For large enough instances, data has to reside on disk and the resulting I/O-operations that ensue become the bottle-neck.

Y.-F. Chen et al. (Eds.): VMCAI 2026, LNCS 16417, pp. 359–382, 2026.
https://doi.org/10.1007/978-3-032-15700-3_17

Adiar [49] is a BDD package written in C++ based on the ideas of Lars Arge [5]: the depth-first recursive algorithms are replaced with iterative algorithms. Here, one or more priority queues reorder the execution of recursive calls such that they are synchronised with a level-by-level traversal of the inputs. This makes Adiar's algorithms, unlike the conventional recursive implementations, optimal in the I/O-model [1] of Aggarwal and Vitter [5,6]. In turn, this enables it to manipulate BDDs beyond the reach of conventional BDD packages at a negligible cost to its running time [49].

Yet, the ideas in [5,6,49] only provide a translation of the simplest BDD algorithms, which do not recurse on the result of other recursive calls. This does not provide a way to translate the more complex BDD algorithms that recurse on intermediate recursion results, e.g. multi-variable quantification. Hence, until this work, Adiar could not easily be used for solving Quantified Boolean formulæ (QBF). Furthermore, game solving and symbolic model checking has until now been out of reach for Adiar.

1.1 Contributions

In Sect. 3, we introduce the notion of *nested sweeping* to provide a framework on which these more complex BDD operations can be implemented. Here, an *outer* bottom-up sweep accumulates the results from multiple nested *inner* sweeps. With this framework in hand, we implement an I/O-efficient multi-variable quantification akin to the one in conventional BDD packages. Furthermore, we identify in Sect. 3.2 optimisations for the nested sweeping framework. The full paper [46] also includes optimisations specific to the quantification operation and an overview of the implementation in Adiar. Section 4 shows that nested sweeping improves the running time in practice by a factor of 1.7 when solving QBF-encodings of two-player games and when reasoning about the transition system in Conway's Game of Life [23]. We compare our approach to related work in Sect. 5 and provide our conclusions and future work in Sect. 6.

2 Preliminaries

2.1 The I/O-Model

Aggarwal and Vitter introduced the I/O-model [1] to analyse the cost of transferring data to and from a slow storage device. Here, computations can only operate on data that resides in *internal* memory, e.g. the RAM, with a finite size of M. Hence, if the input of size N (or some intermediate result) exceeds M then it needs to be transferred to and from *external* memory, e.g. the disk. Yet, each such data transfer (I/O) moves an entire consecutive block of B elements; an algorithm's I/O-complexity is the number of I/Os it uses.

One needs $\mathrm{scan}(N) \triangleq N/B$ I/Os to linearly scan through a consecutive list of N elements in external memory [1]. Assuming $N > M$, one needs to use $\Theta(\mathrm{sort}(N))$ I/Os to sort N elements, where $\mathrm{sort}(N) \triangleq N/B \cdot \log_{M/B}(N/B)$ [1]. Furthermore, one can design an I/O-efficient priority queue capable of doing

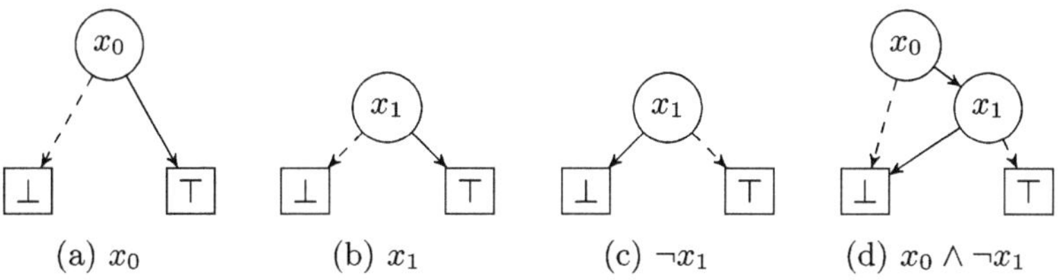

(a) x_0 (b) x_1 (c) $\neg x_1$ (d) $x_0 \wedge \neg x_1$

Fig. 1. Examples of Reduced Ordered Binary Decision Diagrams. Terminals are drawn as boxes surrounding their Boolean value. Internal nodes are drawn as circles and contain their decision variable. Arcs to the *high* and *low* child are respectively drawn solid and dashed.

N insertions and deletions in $\Theta(\mathrm{sort}(N))$ I/Os [4]. For simplicity, we overload $\mathrm{scan}(N)$ to be N and $\mathrm{sort}(N)$ to be $N \log_2 N$ when referring to an algorithm's time complexity rather than its I/O complexity.

Intuitively, an algorithm is I/O-inefficient if it uses an entire I/O to retrieve a block but does not make use of a significant portion of the B elements within. That is, random access can result in N I/Os. For all realistic values of N, M, and B, this is several magnitudes larger than both $\mathrm{scan}(N)$ and $\mathrm{sort}(N)$.

2.2   Binary Decision Diagrams

As shown in Fig. 1, a Binary Decision Diagram [12] (BDD) (based on [2,29]) represents an n-ary Boolean function as a singly-rooted directed acyclic graph (DAG). Each of its two sinks, refered to as *terminals*, contain one of the two Boolean values, $\mathbb{B} = \{\top, \bot\}$. These represent the function's output values. An internal BDD node, v, is associated in $v.\mathtt{var}$ with a Boolean input variable x_i. Furthermore, it has two BDD nodes as children, $v.\mathtt{low}$ and $v.\mathtt{high}$. These three values in f encode the ternary if-then-else $v.\mathtt{var}$? $v.\mathtt{high}$: $v.\mathtt{low}$. What are colloquially referred to as BDDs are in fact *Reduced Ordered* Binary Decision Diagrams (ROBDDs). An Ordered BDD (OBDD) restricts each variable to occur at most once on each path from the root to a terminal and to occur according to a certain order, π. This gives rise to a levelisation of the OBDD where each level, ℓ, is associated with an input variable, x_i. For sake of simplicity, we assume that π is the identity order. A *Reduced* OBDD further restricts the DAG such that (1) no nodes are duplicates of another and (2) no node is redundant, i.e. $v.\mathtt{high} = v.\mathtt{low}$. Assuming the variable ordering, π, is fixed, ROBDDs are a unique canonical form of the Boolean function it represents.

Quantification Algorithm. The levelisation of OBDDs allows the recursive BDD algorithms to both be efficient and elegant. For example, the **or** operation works by a product construction of the two input BDDs. Here, each node of the output BDD simulates, according to π, the decision(s) taken on the shallowest BDD node(s) in the product of nodes from the input.

```
1   exists(v, X)
2       if  v = ⊥ ∨ v = ⊤
3           return v
4       exi0 ← exists(v.low, X)
5       exi1 ← exists(v.high, X)
6       if v.var ∉ X
7           return Node { v.var, exi0, exi1 }
8       return or(exi0, exi1)
```

Fig. 2. A recursive multi-variable **exists** operation.

$$[\{(0,0), \bot, (1,0)\};\ \{(1,0), \top, \bot\}]$$

(a) Node-based representation of $x_0 \wedge \neg x_1$ (Fig. 1d).

$$[(0,0) \rightarrow (1,0);\ (0,0) \dashrightarrow (1,1);\ (1,0) \rightarrow \bot;\ (1,0) \dashrightarrow \top;\ (1,1) \dashrightarrow \top;\ (1,1) \rightarrow \top;\ ]$$

(b) Arc-based representation of the **or** of $x_0 \wedge \neg x_1$ (Fig. 1d) and x_1 (Fig. 1b).

Fig. 3. BDD Representations in Adiar.

Since $(\exists x : \phi) \equiv \phi[\top/x] \vee \phi[\bot/x]$, the **or** operation can be used as the basis for an existential quantification ($\exists$) for a set of input variables, $X = \{x_i, x_j, \ldots, x_k\}$. As shown in Fig. 2, if v is a terminal then this (sub)BDD depends on none of the to be quantified variables. Otherwise, both its children are resolved recursively into intermediate results, exi0 and exi1. If the decision variable of the root, v.**var**, should not be quantified, a new node with variable v.**var** is created from the two recursive results. Otherwise, exi0 and exi1 are instead combined (recursively once more) with a nested **or** operation.

Similarly, one can implement a universal quantification ($\forall$) by use of a nested **and** operation. For clarity, our contributions in Sect. 3 are only phrased with respect to the **exists** operation. But, everything that follows also applies to **forall** by replacing **or** with **and**.

2.3   I/O-Efficient BDD Manipulation

The Adiar [49] BDD package builds on top of Lars Arge's ideas [5,6] on how to improve the I/O complexity of BDD manipulation. To not introduce random access, Adiar does not use any hash tables nor recursion for its BDD manipulation. As a result, different BDD objects do not share common subtrees in Adiar. For the same reason, it neither uses pointers to traverse its BDDs. Instead, every BDD node v is uniquely identified by a pair $(v.\textbf{var}, v.\texttt{id})$ where $v.\texttt{id}$ is v's index on level $v.\textbf{var}$. Lexicographically, this *unique identifier* (uid) imposes a total ordering of all BDD nodes such that they follow the variable ordering. Yet, the uid does not specify the exact index where one can find the BDD node. For example, the BDD for $x_0 \wedge \neg x_1$ in Fig. 1d is represented in Adiar as the list of nodes in Fig. 3a: every node is a 3-tuple with its uid followed by the unique identifier of its low and its high children.

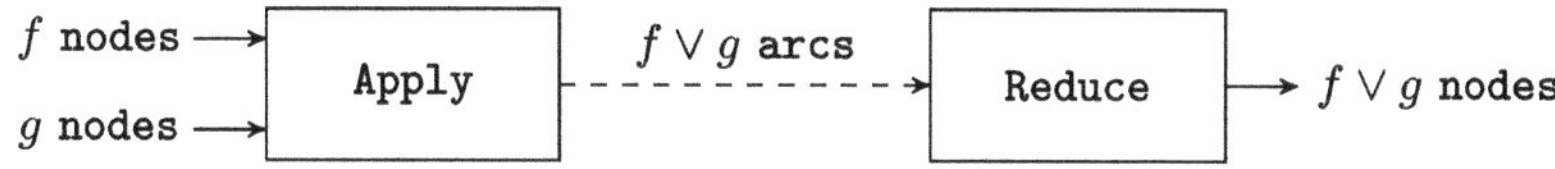

Fig. 4. The Apply–Reduce pipeline of or in Adiar.

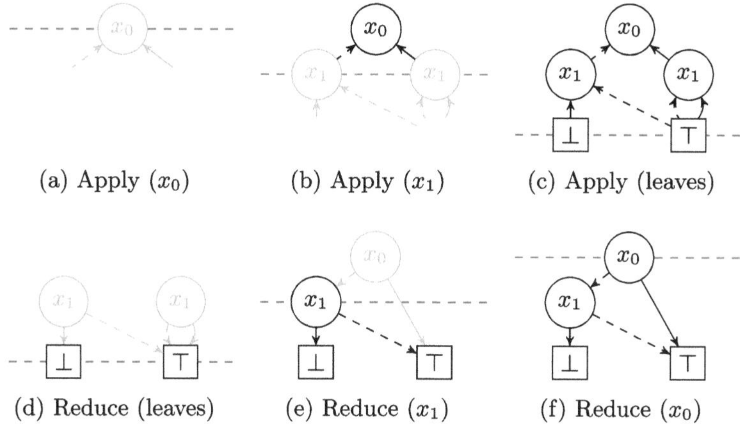

(a) Apply (x_0) (b) Apply (x_1) (c) Apply (leaves)

(d) Reduce (leaves)   (e) Reduce (x_1) (f) Reduce (x_0)

Fig. 5. Step-by-step example of the or of x_1 (Fig. 1b) and $x_0 \wedge \neg x_1$ (Fig. 1d) with time-forward processing; subfigures show the state after processing each level. Arcs in gray are pushed to the algorithm's priority queue whereas the ones in black have been written to the output file.

As depicted in Figs. 4 and 5, the previous BDD operations in Adiar, such as or, process a BDD with two sweeps. Both sweeps use *time-forward processing* [4, 17] to achieve their I/O-efficiency: computation is deferred with one or more priority queues until all relevant data has been read. During the first sweep, the *Apply*[1], the entire recursion tree is unfolded top-down. Here, the priority queues also double as a computation cache [10, 38] by merging separate paths to the same recursion target. Hence, the resulting output is in fact not a tree but a DAG. Yet, it is only an OBDD and needs to be reduced. To do so, Adiar uses an I/O-efficient variant of the original bottom-up Reduce algorithm by Bryant [4, 12]. Here, a priority queue is used to forward the uid of reduced nodes t' in the final ROBDD to their to be reduced parents s in the intermediate OBDD. Yet, to know the parents s, the Reduce needs the intermediate OBDD to be transposed, i.e. the DAG's edges to be reversed. Luckily, the Apply sweep outputs its arcs (directed edges) sorted by their target[2]. This effectively transposes the OBDD and so no extra work is needed for the Reduce [5, 49]. For example, the or of Figs. 1b and

---

[1] Similar to [43], we refer to all top-down manipulating sweeps as Apply, e.g. not, or, and if-then-else. This even includes identity which merely reverses the edges.

[2] In [49], arcs to terminals are actually output to a separate file sorted by their source. This is merely to improve performance. Hence, we will ignore this detail.

1d creates the unreduced BDD in Fig. 5c with the arc-based representation in Fig. 3b.

The I/O and time complexity of this Apply–Reduce tandem is $\mathcal{O}(\mathrm{sort}(N + T))$, where N is the size of the input(s) and T is the size of the unreduced output of the Apply sweep [49]. To catch up with conventional implementation's performance, major efforts have been dedicated to improve on this foundation.

Levelised Cuts [51]. The arcs placed in the above-mentioned priority queues correspond to graph cuts in the (R)OBDDs. In particular, these cuts correspond to the border between the processed and to be processed nodes. Since the priority queues induce a levelised processing order, the shape of these cuts match the (R)OBDD's levels.

This ensures that the maximum size of the priority queues is bounded by the maximum levelised cut in the input. Hence, sound upper bounds on these cuts can in turn be used to determine a priori whether a faster internal-memory priority queue can be used.

In practice, this improves performance for smaller and moderate instances.

Levelised Random Access [50]. Orthogonally, an Apply sweep for a product construction, e.g. an **or**, can be simplified if one of its inputs is narrow, i.e. each level fits into internal memory. In this case, one can load each level in its entirety into internal memory. This provides random access to all of its nodes on said level, making one of two priority queues within the Apply (as in [49]) obsolete.

In practice, this improves performance for larger instances.

3   I/O-Efficient Multi-variable Quantification

Our previous work in [49] only covers simple BDD operations without any data-dependencies in its recursion, e.g. the **or**. Yet, this does not cover the **exists** in Fig. 2, where the nested call to **or** on line 8 depends on the recursions from lines 4 and 5.

To address this, we introduce the *nested sweeping* framework. As shown in Figs. 6 and 7, nested sweeping wraps the algorithm(s) depicted in Fig. 4: after transposing the input in an initial Apply sweep, a single *outer* Reduce sweep

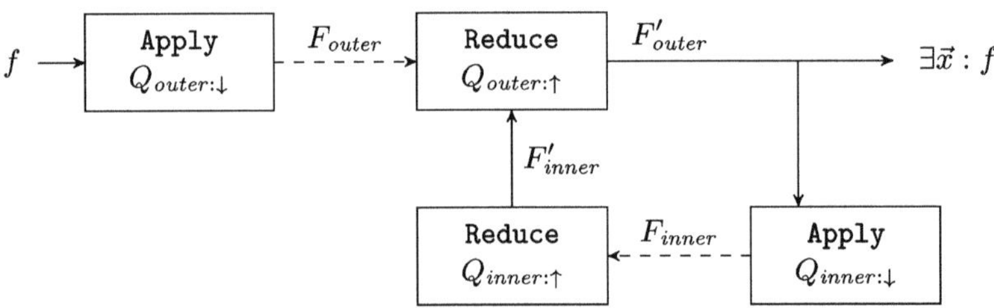

Fig. 6. The Apply–Reduce pipeline of **exists** with Nested Sweeping. F_* are files created by the respective sweep while $Q_{*:*}$ are the priority queues they each use.

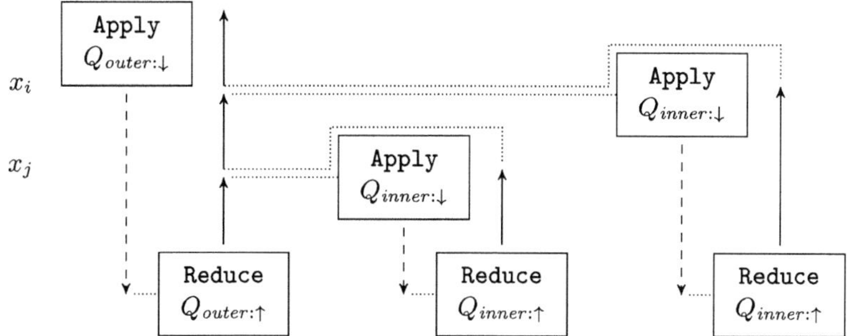

Fig. 7. Sweep direction (solid/dashed) and control-flow (dotted) of Nested Sweeping. The y-axis corresponds to the levels within the BDD. Variables x_i and x_j induce nested sweeps; for **exists**, they are to be quantified variables.

accumulates the result of multiple *inner* Apply–Reduce sweeps. For **exists**, the inner Apply is an **or** sweep.

More precisely, nested sweeping consists of the four phases described below and depicted step-by-step in Fig. 9.

Outer Apply: As shown in Fig. 8, inputs are combined (and possibly manipulated) in an Apply sweep into a single file, F_{outer}. This transposes and merges the inputs such that they are of the form needed by the Reduce of [49].

In the case of **exists**, this is merely a simple transposition of f. In the full paper [46], we explore ways in which this phase can also do double duty for partially resolving the quantification computation.

Outer Reduce: As in [49], each level of F_{outer} is reduced bottom-up by having a priority queue, $Q_{outer:\uparrow}$, forward the information about reduced nodes, t', to their unreduced parents, s. The reduced output is pushed into a new file, F'_{outer}.

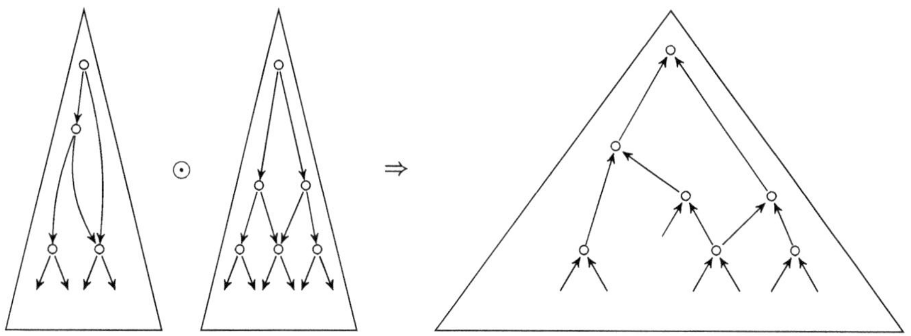

Fig. 8. Outer Apply one (or more) input BDD(s) are processed in a top-down sweep to create a single transposed BDD.

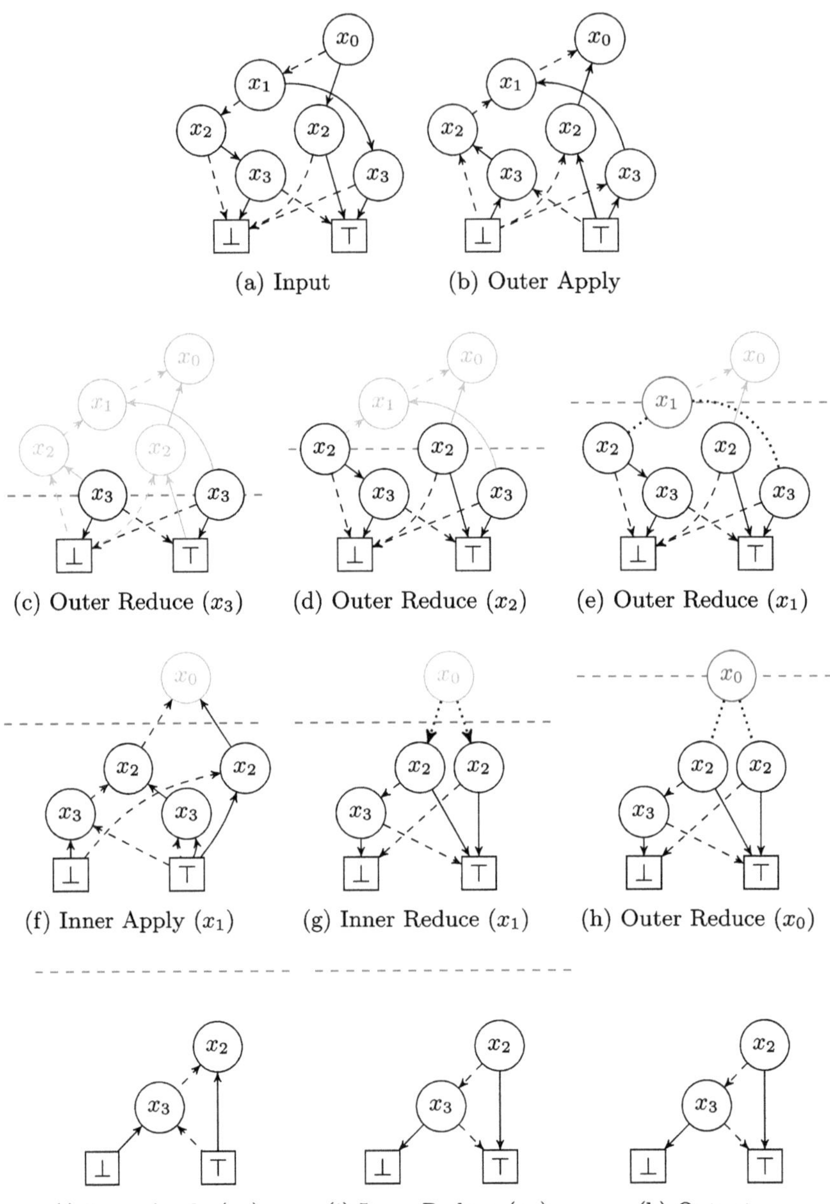

(a) Input   (b) Outer Apply

(c) Outer Reduce (x_3)   (d) Outer Reduce (x_2)   (e) Outer Reduce (x_1)

(f) Inner Apply (x_1)   (g) Inner Reduce (x_1)   (h) Outer Reduce (x_0)

(i) Inner Apply (x_0)   (j) Inner Reduce (x_0)   (k) Output

Fig. 9. Step-by-step example of existential quantification of x_0 and x_1 with nested sweeping; each subfigure shows the state after each respective step. The grey nodes depict unreduced nodes created in 9b while dashed red lines show the latest resolved level in the outer Reduce. The red nodes and dotted lines in 9e and 9h depict Case 3 of the outer Reduce. The dotted arcs in 9g depict Case 3 of the inner Reduce. (Color figure online)

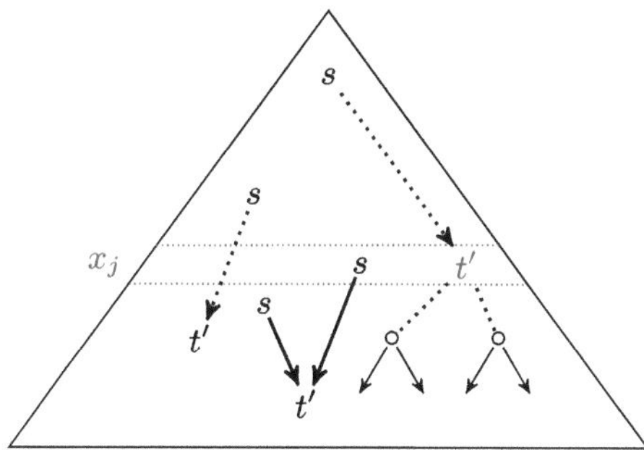

Fig. 10. Outer Reduce: solid arcs stay in $Q_{outer:\uparrow}$ (Case 2a) while dotted arcs are turned into requests for $Q_{inner:\downarrow}$ (Cases 2b on the left and 1 on the right).

Let x_j be the next level that needs a nested sweep. For **exists**, x_j is the largest still to be quantified variable in X. As visualised in Fig. 10, the logic of [49] is extended as follows:

1. If the current level is x_j, each arc $s \longrightarrow t'$ to a reduced node t' at this level is turned into a request and placed in a second priority queue, $Q_{inner:\downarrow}$.
   For **exists**, the requests are of the form $s \longrightarrow (t'.\texttt{low}, t'.\texttt{high})$.
2. If the current level is deeper than x_j, nodes are reduced as in [49] with a caveat: whether the arc $s \longrightarrow t'$ to the reduced node t' is placed in $Q_{outer:\uparrow}$ or in $Q_{inner:\downarrow}$ depends on the level of the unreduced parent s as follows:
   (a) If $x_j \leq s.\texttt{var}$, i.e. s is as deep or deeper than level x_j, then $s \longrightarrow t'$ is placed in $Q_{outer:\uparrow}$ as normal.
   (b) Otherwise, i.e. if $s.\texttt{var} < x_j$, $s \longrightarrow t'$ is placed in $Q_{inner:\downarrow}$ instead.

For **exists**, Case 1 matches the invocation of **or** on line 8 of Fig. 2 whereas 3 is the return with an unquantified variable on line 7.

When level x_j has finished processing, $Q_{inner:\downarrow}$ is populated with all requests that span across level x_j. Now, the inner Apply sweep is invoked.

Inner Apply: As depicted in Fig. 11a, starting with the requests in $Q_{inner:\downarrow}$, the reduced nodes, t', placed in F'_{outer} by the outer Reduce sweep, are processed with an Apply sweep from [49]. The intermediate unreduced result is placed in a new file, F_{inner}.

For **exists**, this sweep is the execution of the **or** on line 8 of Fig. 2. While the algorithms in [49] were only applied to a single BDD, they can also be applied to an entire forest; this merely requires prepopulating their priority queue with recursion requests to each root. Hence, we can reuse the previous top-down algorithms from [49] as is.

Inner Reduce: After the inner Apply sweep, F_{inner} is reduced in another bottom-up Reduce sweep of [49]. This creates the reduced nodes t'' placed in a new file

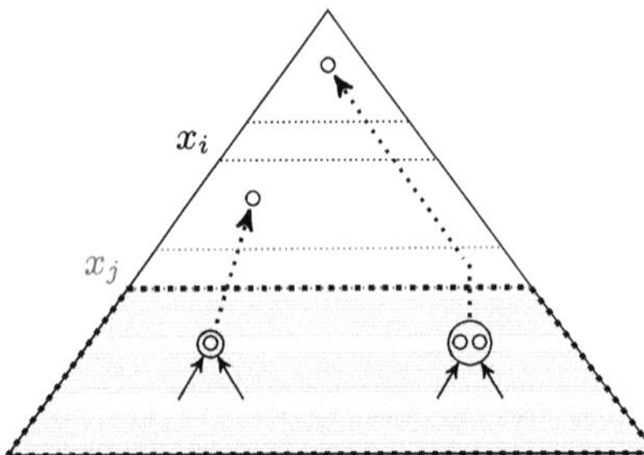

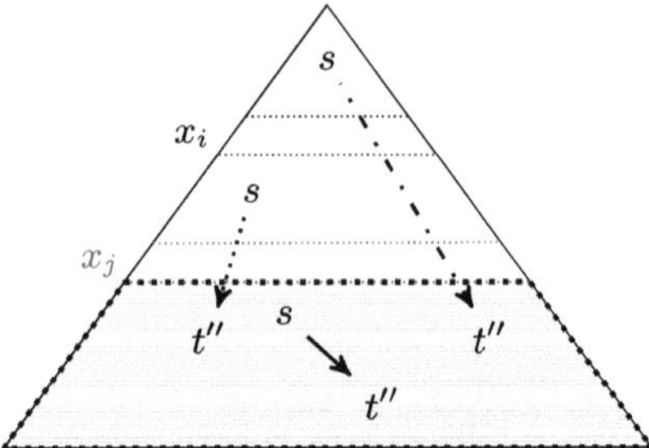

(a) Inner Apply: starting with multiple root requests (dotted), F_{inner} is constructed with the to-be preserved subtrees (left) together with new nodes that are products of previous ones (right).

(b) Inner Reduce: arcs below x_j stay within the inner sweep (Case 1, solid). Arcs that cross x_j are given back to the outer (Case 2, dotted) or to the next inner sweep (Case 3, dash dotted).

Fig. 11. Visualization of the Inner Apply and the Inner Reduce.

F'_{inner}. Let x_i be the next level above x_j that also needs a nested sweep. For exists, x_i is the largest variable smaller than x_j that also needs to be quantified. The arc $s \longrightarrow t''$ is placed in a priority queue, $Q_{inner:\uparrow}$, as follows.

1. If $x_j < s.\mathtt{var}$, i.e. the parent s is below level x_j, then $s \longrightarrow t''$ is forwarded within this inner Reduce sweep's priority queue, $Q_{inner:\uparrow}$.
2. If $s.\mathtt{var} \in [x_i, x_j]$, i.e. the parent s is between level x_i and x_j then $s \longrightarrow t''$ is given back to the outer sweep, $Q_{outer:\uparrow}$. This matches Case 2a in the outer Reduce.
3. If $s.\mathtt{var} < x_i$, i.e. the parent s is above x_i, then $s \longrightarrow t''$ is placed into $Q_{inner:\downarrow}$ to prepare the next invocation of an inner Apply sweep. This matches Case 2b in the outer Reduce.

The three cases above are depicted in Fig. 11b. For exists, Cases 2 and 3 are equivalent to the return from or back to exists. Case is equivalent to a return statement within the or's own recursion. Case 3 is needed to match 2b with x_j replaced with x_i.

Finally, F'_{inner} replaces F'_{outer} and control returns to the outer Reduce sweep to proceed with the levels above x_j.

As shown in Fig. 9, F_{inner} from the inner Apply sweep can be thought of as overlayed on top of F_{outer} from the outer Apply sweep; together they produce a valid (but unreduced) OBDD. The outer and inner Reduce sweeps work together to reduce this into a single file, F'_{outer}. When no more levels, x_j, need to be processed and the outer Reduce sweep has finished processing, then F'_{outer} contains the final reduced BDD of all nested operations.

As mentioned above, the inner Apply sweep needs its priority queue $Q_{inner:\downarrow}$ to be prepopulated with all relevant roots. This is done in Cases 1 and 2b in the outer and Case 3 in the inner Reduce sweeps.

Since the result of the inner Apply and Reduce sweeps replaces the entire set of nodes in F'_{outer}, the priority queue $Q_{inner:\downarrow}$ not only needs to be populated with requests for the nodes that need to be changed but also with requests for the nodes one wishes to keep (see also Figs. 9e and 11a). This makes the inner Apply sweep not only compute the desired result but also act as a mark-and-sweep garbage collection. On the first glance, these additional non-modifying requests may seem too costly – especially if most requests do not modify subtrees. In practice, 33.3% of all requests created throughout our benchmarks in Sect. 4 are subtree modifying. For each benchmark instance, 23.0% of all requests modify subtrees on average (median 35.6%). That is, a reasonable number of all requests (and hence BDD nodes processed) change the subgraph in F'_{outer}.

3.1   Complexity of Nested Sweeping

As mentioned in the description of the outer Apply sweep, nested sweeping works for multiple inputs. In this work, it suffices to assume it only has to deal with a single BDD f of N nodes as also depicted in Figs. 6 and 9.

Lemma 1. *A single BDD f with N nodes can be transposed in $\Theta(sort(N))$ I/Os and time and $\Theta(N)$ space.*

Proof. In $\Theta(\mathrm{scan}(N))$ I/Os and time iterate over and split all nodes v in-order into the two arcs $v.\mathtt{uid}\text{-}\rightarrow v.\mathtt{low}$ and $v.\mathtt{uid}\longrightarrow v.\mathtt{high}$. Sort these $2N$ arcs on their target using $\Theta(sort(N))$ I/Os and time and linear space to transposes them.

Lemma 2. *Ignoring the work done within the inner sweeps, the outer Reduce sweep costs $\Theta(sort(N))$ I/Os and time and requires $\mathcal{O}(N)$ space.*

Proof. This follows from the complexity of the Reduce algorithm in [49] (based on [6]) and the constant extra time spent for each of the N' nodes to resolve the additional logic in Cases 1 and 2 of the outer Reduce sweep.

By combining Lemmas 1 and 2 together with the fact that the last invocation of the inner Apply and Reduce sweeps constructs, together with the outer Reduce sweep, the output of size T, we obtain the following lower bound on the complexity of nested sweeping.

Corollary 1. *Nested Sweeping uses $\Omega(N+T)$ space and $\Omega(sort(N+T))$ time and I/Os where N and T are the size of the input and output, respectively.*

In particular for the $\mathtt{exists}$ BDD operation, let T_j be the size of F'_{outer} when the inner Apply sweep is invoked at level x_j.

Lemma 3. *A single invocation of the inner Apply and Reduce sweeps at x_j costs $\Theta(sort(N'+T_j^2))$ I/Os and time and uses $O(N'+T_j^2)$ space.*

Proof. As in [49], the algorithm's complexity depends on the number of elements placed in the priority queues [4]. In particular, a single nested $\mathtt{or}$ sweep deals with up to $2N$ arcs from F_{outer}. On top of these, it also processes up to $2T_j^2$ arcs created during the product construction of F'_{outer}.

Since nested sweeping closely simulates the (parallelised) recursive BDD algorithm in Fig. 2, one should expect it achieves, similar to the algorithms in [49], major improvements in the number of I/Os at the cost of a log-factor in the running time when compared to the conventional recursive algorithms. This is indeed the case.

Proposition 1. *Quantification of a set of variables, X, is computable with nested sweeping in $O(sort(N^{2^{|X|}}))$ I/Os and time and $O(N^{2^{|X|}})$ space.*

Proof. Due to Lemma 1, F_{outer} from the outer Apply sweep has up to $2N$ arcs. This is also the size of F'_{outer} without any inner sweeps. Each inner Apply and Reduce sweep may increase the size of F'_{outer} quadratically. The result follows from Lemmas 3 to 2. ∎

Asymptotically, this is not an improvement over just quantifying each variable one-by-one using the algorithm already proposed in the full version of [49]. Yet, doing so would involve $2|X|$ sweeps over *all* levels of the input whereas, as highlighted in Figs. 7 and 9, nested sweeping only processes levels below each quantified variable. This difference is also evident in practice: throughout our benchmarks in Sect. 4, when quantifying with nested sweeping rather than each variable independently, the total number of requests processed with the **or** operation decreases by 13.9% while the share of 2-ary product constructions increases from 57.3% to 66.6%.

3.2   Optimisations for Nested Sweeping

While nested sweeping as described above is an improvement over previous work in [49], there are multiple avenues to further improve its performance in practice.

Bail-Out of Inner Sweep: There is no need for the outer Reduce sweep to invoke the inner sweeps if $Q_{inner:\downarrow}$ only contains requests that preserve subtrees, i.e. if Case 1 in the outer Reduce did not create any requests that manipulate the accumulated OBDD in F'_{outer}. On level x_j, such requests can stem from a redundant node t' being suppressed. For **exists**, this may also occur due to either t'.**low** or t'.**high** being the $\bot$ terminal, which is neutral for the **or** operation, or being $\top$, which is short-circuiting it.

In this case, the entire content of $Q_{inner:\downarrow}$ can be redistributed between $Q_{outer:\uparrow}$ and $Q_{inner:\downarrow}$ for the next deepest to be quantified level, x_i. After doing so, the outer Reduce sweep can immediately proceed processing the next level.

For **exists**, any short-circuiting by the **or** operation in Case 1 of the outer Reduce sweep can kill off some subtrees in F'_{outer}. In this case, one cannot skip the last invocation of the inner sweeps. Otherwise, the final result F'_{outer} could include dead nodes. Yet, even so, one can instead of the expensive top-down algorithm, e.g. **or** for **exists**, invoke the inner Apply sweep with a much simpler (and therefore faster) mark-and-sweep algorithm.

In practice, 75.6% of all nested sweeps in our benchmarks (see Sect. 4 for their presentation) are skippable. For each benchmark, between 6.8% and 93.5% of all nested computations were skipped with an average of 59.2% (median of 81.0%). The number of nested levels depends on the problem domain and its instance. For the Garden of Eden (GoE) benchmark, only 26.8% of all nested computations were skipped on average (median 29.0%), whereas 82.7% (median 84.3%) of all levels of the Quantified Boolean Formulas (QBFs) could be skipped.

Root Requests Sorter: Instead of making the outer Reduce sweep push requests directly into $Q_{inner:\downarrow}$, it can push it into an intermediate list of requests, $L_{outer:\downarrow}$. The content of $L_{outer:\downarrow}$ is sorted using the same ordering as $Q_{inner:\downarrow}$ as the inner Apply sweep is invoked, to then merge it on the fly with the inner Apply sweep's priority queue. This allows one to postpone initialising this priority queue until the inner Apply sweep is invoked. This has multiple benefits:

- $Q_{inner:\downarrow}$, resp. $Q_{inner:\uparrow}$, only exists and uses internal memory during the inner Apply sweep, resp. the inner Reduce sweep. Hence, the memory otherwise dedicated to $Q_{inner:\downarrow}$ can be used in the outer Reduce sweep for the Reduce's per-level data structures in [49]. Furthermore, this also increases the amount of space available to the inner Reduce sweep. Hence, this ought to improve the running time of both the inner and the outer Reduce sweeps.
- If $Q_{inner:\downarrow}$, resp. $Q_{inner:\uparrow}$, is initialised for each inner Apply sweep, resp. inner Reduce sweep, then the monotonic and faster *levelised priority queue* in the full version of [49] can be used instead of a regular non-monotonic priority queue.
- Levelised cuts [51] bound the size of the priority queues in each individual inner Apply and Reduce sweep. Hence, for each nested sweep, one can, if it is safe to do so, replace $Q_{inner:\downarrow}$ and/or $Q_{inner:\uparrow}$ with a faster internal memory variant.
- Levelised random access [50] changes the order in which nodes are resolved on each level. This has to be accommodated for in the sorting predicate in $Q_{inner:\downarrow}$. Hence, $L_{outer:\downarrow}$ allows for this optimisation to be applied for each invocation of the inner Apply sweep depending on the width of F'_{outer}.

Furthermore, levelised cuts not only bound the size of $Q_{outer:\uparrow}$ in the outer Reduce sweep but also the size of $L_{outer:\downarrow}$. Hence, while deciding whether $Q_{outer:\uparrow}$ fits into memory, one can also decide whether $L_{outer:\downarrow}$ does.

All in all, this allows the optimisations in [50,51] to be applied on a sweep-by-sweep basis. In practice, if one neither uses faster internal-memory variants of $Q_{inner:\downarrow}$ and $Q_{inner:\uparrow}$ nor levelised random access, then Adiar needs a total of 32.1 h to solve 145 out of the 147 benchmarks in Sect. 4. Using these two optimisations shaves 13.0 h off the total computation time (speedup of 1.68). For each individual instance, this improves Adiar's performance between a factor of 1.07 and 5.05 (1.77 on average). Furthermore, without $L_{outer:\downarrow}$, the exponential blow-up in Proposition 1 implies $Q_{inner:\downarrow}$ would almost always have to use external memory. As the optimisations in [49–51] would then not be applicable, one would expect a slowdown of several orders of magnitude similar to [51].

4   Experimental Evaluation

We have implemented the nested sweeping framework in Sect. 3 in Adiar v2.0. To evaluate the impact of this addition, we have run experiments aiming at answering the following three research questions:

1. How does nested sweeping compare to the repeated use of the single-variable quantification from the full version of [49]?
2. How does Adiar with nested sweeping compare to the external memory BDD package, CAL [43]?
3. How does Adiar with nested sweeping compare to conventional BDD packages that use depth-first recursion and memoisation [9,20,25,31,52]?

As we are only concerned with the design of BDD algorithms, not when and how to use them, we will not compare to non-BDD based approaches.

4.1   Benchmarks

We have implemented the following two benchmarks that rely on multi-variable quantification. Similar to [48,49], all benchmarks have been implemented on top of C++-templated adapters for each BDD package. This makes each BDD package run the exact same set of operations without introducing any indirection. The source code for all benchmarks can be found at the following url:

github.com/ssoelvsten/bdd-benchmark

QBF Solving: Given a Quantified Boolean Formula (QBF) in the QCIR [54] format, each gate of the given circuit is recursively transformed into a BDD. For inputs, we use the 102 encodings from [44] of 2-player games on a grid. In our experience, the symbolic style of these inputs makes them well suited to be solved with BDDs. Hence, they provide a typical use-case of quantification in BDDs. Furthermore, though these inputs are not in CNF they are in prenex form. In practice, resolving these prenex quantifications at the end is computationally much more expensive than computing the to be quantified circuit, i.e. the *matrix*.

Based on preliminary experiments, we use a variable order based on a depth-first traversal of the given circuit. In the prenex, we merge adjacent blocks with the same quantifier to increase the number of concurrently quantified variables.

Garden-of-Eden: A *Garden-of-Eden* [39] (GoE) is any configuration of a cellular automaton without a predecessor. For Conway's Game of Life on a grid of size $n_r \times n_c$, this can be encoded as a relation with $(n_r + 2) \cdot (n_c + 2)$ *previous*-state variables, $\vec{x}$, and $n_r \cdot n_c$ *next*-state variables, $\vec{x'}$. Next-state variables, $x_i' \in \vec{x'}$, in the BDD follow a row-major order. Previous-state variables, $x_i \in \vec{x}$, are interleaved to directly precede their respective next-state variable, x_i'. A GoE corresponds to an unsatisfying assignment to $x_i' \in \vec{x'}$ after having quantified all previous-state variables, $\vec{x}$.

Recent results show there exists no GoE for Conway's Game of Life of size 8×8 or smaller [8]. Hence, the BDD for the $n_r \times n_c \leq 8 \times 8$ sized transition relation will collapse to $\top$ after existential quantification. Yet, the row-major encoding of the transition relation itself only requires a polynomially sized BDD. Hence, the complexity of this problem manifests during the existential quantification.

4.2 Hardware and Settings

As in [48–51], we have run our experiments on the *Grendel* cluster at the Centre for Scientific Computing Aarhus. In particular, we ran both benchmarks on machines with 48-core 3.0 GHz Intel Xeon Gold 6248R processors, 384 GiB of RAM, 3.5 TiB of SSD disk, and run Rocky Linux (Kernel 4.18.0–513). All code was compiled with GCC 10.1.0 or `rustc` 1.72.1. Each BDD package was given $\frac{9}{10}$th of the available RAM, i.e. 345 GiB, leaving $\frac{1}{10}$th to other data structures and the operating system. Next to that, the BDD packages use a single thread and their default/recommended settings.

Note that these machines have vasts amounts of memory. This is to ensure that depth-first implementations are not slowed down by external factors. If less memory is available, then depth-first implementations would have to run multiple garbage collections to stay within the memory limits (cf. the largest instances solved by BuDDy [31] in Fig. 14a). This, in turn, clears their memoisation tables and forces them to recompute previous results. Furthermore, this large amount of memory ensures they can solve larger problems without using the swap partition. If they had to use it then they would slow down by about two orders of magnitude (see the full paper of [49] for an example). Hence, machines of this scale allow us to measure the algorithms' running time without the noise otherwise introduced by their execution environment. Finally, this biases the running time in favour of the depth-first implementations, which in turn makes the numbers we report on Adiar's relative performance close to the worst-case.

Adiar does not yet support dynamic variable reordering. Hence, we have disabled reordering for all other BDD packages as part of these experiments.

4.3 Experimental Results

The computing cluster's scheduler does not let many long-running jobs run concurrently. To obtain all 1176 data points reported below within only a few months, we had to place each of the 147 instances in buckets of instances with a common timeout. In particular, an instance is placed in the bucket with the smallest timeout that is four times larger than Adiar needed during preliminary experiments. That is, a BDD package timing out should only be understood as it (possibly) being considerably slower than Adiar.

Depending on an instance bucket placement, running time measurements were made 1 to 3 times. Due to node failures on the cluster, Adiar with nested sweeping was run once more, resulting in its measurements being repeated on many instances 4 times. On average, all data points had 3.0 measurements.

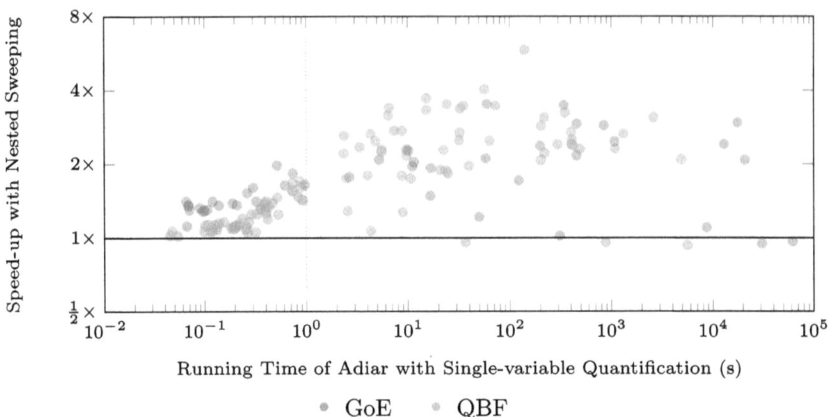

Fig. 12. Relative time ($T_{\mathrm{old}}/T_{\mathrm{new}}$) of quantification with nested sweeping (T_{new}) compared to the previous repeated single-variable quantification (T_{old}).

Similar to [48,49,51], we report for each benchmark the minimum time recorded as it is the measurement with the least noise [16]. Average ratios are aggregated using the geometric mean.

Adiar needs less than 1 s to solve 27 out of the 45 GoE instances, resp. 50 out of the 102 QBF instances. As will become evident later with Figs. 13 and 14, this makes them so small that they are not within the current scope of Adiar. For completeness, we still show and discuss these results.

RQ 1: Improvement by Nested Sweeping. Figure 12 shows the speed-up of using Adiar with nested sweeping relative to quantifying each variable individually. Across all instance sizes, nested sweeping is in general an improvement in performance. We have recorded a slowdown of up a factor of 1.05 for 5 instances. Yet, we also recorded speed-ups up to a factor of 5.88 for the 142 remaining instances. On average, performance improves by a factor of 1.7 for both QBF and GoE. The total computation time was decreased by 21% from 49.4 h to 39.1 h.

RQ 2: Comparison to CAL. To the best of our knowledge, CAL [43] (based on [7,40]) is the only other BDD package also designed to manipulate BDDs larger than main memory. To do so, it uses breadth-first algorithms that should work well with BDDs stored on disk via the operating system's swap memory [43]. CAL also includes algorithms to support multi-variable quantification [43]. For more details, see [43] and Sect. 5. The machines for our experiments provide a 48 GiB swap partition, i.e. a 12.5% increase in available space.

Preliminary experiments indicated CAL's breadth-first algorithms are much slower than Adiar's time-forward processing. Hence, we multiplied the timeout for CAL by a factor of 3. But as is evident in Fig. 13, this increase turned out to still overestimate CAL's performance on larger instances. Hence, the running

Table 1. Time taken and the average ratio between Adiar and CAL for the 124 commonly solved instances. The average (geometric mean) pertains only to instances where Adiar needed at least 1 s to solve them. Ratios larger than 1.00 means Adiar is faster.

	Time		# Solved		Avg. Ratio (1+s)	
	• GoE	• QBF	• GoE	• QBF	• GoE	• QBF
Adiar	7431.9 s	688.0 s	45	102	–	–
CAL	184688.3 s	295660.0 s	38	86	5.0	25.2

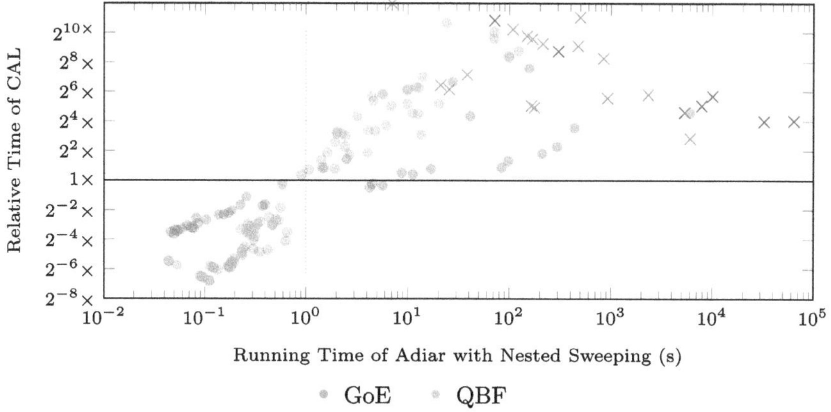

Fig. 13. Relative performance ($T_{\mathrm{CAL}}/T_{\mathrm{Adiar}}$) of CAL ($T_{\mathrm{CAL}}$) compared to Adiar with Nested Sweeping (T_{Adiar}). Time-/Memouts are marked as crosses.

times and averages in Fig. 13 and Table 1 pertain only to the 124 instances which CAL can solve within the given RAM, SWAP, and the time limits.

Even though this discards the instances where CAL struggles, i.e. the data points that remain are in CAL's favour, Fig. 13 shows Adiar heavily outperforms CAL for instances where Adiar takes 1 s or longer to solve. Where CAL uses 133.4 h to solve 124 instances, Adiar, by solving them in only 2.3 h, is 59.1 times faster. On these larger instances, CAL is on average 14.7 times slower than Adiar. As is evident in Fig. 13 and Table 1, Adiar especially outperforms CAL on the QBF benchmark. For example, the largest difference was measured for the `hex/hein_15_5x5-13` QBF instance, where CAL is 1081 times slower than the 71.2 s Adiar needs to solve it.

CAL is considerably faster for the instances where Adiar takes less than 1 s to solve. At this small scale, CAL uses the conventional depth-first algorithms [51] to sidestep the performance issues of its external memory breadth-first algorithms. Doing the same for Adiar is still left as future work [50,51].

RQ 3: Comparison To Depth-First Implementations. We have compared the performance of Adiar with BuDDy 2.4 [31], CUDD 3.0.0 [52], LibBDD 0.5 [9], OxiDD 0.6 [25], and Sylvan 1.8.1 [20]. Their individual performance relative to

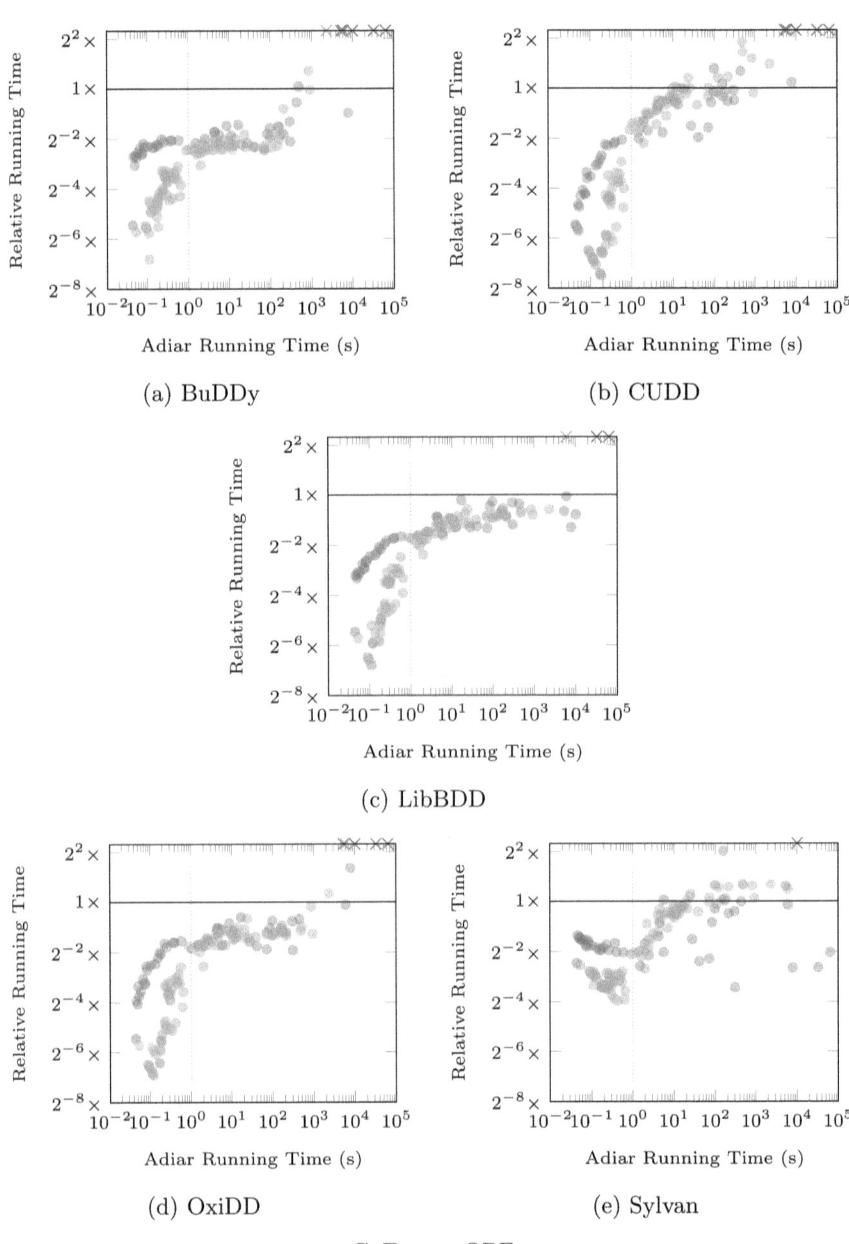

Fig. 14. Relative performance of depth-first implementations compared to Adiar with nested sweeping. Time-/Memouts are marked as crosses.

Adiar is shown in Fig. 14. Out of the 147 instances, 140 are solved by all depth-first BDD packages, i.e. the remaining 7 instances have at least one BDD package running out of memory (MO) or time (TO). Adiar solves all of them. Running out of time is most likely due to repeated need for garbage collection, which essentially is equivalent to an MO. Yet for fairness, Table 2 shows the total time for these 140 commonly solved instances. The average ratio, on the other hand, pertains to all instances solved by the respective BDD package.

The relative running time of BuDDy in Fig. 14a and OxiDD in Fig. 14d shows that Adiar's performance can be divided into three categories: the *small* instances that takes Adiar less than 1 s to solve but is out of its (current) scope, the *moderate* instances where Adiar needs between 1 and 10^3 s to solve and it is up to a constant factor of 4 slower than other BDD packages, and the *large* instances beyond 10^3 s where other BDD packages slow down compared to Adiar due to limited internal memory and repeated garbage collection. While the distinction is not as clear for CUDD in Fig. 14b, LibBDD in Fig. 14c, and Sylvan in Fig. 14e, they also follow the same trend. This relative performance is similar to our previous results in [48–51]. That is, nested sweeping allows Adiar to compute quantifications at no additional cost to our previous work.

As shown in Table 2, Adiar solves the 40 common GoE instances in 2.7 h and the 100 common QBF instances in 1.25 h. This makes it as fast as CUDD for both sets of benchmarks and faster than OxiDD on GoE and Sylvan on QBF. BuDDy is faster for both benchmarks but its use of 32 bits indices limits the maximum BDD size it supports and hence the number of instances it can solve. As shown in Fig. 14c, LibBDD is the only BDD package that is consistently faster than Adiar (ignoring its three MOs). Most likely, this is due to its lack of a shared unique node table. This improves its cache locality and removes the need for expensive garbage collections. Hence, LibBDD can either fit its BDD computations into the internal memory (and it is faster than Adiar) or it aborts.

As is evident from Fig. 14e, Sylvan is comparatively good at some of the larger GoE instances, thereby beating all other BDD packages in the total time

Table 2. Total time needed by Adiar and conventional depth-first implementations to solve the 140 commonly solved instances. The average (geometric mean) covers all instances that were commonly solved by all BDD packages and where Adiar needed at least 1 s to solve. Ratios smaller than 1.00 means Adiar is slower.

	Time		# Solved		Avg. Ratio (1+s)	
	• GoE	• QBF	• GoE	• QBF	• GoE	• QBF
Adiar	9655.7 s	4499.4 s	45	102	–	–
BuDDy	4725.5 s	3793.1 s	40	100	0.30	0.25
CUDD	10892.8 s	4591.9 s	40	101	0.61	0.75
LibBDD	4365.7 s	2687.3 s	43	101	0.54	0.45
OxiDD	21223.9 s	2379.6 s	41	101	0.48	0.39
Sylvan	2925.4 s	5841.4 s	44	102	0.46	0.70

to solve its subset of the GoE benchmarks. Furthermore, it is the only other BDD package able to solve all QBF instances within the given time limit. This is in parts thanks to its small memory footprint per BDD node [20]. Yet, Sylvan requires a total of 5.0 h to solve all 102 QBF instances whereas Adiar only needed 3.6 h, making Sylvan a factor of 1.4 times slower than Adiar. If Sylvan was given access to more than a single thread, then it would, of course, be expected to be at least as fast if not much faster than Adiar [20].

Considering the 140 commonly solved instances are, by definition, not the main focus of the algorithms in Adiar and that the experiments have been designed in favour of the conventional BDD packages, it is not surprising that Adiar does not greatly outshine the other BDD packages. Even so, it is faster than some implementations in some cases – despite storing BDDs on disk. Most importantly, it solves more instances than any other BDD package, witnessed by the timeouts in Fig. 14.

5   Related Work

Many other implementations of BDDs also support quantification of multiple variables [9,20,25,31,43,52]. All these are based on a nested (inner) operation being accumulated in an (outer) traversal of the input; the nested sweeping framework achieves the same within the time-forward processing paradigm [4, 17].

The CAL [43] BDD package (based on [7,40]) is to the best of our knowledge the only implementation of BDDs also designed to compute on BDDs whose size exceed main memory. To do so, it uses breadth-first algorithms that are resolved level by level. For each level it still follows the conventional approach: a unique node table is used to manage BDD nodes while a polynomial running time is guaranteed by use of a memoisation table. These per-level hash tables, both in theory and in practice, put an upper bound on the maximum BDD width that CAL can support with a certain amount of internal memory [6].

To skip redundant computations, its quantification operation also required additional ideas particular to the design of CAL. By the nature of nested sweeping, our proposed algorithm is, unlike CAL, not able to skip redundant computations. Furthermore, the lack of a unique node table in Adiar requires our algorithms to retraverse and copy the subtrees that are unchanged. Even so, as evident in Sect. 4, Adiar with nested sweeping outperforms CAL by up to several orders of magnitude. Moreover, the I/O-efficient approach in [6,49], and by extension the ones in this work, are, unlike CAL, I/O efficient despite a BDD's level is wider than main memory.

6   Conclusions and Future Work

Each sweep in [49] is independent of the others. Using only this approach, one can only quantify a single variable a time but not multiple at once. In this work, we enable multi-variable quantification with the *nested sweeping* framework. Here, multiple sweeps work together: each sweep forwards information within priority queues to itself, its parent, or its child in a recursion stack.

In practice, nested sweeping has improved the total time that Adiar needs to solve our quantification benchmarks by 21%. On average, it improves the running time of each instance by a factor of 1.7. This allows us to extend our previous results in [48–51] to Adiar's quantification operations: ignoring small instances, Adiar is at most 4 times slower than conventional depth-first implementations. Adiar even outperforms depth-first implementations as they get closer to the limits of internal memory. As Adiar's nested sweeping algorithms are implemented on-top of the I/O-efficient data structures that were also used in [48–51], its performance is unaffected by a limited internal memory [49]. For example, whereas CUDD [52] could solve 141 out of our 147 benchmark instances in 5.6 h, Adiar needed only 4.6 h to do the same and it could also solve the remaining 6 instances. Adiar is also faster, often by several orders of magnitude, than the only other external memory BDD package, CAL [43].

The nested sweeping framework has already been generalised to pave the way for the implementation of other multi-recursive BDD operations. We have used this to implement the relational product [47]; doing so requires several additional optimisations for its variable relabelling and its combined **and-exists** [47]. In [47], we also provide an evaluation on a large collection of symbolic model checking experiments. We also hope to use nested sweeping for functional composition and as the foundation for novel I/O-efficient variable reordering procedures [45]. Finally, nested sweeping opens up the possiblity to create an I/O-efficient implementation of other types of decision diagrams. For example, both Quantum Multiple-valued Decision Diagrams [36] and Polynomial Boolean Rings [11] require nested sweeps to implement their multiplication operations.

Acknowledgements. Thanks to the Centre for Scientific Computing, Aarhus, for running our benchmarks on the Grendel cluster and thanks to Marijn Heule and Randal E. Bryant for suggesting the Garden of Eden problem and their ideas on how to encode it.

Data Availability Statement. The source code for all our benchmarks is available on the following GitHub repository at commit **51b1375**: github.com/ssoelvsten/bdd-benchmark The source code is also available at doi:10.5281/zenodo.4718224 on Zenodo. The raw data and its analysis is available at doi:10.5281/zenodo.17054026.

References

1. Aggarwal, A., Vitter, J.S.: The input/output complexity of sorting and related problems. Commun. ACM **31**(9), 1116–1127 (1988). https://doi.org/10.1145/48529.48535
2. Akers, S.B.: Binary decision diagrams. IEEE Trans. Comput. **C-27**(6), 509–516 (1978). https://doi.org/10.1109/TC.1978.1675141
3. Amparore, E.G., Donatelli, S., Gallà, F.: starMC: an automata based CTL* model checker. PeerJ Comput. Sci. **8**, e823 (2022)

4. Arge, L.: The buffer tree: a new technique for optimal I/O-algorithms. In: Akl, S.G., Dehne, F., Sack, J.-R., Santoro, N. (eds.) WADS 1995. LNCS, vol. 955, pp. 334–345. Springer, Heidelberg (1995). https://doi.org/10.1007/3-540-60220-8_74

5. Arge, L.: The I/O-complexity of ordered binary-decision diagram manipulation. In: Proceedings of International Symposium on Algorithms and Computations, ISAAC'95, pp. 82–91 (1995)

6. Arge, L.: The I/O-complexity of ordered binary-decision diagram manipulation. In: Efficient External-Memory Data Structures and Applications (PhD Thesis), pp. 123–145. Aarhus Universitet, Datalogisk Institut (1996)

7. Ashar, P., Cheong, M.: Efficient breadth-first manipulation of binary decision diagrams. In: IEEE/ACM International Conference on Computer-Aided Design (ICCAD), pp. 622–627. IEEE Computer Society Press (1994). https://doi.org/10.1109/ICCAD.1994.629886

8. Beer, R.D.: Cultivating the garden of Eden. arXiv (2022). https://doi.org/10.48550/arXiv.2210.07837

9. Beneš, N., Brim, L., Kadlecaj, J., Pastva, S., Šafránek, D.: AEON: attractor bifurcation analysis of parametrised Boolean networks. In: Lahiri, S.K., Wang, C. (eds.) CAV 2020. LNCS, vol. 12224, pp. 569–581. Springer, Cham (2020). https://doi.org/10.1007/978-3-030-53288-8_28

10. Brace, K.S., Rudell, R.L., Bryant, R.E.: Efficient implementation of a BDD package. In: 27th Design Automation Conference (DAC), pp. 40–45. Association for Computing Machinery (1990). https://doi.org/10.1109/DAC.1990.114826

11. Brickenstein, M., Dreyer, A.: PolyBoRi: a framework for Gröbner-basis computations with Boolean polynomials. J. Symb. Comput. **44**(9), 1326–1345 (2009). https://doi.org/10.1016/j.jsc.2008.02.017

12. Bryant, R.E.: Graph-based algorithms for boolean function manipulation. IEEE Trans. Comput. **C-35**(8), 677–691 (1986)

13. Bryant, R.E., Biere, A., Heule, M.J.H.: Clausal proofs for pseudo-boolean reasoning. In: TACAS 2022. LNCS, vol. 13243, pp. 443–461. Springer, Cham (2022). https://doi.org/10.1007/978-3-030-99524-9_25

14. Bryant, R.E., Heule, M.J.H.: Dual proof generation for quantified boolean formulas with a BDD-based solver. In: Platzer, A., Sutcliffe, G. (eds.) CADE 2021. LNCS (LNAI), vol. 12699, pp. 433–449. Springer, Cham (2021). https://doi.org/10.1007/978-3-030-79876-5_25

15. Bryant, R.E., Heule, M.J.H.: Generating extended resolution proofs with a BDD-based SAT solver. In: TACAS 2021. LNCS, vol. 12651, pp. 76–93. Springer, Cham (2021). https://doi.org/10.1007/978-3-030-72016-2_5

16. Chen, J., Revels, J.: Robust benchmarking in noisy environments. arXiv (2016). https://arxiv.org/abs/1608.04295

17. Chiang, Y.J., Goodrich, M.T., Grove, E.F., Tamassia, R., Vengroff, D.E., Vitter, J.S.: External-memory graph algorithms. In: Proceedings of the Sixth Annual ACM-SIAM Symposium on Discrete Algorithms, SODA '95, pp. 139–149. Society for Industrial and Applied Mathematics (1995)

18. Ciardo, G., Miner, A.S., Wan, M.: Advanced features in SMART: the stochastic model checking analyzer for reliability and timing. SIGMETRICS Perf. Eval. Rev. **36**(4), 58–63 (2009)

19. Cimatti, A., Clarke, E., Giunchiglia, F., Roveri, M.: NuSMV: a new symbolic model checker. Int. J. Softw. Tools Technol. Transfer **2**, 410–425 (2000). https://doi.org/10.1007/s100090050046

20. van Dijk, T., van de Pol, J.: Sylvan: multi-core framework for decision diagrams. Int. J. Softw. Tools Technol. Transfer 1–22 (2016). https://doi.org/10.1007/s10009-016-0433-2
21. Fried, D., Tabajara, L.M., Vardi, M.Y.: BDD-based boolean functional synthesis. In: Chaudhuri, S., Farzan, A. (eds.) CAV 2016. LNCS, vol. 9780, pp. 402–421. Springer, Cham (2016). https://doi.org/10.1007/978-3-319-41540-6_22
22. Gammie, P., van der Meyden, R.: MCK: model checking the logic of knowledge. In: Alur, R., Peled, D.A. (eds.) CAV 2004. LNCS, vol. 3114, pp. 479–483. Springer, Heidelberg (2004). https://doi.org/10.1007/978-3-540-27813-9_41
23. Gardner, M.: The fantastic combinations of John Conway's new solitaire game "life". Sci. Am. **223**(4), 120–123 (1970). https://doi.org/10.1038/scientificamerican1070-120
24. He, L., Liu, G.: Petri net based symbolic model checking for computation tree logic of knowledge. arXiv (2020). https://arxiv.org/abs/2012.10126
25. Husung, N., Dubslaff, C., Hermanns, H., Köhl, M.A.: OxiDD: a safe, concurrent, modular, and performant decision diagram framework in Rust. In: Tools and Algorithms for the Construction and Analysis of Systems (TACAS'24). LNCS, vol. 14572. Springer, Heidelberg (2024). https://doi.org/10.1007/978-3-031-57256-2_13
26. Kant, G., Laarman, A., Meijer, J., van de Pol, J., Blom, S., van Dijk, T.: LTSmin: high-performance language-independent model checking. In: Baier, C., Tinelli, C. (eds.) TACAS 2015. LNCS, vol. 9035, pp. 692–707. Springer, Heidelberg (2015). https://doi.org/10.1007/978-3-662-46681-0_61
27. Karplus, K.: Representing boolean functions with if-then-else DAGs. Technical report. University of California at Santa Cruz, USA (1988)
28. Klarlund, N., Rauhe, T.: BDD algorithms and cache misses. In: BRICS Report Series, vol. 26 (1996). https://doi.org/10.7146/brics.v3i26.20007
29. Lee, C.Y.: Representation of switching circuits by binary-decision programs. Bell Syst. Techn. J. **38**(4), 985–999 (1959). https://doi.org/10.1002/j.1538-7305.1959.tb01585.x
30. Lin, Y., Tabajara, L.M., Vardi, M.Y.: ZDD boolean synthesis. In: TACAS 2022. LNCS, vol. 13243, pp. 64–83. Springer, Cham (2022). https://doi.org/10.1007/978-3-030-99524-9_4
31. Lind-Nielsen, J.: BuDDy: a binary decision diagram package. Technical report, Department of Information Technology, Technical University of Denmark (1999)
32. Lomuscio, A., Qu, H., Raimondi, F.: MCMAS: an open-source model checker for the verification of multi-agent systems. Int. J. Softw. Tools Technol. Transfer **19**(1), 9–30 (2015). https://doi.org/10.1007/s10009-015-0378-x
33. Madsen, M., Van de Pol, J.: Polymorphic types and effects with Boolean unification. Proc. ACM Program. Lang. **4**(OOPSLA) (2020). https://doi.org/10.1145/3428222
34. Madsen, M., Van de Pol, J., Henriksen, T.: Fast and efficient boolean unification for Hindley-Milner-style type and effect systems. Proc. ACM Program. Lang. **7**(OOPSLA 2) (2023). https://doi.org/10.1145/3622816
35. Michaud, T., Colange, M.: Reactive synthesis from LTL specification with Spot. In: 7th Workshop on Synthesis, SYNT@ CAV, vol. 5 (2018)
36. Miller, D., Thornton, M.: QMDD: a decision diagram structure for reversible and quantum circuits. In: 36th International Symposium on Multiple-Valued Logic, pp. 30–36 (2006). https://doi.org/10.1109/ISMVL.2006.35
37. Minato, S.i., Ishihara, S.: Streaming BDD manipulation for large-scale combinatorial problems. In: Design, Automation and Test in Europe Conference and Exhibition, pp. 702–707 (2001). https://doi.org/10.1109/DATE.2001.915104

38. Minato, S.i., Ishiura, N., Yajima, S.: Shared binary decision diagram with attributed edges for efficient Boolean function manipulation. In: 27th Design Automation Conference, pp. 52–57. Association for Computing Machinery (1990). https://doi.org/10.1145/123186.123225
39. Moore, E.F.: Machine models of self-reproduction. In: Proceedings of Symposia in Applied Mathematics, vol. 14, pp. 17–33 (1962). https://doi.org/10.1090/psapm/014
40. Ochi, H., Yasuoka, K., Yajima, S.: Breadth-first manipulation of very large binary-decision diagrams. In: International Conference on Computer Aided Design (ICCAD), pp. 48–55. IEEE Computer Society Press (1993). https://doi.org/10.1109/ICCAD.1993.580030
41. Pastva, S., Henzinger, T.: Binary decision diagrams on modern hardware. In: Conference on Formal Methods in Computer-Aided Design, pp. 122–131 (2023)
42. Renkin, F., Schlehuber-Caissier, P., Duret-Lutz, A., Pommellet, A.: Dissecting ltlsynt. Formal Methods Syst. Des. **61**, 248–289 (2022). https://doi.org/10.1007/s10703-022-00407-6
43. Sanghavi, J.V., Ranjan, R.K., Brayton, R.K., Sangiovanni-Vincentelli, A.: High performance BDD package by exploiting memory hierarchy. In: 33rd Design Automation Conference (DAC), pp. 635–640. Association for Computing Machinery (1996). https://doi.org/10.1145/240518.240638
44. Shaik, I., Van de Pol, J.: Concise QBF encodings for games on a grid (extended version). arXiv (2023). https://doi.org/10.48550/arXiv.2303.16949
45. Sølvsten, S.C.: I/O-efficient Symbolic Model Checking. Phd thesis, Aarhus University (2025). https://soeg.kb.dk/permalink/45KBDK_KGL/1pioq0f/alma99126389524805763
46. Sølvsten, S.C., van de Pol, J.: Multi-variable quantification of BDDs in external memory using nested sweeping (extended paper). arXiv (2024). https://arxiv.org/abs/2408.14216
47. Sølvsten, S.C., van de Pol, J.: Symbolic model checking in external memory. arXiv (2025). https://arxiv.org/abs/2505.11229
48. Sølvsten, S.C., Van de Pol, J.: Adiar 1.1: Zero-suppressed decision diagrams in external memory. In: NASA Formal Methods Symposium. LNCS, vol. 13903, Springer, Heidelberg (2023). https://doi.org/10.1007/978-3-031-33170-1_28
49. Sølvsten, S.C., de Pol, J., Jakobsen, A.B., Thomasen, M.W.B.: Adiar binary decision diagrams in external memory. In: TACAS 2022. LNCS, vol. 13244, pp. 295–313. Springer, Cham (2022). https://doi.org/10.1007/978-3-030-99527-0_16
50. Sølvsten, S.C., Rysgaard, C.M., Van de Pol, J.: Random access on narrow decision diagrams in external memory. In: Model Checking Software, pp. 137–145. Springer, Heidelberg (2024). https://doi.org/10.1007/978-3-031-66149-5_7
51. Sølvsten, S.C., Van de Pol, J.: Predicting memory demands of BDD Operations using maximum graph cuts. In: André, É., Sun, J. (eds.) Automated Technology for Verification and Analysis. Lecture Notes in Computer Science, vol. 14216, pp. 72–92. Springer, Heidelberg (2023). https://doi.org/10.1007/978-3-031-45332-8_4
52. Somenzi, F.: CUDD: CU decision diagram package, 3.0. Technical report, University of Colorado at Boulder (2015)
53. Van Dijk, T., Van Abbema, F., Tomov, N.: Knor: reactive synthesis using oink. In: Tools and Algorithms for the Construction and Analysis of Systems. pp. 103–122. Springer, Heidelberg (2024). https://doi.org/10.1007/978-3-031-57246-3_7
54. QBF Gallery 2014: QCIR-G14: a non-prenex non-CNF format for quantified Boolean formulas (2014)

Probabilistic Verification for Modular Network-on-Chip Systems

Nick Waddoups[1]([✉]) [iD], Jonah Boe[2] [iD], Arnd Hartmanns[3] [iD], Prabal Basu[4] [iD], Sanghamitra Roy[1] [iD], Koushik Chakraborty[1] [iD], and Zhen Zhang[1] [iD]

[1] Utah State University, Logan, UT, USA
{nick.waddoups,sanghamitra.roy,koushik.chakraborty,
zhen.zhang}@usu.edu
[2] Hill Air Force Base, Davis County, UT, USA
[3] University of Twente, Enschede, The Netherlands
a.hartmanns@utwente.nl
[4] Cadence Design Systems, San Jose, CA, USA

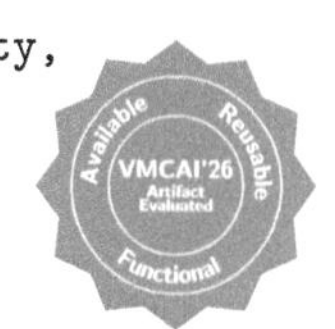

Abstract. Quantitative verification can provide deep insights into reliable Network-On-Chip (NoC) designs. It is critical to understanding and mitigating operational issues caused by power supply noise (PSN) early in the design process: fluctuations in network traffic in modern NoC designs cause dramatic variations in power delivery across the network, leading to unreliability and errors in data transfers. Further complicating these challenges, NoC designs vary widely in size, usage, and implementation. This case study paper presents a principled, systematic, and modular NoC modeling approach using the MODEST language that closely reflects the standard hierarchical design approach in digital systems. Using the MODEST TOOLSET, functional and quantitative correctness was established for several NoC models, all of which were instantiated from a generic modular router model. Specifically, this work verifies the functional correctness of a generic router, inter-router communication, and the entire NoC. Statistical model checking was used to verify PSN-related properties for NoCs of size up to 8×8.

1 Introduction

As modern digital systems grow in complexity, there is a need for efficient on-chip communications as the traditional shared data bus becomes overloaded when more sub-systems are added to a chip. *Network-on-Chip* (NoC) designs

N. Waddoups and Z. Zhang gratefully acknowledge the support of the ECE Department at Utah State University. A. Hartmanns was supported by the EU's Horizon 2020 R&I programme under MSCA grant 101008233 (MISSION), the Interreg North Sea project STORM_SAFE, and NWO VIDI grant VI.Vidi.223.110 (TruSTy). S. Roy and K. Chakraborty were supported in part by National Science Foundation (NSF) grant CNS-2106237. Any opinions, findings, and conclusions or recommendations expressed in this material are those of the author(s) and do not necessarily reflect the views of the NSF.

are a widely-used alternative in computer chips that improve communication throughput and efficiency of a system by allowing more parallel communication.

NoCs have attracted substantial research attention for improving both reliability and performance, especially in complex designs such as System-on-Chips (e.g., [32,48]). As modern semiconductor production integrates more subsystems on a single chip, from standard processing units to specialized hardware accelerators [40], NoCs are increasingly being adopted. However, their unique design challenges require correctness guarantees, particularly in safety-critical systems where faults can cause catastrophic failure [6,26].

One major challenge in NoC designs is *power supply noise* (PSN), which occurs with sudden shifts in traffic volume across the network. These shifts lead to fluctuations in the power drawn by each router and can result in data corruption or loss. Previous work [5] shows that PSN is magnified by the shift to smaller chip technology, with PSN increasing from 40% of the supply voltage to around 80% when moving an 8×8 NoC from 32 nm to 14 nm technologies.

Computer chips are designed in stages, starting with an architectural level specification, then moving onto an implementation in a hardware description language, and finishing with a physical layout before manufacturing. Design changes later in the cycle are significantly more costly than they would be if made earlier in the cycle. Incorporating probabilistic verification during the architectural stage of the NoC design flow allows for *early* quantification of PSN in order to mitigate issues. While formal verification of NoCs is an active area of study, there is only limited work on probabilistic verification of small-scale NoCs [34,38]. We argue that probabilistic verification is *required* to thoroughly check for PSN-related issues, as they are highly dependent on the inherent stochastic nature of network packet generation patterns.

Building on the second author's M.S. thesis work [7], this paper has the following major contributions:

1. A highly modular NoC model, written in the probabilistic modeling language MODEST [8,17], to allow easy construction of arbitrary-size NoCs and characterization of PSN (Sect. 4). Previous work [38] presents only probabilistic verification of a small-scale 2×2 NoC, modeled in a monolithic, non-modular manner, which does not scale. The parameterizable nature of our modular model enables conventional and probabilistic model checking of diverse NoC implementations.
2. Verification of functional correctness of a generic router, inter-router communication, and full NoC model (Sect. 5).
3. Comparative model-checking: By reconfiguring our modular 2×2 NoC model into one equivalent to the earlier monolithic 2×2 NoC [38], we identified a deep-buried discrepancy in the modeling of packet consumption (Sect. 7). After resolving the discrepancies, our model successfully reproduced the results from [38], validating our unique modular modeling approach.
4. Probabilistic verification of PSN-related properties for NoCs of size up to 8×8, demonstrating the scalability and flexibility of the modular approach, while uncovering PSN characteristics not captured in previous work [38].

5. An analysis of the impacts of routing algorithms, demonstrating the frameworks ability to characterize PSN effects and reveal routing and scheduling strategies to improve reliability early in the design process.

2   Related Work

Probabilistic Modeling Checking. Research and development of probabilistic verification tools has been carried out for decades, stemming from research in the 1980s [27]. Modern research has produced probabilistic verification tools such as MODEST [20], PRISM [29], STORM [23], among others [18,24,39,44]. They have been widely applied to fields from flight controllers [52] to hardware security and reliability [15,28,36,37] to synthetic biology [10,31,35,46]. The MODEST TOOLSET can analyze models with quantitative data [20] and includes a built-in statistical model checker [9] used to generate PSN probabilities in this work.

NoC Verification. Previous NoC verification work mainly focuses on functional correctness of NoCs. It has been used to verify the correct routing behavior [41, 47], check security properties [42], and evaluate performance [1]. These works have led to a *qualitative* understanding of NoCs, but cannot fully quantify critical properties involving reliability.

Probabilistic NoCs in Modest. Previous probabilistic verification for NoCs [34, 38] was performed using MODEST. [38] developed a 2×2 NoC to quantify the effects of PSN using probabilistic model checking. This model was developed as a proof of concept, but was modeled in a *monolithic* fashion where all behavior and state variables were explicitly declared. This style of modeling is not scalable, and mistakes are easy to make. The goal of this work is to significantly enhance the scalability and usability of the concepts in [38] to characterize PSN in NoCs.

3   Preliminaries

3.1   NoC Architecture

The NoC architecture presented in this paper is a variable-size square mesh network composed of individual routers. Figure 1a shows a 2×2 NoC composed of four individual routers and Fig. 1b shows an individual router.

A 2×2 NoC is made up of four routers in a grid pattern. Each router is responsible for receiving packets and routing them to their destination. Routers are denoted using r_i, where i is the router ID.

A router contains first-in-first-out (FIFO) fixed-size *buffers* to store incoming network packets and *channels* to send packets. Buffers are represented by small blocks ("▢▢▢") and channels by arrows ("→") in Fig. 1. Additionally, "×" indicates that the buffer and channel on that side of the router are not connected which occurs when a router is positioned at the edge of the network.

A specific buffer in a router is addressed as r_i^b, where b is the buffer label and i is the router ID. The buffer is described either by its position relative to a router, such as r_i^{North}, or generally, such as r_i^{Source}. In digital design, a fixed-size FIFO buffer has queue-like semantics where elements are inserted in the back and popped from the front. In our models, a channel represents physical wires that carry network data from one router to another.

Packets are used to communicate data between routers in a NoC, similar to packets in a traditional computer network. Starting at one router, a packet is moved from router to router through the network until it reaches its destination. Packets are introduced into and removed from the NoC by a *processing element* (PE) connected to each router via the local buffer and channel. A PE could be a CPU, accelerator core, memory cache, or other digital circuit. To move across the network, packets are (1) popped from the front of a buffer by a router, (2) routed through a channel towards their destination, and (3) pushed into the buffer connected to the channel.

As a digital circuit, NoCs are synchronized by a global clock. In our model, each channel can only be used once per clock cycle to send a packet. Additionally, in our model, each buffer can only accept one new packet per clock cycle. The sending of packets happens simultaneously (i.e., during the same clock cycle) between all NoC routers.

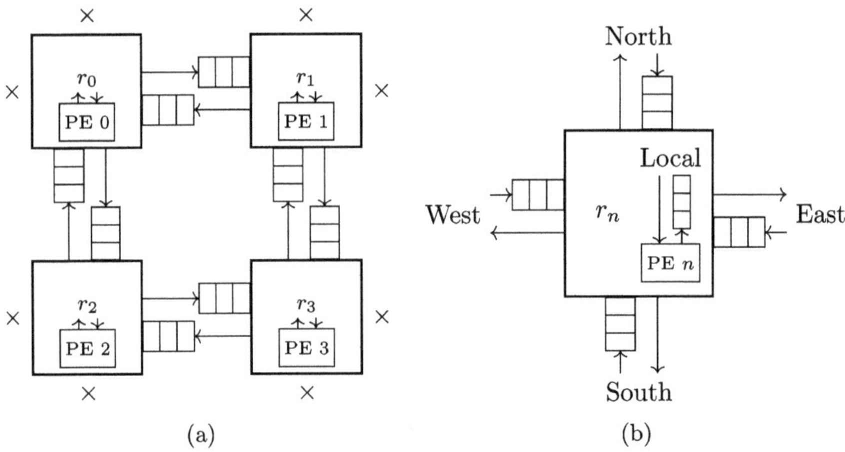

Fig. 1. A 2×2 NoC architecture and an individual router.

In digital design, a packet is often split up into smaller components known as *flits* [5] due to limitations of channel bit-width. When a packet is transferred from one router to another, it is broken up into flits that are transmitted contiguously and sequentially over a channel, with one flit transferred per clock cycle. The number of flits in a packet is determined by the bit width of the channel, the size of the packet in bits, and the protocol being used.

The NoC model in this paper assumes a *single-flit packet* that can be transferred between routers in one clock cycle for the following reasons:

1. This work is targeted at characterizing PSN *early* in the design cycle, before the implementation of the NoC is complete. As such, the channel width and packet size may not be known when this modeling is intended to take place.
2. This work focuses on characterizing PSN based on high-level traffic patterns, routing algorithms, and NoC size. As such, our main consideration is to look at the high-level of packet traffic, not the individual movements of flits in the system. The abstraction presented in this work forgoes verification of circuit specific details for scalability and flexibility, which is desirable early in the design cycle. For more in-depth methods that look at characterizing PSN later in the design cycle, see [5].
3. Using single-flit packets greatly improves the tractability of the state space for verification. Roughly, a single-flit packet model has a state space complexity of $O(S^n)$, where S is the number of states for each buffer, and n is the number of buffers. Modeling multi-flit packets introduces many other possible states for each buffer, which would worsen the state space explosion problem and render verification intractable for the model.

Because we assume a single-flit packet, *packets* and *flits* are equivalent in our model as they each carry only a small piece of data and can be transferred from router to router in one clock cycle.

3.2 Routing Algorithm

Similar to computer networking, the path a flit takes to get from the sender to the receiver depends on the implemented routing protocol. The NoC design presented in this work allows *any* NoC routing algorithm to be modeled.

Results shown in this paper use the *X-Y routing algorithm*: It moves flits first horizontally across the NoC until they are in the same column as the destination router, and then moves them vertically until they reach their destination. This routing algorithm was chosen because it is deterministic, deadlock free, and is the same algorithm used in [38]. Additionally, the modularity of this work provides a framework for implementing other routing algorithms. More details about our implementation of the X-Y routing algorithm are given in [51].

3.3 Router Arbitration

An important aspect of the router is the order in which each router's buffers are *serviced*, i.e. when a flit is removed from a buffer and sent through a channel to a destination. This order determines a router's fairness and is also critical to ensuring correct operation. An incorrect algorithm may deadlock, starve a particular router, or allow multiple flits to be routed per clock cycle across a single channel. We discuss routing correctness in Sect. 4.

Our NoC modeling framework uses CTL formulas of the form "forall globally" ($\forall\Box$) and "exists eventually" ($\exists\Diamond$) to ensure the functional correctness of a single router model and the composition to form an $n \times n$ NoC. *Probabilistic CTL* (PCTL) [19] is used to quantitatively check PSN-related properties. The PCTL syntax relevant to this work is $\mathbf{P}_{\sim q}(\lozenge^{\leqslant K}\Psi)$. That is, whether the probability that the state formula Ψ is eventually true within K steps lies within the interval specified by $\sim q$, where $\sim \in \{<, \leqslant, \geqslant, >\}$ and $q \in [0, 1]$. Additionally, the probabilistic operator $\mathbf{P}_{\sim q}$ can be replaced by $\mathbf{P}_{=?}$, indicating that the actual probability of the path formula $\lozenge^{\leqslant K}\Psi$ is of interest. For a complete definition of PCTL's syntax and semantics, see [30].

3.4 Power Supply Noise (PSN)

PSN has two components in a NoC: *resistive noise* – the product of the inherent resistance R of a circuit and the current I drawn, and *inductive noise* – the rate of change of current $\frac{\Delta I}{\Delta t}$ through the inherent inductance L of the circuit. The voltage drop due to PSN can radically degrade the delay of various on-chip circuit components, causing *timing errors*[1] in the network. Existing approaches to mitigate PSN are a far cry from a truly reliable NoC design paradigm, due to the lack of worst-case peak PSN guarantees [5,43].

Instead of directly measuring I, the model measures the activity of the router. Router activity is determined by the number of packets a router moves during a single clock cycle, ranging from zero (no packets) to five (all buffers used). This abstraction is supported by [5], which shows that high activity correlates to higher current, and an abrupt change in router activity leads to a high rate of current change. Like [34,38], PSN is tracked over clock cycles by introducing two counters, *resistiveNoise* and *inductiveNoise*. These counters represent the number of clock cycles where the router activity is over a user-specified router activity threshold T, where $T \in [0, 5]$.

Property 1 below is used to check the probability that the *resistiveNoise* counter will eventually exceed some threshold K within N clock cycles. Property 2 shows a similar property for determining the *inductiveNoise* probability within N clock cycles. Note that clock cycle progress is a *reward annotation* to certain transitions, i.e., accumulate(clk), instead of being encoded in the structure of an expanded state space.

$$\mathbf{P}_{=?}(\lozenge^{[\text{accumulate}(clk)\leqslant N]} \; resistiveNoise \geqslant K) \tag{1}$$

$$\mathbf{P}_{=?}(\lozenge^{[\text{accumulate}(clk)\leqslant N]} \; inductiveNoise \geqslant K) \tag{2}$$

This work characterizes PSN using the same method as [34,38]. This method abstracts away circuit-specific details in order to characterize PSN at a behavioral level. This is done because (1) early in the design process, engineers are often more concerned with high-level behavior, (2) circuit-level netlist models are not available early in the design process, and (3) once made available, circuit-level netlist models are intractable for formal verification. This abstraction to a

---

[1] A timing error is a fault caused when a pipe stage delay exceeds the clock period, leading to an erroneous value appearing in the network. In practice, this would appear as corrupted, dropped, or spurious data.

behavioral characterization of PSN in NoCs cannot be directly converted or compared to measured PSN on a digital circuit. However, the correlation between router activity and PSN shown in [5] indicates that the high-level abstraction used in this work enables valid characterization of PSN early in the design stage, providing key insights into the expected profile of PSN in the design.

3.5 Statistical Model Checking (SMC)

SMC is a quantitative verification approach that enables verifying the probability that a system model satisfies some property of interest. SMC uses data collected from random simulation runs to approximate the probabilistic properties of a model [33]. As a result, the confidence level of these approximations depends on the number of runs executed, with more executions resulting in higher confidence. Models containing rare events may take hundreds of thousands of runs to achieve accurate results. This work uses the SMC tool `modes` [9], part of the MODEST TOOLSET [20]. While the MODEST TOOLSET also contains tools for probabilistic model checking [2] (which uses exhaustive state space exploration), this work primarily uses SMC for its unparalleled scalability.

3.6 The Modest Toolset

The presented NoC architecture, routing algorithms, and properties are modeled and specified using the MODEST formal modeling language. Properties are then checked with the MODEST TOOLSET. While a comprehensive description of MODEST is outside the scope of this work, some concepts critical to the function of the model are outlined here. The `process` construct in MODEST models an asynchronous process and may contain local variables, atomic assignments, and synchronizing actions. Processes can be composed using either *sequential* or *parallel* composition. Parallel processes are synchronized through *actions*, which act as a synchronization barrier in a parallel composition where all participating processes must synchronize before performing the next step. A short example of `process` and composition syntax is given in Code Segment 1. We refer the interested reader to [22] for a tutorial and reference for MODEST.

The semantics of MODEST consists of two components: (1) a *symbolic semantics* [8, Sect. 4.2] that maps syntax to a symbolic *stochastic timed automaton* (STA) and (2) a *concrete semantics* [8, Sect. 5] that transforms the STA into a *timed probabilistic transition system* (TPTS), a form of uncountable-state Markov decision process. In our work, we use a subset of MODEST where the concrete semantics yields a *discrete-time Markov chain* (DTMC).

3.7 Rationale for Choosing MODEST

MODEST is our preferred modeling language and toolset for this work, as it is the *only* mature probabilistic formal modeling language that supports flexible data structures for modular modeling, such as arrays, linked lists, and structs.

These data structures are *not* supported by the other mature probabilistic modeling language, that of PRISM [29], yet they are key to our goal of developing a *highly modular* NoC model.

While the NoC model analyzed in this work is synchronous, MODEST and PRISM are inherently asynchronous. To resolve this, we can use the synchronizing actions offered by both MODEST and PRISM that force specified parts of the model to be synchronous. Other naturally synchronous formal modeling languages such as nuXmv [12] or Lustre [11,13] lack the required support for specifying models with probabilistic characteristics.

Finally, the MODEST TOOLSET has been actively maintained for many years and provides a suite of tools for statistical and probabilistic model checking of PCTL properties, with recently added support for checking CTL properties.

```
process P0() { /*...*/ } // process declaration
par { :: P0() :: P1() } // parallel composition
P0(); P1(); // sequential composition
```

Code Segment 1. Process and Parallel Composition Syntax in MODEST

4   Modular Design of the Formal NoC Model

We provide a framework for creating arbitrary $n{\times}n$ two-dimensional mesh NoCs in the MODEST language. The NoC model is a composition of instances of a router process, modeled using the **process** construct in MODEST. This modular style was influenced by modern digital design using *hardware description languages* such as Verilog [45] or VHDL [49], where modular design is essential for scalable designs and commonly used across all hardware designs. A modular design is defined by the clean separation of circuitry between different modules, where each module has a well defined interface that can be connected to other modules. Our modular design of the NoC model provides scalable results beyond that of previous works and more closely aligns with digital design practices.

This work presents a NoC model with easy customization of design parameters, including topological size, flit injection pattern, buffer size, and routing algorithm. The PSN characterization and verification of functional correctness is specified as properties *independent* of the specific model allowing for comparison of PSN characterization between models. Parameters such as topological size, buffer size, and PSN thresholds can be easily updated without knowledge of the MODEST language. Customization to behavior (e.g. packet injection pattern, routing algorithm, or priority arbitration) require writing MODEST code.

The full specification of the NoC model is too complex to fit within the page limit of this work. For this purpose, [51] has been made available with more details, examples, and results that support this work.

4.1   NoC Construction Using Parallel Composition

An $n \times n$ NoC is described in our framework as the parallel composition of n^2 router processes and a clock process. For example, the 2×2 NoC shown in Fig. 1a is a parallel composition of four routers and a clock, i.e., $r_0 \parallel r_1 \parallel r_2 \parallel r_3 \parallel Clock$. Code Segment 2 shows the parallel composition in MODEST, where each router is given its unique ID as input parameter with ID $\in [0, n^2 - 1]$. The router process demonstrates how a modular design improves the scalability of formal NoC models, as constructing a NoC of arbitrary size is accomplished by instantiating more routers.

```
par { :: Clock() :: Router(0) :: Router(1) :: Router(2) :: Router(3)}
```

Code Segment 2. Creation of a 2×2 NoC in MODEST

4.2   Router Process

The `Router` process models the generic router model in Fig. 1b. This router has four input buffers and four output channels for interfacing with neighboring routers and a local input buffer for introducing new packets. The local buffer is the only way new packets are introduced into the NoC. The router model can be instantiated many times with a unique ID (Code Segment 2).

Each router maintains variables that define the state of the NoC, such as the elements in each buffer, previously serviced buffers, and various boolean flags that indicate what actions a buffer can take. These variables are stored in a global array of router objects. When instantiating a router process, the given `id` acts as an index into the array of router objects. These router objects cannot be completely default initialized due to design choices of MODEST, but a script is provided to automate this process as described in [51, Sec. A.5].

The buffers are modeled as a functional programming language-style *list*. This implementation efficiently models a FIFO queue, and allows for buffer size to be easily specified by a single global constant.

As shown in Code Segment 3, the router process is divided into five major sub-processes: generating new flits for the local buffer, preparing the router for flit advancement, advancing flits to their respective output channels, updating the buffer scheduling priority, and updating the PSN variables. After the `nextClockCycle` action, the router process calls itself recursively, which models the repeated execution of the same five sub-processes in every clock cycle.

Although the modular style allows for scalability, code reuse, and customization, it also initially had synchronization issues between routers that were not present in the previous work [38] because [38] did not use parallel composition to model the NoC. The parallel composition shown in Code Segment 2 is asynchronous, which means that each router instance executes its actions at its own speed without waiting for other routers.

```
process Router(int id) {
  GenerateFlits(id); PrepRouter(id);
  AdvanceRouter(id); UpdatePiority(id);
  nextClockCycle; /* synchronizing clock action */
  Router(id) /* recursive call models next clock cycle */ }
```

Code Segment 3. Router Process in MODEST

Two key issues presented themselves when initial verification attempts were performed on the NoC. First, the asynchronous composition meant that there was no guarantee that all routers would be synchronized. Because of this, one router could get ahead of its neighbors, effectively allowing multiple steps to be taken per clock cycle, which is not possible in a synchronous digital system. This error was fixed in [7] by introducing a global synchronization action, similar to `nextClockCycle` in Code Segment 3.

In MODEST, synchronization actions act as a barrier that every process must reach before proceeding. When the `nextClockCycle` action is reached, every process must wait for every other process to also reach the `nextClockCycle` action before proceeding. This prevents routers from performing multiple steps per clock cycle, and ensures that per-cycle PSN results are correct.

Unfortunately, a single `nextClockCycle` action is not sufficient to prevent all synchronization errors in the router. While the `nextClockCycle` action synchronizes the `Clock` and `Router` processes, errors can still occur where one flit passes through multiple routers in one clock cycle. This is a write-before-read conflict, where one router may call `AdvanceRouter` before another has called `PrepRouter`, and is not a valid NoC behavior.

For example, if r_0 has a flit destined for r_3 in the 2×2 NoC depicted in Fig. 1a, it should take a minimum of two clock cycles to reach its destination. On clock cycle 1, the flit is sent from r_0 to r_1. Then, on clock cycle 2, it is sent from r_1 to r_3. However, if r_0 executes `AdvanceRouter` before r_1 executes `PrepRouter`, the flit could travel from r_0 to r_1 to r_3 in a single clock cycle.

A second synchronization action `sync` was inserted between the `PrepRouter` and `AdvanceRouter` processes in [7] and this successfully prevented write-before-read conflicts. However, when performing functional verification on the resulting model, the state space would explode due to the need to store states resulting from interleavings of each router *independently* executing each sub-process. These interleavings were not desired as they would not occur in a synchronous digital system. In this work, we inserted *fine-grained synchronization actions* into every sub-process so that all routers in the parallel composition executed in lock step. This model update not only alleviates state space explosion, but also more faithfully models synchronous digital hardware, where assignments across all components execute in lock step during every clock cycle.

4.3   Synchronizing Actions

The MODEST language is inherently asynchronous while most digital systems are synchronous with a global clock synchronizing all the updates in lock-step. To ensure that our MODEST model correctly models a synchronous system, we carefully employ fine-grained synchronization actions among all processes. In MODEST, each assignment block – assignments within {= =} braces – executes atomically. As described in this section, the NoC model is broken up into many processes, where each process handles one element of the model. Parallel composition of these processes produces one single model that acts as a NoC. In parallel assignment in MODEST, each unique interleaving of assignment blocks is explored. This produces many interleavings of actions, which is useful for many problems that MODEST can model. However, when modeling synchronous digital designs, we are not concerned with interleaving states, we are only concerned with the changes that occur with the synchronous clock.

Code segment 4 shows an example of a process that updates an element in array x based on the input id. The possible states including interleavings for these assignments are shown in Fig. 2a. If this were modeling a digital circuit, and x was a digital register block that updated with the clock, states s_1 and s_2 would never be observed due to the synchronous semantics of the digital system, as shown in the clock diagram in Fig. 2b. In practice, this problem of unnecessary interleaving causes an exponential explosion of state space that makes the modular model challenging to verify with the available hardware. This issue was not a problem for previous work [38] because that model was not made of modular processes composed together, but was rather represented by a large process that orchestrated each part of the model and updates occurred in the same assignment block.

```
int x[] = {0, 0};
process setTo5(int id) {{= x[id] = 5; =}}
par { :: setTo5(0) :: setTo5(1) }
```

Code Segment 4. Parallel Interleaving Example

To resolve this issue we use the synchronizing actions provided by MODEST. Actions provide a synchronization method where parallel assignments happen in lockstep with one other. For a more in-depth explanation of actions in MODEST, see [22]. Adding a clock action that models the synchronous nature of digital circuits eliminates states s_1 and s_2, and we are left with a single state transition that acts akin to the synchronous digital system as shown in Fig. 2c. An alternative would be to apply probabilistic partial order reduction [3,4,14] to pick *one interleaving* of the independent assignments; by exploiting the system's synchronous nature, we achieve a stronger reduction that collapses the entire interleaving into *one transition* synchronizing all the involved parallel processes.

To synchronize the parallel composition of router processes, each assignment in each sub-process called in router has a unique synchronizing action associated

```
action clock; int x[] = {0, 0};
process setTo5(int id) { clock{= x[id] = 5; =} }
par { :: setTo5(0) :: setTo5(1) }
```

Code Segment 5. Synchronizing Parallel Interleavings with a Clock Action

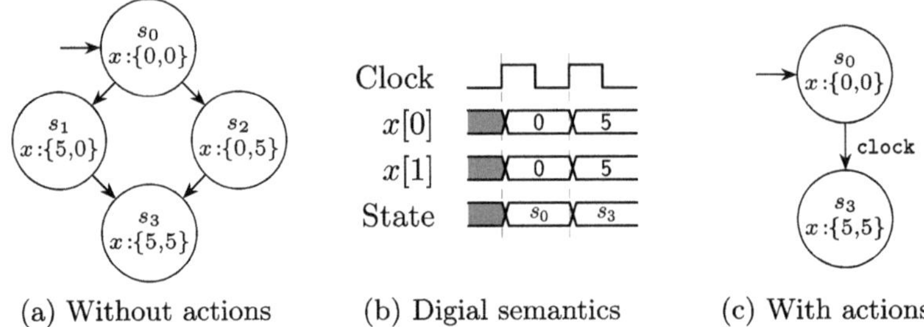

(a) Without actions   (b) Digial semantics   (c) With actions

Fig. 2. Adding Actions to Reduce Unnecessary State Space.

to it. This ensures that in each state update the synchronous semantics of digital circuits is correctly represented and, like Fig. 2 shows, helps to reduce the state space and improve verification time. Code Segment 3 shows the relevant sub-processes to be synchronized. Code Segment 6 shows an example of how each sub-process of **Router** uses a unique action to ensure that the assignments are synchronized. The effect of these actions is that each of the routers update their state together, and no interleaving is allowed. Figure 3 shows an example trace of the state updates of an $n \times n$ NoC, where each router updates in sync with the other routers.

```
action gf, pr, ar, up;
process GenerateFlits(int id) { gf {= /*...*/ =} }
process PrepRouter(int id)    { pr {= /*...*/ =} }
process AdvanceRouter(int id) { ar {= /*...*/ =} }
process UpdatePiority(int id) { up {= /*...*/ =} }
```

Code Segment 6. Use of Synchronizing Actions in Router Processes

4.4   Clock Process

A clock process (Code Segment 7) is composed with the router instances to count the elapsed clock cycles (Code Segment 2). The clock count is used to generate the PSN results (Sect. 8) and to give context to different flit injection strategies. MODEST allows for variable assignment to occur with a synchronization action (i.e. **nextClockCycle**), which is used to increment the clock count.

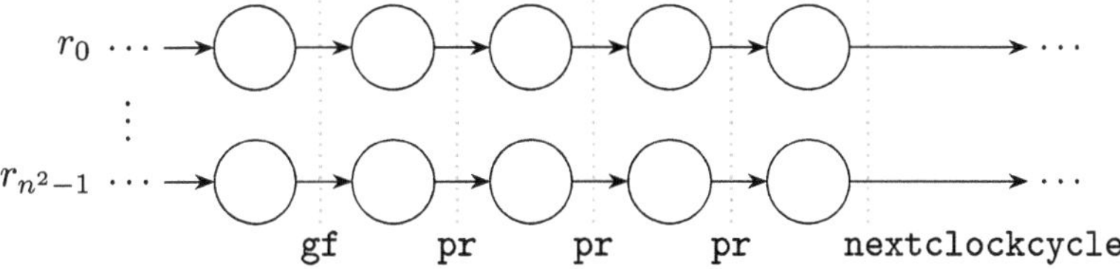

Fig. 3. Impact of Synchronizing Actions in a Parallel Composition of n^2 Routers.

```
process Clock() { nextClockCycle{= clk = (clk + 1) =}; Clock()}
```

Code Segment 7. Clock Process in MODEST

4.5   Flit Generation Algorithm

The modular `GenerateFlits` process allows for easy customization of network traffic patterns in order to measure the PSN impact of different flit injection strategies. This process has one input parameter `id` and may add a flit to r_{id}^L. This simulates the generation of network traffic in the NoC. Determining the traffic generation pattern needs to be done after observing patterns in real-world or simulated environments, as the generation pattern will vary based on application and use-case.

Results in this work use the periodic flit generation pattern from [38] with a 3/10 duty cycle. In this pattern, each router has a new flit synchronously added to the local buffer during the first 3 out of every 10 clock cycles. Destinations for each new flit are chosen according to a uniform distribution of all other routers in the NoC. While this specific distribution and generation pattern were chosen for direct comparison with [38], the modular framework allows arbitrary generation patterns and destination distributions. [51, Sec. A.4] gives further examples.

4.6   Routing Algorithm

The *routing algorithm* used by each router determines how incoming flits should be treated. This is another modular process with input parameter `id` that can be adjusted to model different routing algorithms in the NoC. Due to the page limit, the full details and semantics of the `AdvanceRouter` process are included in the [51, Sec. A.4].

This work implements the X-Y routing algorithm, as described in Sect. 3, because it is deterministic and deadlock free. X-Y routing was used in [34,38] and a goal of this work was to reproduce results from those papers. Additionally, 7 of the 16 real-world NoCs surveyed in [25] used X-Y (or Y-X) routing.

4.7   Priority Tracking and Updates

During routing, flits are passed between buffers on channels. Since a channel can be used *only once* per clock cycle, conflicts may arise when multiple flits need to use the same channel. To ensure that all buffers are *serviced*, flags are set when a buffer cannot be serviced due to conflicts. A buffer is *serviced* when a flit is removed from the front and removed or sent down a channel to another buffer of the neighboring router.

The `UpdatePriority` sub-process uses these flags to generate the priority of buffers to service for the next cycle. The order of buffers to be serviced next is stored in an array and used during the next clock cycle as the initial order for flits to be sent. As part of the modularity of this design, the priority algorithm used for each buffer can be reconfigured in the model to meet the design goals. This work uses a round-robin process to ensure that each buffer is serviced fairly.

4.8   Noise Tracking

PSN is characterized by tracking the activity level of each router in the network every clock cycle. As discussed in Sect. 3.4, resistive noise is related to the activity during the current clock cycle, and inductive noise is related to the change in activity over two consecutive clock cycles. Further discussion of PSN characterization is given in Sect. 6.

5   Correctness of the Modular NoC

MODEST uses the NoC model we provide to explore the complete state-space, and then the MODEST TOOLSET's `mcsta` model checker [21] uses $\forall\square$ and $\exists\lozenge$ CTL formulas to verify the correctness of a NoC design. Verification of these properties is carried out on a generic router and inter-router communications. In this section, i and j represent valid router IDs in a NoC and $j \twoheadrightarrow r_i^L$ denotes a packet destined for j being added to the *Local* buffer of r_i.

5.1   Flit Generation Verification

As discussed in Sect. 4.5, the flit generation pattern for a given NoC design can be easily modified. Thus verifying a unique flit generation pattern requires a verification strategy suited to each pattern, so we do not present a general method for verifying the correctness of *all* flit generation patterns.

However, one property that all flit generation patterns must demonstrate across all paths is that a flit destined for the router they are originating from should not be generated as shown in Property 3.

$$\forall i : \forall\square(\neg(i \twoheadrightarrow r_i^L)) \tag{3}$$

While we do not present a method for verifying all possible flit generation patterns, we do demonstrate a possible approach to such verification. The flit generation algorithm used in [38] assigns destinations using a uniformly random distribution. We can verify that there exists an execution path in which a flit is sent from router i to router j for all routers i, j using the following CTL property:

$$\forall i : \forall j : j \neq i \implies \exists\lozenge(j \twoheadrightarrow r_i^L) \tag{4}$$

5.2   Scheduling Priority Verification

The implementation of this scheduler relies on a list of buffers that have been serviced or are waiting for service. It is important that this list is coherent, i.e. that the list always contains all buffers and no duplicate entries. Property 5 shows the CTL formula for this property, where i is a placeholder for router ID.

$$\forall i : \forall\square(local \in r_i.priorityList \wedge north \in r_i.priorityList$$
$$\wedge \ east \in r_i.priorityList \wedge south \in r_i.priorityList \tag{5}$$
$$\wedge \ west \in r_i.priorityList) \wedge \text{length}(priorityList) = 5$$

5.3   Flit Propagation Verification

The real-world NoC design deploys buffers of bounded sizes, but they are modeled as an unbounded queue in MODEST. This design choice is used to facilitate the instantiation of a buffer of arbitrary size, but makes it possible for a bounded buffer to exceed its specified size in the model. Therefore, we must specify a property to check that all the buffers in a NoC never exceed the user-specified size. This property should also check that a flit is never accidentally sent to a full buffer, as that would increase the length of the buffer by one. In CTL, this is

$$\forall i, j : \forall\square(\text{length}(r_i.buffer_j) \leqslant BUFFER\_SIZE) \tag{6}$$

Additionally, each channel should only admit at most one flit during a clock cycle to correctly model a synchronous NoC. We verify this property by adding a counter to each channel that tracks how many times it was used during a single clock cycle. It is incremented when a flit is sent across the channel during the `AdvanceRouter` process, and is reset back to zero in the `UpdatePriority` process. The CTL property to check that no counter is ever greater than one is

$$\forall i, j : \forall\square(r_i.channel_j.used\_count \leqslant 1) \tag{7}$$

5.4   Improvement Over Previous Works

Previous works such as [7,38] were not able to perform CTL model checking as `mcsta` did not support CTL then. Instead, they encoded functional correctness using the PCTL syntax and checked if the probability of PCTL encodings of Properties 3-7 were 0.0 or 1.0, which may not be the same for certain probabilistic loops, and requires less scalable model checking methods.

Additionally, due to the many unnecessary interleavings generated by the lack of true synchronization as discussed in Sect. 4.2, previous models were only able to be checked up to a small number of clock cycles, effectively performing bounded model checking. Our work checks an unbounded number of cycles for the 2×2 NoC. Checking 3×3 however still runs out of memory; future work can look at abstracting parts of the modular model to scale up verification.

6   PSN Probability

As discussed in Sect. 3.4, this work measures PSN through an behavioral description instead of circuit level details such as resistance, inductance, and current. This section outlines our method for calculating PSN probabilities and how it is customized to provide more information.

Resistive noise occurs when activity occurs in a router during a single clock cycle. Each router maintains a local `activity` counter to track events that occur during a single clock cycle. The counter `activity` increases when a packet is sent or received. This activity level translates into a resistive noise event when it crosses the activity threshold K (i.e., `ACTIVITY_THRESH`) set by the user. We then use Property 1 to determine the cumulative probability that N resistive noise events occur in k clock cycles. N is set by the user as `RESISTIVE_THRESH`.

Inductive noise occurs when the activity level in a router changes between two clock cycles. Each router also maintains a local `lastActivity` variable that stores the activity of the previous clock cycle. When the difference between `activity` and `lastActivity` is greater than the activity threshold K, it generates an inductive noise event. We then use Property 2 to determine the cumulative probability that M inductive noise events occur. M is set by the user as `INDUCTIVE_THRESH`.

By default, the PSN characterization in this model is a global characterization, where PSN is considered for all routers in the network. However, since each router maintains a local `activity` and `lastActivity` variable, it is easy to extract PSN probabilities for a single router or group of routers instead of all routers, as shown in Sect. 7.

7   Results and Discussion

The results presented in this work were generated on a machine using a 12-core AMD Ryzen Threadripper Processor (at 3.5 GHz), 132 GB of memory, and Ubuntu Linux version 22.04 LTS. MODEST TOOLSET version 3.1.290 was used. All formal models, plots, and scripts are publicly available at [16]. The runtime results are presented in Table 1.

7.1   CTL Verification Results for a 2×2 NoC

CTL model checking results for the 2×2 NoC were generated using `mcsta`, the model checking engine in MODEST. In addition, the `chainopt` option was used to reduce the state space by collapsing chains of states connected by probability-1 transitions. All properties specified in Sect. 5 were verified on the 2×2 model.

7.2   PSN Results

The results of the PSN PCTL properties were captured using `modes`, the statistical model checker in MODEST. SMC was used due to memory constraints for

Table 1. Runtime Results

Size	Parameter	Tool	Confidence Interval	Runtime (HH:MM:SS)	Memory (MB)	States
2×2	Functional Properties	mcsta	–	00:10:31	9844	3446073
2×2	Resistive PSN	modes	95%	00:06:11	105	–
2×2	Inductive PSN	modes	95%	00:56:08	105	–
3×3	Resistive PSN	modes	95%	00:03:07	129	–
3×3	Inductive PSN	modes	95%	00:14:13	131	–
4×4	Resistive PSN	modes	95%	00:02:42	164	–
4×4	Inductive PSN	modes	95%	00:16:13	172	–
8×8	Resistive PSN	modes	95%	00:05:23	545	–
8×8	Inductive PSN	modes	95%	00:21:08	565	–
Total				02:14:36		

models larger than 2×2. In `modes`, the `max-run-length` option is set to zero to allow longer simulation runs (as `modes` by default aborts when it encounters a simulation run longer than 10,000 transitions, as a heuristic to avoid hanging on unintended infinite paths – which our model does not have).

The main goal of this work was to create a modular NoC framework that is more flexible and scalable than the previous work. However, in order to reproduce and compare the probabilistic verification results between this work and similar previous work [38], we configure the NoC model to match the previous ones. Therefore, this model uses a buffer length of four, the X-Y routing algorithm, an activity threshold of three, and the 3/10 flit injection pattern (detailed in Sect. 4.5) unless otherwise stated.

2×2 **NoC Verification Results.** Figure 4 shows the PSN results for a 2×2 NoC for our modular design. Each plot shows the *cumulative distribution functions* (CDFs) of the probability of PSN events greater than or equal to the amount shown in the legend. Both resistive and inductive noise display a step-like growth corresponding to the 3/10 flit injection pattern detailed in Sect. 4.5.

7.3   Comparison to Previous 2×2 NoC Model

The 2×2 model results in [38] were regenerated using the `modes` SMC tool and compared the modular 2×2 results of this work in order to reproduce their findings. A similar comparison of a central router from [34] to the modular model is given in [7].

The initial results of the modular design were quite different from [38]. Because the modular model had been proved functionally correct by passing the CTL properties detailed in Sect. 5, we determined the discrepancies derived either from architectural differences or from an error in previous work.

We used comparative model checking to narrow down the differences between the two models and discovered two architectural differences. The initial modular implementation did not generate new flits on clock cycle 0, while the previous

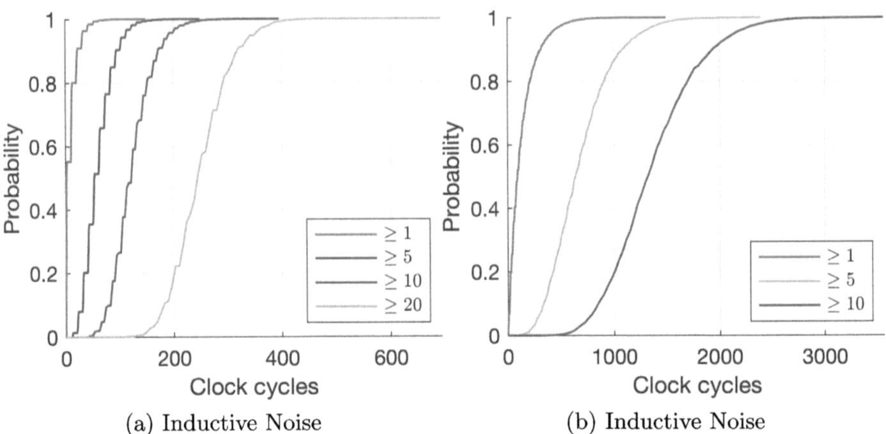

(a) Inductive Noise   (b) Inductive Noise

Fig. 4. PSN for 2×2 Modular Model.

monolithic model of [38] did. We corrected this by rearranging the sub-processes shown in Code Segment 3 to occur before the synchronizing clock action.

Additionally, the modular model allows multiple flits to be consumed during a single clock cycle, while the monolithic model allowed only a single flit to be consumed per cycle. For example, if two flits destined for r_0 arrived at r_0 on the same clock cycle in different channels, the modular model would consume both flits, while the monolithic model would consume only one. While both designs are valid, the reality is that the hardware attached to each router is unknown. It is therefore *assumed* in this work that the hardware attached to each router is capable of consuming multiple flits per cycle, and we adjusted the monolithic model to accept multiple flits per clock cycle.

We performed comparative model checking by assuming that both models were correct, and then generating the PSN characterization for each model and comparing the output. If the output differs, at least one of the models is faulty as they should be equivalent. Determining which model is faulty is done by simulating traces where the correct output is known on each model. Whichever model returns the incorrect output is the faulty model, and the trace can be analyzed to determine where the error originated. This process is repeated iteratively until the models match. We used this method because we could not directly compare states due to the different implementations of each model. Matching results after comparative-model checking was completed is shown in Fig. 5.

7.4   PSN Results for Larger NoCs

The advantage of a modular NoC model is the ability to scale it with ease. By instantiating 9, 16, and 64 routers for 3×3, 4×4, and 8×8 NoCs, respectively, we are quickly able to scale our PSN analysis. Figure 6 shows the inductive noise CDFs for these sizes. Resistive noise CDFs are not shown here due to page limitations, but are shown in [51, Sec. A.1].

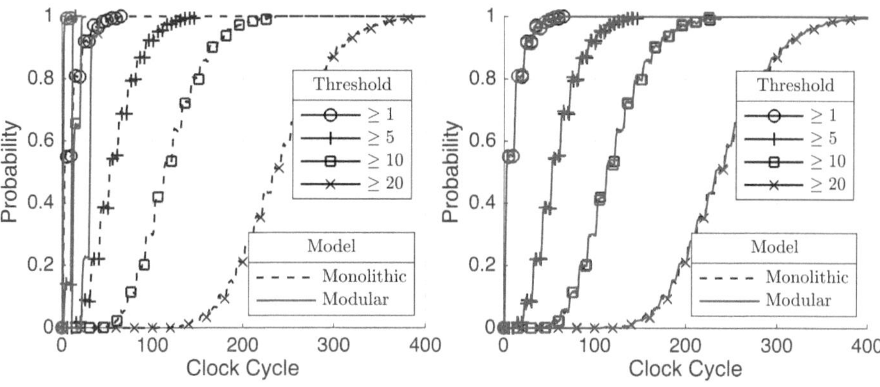

(a) Before Comparative Model Checking   (b) After Comparative Model Checking

Fig. 5. Modular and Monolithic 2×2 NoC Resistive Noise Comparison.

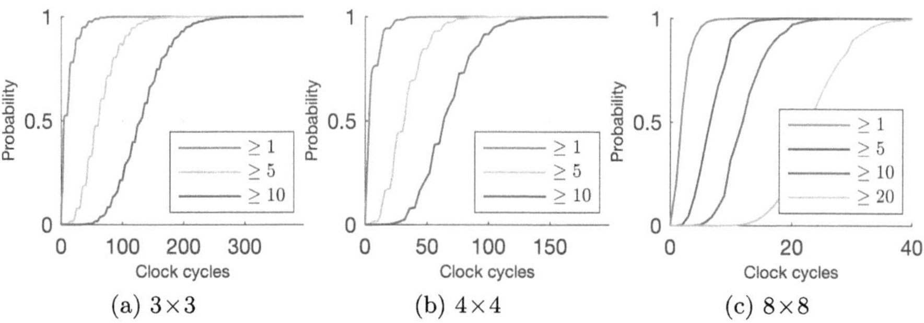

(a) 3×3 (b) 4×4 (c) 8×8

Fig. 6. Inductive Noise CDF for 3×3, 4×4, and 8×8 Modular Model.

Similar to the 2×2 results in Fig. 4, the 3×3 and 4×4 results show a pronounced pattern of steep slopes followed by rough edges which is a result of our chosen 3/10 flit injection pattern. As expected, compared to a 2×2 NoC, the likelihood of PSN events is much higher and occurs sooner in the 3×3 and 4×4 NoCs due to the higher number of routers and network traffic. Inductive noise events are likely to happen sooner in the 8×8 NoC compared to smaller models, and the pronounced steps every 3/10 cycles are gone. This is because packets are much more likely to persist in the 8×8 NoC, as packets in the 8×8 are likely to travel farther.

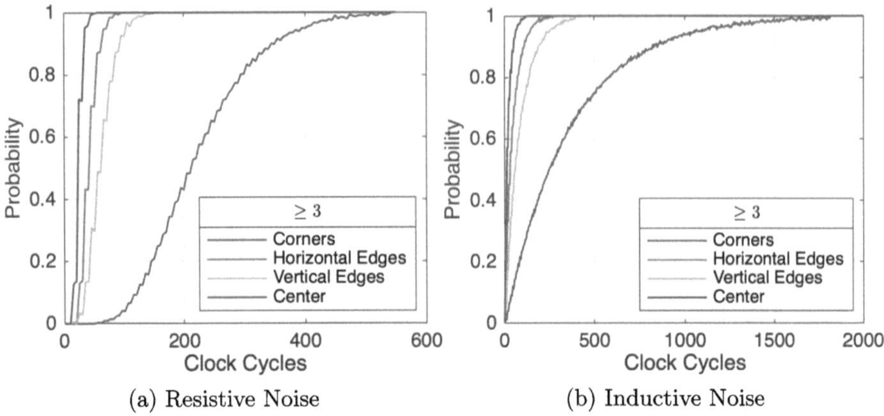

(a) Resistive Noise   (b) Inductive Noise

Fig. 7. Router Specific PSN for 3×3 Modular Model.

7.5 PSN for Specific Routers

One difference between a 2×2 and 3×3 NoC are the different routing connections used. A 2×2 NoC, such as the one shown in Fig. 1a, only uses *corner routers* with two neighbors. A 3×3 configuration has *corner routers* with two neighbors, *edge routers* with three neighbors, and a *central router* with four neighbors.

The modular NoC model allows us to obtain insights that were impossible to obtain from the monolithic design style [34,38]. Rather than checking PSN globally, we can inspect the `activity` and `lastActivity` of an individual router to get a CDF for PSN for that specific router. The results of PSN resulted from different router types are shown in Fig. 7. PSN results were generated for each router type with an `ACTIVITY_THRESH` of 3.

The router-specific results in Fig. 7 show that the PSN increases toward the center of the network. Although the corner routers experienced a very gradual increase in PSN with respect to time, the center router experienced high levels of PSN almost immediately. This is because the central router has more active buffers than the others, and consequently, it experiences higher PSN. However, the horizontal edge routers experienced a higher PSN than those on the vertical edges, regardless of having the same number of buffers. This is a result of the X-Y routing causing more traffic through those areas. Therefore, placing most traffic-prone neighbors on the NoC perimeter may reduce PSN as the effects of PSN are less prevalent on corner and edge routers.

This ability to check PSN for specific parts of a NoC can be used to provide valuable information on how the chosen routing algorithm and the flit injection pattern affect PSN. For example, if a new routing algorithm was designed to minimize traffic to the central router of a 3×3 NoC, our modular model could verify the effects on PSN that the new algorithm would have compared to the current X-Y routing algorithm. This would allow routing algorithm design to occur early in the design cycle, leading to more effective designs.

Previously, the largest topology achieved for a probabilistic NoC model was a 2×2 network. As mentioned, 2×2 routers would only have three input buffers (two neighbor buffers and a local buffer). With the PSN threshold set to three, this inherently results in lower PSN between routers with higher conflict rates, because if even one buffer goes unserviced in a cycle, there can be no PSN on that router. PSN characterization for 2×2 networks does not scale to larger NoCs, making it difficult to make any particular recommendation to chip designers based on 2×2 findings alone.

In the work of [38], it is concluded that PSN can be reduced by injecting delays between flits. While this is true, this work affirms that PSN is difficult to completely eradicate using the flit injection pattern alone. The larger a NoC becomes, the longer it takes on average for flits to propagate through the system. Injecting flits in only a single clock cycle can still result in PSN. In addition to reducing the flit injection rate, this work recommends grouping routers into high-traffic regions, with the highest traffic connections reserved for vertical neighbors. In this way, flits can propagate through the NoC more quickly. This recommendation, in conjunction with keeping heavy traffic on the perimeter of the NoC, may result in even lower traffic through the central routers.

8   Conclusion

This paper presents a case study on the development of an easy-to-use, scalable, and modular formal NoC model using the MODEST modeling language. The paper describes the CTL properties to verify the correctness of the model and the PCTL properties to quantify PSN. It also describes the diagnoses of a discrepancy in the quantification of PSN in a 2×2 NoC while attempting to reproduce a previous work [38], which revealed an inaccurate modeling of the flit consumption behavior in [38]. In addition, results from statistical model checking on the 3×3, 4×4, and 8×8 NoCs suggested network flit scheduling schemes to minimize PSN. While this work focuses on NoCs sized up to 8×8, it can be easily scaled to larger models. The modular design of the model makes it ideal for examining NoC designs early in the design cycle. Doing so may help find potential errors sooner and may encourage designs that are more robust in their function. Future work includes scaling the CTL correctness verification to arbitrarily sized models and exploring a wide range of flit injection patterns and routing algorithms.

Data Availability Statement. A docker environment with the tools, models, and results of this paper is available on **Zenodo** [50]. Additionally, the models are available on **GitHub** [16].

References

1. Alhubail, L., Bagherzadeh, N.: Power and performance optimal noc design for cpu-gpu architecture using formal models. In: 2019 Design, Automation & Test in Europe Conference & Exhibition (DATE), pp. 634–637 (2019). https://doi.org/10.23919/DATE.2019.8714769
2. Baier, C., de Alfaro, L., Forejt, V., Kwiatkowska, M.: Model checking probabilistic systems. In: Handbook of Model Checking, pp. 963–999. Springer, Cham (2018). https://doi.org/10.1007/978-3-319-10575-8_28
3. Baier, C., D'Argenio, P.R., Größer, M.: Partial order reduction for probabilistic branching time. In: Cerone, A., Wiklicky, H. (eds.) Proceedings of the Third Workshop on Quantitative Aspects of Programming Languages (QAPL 2005). Electronic Notes in Theoretical Computer Science, vol. 153, pp. 97–116. Elsevier (2005). https://doi.org/10.1016/J.ENTCS.2005.10.034
4. Baier, C., Größer, M., Ciesinski, F.: Partial order reduction for probabilistic systems. In: 1st International Conference on Quantitative Evaluation of Systems (QEST 2004), pp. 230–239. IEEE Computer Society (2004). https://doi.org/10.1109/QEST.2004.1348037
5. Basu, P., Shridevi, R.J., Chakraborty, K., Roy, S.: IcoNoClast: tackling voltage noise in the NoC power supply through flow-control and routing algorithms. IEEE Trans. VLSI Syst. **25**(7), 2035–2044 (2017)
6. Becchu, N.K.R., Harishchandra, V.M., Yernad Balachandra, N.K.: System level fault-tolerance core mapping and fpga-based verification of noc. Microelectron. J. **70**, 16–26 (2017). https://doi.org/10.1016/j.mejo.2017.09.010. https://www.sciencedirect.com/science/article/pii/S0026269217302884
7. Boe, J.W.: Probabilistic Verification for Modular Network-on-Chip Systems. Master's thesis, Utah State University (2023). https://doi.org/10.26076/4f5a-4a68. https://digitalcommons.usu.edu/etd/8763
8. Bohnenkamp, H.C., D'Argenio, P.R., Hermanns, H., Katoen, J.P.: MoDeST: a compositional modeling formalism for hard and softly timed systems. IEEE Trans. Softw. Eng. **32**(10), 812–830 (2006). https://doi.org/10.1109/TSE.2006.104
9. Budde, C.E., D'Argenio, P.R., Hartmanns, A., Sedwards, S.: A statistical model checker for nondeterminism and rare events. In: Beyer, D., Huisman, M. (eds.) TACAS 2018. LNCS, vol. 10806, pp. 340–358. Springer, Cham (2018). https://doi.org/10.1007/978-3-319-89963-3_20
10. Buecherl, L., et al.: Stochastic hazard analysis of genetic circuits in iBioSim and STAMINA. ACS Synth. Biol. **10**(10), 2532–2540 (2021). https://doi.org/10.1021/acssynbio.1c00159. pMID: 34606710
11. Caspi, P., Pilaud, D., Halbwachs, N., Plaice, J.A.: Lustre: a declarative language for real-time programming. In: Proceedings of the 14th ACM SIGACT-SIGPLAN Symposium on Principles of Programming Languages, POPL '87, pp. 178–188. Association for Computing Machinery, New York (1987). https://doi.org/10.1145/41625.41641. https://doi-org.dist.lib.usu.edu/10.1145/41625.41641
12. Cavada, R., et al.: The NUXMV symbolic model checker. In: Biere, A., Bloem, R. (eds.) CAV 2014. LNCS, vol. 8559, pp. 334–342. Springer, Cham (2014). https://doi.org/10.1007/978-3-319-08867-9_22
13. Champion, A., Mebsout, A., Sticksel, C., Tinelli, C.: The kind 2 model checker. In: International Conference on Computer Aided Verification (2016). https://api.semanticscholar.org/CorpusID:11582048

14. D'Argenio, P.R., Niebert, P.: Partial order reduction on concurrent probabilistic programs. In: 1st International Conference on Quantitative Evaluation of Systems (QEST 2004), pp. 240–249. IEEE Computer Society (2004). https://doi.org/10.1109/QEST.2004.1348038

15. Fang, L., Yamagata, Y., Oiwa, Y.: Evaluation of A resilience embedded system using probabilistic model-checking. In: Pang, J., Liu, Y. (eds.) Proceedings Third International Workshop on Engineering Safety and Security Systems, ESSS 2014, Singapore, Singapore, 13 May 2014. EPTCS, vol. 150, pp. 35–49 (2014). https://doi.org/10.4204/EPTCS.150.4

16. Artifact repository for "probabilistic verification for modular network-on-chip systems" (2025). https://github.com/formal-verification-research/VMCAI26_Modular_NoC_Artifact. Accessed 16 Sept 2025

17. Hahn, E.M., Hartmanns, A., Hermanns, H., Katoen, J.P.: A compositional modelling and analysis framework for stochastic hybrid systems. Formal Methods Syst. Des. **43**(2), 191–232 (2013). https://doi.org/10.1007/S10703-012-0167-Z

18. Hahn, E.M., Li, Y., Schewe, S., Turrini, A., Zhang, L.: ISCASMC: a web-based probabilistic model checker. In: Jones, C., Pihlajasaari, P., Sun, J. (eds.) FM 2014. LNCS, vol. 8442, pp. 312–317. Springer, Cham (2014). https://doi.org/10.1007/978-3-319-06410-9_22

19. Hansson, H., Jonsson, B.: A logic for reasoning about time and reliability. Formal Aspects Comput. **6**(5), 512–535 (1994). https://doi.org/10.1007/BF01211866

20. Hartmanns, A., Hermanns, H.: The modest toolset: an integrated environment for quantitative modelling and verification. In: Ábrahám, E., Havelund, K. (eds.) TACAS 2014. LNCS, vol. 8413, pp. 593–598. Springer, Heidelberg (2014). https://doi.org/10.1007/978-3-642-54862-8_51

21. Hartmanns, A., Hermanns, H.: Explicit model checking of very large MDP using partitioning and secondary storage. In: Finkbeiner, B., Pu, G., Zhang, L. (eds.) Automated Technology for Verification and Analysis - 13th International Symposium, ATVA 2015, Shanghai, China, October 12-15, 2015, Proceedings. Lecture Notes in Computer Science, vol. 9364, pp. 131–147. Springer, Heidelberg (2015). https://doi.org/10.1007/978-3-319-24953-7_10

22. Hartmanns, A., Hermanns, H.: A modest markov automata tutorial. In: Krötzsch, M., Stepanova, D. (eds.) Reasoning Web. Explainable Artificial Intelligence - 15th International Summer School 2019, Bolzano, Italy, 20–24 September 2019, Tutorial Lectures. Lecture Notes in Computer Science, vol. 11810, pp. 250–276. Springer, Heidelberg (2019). https://doi.org/10.1007/978-3-030-31423-1_8

23. Hensel, C., Junges, S., Katoen, J.-P., Quatmann, T., Volk, M.: The probabilistic model checker STORM. Int. J. Softw. Tools Technol. Transfer 1–22 (2021). https://doi.org/10.1007/s10009-021-00633-z

24. Jeppson, J., et al.: STAMINA in C++: modernizing an infinite-state probabilistic model checker. In: Jansen, N., Tribastone, M. (eds.) Quantitative Evaluation of Systems - 20th International Conference, QEST 2023, Antwerp, Belgium, 20–22 September 2023, Proceedings. Lecture Notes in Computer Science, vol. 14287, pp. 101–109. Springer, Heidelberg (2023). https://doi.org/10.1007/978-3-031-43835-6_7

25. Jerger, N.E., Krishna, T., Peh, L.S.: On-chip networks. In: Synthesis Lectures on Computer Architecture, Springer, Cham (2017). https://doi.org/10.1007/978-3-031-01755-1

26. Jiang, S.Y., Luo, G., Liu, Y., Jiang, S.S., Li, X.T.: Fault-tolerant routing algorithm simulation and hardware verification of noc. IEEE Trans. Appl. Supercond. **24**(5), 1–5 (2014). https://doi.org/10.1109/TASC.2014.2346484

27. Katoen, J.P.: The probabilistic model checking landscape. In: 2016 31st Annual ACM/IEEE Symposium on Logic in Computer Science (LICS), pp. 1–15 (2016)
28. Kumar, J.A., Vasudevan, S.: Automatic compositional reasoning for probabilistic model checking of hardware designs. In: QEST 2010, Seventh International Conference on the Quantitative Evaluation of Systems, Williamsburg, Virginia, USA, 15–18 September 2010, pp. 143–152. IEEE Computer Society (2010). https://doi.org/10.1109/QEST.2010.25
29. Kwiatkowska, M., Norman, G., Parker, D.: PRISM 4.0: verification of probabilistic real-time systems. In: Gopalakrishnan, G., Qadeer, S. (eds.) CAV 2011. LNCS, vol. 6806, pp. 585–591. Springer, Heidelberg (2011). https://doi.org/10.1007/978-3-642-22110-1_47
30. Kwiatkowska, M., Norman, G., Parker, D.: Stochastic Model Checking, pp. 220–270. Springer, Heidelberg (2007). https://doi.org/10.1007/978-3-540-72522-0_6
31. Lakin, M., Parker, D., Cardelli, L., Kwiatkowska, M., Phillips, A.: Design and analysis of DNA strand displacement devices using probabilistic model checking. J. R. Soc. Interface 9 (2012). https://doi.org/10.1098/rsif.2011.0800
32. Lecler, J.J., Baillieu, G.: Application driven network-on-chip architecture exploration and refinement for a complex soc. Design Autom. Emb. Sys. **15**, 133–158 (2011). https://doi.org/10.1007/s10617-011-9075-5
33. Legay, A., Delahaye, B., Bensalem, S.: Statistical model checking: an overview. In: Barringer, H., et al. (eds.) RV 2010. LNCS, vol. 6418, pp. 122–135. Springer, Heidelberg (2010). https://doi.org/10.1007/978-3-642-16612-9_11
34. Lewis, B., Hartmanns, A., Basu, P., Jayashankara Shridevi, R., Chakraborty, K., Roy, S., Zhang, Z.: Probabilistic verification for reliable network-on-chip system design. In: Larsen, K.G., Willemse, T. (eds.) FMICS 2019. LNCS, vol. 11687, pp. 110–126. Springer, Cham (2019). https://doi.org/10.1007/978-3-030-27008-7_7
35. Madsen, C., Myers, C.J., Roehner, N., Winstead, C., Zhang, Z.: Utilizing stochastic model checking to analyze genetic circuits. In: 2012 IEEE Symposium on Computational Intelligence in Bioinformatics and Computational Biology (CIBCB), pp. 379–386 (2012). https://doi.org/10.1109/CIBCB.2012.6217255
36. Maes, R.: An accurate probabilistic reliability model for silicon PUFs. In: Bertoni, G., Coron, J.-S. (eds.) CHES 2013. LNCS, vol. 8086, pp. 73–89. Springer, Heidelberg (2013). https://doi.org/10.1007/978-3-642-40349-1_5
37. Mundhenk, P., Steinhorst, S., Lukasiewycz, M., Fahmy, S.A., Chakraborty, S.: Security analysis of automotive architectures using probabilistic model checking. In: Proceedings of the 52nd Annual Design Automation Conference, p. 38. ACM (2015)
38. Roberts, R., et al.: Probabilistic verification for reliability of a two-by-two network-on-chip system. In: Lluch Lafuente, A., Mavridou, A. (eds.) FMICS 2021. LNCS, vol. 12863, pp. 232–248. Springer, Cham (2021). https://doi.org/10.1007/978-3-030-85248-1_16
39. Roberts, R., Neupane, T., Buecherl, L., Myers, C.J., Zhang, Z.: STAMINA 2.0: improving scalability of infinite-state stochastic model checking. In: Finkbeiner, B., Wies, T. (eds.) VMCAI 2022. LNCS, vol. 13182, pp. 319–331. Springer, Cham (2022). https://doi.org/10.1007/978-3-030-94583-1_16
40. Royannez, P., et al.: 90 nm low leakage soc design techniques for wireless applications. In: ISSCC. 2005 IEEE International Digest of Technical Papers, Solid-State Circuits Conference, 2005, vol. 1, pp. 138–589 (2005). https://doi.org/10.1109/ISSCC.2005.1493907

41. Salamat, R., Khayambashi, M., Ebrahimi, M., Bagherzadeh, N.: A resilient routing algorithm with formal reliability analysis for partially connected 3d-nocs. IEEE Trans. Comput. **65**(11), 3265–3279 (2016). https://doi.org/10.1109/TC.2016.2532871
42. Sepulveda, J., Aboul-Hassan, D., Sigl, G., Becker, B., Sauer, M.: Towards the formal verification of security properties of a network-on-chip router. In: 2018 IEEE 23rd European Test Symposium (ETS), pp. 1–6 (2018). https://doi.org/10.1109/ETS.2018.8400692
43. Shridevi, R.J., Ancajas, D.M., Chakraborty, K., Roy, S.: Tackling voltage emergencies in noc through timing error resilience. In: ISLPED, pp. 104–109 (2015)
44. Song, S., Sun, J., Liu, Y., Dong, J.S.: A model checker for hierarchical probabilistic real-time systems. In: Madhusudan, P., Seshia, S.A. (eds.) CAV 2012. LNCS, vol. 7358, pp. 705–711. Springer, Heidelberg (2012). https://doi.org/10.1007/978-3-642-31424-7_53
45. IEEE Standard for SystemVerilog–Unified Hardware Design, Specification, and Verification Language (2018). https://doi.org/10.1109/IEEESTD.2018.8299595
46. Taylor, L., Israelsen, B., Zhang, Z.: Cycle and commute: rare-event probability verification for chemical reaction networks. In: Nadel, A., Rozier, K.Y. (eds.) Proceedings of the 23rd Conference on Formal Methods in Computer-Aided Design – FMCAD 2023, pp. 284–293. TU Wien Academic Press (2023). https://doi.org/10.34727/2023/isbn.978-3-85448-060-0_37
47. Taylor, L., Zhang, Z.: Scaling up livelock verification for network-on-chip routing algorithms. In: Finkbeiner, B., Wies, T. (eds.) VMCAI 2022. LNCS, vol. 13182, pp. 378–399. Springer, Cham (2022). https://doi.org/10.1007/978-3-030-94583-1_19
48. Tsai, L., Hu, C.: Networks on chips: structure and design methodologies. J. Electric. Comput. Eng. **2012** (2012). https://doi.org/10.1155/2012/509465
49. IEEE Standard for VHDL Language Reference Manual (2019). https://doi.org/10.1109/IEEESTD.2019.8938196
50. Waddoups, N., et al.: Artifact for "probabilistic verification for modular network-on-chip systems artifact" submitted to vmcai26 (2025). https://doi.org/10.5281/zenodo.17247418
51. Waddoups, N., et al.: Probabilistic verification for modular network-on-chip systems (extended version) (2025). https://arxiv.org/abs/2511.13890
52. Wang, L., Cai, F.: Reliability analysis for flight control systems using probabilistic model checking. In: 2017 8th IEEE International Conference on Software Engineering and Service Science (ICSESS), pp. 161–164 (2017). https://doi.org/10.1109/ICSESS.2017.8342887

Author Index

A
An, Jie   173
Andreotti, Bruno   1

B
Bae, Kyungmin   335
Barbosa, Haniel   1
Barrett, Clark   99
Basu, Prabal   383
Bertrand, Nathalie   21
Boe, Jonah   383
Bonakdarpour, Borzoo   263
Boyer, Benoît   287

C
Cao, Xiuqing   82
Chakraborty, Koushik   383
Cho, Youngchan   44
Constable, Scott   147

D
Dam, Mads   147

E
Erhard, Julian   309

F
Faissole, Florian   287
Fisman, Dana   58

G
Guanciale, Roberto   147
Guo, Rui   82

H
Hartmanns, Arnd   383
Hélouët, Loïc   21

I
Isac, Omri   99

Izsak, Noa   58

K
Katz, Guy   99

L
Lefaucheux, Engel   21
Lewis, Marco   125
Li, Yang   82
Li, Yong   173
Lin, Wang   82
Lindner, Andreas   147

M
Meng, Junjie   173
Monat, Raphaël   197

N
Nemati, Hamed   147

O
Onderka, Jan   237

P
Palmskog, Karl   147
Paparazzo, Luca   21

R
Rafieioskouei, Arshia   263
Rand, Robert   44
Ratschan, Stefan   237
Refaeli, Idan   99
Remil, Naïm Moussaoui   212
Rogale, Kenneth   263
Roy, Sanghamitra   383

S
Sankur, Ocan   287
Schwarz, Michael   309
Sølvsten, Steffan Christ   359

Son, Byoungho   335

T
Turrini, Andrea   173

U
Urban, Caterina   212

V
Valiron, Benoît   125

van de Pol, Jaco   359

W
Waddoups, Nick   383
Wu, Haoze   99

Z
Zhang, Miaomiao   173
Zhang, Zhen   383